DATE DUE

Western Lands and Waters Series
XXVI

"San Francisco Speculators"
J. Ross Browne, *Harper's New Monthly*, January 1861

BONANZAS & BORRASCAS

GOLD LUST

&

SILVER SHARKS

1848–1884

RICHARD E. LINGENFELTER

THE ARTHUR H. CLARK COMPANY

An imprint of the University of Oklahoma Press

Norman, Oklahoma

2012

Also by Richard E. Lingenfelter

Bonanzas & Borrascas: Copper Kings & Stock Frenzies, 1885–1918 (2012)

The Mining West: A Bibliography & Guide to the Literature & History of Mining in the American & Canadian West (2002)

Death Valley & the Amargosa: A Land of Illusion (1986)

Steamboats on the Colorado River, 1852–1916 (1978)

The Hardrock Miners: A History of the Mining Labor Movement in the American West, 1863–1893 (1974)

Songs of the American West (1968), with Richard Dwyer and David Cohen

First through the Grand Canyon (1958)

Library of Congress Cataloging-in-Publication Data

Lingenfelter, Richard E.

  Bonanzas & borrascas : gold lust and silver sharks, 1848–1884 / Richard E. Lingenfelter.

    p. cm. — (Western lands and waters series ; v. 26)

  Includes bibliographical references and index.

  ISBN 978-0-87062-405-6 (hardcover : alk. paper) 1. Mineral industries—West (U.S.)—History—19th century. 2. Mineral industries—Economic aspects—United States—History—19th century. 3. Copper mines and mining—West (U.S.)—History—19th century. 4. Gold mines and mining—West (U.S.)—History—19th century. 5. Silver mines and mining—West (U.S.)—History—19th century. 6. West (U.S.)—History—19th century. 7. Speculation—United States—History—19th century. 8. Investments—United States—History—19th century. I. Title. II. Title: Bonanzas and borrascas. III. Series.

  HD9506.U63W354 2012

  338.20978′09034—dc23

2011024556

Bonanzas & Borrascas: Gold Lust and Silver Sharks, 1848–1884
is Volume 26 in the Western Lands and Waters Series.

To Naomi

Of all those expensive and uncertain projects, which bring bankruptcy on the greater part of the people who engage in them, there is none perhaps, more perfectly ruinous than the search after new silver and gold mines. It is perhaps the most disadvantageous lottery in the world.

ADAM SMITH
AN INQUIRY INTO THE NATURE AND
CAUSES OF THE WEALTH OF NATIONS (1776)

The investor in a mine has two risks to encounter: first, the character of the mine itself; second, the character of its management, which is the greater risk of the two.

PETER BURNETT
RECOLLECTIONS AND OPINIONS
OF AN OLD PIONEER (1880)

The secret of great fortunes without apparent cause is a crime forgotten, because it was so neatly done. (Le secret des grandes fortunes sans cause apparente est un crime oublié, parce qu'il a été proprement fait.)

HONORÉ DE BALZAC
LE PÈRE GORIOT (1834)

CONTENTS

Illustrations 11

Preface 13

1. Mines and Money 19

2. The Gold Lust 43

3. The Silver Rage 85

4. The Diamond Dream 155

5. The Big Bonanza Boom 207

6. The Silver Sharks 273

Appendix: Western Mining Dividends, 1803–1884 355

Abbreviations Used in the Notes 359

Notes 361

Index 437

ILLUSTRATIONS

San Francisco Speculators . . . . *Frontispiece*

Map of the Mining West in the 1850s 44

"To California on Shares" 47

Compagnie Française & Américaine de San-Francisco . . 50

Philadelphia and California Mining Company 59

Darius Ogden Mills 72

Tuolumne County Water Company 75

Map of the Mining West in the 1860s 84

Gould & Curry Silver Mining Company 87

Grosh Consolidated Gold and Silver Mining Company . . 94

The Silver Mania at San Francisco 105

Mariposa Company 110

Bodie Bluff Consolidation Mining Company 127

New Mexico Mining Company 134

William Sharon 143

Sutro Tunnel Company 148

Map of the Mining West in the 1870s 156

Emma Silver Mining Company, Limited 165

Ontario Silver Mining Company 172

Maxwell Land Grant and Railway Company 181

Belcher Silver Mining Company 196

John W. Mackay 209

Consolidated Virginia Mining Company 210

San Francisco Stock and Exchange Board 214

Ralston Panic at the Bank of California, August 26, 1875 . . 227

George Hearst 244

Homestake Mining Company 247

Justice Mining Company 255

"The Curse of California" 262

Map of the Mining West in the 1880s 272

The American Mining Stock Exchange, No. 63 Broadway . . 275

Chrysolite Silver Mining Company 280

The Wild-Cat Mining Swindle 289

Atlantic-Pacific Railway Tunnel Company 301

Cataract and Wide West Gravel Mining Company . . . 319

Edison Ore Milling Company, Limited 322

Quicksilver Mining Company 329

The Panic in Wall Street on the Morning of May 14, 1884 . . 351

PREFACE

BONANZAS & BORRASCAS is the saga of mines and money in the
American West from the opening of the first copper bonanza in
New Mexico, through the great California gold rush and the silver
rushes that followed, to the end of the copper frenzy at the close of World
War I. It is an epic tale of rich, profitable bonanzas and poor, profitless
borrascas, of investors and speculators and their fortunes and misfortunes, and of all the enormous wealth in gold, silver, and copper they
produced, wealth that spurred the economy, attracted myriad argonauts
and settlers, and transformed the West and the nation. Throughout it
traces the intertwined evolution of western mining practice and finance,
driven not only by new discoveries and new technologies, but by the ebb
and flow of the economy and the demands of the nation.

This book moves from the early years, when western investors and speculators dominated both the mines and markets, to a watershed change in
the early 1880s after California's new constitution cracked down on stock
speculation, banning margin trading, and drove San Francisco's mining
sharks to New York. The story breaks at that watershed. Thereafter, as
revealed in the companion volume to this, *Bonanzas & Borrascas: Copper
Kings and Stock Frenzies, 1885–1918*, easterners bought control of most of
the large mines to further exploit eastern markets for even bigger profits
and losses. Underlying all this at the same time, developing technology in
the western mines also expanded beyond the working of high-grade lodes
of gold and silver, to the opening of much larger, lower-grade deposits
predominantly of copper, which rapidly became the leading metal as the
electrification of the nation drove up the demand and prices.

This is especially a stockholder's-eye view of western mining, and a
story of great expectations. It tells both of dazzling profits and crushing

losses. It is the saga of those ambitious argonauts who left their families, farms, and factories in the East to seek their fortunes in the frantic rushes to the West, and of the triumph of the shrewd and fortunate, like Darius Mills, George Hearst, and John Mackay, who invested in promising prospects that exceeded their wildest dreams and made them some of the richest men in America. And it is the tale of countless others, all those armchair argonauts—those more encumbered brothers and sisters—held back by circumstance or duty, who longed just as feverishly for their own chance to strike it rich. They gambled their savings, some even their homes, for shares of stock in such dreams. Many did win rich returns, but most lost everything in the resurgent orgies of mining stock speculation that seized the nation in every new wave of prosperity.

Yet many who bought shares from wildcat promoters still had faith in their bogus promises, or at least were not yet ready to admit that they had been conned. So even after all work was abandoned, they still held on to their gilt-edged shares, like perpetual lottery tickets, till the day they died, still hoping against hope that they might yet draw the prize, each brightly gilded share a tantalizing promise of "pie in the sky bye and bye."

It was, of course, the very success of the western mines that lured most investors and speculators; all the titillating tales of fabulous fortunes from the western bonanzas excited a swarm of raw passions in many who heard or read them—envy of the quick fortunes of others, greed for a share of such riches, and a panicky fear of missing out on the chance of a lifetime. This lust for quick wealth drove them to buy shares in almost any company that promised a piece of the profits of the western mines. Some paid off, but most of those companies were mere shimmering fantasies concocted by obliging "mining sharks" seeking their own quick fortunes in exploiting the "shining marks." For "swindling is demand-determined," as economist Charles Kindleberger succinctly observed; "in a boom, fortunes are made, individuals wax greedy, and swindlers come forward to exploit that greed."[1]

Yet those most driven by greed were the directors and managers of some of the biggest and most profitable mines, who had already gained startling fortunes but still wanted more. So some of the greatest stockholder losses came from insider looting and market manipulation of those mines. For as California's pioneer banker and governor Peter Burnett long ago warned, mining investors faced two basic risks: "first, the character of the mine itself; second, the character of its management, which is the greater risk of the two." This history, then, follows the money trail from both the rich pockets of ore and the bulging pockets of investors and speculators through the mills and the markets in the exploitation of both the mines and the public.[2]

In all, the saga ties together the fortunes of the East and the West, exploring not only the impact of eastern investors and speculators on the western mines but the generally unrecognized impact of the western mines on Wall Street and Washington. For profits from Nevada's Comstock Lode built New York's "first skyscraper," the Mills Building, across the street from the New York Stock Exchange. The scandalously manipulated shares of the western giants Amalgamated Copper and Asarco made up more than half of the annual trading of all the industrial shares on that exchange for years, and they were the driving factors in the early Dow Jones Industrial Averages. The calamitous collapse of a Colorado mining shark, Ferdinand Ward, brought on the Wall Street Panic of 1884 and wiped out his partners, former president Ulysses S. Grant and his sons. The bitter fight between Henry Rogers and Fritz Heinze over the copper in Butte, Montana, triggered the Panic of 1907, which led to the creation of the Federal Reserve Board. And a former mining shark and eminent corporate lawyer, Samuel Untermyer, waged an unrelenting fight for market reform and regulation that culminated in the formation of the Securities and Exchange Commission, after the devastating crash of 1929 and the defeat of another former mining operator, President Herbert Hoover.

If the money made sometimes seems too trifling to tweak your own greed, remember that it is all tallied in the long-gone dollars of the latter half of the nineteenth and the early twentieth centuries. Those were times when the price of gold was fixed at $20.67 an ounce, not the $1,000 or more that it is today, and the average wage in the country was only $1 a day, or $300 a year, while the average today is above $30,000. So to get a better feel for the excitement of the times, just add a couple of zeros to those early dollars and look at the numbers again, for the tens of millions then are the billions of today.[3]

Despite the enormous impact of mining on the West and the nation, only parts of its history have been told. Rodman Paul's great classic, *Mining Frontiers of the Far West, 1848–1880*, traced just the beginnings of gold and silver mining, while William Greever's *Bonanza West* focused more on the social aspects of the mining rushes before 1900. Charles Hyde alone has followed the full course of copper mining in America. Others have studied more limited economic aspects. The most notable are Clark Spence's pioneering work on British investment in the mining frontier from 1860 to 1901, Joseph King's study of Colorado mining finance from 1859 to 1902, Marian Sears's survey of the mining stock exchanges from 1860 to 1930, and Richard Peterson's study of the roots and practices of some of the Bonanza Kings, as well as Dan Plazak's recent, entertaining accounts of some popular mining frauds. Many other regional and corporate studies are noted in the author's *The Mining West: A Bibliography & Guide to the History & Literature of Mining in the American & Canadian West.*[4]

This, then, is the first broad history of mining in the American West during its great heyday to the end of World War I, before it was overtaken by oil. It is also the first deep exploration into the maze of Western mining finance, following the fortunes and misfortunes of stockholders in the hundred or more most profitable bonanzas and almost as many borrascas.

The tangled paths are revealed through a wide range of sources—the popular and the financial press, the mining and muckraking magazines, the company prospectuses and reports, stock prices and dividends, stockholders' suits and investigating committees, government reports and investigations, and personal reminiscences, letters, ledgers, and diaries. They are also illuminated by the stock certificates themselves, both of the great mining bonanzas and the scandalous mining humbugs, those gilded securities that were the very vehicles of the financial saga, those ever-enticing tickets to the lottery.

The quest has been aided immeasurably by the advent of extensive, searchable online archives, which make it possible to probe much deeper than was ever practical before. By far the most useful of these have been the newspaper databases of the New York *Times* and *Tribune*, the *Wall Street Journal*, the Washington *Post*, the Chicago *Tribune*, the San Francisco *Chronicle*, the Los Angeles *Times*, and numerous popular magazines, all offered by ProQuest, plus the London *Times* and many in the 19th Century U.S. Newspapers series offered by Gale. I also drew on a rich sampling of western and eastern papers from Ancestry.com, California

Digital Newspaper Collection, Chronicling America, Colorado's Historic Newspaper Collection, Newsbank, NewspaperArchive, and Utah Digital Newspapers, as well as the broad book and serial databases of Google Books.

Finally, I am personally indebted to Lynda Claassen, Deborah Cox, and Elliot Kanter of the Geisel Library of the University of California, San Diego; to Dave Beach, Guy Cifre, Jeff Daly, Wendell Hammon, Fred Holabird, Bob Kerstein, George LaBarre, Brian Levine, Homer Milford, Geff Pollock, Ken Prag, Rick Rothschild, Bob Spude, Mike Veissid, Scott Winslow, and Sally Zanjani; especially to Bob Clark and Steven Baker of the University of Oklahoma Press, plus copyeditor Margery Tippie—but most of all to my wife, Naomi.

La Jolla, California
April 5, 2011

1

MINES AND MONEY

MINES MADE MONEY and money made mines. Together they laid the foundation of much of the American West and enriched the nation. The mixing of mines and money was driven not just by the discovery of new bonanzas but also by new technologies, shifting market demands, waves of prosperity and depression, and all the raw passions of human nature.

The interplay of mines and money in the Americas started with Cristoforo Colombo, and the Spanish *conquistadores* eventually carried it into the Southwest. There it moved slowly for centuries until the sudden demand for copper in the Napoleonic Wars led to the opening of the first western bonanza in New Mexico in 1803. Then, in 1848, the chance discovery of gold in California unleashed a flood of riches that startled the world and launched a great rush of money into western mining investments and speculation that has surged and crested in great waves and countless ripples for over a century and a half. But unlike the mining rushes themselves, which drew people west in times of economic depression, the mining stock manias of investment and speculation drew money in times of prosperity.

For the ever-swelling flow of wealth from the western mines, the seemingly endless news of exciting new bonanzas, and the steady evolution of mining technology always offered rich enticements and ready opportunities for investment and speculation whenever prosperous times arose. What made the difference between rich, profitable bonanzas and poor, profitless borrascas was, of course, the value per ton of the rock, or ore, but what constituted ore was ever changing and ever growing. Ore is by definition economically profitable rock. Ore is not an inherent mineral property. It depends not only on the natural concentration of minerals in the rock but also on the state of technology that sets the cost of its extraction and refining, and on the social and economic demands that set its marketable price.

The natural concentration that produces metallic ores comes primarily from minerals dissolved in hot solutions, thought to be predominantly hot ground waters heated by deep, intruding magma that fractures the overlying rock. These hot, mineral-laden waters rise along the deep fractures and through more porous near-surface sediments, and as they cool in colder rocks above, the dissolved minerals precipitate out as concentrates of metals, quartz and other crystals in relatively thin veins, or lodes, in the fractures and in broader, thicker beds in the near-surface sediments. Later weathering and erosion can further enrich heavy minerals at the surface, which not only help lead prospectors to the underlying ore but also help mislead investors, since richer surface showings can greatly exaggerate the apparent value of the underlying ore.[1]

Moreover, since the total amount of available ore increases dramatically as the concentrated ore value per ton decreases, the overall mineral production has steadily risen as the workable grade of ore has fallen. In the history of producing mines, this also means that those with the very richest ores, those that excited the greatest public interest, were often among the shortest-lived producers, because such rich ore deposits were likely to be so small that they were rapidly worked out, typically in less than a year. That fact, in turn, excited a rapacious greed in the promoters and managers to rip out all they could and unload the carcass on others. Only in the development of larger, lower-grade ore deposits with workable lifetimes of a decade or more did outside investors stand a better chance. These, of course, were also the biggest and most efficient mines that could patiently work the low-grade ores and still make a profit of 20 percent or more. But many of the biggest bonanzas still fell prey to a myriad of management manipulations and frauds.[2]

For most of the nineteenth and early twentieth centuries the price of gold was set at $20.67 a troy ounce, so gold ore was determined primarily by its recovery costs. The richness of paying rock has steadily dropped from hundreds of dollars, or tens of ounces of gold, per ton of rock to current values of less than a tenth of an ounce per ton, as the technology evolved from solitary panning in the placers to gigantic hydraulic and dredging operations, and from rich surface diggings in the lodes to ever-deeper hard rock and open pit operations, coupled with ever-more-efficient refining processes moving from mercury amalgamation to cyanidization. Only in the last quarter of the twentieth century did the ending of the "gold standard" finally set free the price of gold, which quickly rocketed to several hundred dollars an ounce. That triggered the "invisible gold rush" to very low grade ores and raised western gold production to its highest levels in history.[3]

Other ores, particularly those of silver, copper, and uranium, which outside of gold have most attracted investors, followed more erratic courses as the market prices rose and fell. Silver prices, roughly fixed for a time at the monetary ratio of one-sixteenth that of gold, supported a boom in western silver mining that lasted until the 1870s, when over-production drove prices down after they were cut loose with the demon-etization of silver. Copper prices fluctuated wildly over the years, and the coming of cheaper rail transportation opened many new mines. Then, with the burgeoning demand driven by the electrification of America, western copper production took off in the 1880s and soon dominated western metal production. Uranium boomed briefly in the 1950s with the demands of the Cold War arsenal and the fledgling nuclear power industry. Surges in the mining of each of these metals stirred new waves of mining stock mania.

For a time each successive wave seemingly surpassed those before until the great boom at the turn of the twentieth century, when more than four million people, roughly one out of every dozen adults, risked at least part of their savings on the dream of wealth in the western mines. But by the First World War the rate of discovery of exciting new bonanzas had plummeted and the growing outrage over mining frauds had finally forced a federal crackdown, so the epidemic of mining stock manias at last subsided. Their place was quickly taken, however, by excitements in oil, automobiles, radios, and countless other new markets, all the way up to the Internet, that have subsequently enticed a major share of investors, speculators, and swindlers as well.[4]

BUT FAR AHEAD of most of the technological and economic revolu-tions that changed the nature of mining investments and speculations came the social and political revolutions that changed the nature of the community of investors and speculators. This community grew from the early, small royal circle of regents and court favorites, through participating shareholders of a growing merchant class, and then, with the rise of the middle class and the beginnings of mass mining stock manias in the nineteenth century, to the populace at large. With this change came also a shift in the form of the companies, from the first royally chartered ventures, granted broad monopolies to simply seize and exploit all lands "not actually possessed by any Christian prince or people," to the later, much more limited joint-stock companies, incorporated to buy and work specific mines. With these changes too came a shift in the schemes and strategies of promoters and their relations with investors, moving from an occasionally desperate need to satisfy the expectations of a few very powerful investors, who could hold the power of life and death, to the only fleeting need to raise hopes while unloading shares on a large flock of virtually powerless investors, who could rarely mount an effective recourse in the courts.[5]

Speculation in the mineral riches of the Americas, in fact, began with the European conquest, driven by a lust for quick profits in gold and silver, and it continued hand in hand with that bloody struggle on through the nineteenth century in the mines of the American West. Right from the start that exuberant Genoan, Cristoforo Colombo, had even tried to excite the lust of the Spanish monarchs, Isabella and Ferdinand, for nothing less than the biblical mines of Solomon and Ophir in order to open their purse and win their support for his speculative search for a westerly route to the Orient. But in spite of his erroneous arguments about the nearness of Asia, it was just the chance of great gains for very little risk that finally won the monarchs over, ready for rich new lands to conquer after their triumphant reconquest of Spain from the Moors. On April 17, 1492, they formally agreed to finance the venture, taking for themselves all the lands and 90 percent of all the gold and other treasures that he might gain. But they offered Colombo the grand titles of admiral, viceroy, and governor-general of these lands and the remaining 10 percent share of the gold and treasures tax free. The total investment of the crown was indeed quite modest, amounting to only about 400 ounces of gold. That had the buying power of roughly $40,000 in the nineteenth century and was small indeed compared to later western mining speculations. Still, it was enough to pay for all the provisions and crew, while at no cost to themselves; the monarchs simply ordered the city of Palos to provide two of the three ships that Colombo wanted, as punishment for some royal displeasure. The admiral himself had to borrow about seventy ounces of gold, or $7,000, to charter a third, the *Santa Maria*, for his flagship.[6]

Yet, small as their investment was, Colombo had so exaggerated the expected profits that when the islands he first found delivered only handfuls of gold trinkets, he nonetheless told the crown that he'd found "great mines of gold" that could "give them as much gold as they want." And when he founded the first colony, Hispaniola, a year later, he and his two brothers turned it into a deadly hell trying to maintain that illusion. He demanded on pain of death a quarterly tribute of roughly an ounce of gold from every adult islander, as he madly tried to satisfy all the avaricious expectations, not only of his royal backers, but also those of his gold-seeking colonists as well as himself. Even then he fell far short, and after seven years he was finally removed as governor. The islanders meanwhile were decimated by European diseases, which killed well over 100,000 souls, before Spanish miners brought in slaves and at last managed to dig out an estimated 30,000 ounces of gold per year. Of this the crown at first took a half, but when that proved too punitive to mining investors they cut back to only a fifth, the *quinto real*, which became the standard tax, or "royalty," after 1504. This alone was a magnificent, if blood-stained, return on the crown's original investment, amounting to a fifteenfold profit per year. But this was nothing compared to the incredible additional profits that soon came from the conquests of the Aztecs and the Incas. In fact, this venture is very likely the most profitable investment ever made, as the mines of the Americas would make Spain for a time the richest power on earth. Colombo, too, though smarting from his removal, still collected his tenth of the crown's share of the gold, which paid him a handsome profit, and he claimed to have left some $700,000 in gold in Hispaniola, against which he freely borrowed further cash. But he still wanted more and hounded the court almost to the day he died, in 1506, for a further third of all the colonial trade from a fancied tax he claimed he had the right to collect as Admiral of the Ocean Sea.[7]

THE LURE OF REAL GOLD, however, once proven in Hispaniola, soon excited a ravenous army of "speculators and adventurers, who swarmed, like so many famished harpies, in the track of discovery." In the ensuing century they swarmed all over the Americas, organizing over a hundred companies to complete the conquest of its people and seize their land and their riches. Unlike Colombo's original venture, nearly all of these companies were financed by private speculators instead of the monarchy. The crown simply licensed the ventures and took its royal fifth of the proceeds, leaving all the risk to the speculators. These companies generally consisted of a few well-to-do and well-connected promoters, who put up most of the cash investment, and a few hundred hopeful adventurers, who became active shareholders, willing to risk their very lives for a share of the hoped-for riches of the conquest. Most were after the quick profits of gold and other treasures, but many also sought longer-term profits from grants of land and the labor of those who lived on it. By far the most profitable of these ventures were those of Cortés and Pizarro in seizing the truly fabulous riches of Mexico and Peru.[8]

When the first bits of gold were brought to Cuba from the still barely explored mainland in 1518, they excited fresh rumors of a rich empire to the west. Governor Diego Velásquez lacked a royal license for its conquest, but he promptly readied half a dozen ships for a trading venture to explore the coast. His two closest advisers, however, his colonial secretary, Andrés de Duero, and the royal accountant, Amador de Lares, conspired to get him to appoint as captain of the fleet an avaricious young speculator, Hernán Cortés, who had secretly agreed to pursue the conquest even without license and share equally with them whatever he gained. Cortés had dropped out of college to join the rush to the New World and had made a good income from land grants and gold concessions in Cuba and Hispaniola. But he had seen Velásquez rise quickly to wealth and power in the conquest of Cuba just seven years before, and he was eager to follow the same course. So he mortgaged his estate for 8,000 pesos de oro, about $90,000 in nineteenth-century dollars, to buy arms and other supplies, claiming he put up two-thirds of the cash outlay for the venture, and he hastily recruited about five hundred adventurers for shares. When Velásquez realized what was happening, he tried to stop it, but Cortés was forewarned by his partners and quickly set sail.[9]

He established Vera Cruz as a base on the mainland, and recruiting several thousand more coastal people, previously conquered by the Aztecs, he moved on to the great city of Tenochtitlán, the seat of the Aztec empire. The emperor Moctezuma tried to buy off Cortés and his followers with fabulous treasures of gold and silver that far surpassed even their wildest dreams. But such riches only further excited their greed and steeled their determination. After the murder of the emperor and the bloody slaughter of thousands, when the Aztecs briefly drove the invaders out of the city in panicky retreat, Cortés, refortified and reinforced with the aid of Duero and Lares, was finally victorious. He had at last gained for Ferdinand's successor, Carlos I, such riches as Colombo had only talked of, and he was richly rewarded with the governorship of what was then proclaimed New Spain.[10]

Although much of the original treasure was apparently lost in the muddy depths of the lake surrounding the city during the retreat, the Spaniards still got away with around $4,000,000 worth of gold and jewelry. But after the crown and Cortés each took a fifth and more, and he gave generous shares to his partners and the governor, who had put up the other third of the costs, plus favored officers, there was little left for all the other adventurous shareholders, who had joined for promised fortunes. They were outraged when they were each offered a paltry $1,200 and found that their combined share amounted to less than 10 percent of the total! Cortés, however, had become a thoroughly modern executive, grabbing for himself nearly a thousand times as much as he gave to most of his fellow adventurers. To quiet their protests, he bought off their leaders and with "honeyed words" promised the rest that this was only the first taste of riches and that they would soon all be rich lords of the lands and the mines, which would indeed pay the crown and later investors fabulous profits for centuries. Many had also already helped themselves to the treasure in the panic of the retreat, though some had died for their greed, dragged down under the weight of it. Despite such disputes, the wonderful riches of the Aztecs further fanned the flames of conquest and fed the fevers of an ever-growing army of adventurers.[11]

Even more spectacular and contentious was Pizarro's venture in Peru. Francisco Pizarro, a doggedly ambitious, but illiterate, son of an infantry captain, was in his early twenties when he came to New World in 1501 to join the ranks of its soldiers of fortune. He went with Vasco Núñez de Balboa to Panama a dozen years later, hastily switching allegiances after the jealous new governor, Pedro Arias de Avila, executed Balboa on trumped-up charges of treason. Finally in 1522, fired by the stories of Cortés's rich conquest and rumors of the riches of the Inca empire to the south, Pizarro, by then in his mid-forties, joined with another aging adventurer, Diego de Almagro, and the older, wealthier, and more influential chief justice Gaspar de Espinosa in a partnership for the conquest of the Incas. Espinosa, who was said to have already seized over $900,000 in gold and two thousand slaves in brutal expeditions against the people of Panama, agreed to put up nearly a third of that amount as the major backer of the venture. Pizarro would lead the assault and Almagro would provide support, and they would share equally all the treasure, land, and slaves that they seized. With the approval of the governor, who was promised a cut, they recruited a couple hundred adventurer-shareholders and were soon under way, but the conquest would be a long time coming.[12]

After six years and two aggressive but failed expeditions, a new governor balked at allowing them another attempt. At the same time Pizarro wanted direct control, so he went to Spain, and in 1529 he won a royal license that gave him all the power as both governor and captain general of the future province of Peru. When Pizarro returned to Panama, Almagro, furious at being excluded from power, was further enraged by the arrogant behavior of Pizarro's half-brother Hernando, who had returned with him to share in the spoils of the conquest. Only through Espinosa's impassioned pleas and a new contract, confirming once again that all three would still share equally, was the dissension at least temporarily smoothed over, and early in 1531 they finally launched their third expedition. This time Pizarro, feigning only peaceful intentions, started with a force of 160 soldier-shareholders and a large following of coastal people into the heart of the Inca empire, where they were soon met by the new young emperor, Atahuallpa, at the city of Cajamarca in November of 1532. There Pizarro lured the Incas into a trap, where more than a thousand were slaughtered and the emperor himself was seized for ransom. For his freedom Atahuallpa offered to fill the room where he was being held with gold as high as he could reach. Pizarro eagerly accepted, and the Inca dispatched couriers who brought back dazzling treasures worth well over $15,000,000 in nineteenth-century dollars. But Pizarro then murdered him anyway.[13]

Setting aside the crown's fifth, they immediately divided the rest of the loot. Since Pizarro was already granted control of the empire, he was far more equitable than Cortés and allotted roughly two-thirds of the booty to his fellow adventurers. The sixty cavaliers got a handsome dividend of roughly $100,000 apiece for their share, and the hundred infantrymen remaining got nearly half that. Thus Pizarro generously ensured their continued loyalty in finishing the conquest on which his own rewards rested. He then took only an eighth of the treasure to divide equally among Almagro, Espinosa, and himself, which gave each of them over $600,000, only ten times that of most of his men. And while that repaid Espinosa nearly double, it amounted to only 10 percent interest over ten years, and he surely expected much more, as did Almagro. Pizarro provoked further dissatisfaction by giving Hernando and his two other brothers a share nearly equal to that of his partners, and by refusing to share with Almagro's men, who had arrived just after Atahuallpa's capture. A year later, however, they were all further rewarded after the final conquest of the Inca capital of Cuzco, in which they amassed an enormous new heap of treasure. This totaled somewhere between the $7,000,000 reported to the crown and as much as four times that, based on the reported shares of $70,000 for cavaliers in a similar allotment among a force that now numbered about 480. The official tally would have given the three original partners at least another $300,000, but Almagro was by then much more interested in the division of the land.[14]

When the enormous treasures of the Incas were presented to Carlos I in 1534, Almagro was finally rewarded, after a fashion, with a license to conquer the lands to the south of those already granted to Pizarro. But instead of finally quieting their past difficulties, this sparked a final deadly conflict among the partners, because it was not yet clear in whose territory the great capital of Cuzco lay. The bloody struggle for possession ended with the deaths of all the partners. Espinosa died mysteriously trying to mediate a compromise in 1537; Almagro was captured and killed the following year by Pizarro's brother Hernando, who was sent to prison for twenty years; and Pizarro himself was assassinated in 1541 by Almagro's followers. Nonetheless, the Inca treasures excited all the more the greed of adventurers and speculators, and even these riches were soon surpassed by the opening in 1545 of the incredible silver bonanza of Cerro Rico de Potosí to the south of Cuzco.[15]

LIKE ALL SPECULATIONS, however, most of the schemes of conquest were only marginally profitable at best, and many were highly overrated and proved to be utter losses to their backers. Indeed, one of the more costly failures was that which attracted the first Spanish speculators into what is now the American Southwest. It was the conquest of the seven supposedly fabulous cities of Cibola somewhere north of the frontier of New Spain that lured Francisco Vásquez de Coronado and others. The first notion of Cibola was inspired by a little, old copper bell that a shipwrecked adventurer, Alvar Núñez Cabeza de Vaca, had picked up in 1536, while wandering across the northern frontier of Mexico after an abortive expedition to Florida. The nomadic natives who had given him the bell said it had been traded to them from people to the north who also had plates of copper. They further told him that there were large cities to the north, but what excited him most was the fact that the little bell had been cast in a mold. From this Cabeza de Vaca concluded that they must have a "foundry" to refine and work the metal, and from that it was soon imagined that they must also have great treasures.

The notion of yet another rich empire to the north grew rapidly after the conquest of the Incas to the south, exciting even Cortés to try to search for it. Cortés had been displaced as overlord of New Spain by the appointment of a royal viceroy, and he was again seeking new conquests. He had already launched an expedition that discovered the California peninsula in 1533, naming it for the fictional island of gold ruled by the black Amazon queen Calafia in a popular romance by Garci de Montalvo. But Cortés had lost much of his fortune in that venture, equivalent to about $3,500,000, in a disastrous attempt to establish a colony there two years later, so he was desperately in need of new treasures.[16]

The viceroy of New Spain, Antonio de Mendoza, however, was equally determined to get control of any new empire for himself. So in 1539, while Cortés was sending a naval expedition up the coast in quest of Cibola, Mendoza formed a partnership with Coronado, an aggressive young governor on the northwestern frontier, to organize an overland expedition. They agreed to put up nearly $1,000,000 to help outfit the venture, splitting the cost nearly equally, with Mendoza keeping a controlling 55 percent interest, and to raise the money they both borrowed heavily against their own estates and the prospective profits of the conquest. With three hundred ready adventurers, who joined for a share, and eight hundred native converts, Coronado temporarily subdued the Zuni and other pueblos in what is now New Mexico in the summer of 1540, only to find that their vaunted gold was all a fantasy. To be rid of him, the pueblo dwellers told him stories of the new golden cities of Quivira much farther to the east. So he spent two years wandering around the Southwest and out into the Great Plains before he was kicked in the head by a horse and finally became convinced that there were no more rich empires to be conquered. The failure of the venture left Coronado and many of his fellow adventurers deeply in debt and killed all interest in conquest to the north for half a century. But the discovery of the rich Mexican silver mines at Zacatecas in 1546, right after the opening of those at Potosí, fueled a relentless quest for silver that slowly spread northward again.[17]

Finally, in 1581, a small party of explorers again reached the Pueblos along the Rio Grande, and at one Felipe de Escalante and others were given some tantalizingly rich rock, running as much as half silver. It came from the Cerrillos, the low hills just east of the river, where the Pueblos had mined turquoise and lead for centuries. Reports of more silver deposits followed, and the rich Pueblos themselves were also eyed as profitable fiefdoms for conquest, so in 1583 the Spanish crown opened a license for their "pacification" to competitive bidding under new colonization laws—a process that dragged on for fifteen years. The winning bidder came from a whole new generation of *conquistadores*, a wealthy and well-connected middle-aged silver mine owner, Juan de Oñate, whose father was one of the founders of Zacatecas and whose wife was no less than a granddaughter of Cortés and a great-granddaughter of Moctezuma. Oñate financed the venture at a cost of about $250,000, of which 40 percent was loaned by the crown, and he received the usual powers of governor and captain general of the new province of New Mexico, plus he only had to pay half of the royal fifth of what he took from any mines he might open. In the summer of 1598, accompanied by nearly two hundred Spanish colonists including Escalante, hundreds of natives, and several thousand head of livestock, Oñate at last marched up the Rio Grande and conquered the Pueblos. But his prime interests were revealed when, just nine days after selecting a site for his new capital, he began digging out ore samples from the Cerrillos, and he put another Zacatecan, his cousin Vicente de Zaldívar, in charge of developing the mines. Thus began the first European mining ventures in what is now the United States. But these and other silver lodes that they discovered all proved to be fairly small, and jaded as they were by the riches of Zacatecas, the Spaniards were greatly disappointed. Nonetheless, the mines would still be worked on and off for centuries, a span interrupted only briefly when the Pueblos revolted in 1680 and drove out all the Spanish for a dozen years.[18]

IN THE MEANTIME, Spain's rivals also lusted for a share of the riches of the Americas, but to little avail. The first to try was François I, who, after a humiliating defeat by Carlos I in Europe, commissioned mariner Jacques Cartier to challenge Spain's monopoly of the New World by exploring new lands, where it was said "a great quantity of gold, and other precious things, are to be found," and planting the first French colonists there. For this, François invested about $60,000 in nineteenth-century values, plus two galleons and a crew of fifty convicts from the jails of Brittany. But that still left Cartier and the future colony's governor, Jean-François de la Rocque de Roberval, to raise the remaining $100,000 needed to outfit the venture and the colonists. Between 1534 and 1541 Cartier made three voyages to the new lands. There he discovered the great St. Lawrence and built a fort at Cap-Rouge, near present-day Quebec, where he dutifully, but guardedly, reported the discovery of black slate threaded with veins that shone "like gold and silver" and studded with "stones like diamants" that glistened as if they "were sparkles of fire." But when Roberval and the colonists arrived in 1542, their eager goldsmiths quickly found that all the imagined riches were nothing like the real thing. So after a hard winter of scurvy, increasingly hostile Iroquois, and the deaths of a quarter of the colonists, the lengthy venture was finally abandoned as a total loss.[19]

Next came Elizabeth I, who in a swell of speculation fueled by the looting of Spanish treasure galleons, launched England's first and certainly most curious American mining venture. Without her participation, the merchant Companye of Kathai, promoted by a London speculator, Michael Lok, for about $25,000 in later dollars, had sent an ambitious former pirate, Martin Frobisher, to scout for a northwest passage through the Arctic to the Orient in 1576. Frobisher went only as far as Baffin

Island, where he happily mistook a deep bay, which now bears his name, for the entrance to the great passage. Then, on a tiny coastal island, one of his crew picked up a fascinating "blacke stone" that would divert the whole direction of the company. For Frobisher quickly returned to England, and Lok excited the investors with news that his assayers had found that the "glittering" rock was very rich in gold! His eager assayers claimed it ran as high as twenty-five ounces a ton, equivalent to $2,000 a ton, and Lok claimed that even the worst ore would clear a profit of $800 a ton. When Frobisher then boasted that he could fill all the queen's ships with such ore, Elizabeth immediately took over the company and became its major shareholder. She put up nearly $80,000 to turn it into a grand mining venture, sending over 150 miners to work the "mines" off Baffin Island and a fleet of fifteen ships to haul all the ore. In two more voyages over the next year and a half Frobisher brought back around 1,500 tons of rock for an expected profit of more than $1,000,000. But when the company's new $40,000 smelter on the outskirts of London failed to recover more than a trace of gold, the whole scheme collapsed in 1578, for a total loss of about $400,000. Amid bitter lawsuits and charges of embezzlement and self-deception, Lok was sent to debtor's prison and his principal assayer and smelterman fled the country. Frobisher, however, claiming Lok had deceived him too, returned to privateering to redeem himself by helping Francis Drake take about $10,000,000 in Spanish treasure, and he was later knighted for his part in the defeat of the Armada.[20]

Still hungry for a share of the riches of America, Elizabeth next gave a license to Humphrey Gilbert in 1578 to seize and exploit all lands "not actually possessed by any Christian prince or people," reserving for herself in the Spanish tradition a "fifth part of all the ore of gold and silver which should at any time there be gotten." But after Gilbert failed twice and was then lost at sea, she turned in 1584 to his much younger half-brother Walter Raleigh, whose Roanoke venture also ended in a tragic and profitless disaster. Her successor, James I, had only slightly better luck in 1606, when he gave a charter to a group of London adventurers headed by Thomas Gates to form the Virginia Company to establish a colony and "to dig, mine and search for all manner of gold, silver and copper." They easily sold all one thousand shares for the equivalent of $250,000 and did at least succeed in planting a permanent colony at

Jamestown. But in their first year the struggling colonists were also seized by a golden fantasy. "There was no talke, no hope, no worke, but dig gold, wash gold, refine gold, load gold," a disgusted John Smith wrote, while the colonial commander Christopher Newport encouraged the settlers to gather up a shipload of worthless "glittering dirt" to feed their London investors. It was, of course, just fool's gold, for unlike Spanish America, paying gold would not be found in English America for nearly two centuries. The English settlers meanwhile turned to quests for more utilitarian coal, iron, lead, and copper, but even profitable copper mines still weren't found until a century later in Connecticut and New Jersey.[21]

By then new illusory riches in the American West were helping to lure French investors and speculators into the first and most infamous of stock manias, the "Mississippi Bubble." This complex speculative frenzy, which rocked Paris and the rest of Europe in 1719–1720, grew out of financial revolutionary John Law's grand scheme to revive the debt-ridden economy of France following the death of Louis XIV. After establishing the Banque Générale to issue paper money, Law formed the Compagnie d'Occident, the Company of the West, to exploit the vast Louisiana colony that stretched west from the Mississippi to the Rockies. Among the company's many inducements were fabulous tales of mountains of gold and silver in the great range and a gigantic emerald rock that sat on the banks of the Arkansas River. But the driving force of the speculative fever was Law's subsequent merger of this company with others controlling French trade with China and India, as well as the slave trade with Africa, into a giant conglomerate, the Compagnie des Indes, which then took over French tax collection, the mint, the central bank, and finally the national debt! Stock in this company sky-rocketed from 490 livres a share in May of 1719 to just over 10,000 livres in December, before Law froze the price at 9,000. By then the total market value of shares was close to 5,000,000,000 livres, about $1,250,000,000 in nineteenth-century dollars, and former merchant Daniel Defoe lamented, "If ever a whole Nation was Mad in the World, this is the Time." Even Law began to worry after the accompanying inflation rate reached 23 percent per month, and the following May, when he announced that he would deflate share prices, he was temporarily dumped from the company and the market collapsed. Nonetheless, the company did briefly search for the great emerald and the mountains of gold and silver, but all they ever found was lead in Missouri. Eventually, after the spark of democratic revolution ignited in the English colonies east of the Mississippi in 1775 and finally spread to France in 1789 and its Haitian colony on Hispaniola a decade later, Napoleon Bonaparte, in 1803, hastily sold the Louisiana colony to the new American republic for $15,000,000 to raise quick cash for his European adventures.[22]

ALL THE WHILE to the west the Spanish settlers of New Mexico were working small mines in the Cerrillos and placers to the south, and they pushed the quest for new gold and silver mines north into the San Juans in what would become Colorado and west to the Colorado River. Miners from Sonora also moved north toward the Gila River after a Yaqui Indian discovered enormous slabs of native silver, some weighing as much as a ton. These *planchas de plata* were found in 1736 about forty miles south of Tubac on what Basque miners would christen the Cerro de la San Antonio de la Arizona, the Hill of St. Anthony of the Good Oak, that would give the country its name. Although the planchas proved to be only a shallow surface deposit, miners soon opened deeper silver mines farther north around Tubac and the mission of San Xavier del Bac.[23]

But it was a copper mine, the Santa Rita del Cobre, that proved to be the first big bonanza and paid the first real fortunes to investors in the American West. Its spectacular outcrops of native copper at the foot of the Mogollons, west of the Rio Grande, had been worked along with surrounding turquoise and lead mines by the nearby Mimbres pueblos and might even have been the source of the copper in Cabeza de Vaca's old bell. But long after those pueblos were abandoned, the Dene, or "people," whom others called Apache, or "enemy," invaded the Mogollons just ahead of Coronado. Later Spanish adventurers weren't eager to challenge the Apache for anything as cheap as copper until the crown, to stimulate new production to meet rising demands from the onset of the Napoleonic wars, dramatically raised its price in July of 1799. Then an enterprising lieutenant from Chihuahua, José Manuel Carrasco, who in 1785 had taken part in punitive raids against the Apaches in the area— which was already known as El Cobre, the Copper—returned, and on June 30, 1801, he formally claimed the old diggings. With a well-armed force, he started taking out copper in 1803, but since he lacked the capital to adequately protect and work the mine, he sold out the following year to a prosperous Chihuahua merchant, Francisco Manuel Elguea, who already held a monopoly on the Santa Fe trade and would become the first successful mining investor in the West.[24]

In company with Blas Calvo y Muro, who had discovered gold placers nearby that provided immediate returns, Elguea brought in a large crew of miners, built a fort, and got the governor to provide a garrison for protection. The native copper was so pure that it was easily and cheaply refined. So the royal mint in Mexico City immediately gave him a generous contract for 200,000 pounds of copper a year, at a premium price of 25 cents a pound, and waived the royal tax. With addition commercial sales, he may have mined at least 300,000 pounds a year for a handsome annual profit of over $50,000. Although Elguea died just two years later, his widow's new husband, Francisco Pedro de Guerra, continued to develop the mine by leases, and the family fortune continued to grow. The camp also grew steadily from 160 souls in 1806 to nearly 400 by 1812, as smelting furnaces were built to work the even more abundant copper oxides.[25]

The revolutionary transformation of the Americas, however, soon brought changes to Santa Rita after the overthrow of Spanish colonial rule and the creation of the republic of Mexico in 1823. After the withdrawal of the Spanish garrison at the onset of the revolution, the Apaches began attacking woodcutters, forcing the shutdown of the furnaces for lack of fuel and essentially closing the mines. But the lifting of the ban of foreign investors after the revolution attracted aggressive new speculators, who soon took over the mines. The first was Sylvester Pattie, a hapless Missouri sawmill operator in his early forties, who in a fit of despondency after his wife died had sold his mill for $375 and headed to New Mexico with his twenty-one-year-old son, James, in the summer of 1825 to try his luck in a party of fur trappers. The following summer, after they lost their cache of furs, Guerra's Santa Rita agent, Juan de Onis, hired them for a dollar a day to guard a new gang of woodcutters so the furnaces could be started up again.[26]

The Patties were so successful that, when the trapping season began again in the fall, Onis offered them a five-year lease on some of the mines for only $1,000 a year. Sylvester Pattie and a couple of partners, Nathaniel Pryor and James Kirker, promptly accepted, but his son refused to stay and joined another party to go trapping again. Pattie and the others parlayed the lease into a very profitable enterprise, not only making a handsome profit on the copper, but also on a 200 percent markup on food and supplies for the miners. By the time his son returned in the spring of 1827, broke again after the trappers were ambushed, Pattie and his partners had accumulated over $30,000 in just six months and looked forward to making a fortune of at least $300,000 before the lease expired. To turn an even bigger profit, Pattie asked his son to take the $30,000 to St. Louis to buy provisions more cheaply. But again his son refused to help, so Pattie instead sent a clerk, who absconded with the cash. Meanwhile, Guerra, threatened by new laws expelling nearly all Spanish-born citizens from Mexico, had hoped to sell his interest in the mines to Pattie and his partners. But since they could no longer afford to buy, Pattie and his son turned once more to trapping and headed to California with Pryor. There the luckless pair were arrested as illegal aliens, trapping without a license, and the thoroughly depressed Pattie died, at the age of forty-five, in prison in San Diego. Such was the fate of the first Anglo-American investor in the western mines.[27]

With the aid of James Kirker, however, the Santa Rita bonanza quickly attracted new American investors, who eagerly took up Guerra's interest, together with a lease from Elguea's heirs, late in 1827 and made real fortunes. They were Kirker's friend Robert McKnight and his business partners, Stephen and Andrew Curcier. McKnight, who became the mine manager, had come to New Mexico from St. Louis in 1812 as an impulsive twenty-two-year-old trader, mistakenly thinking that the first stirrings of revolution had overthrown the Spanish. He, too, was arrested as an illegal alien and imprisoned, but he was finally released after eight years and turned to mining in Chihuahua. The Curcier brothers, who apparently put up most of the money, had fled to the United States from the Caribbean during the Haitian revolution, and after twenty years in Philadelphia they came to Chihuahua in 1826, looking for better opportunities after the successful Mexican revolution. Talking the governor of Chihuahua into stationing a garrison at Santa Rita again, and hiring Kirker to help provide additional protection against the Apaches, they resumed large-scale copper mining.[28]

As Elguea had done, they apparently contracted with the new republic's mint in Mexico City, which began striking copper coins in 1828 when the Santa Rita mine reopened, minting $4,700,000 worth before stopping in 1837, when the Apaches finally forced the mine to shut down again. During that time, the camp flourished once more as the Curciers brought in several hundred miners and opened a dozen deep shafts, producing perhaps as much as 1,000 tons of copper a year, if the mine was in fact the main source for the mint. But it all came to an end when the Apaches stopped the supply wagons after their chief and twenty followers were killed in an ambush. By then, however, the three partners had all made their fortunes. Esteban Courcier, as he came to be known, was said to have made $500,000, and acquiring most of Andrew's share after his death, he became one of Chihuahua's wealthiest citizens and the most successful of the pioneering western mining investors. Santa Rita's bonanza, mostly in massive, low-grade "porphyry copper" ores, still had far more fabulous fortunes that would pay billions in dividends to its future investors. It eventually produced some 5,000,000 tons of copper by the end of the twentieth century, worth by then well over $10,000,000,000 and making it the fourth largest copper producer in America.[29]

The other great outcropping of native copper in the West, though not nearly as large as Santa Rita, was the enormous Bonanza vein in the remote headwaters of the Copper River basin beside the glaciers of Mount Wrangell, near the southern coast of Alaska. There, running for many miles and spanning an equal breadth, were whole mountainsides strewn with copper nuggets and rich chunks of ore, all exposed for the picking. For several centuries they had been worked by a small but enterprising band of Dene, called the Mednovsty, or Coppers, by the Russians, and now known as the Ahtna. Settling in the summers at Taral in the canyon of the Copper River about ninety miles above its mouth, and in the fall and winters about sixty miles up its tributary, the Chitina, at a camp close to the bonanza, they controlled access to the treasure. There, too, they also fashioned the copper into a wide variety of axes, bowls, spear points, knives, needles, beads, and ornaments, which they traded throughout Alaska and western Canada, making their *tyone*, or "chief," one of the richest and most powerful men in the Great Northwest. Even after the Russian fur traders came, the skillful and determined tyones managed to conceal the source of their copper and protect their monopoly to the end of the nineteenth century, a truly unprecedented success.[30]

It was a hard fight at first, however, for starting in 1795 one of the fur company managers dispatched a dozen men to seek out and take over the Ahtna's copper. Clearly perceiving the threat, the Ahtna massacred the whole party, but that didn't stop the search. So when Aleksandr Baranov, later head of the Russian-American company, persisted in sending still more search parties over the next ten years, the tyones simply offered guides, who carefully steered the parties away from the secret ground. In the fall of 1804, however, one persistent searcher, Konstantin Galaktionov, apparently discovered the bonanza at last, and as his reward he and his guide were also killed before he could get out to tell. After that Baranov finally backed off, and not wanting to jeopardize the highly profitable fur trade, he simply exiled the three men offered up as the murderers. Even after Alaska became an American possession in 1867, prospecting in the area was courteously but effectively thwarted by the last powerful tyone, Nicolai, until shortly before his death in 1899. By then the steady influx of manufactured goods from the outside had diminished the trade value of his copper, and a throng of gold seekers heading for the Klondike had also brought in a deadly outbreak of smallpox. So on his deathbed that summer he gave a party of prospectors a map and a guide to part of the bonanza in exchange for a cache of food for his hungry family. All the ground for miles around was soon staked, and the great Kennecott mine eventually produced nearly 600,000 tons of copper now worth over \$1,000,000,000.[31]

MEANWHILE, soon after the American Revolution and even before the new nation had begun to expand beyond the Mississippi, the rising expectation of vast mineral riches in the "West" aroused the greed of a twenty-eight-year-old would-be monopolist, Nicholas I. Roosevelt, and a couple of New Jersey copper mining associates, who tried to claim it all. In January of 1796, in the spirit of earlier royal charters, they unblushingly asked the Congress for nothing less than the "exclusive right of searching for and working mines in the North-West and South-West Territory" for as many years as the lawmakers "in their wisdom shall conceive the difficulties, the dangers, the expenses, the importance, and the merit of the undertaking may deserve." In exchange for this indulgence, Roosevelt and his partners offered to pay the government, as a royalty, an unspecified but "reasonable portion of the produce of such mines, pits and quarries" that they might find.[32]

In hopes of gaining the sympathy of the lawmakers by impressing upon them the great risk that the would-be monopolists would be taking, they quoted pioneering economist Adam Smith's new *Inquiry into the Nature and Causes of the Wealth of Nations*. There he had noted that "of all those expensive and uncertain projects, which bring bankruptcy on the greater part of the people who engage in them, there is none perhaps, more perfectly ruinous than the search after new silver and gold mines. It is perhaps the most disadvantageous lottery in the world, or the one in which the gain of those who draw the prizes bears the least proportion to the loss of those who draw the blanks; for though the prizes are few and the blanks many, the common price of a ticket is the whole fortune of a very rich man." The Congress in their wisdom, however, followed instead Smith's damning advice, understandably omitted by the petitioners even though it was the very next sentence, that "projects of mining, instead of replacing the capital employed in them, together with the ordinary profits of stock, commonly absorb both capital and profit. They are the projects, therefore, to which of all others a prudent law-giver, who desired to increase the capital of his nation, would least choose to give any extraordinary encouragement." So the prudent lawgivers, still in an egalitarian mood, simply let the coveted monopoly die a quiet death in the Senate the following year and gave everyone a chance at the prize and the pain.[33]

Just two years later, in 1799, the first paying gold mine in the country was discovered by a twelve-year-old North Carolina farm boy, Conrad Reed. While fishing in a creek, he picked up a seventeen-pound lump of metal that his father used as a doorstop until it was finally found to be gold in 1803. His father then took in three partners on equal shares to put six slaves to work digging for gold after the crops were in. They took out over $14,000 worth in two years and began to excite the interest of outsiders. The most ambitious of these was William Thornton, a forty-seven-year-old Washington physician, head of the U.S. Patent Office, and amateur architect of the original federal Capitol. Hearing of gold nuggets as big as twenty-eight pounds, Thornton rushed to Reed's farm and quickly bought up over 35,000 nearby acres, before most of the owners realized their potential. Early in 1806 he formed the country's first joint-stock, gold mining venture, the North Carolina Gold-Mine Company, with 1,100 shares of $100 each, and an impressive array of trustees that included Thomas Tucker, treasurer of the United States; John Van Ness, president of the Bank of the United States; and Daniel Carroll, the largest land holder in the District of Columbia.[34]

To push the shares, Thornton wrote an enticing prospectus, claiming that his lands held "nearly the whole of the valuable gold mines," and curiously emphasizing that they lay at "precisely . . . the same latitude as some of the richest Spanish mines" in New Mexico! Moreover, he claimed that although mining is frequently risky, "in the present case it cannot be called mining," because "no deep digging or blasting is necessary," and "the gold is pure and requires no process." All that was needed was the labor of "black boys," who would consider gold digging "more as amusement than work." Thus he assured investors that every share would soon sell for profits of "many hundreds per cent." But like the great majority of mining stock ventures that would follow over the next two centuries, those assurances were never realized, and the company folded in the depression of 1808. Even a capital as modest as $110,000 far exceeded the capacity of the fledgling southern placer operations to repay, and it took over twenty years before the accumulated gold from all the mines even matched that amount.[35]

It was the discovery of the sources of the placer gold and the beginning of lode mining along the southern gold belt that finally stirred the first ripple of an American mining stock mania. The first gold lode was found not far from Reed's farm in 1825, and within the next few years the gold belt was traced on down the Piedmont and into the treaty lands of the Cherokee Nation. With the connivance of the popular Democratic president, Andrew Jackson, Georgia seized the Cherokee lands illegally, savagely drove the people from their towns, farms, and plantations, and banished them down the "Trail of Tears" to the west. The state then gave away their lands in a lottery of forty-acre plots, spurring wild speculation in gold lots, which the lucky winners sold for as much as $10,000. By 1828 Piedmont gold production had jumped from $5,000 to nearly $50,000 a year, and by 1830 it was close to $500,000 a year. Moreover, rumors spread that one marvelous mine, the Capps, had yielded 1,600 pounds, or $400,000 worth, of gold in just one week! The gold frenzy had begun, and before it ended in 1837 the southern mines had yielded up over $5,000,000 in gold. In the heady prosperity of the early 1830s, several dozen mining companies were chartered to share in the exploitation of mines from Virginia to Georgia and of investors all the way from Philadelphia to London. Even the Democratic vice president, John C. Calhoun, and the powerful Whig senator Henry Clay joined in the speculation. Most companies had only modest initial offerings of under $100,000, but some reached for as much as $1,500,000. All together they sought to raise well over $10,000,000 from excited investors.[36]

The most extravagant venture was the Mecklenburg Gold Mining Company, chartered in North Carolina in 1830 for $300,000 and financed in London to lease the marvelous Capps and the other rich mines around Charlotte. Its organizer and manager, a flamboyant Italian count, Vincent de Rivafinoli, brought in tons of machinery and a crew of European miners and millmen to direct several hundred rented slaves. He also started a small plantation to feed them all, built himself one of the finest homes in Charlotte, and spent an undue amount of time in New York, where he and a partner opened the first Italian opera house. But he failed to pay the mining investors a profit. By the time they finally got around to "forcibly" removing him in the fall of 1832, however, the company was hopelessly in debt. Rivafinoli, however, would prove to be only the first of an endless line of profligate managers sent forth by hapless investors in American mines.[37]

The next ripple of mining stock mania came during the brief prosperity of 1844–46, when the Michigan "copper fever" seized the Northeast with the opening of the rich mineral lands on the Keweenaw Peninsula in Lake Superior. The Chippewas had long worked the native copper deposits there, and Jesuit missionaries and fur trappers had made them known to Europeans in the seventeenth century. But it was not until 1843 that the Congress bought the mineral lands from the Chippewas and speculators grabbed up leases on nearly four hundred tracts for exploration and mining. Each tract of nine square miles was leased with no down payment for nine years at a rent of 6 percent of the mineral production. Extravagant reports of rich strikes in some tracts, predicted to yield as much as $6,000,000 a year from a single vein, excited even the most conservative investors. Before it was over, a hundred companies had been formed to meet the demand for shares in what was dismissed by some as a "subterranean lottery." The great proliferation of new companies was also greatly aided by the faster and easier new incorporation laws that had replaced the slower and cumbersome special charter laws of most states. In all, these companies offered investors around $10,000,000 in shares, but only a handful of them ever produced any copper.[38]

Even the most productive of the pioneering ventures, the Pittsburgh and Boston Mining Company, which worked the phenomenally rich native copper deposits of the Cliff mine, was slow to pay a profit to its investors. It was organized in May of 1844 by Pittsburgh doctor Curtis Hussey, druggist John Hays, and several friends with an initial investment of $15,000, divided into 6,000 shares of $2.50 each. But they paid out another $18.50 a share, or a total of $111,000 in assessments before they finally started seeing dividends, the first from any of the Lake Superior mines, in May of 1849. Then they got their original investment back within a year, and it was pure profit after that. By the time most of the dividends had been paid out in 1867, they had collected nearly $400 a share, or a total of $2,280,000, making it a truly fabulous investment, paying any of its original investors who held on an average income of 70 percent per annum for twenty-four years![39]

The public, of course, didn't do anywhere near that well, for their own exuberance in the market minimized their profits. Once the Pittsburgh and Boston began paying its annual dividends of $10 a share or more, amounting to a return of more than 50 percent a year, investors clamored to get in, and they quickly drove the shares up to $105 apiece on the Boston Stock Exchange before slacking off. That price just allowed investors to collect at least 10 percent a year from the dividends, which was barely a couple points above what they could draw in interest on secured loans and several points above savings account rates, though at far greater risk. So even though the dividend rate tripled over the next decade, ever-eager investors also tripled the price and kept the profit margin thin. But because their exuberance held down their percentage returns, the investors had to hope that the mine could keep paying such dividends for at least a decade or more just to get their initial investment back, let alone make a profit. Even the Pittsburgh and Boston ore, however, couldn't keep up its high dividends that long, so nearly all of those who bought in after 1858, when the stock first went above $300 a share, never even got their money back! For once the dividends stopped, the stock prices soon collapsed as everyone rushed to get out, suddenly realizing that, unlike many other business investments, nearly all of the assets of mines were in the ore itself, and when that was eventually exhausted, there was little else left. But that was just one of many simple lessons that generation after generation of hopeful new mining investors would have to learn, again and yet again. Once they did, there is little wonder that many sold out to go for quicker speculative profits instead.[40]

Nonetheless, the Pittsburgh and Boston company was a marvelous example of just how big the lottery prizes could be for early investors, and its widely heralded profits would lure many into the market. But what it paid was still far from the instant fortunes that many feverish speculators wanted. So most promoters, in fact, didn't even bother to look for metal outside the pockets of investors, and they took their profits just from talk of dividends. The copper fever soon broke, however, after the government revoked most of the free speculative leases and began selling the land outright at $5 an acre, forcing promoters to pay up front. Still, Michigan's copper mines were truly rich, and with the opening of the enormous Calumet & Hecla bonanza in 1865 they would dominate American copper production until the early 1880s, and they would eventually produce nearly as much as would the great Santa Rita. But in the minds of American investors and speculators, all the copper mines were quickly outshone by rich new mines of gold and silver in the West, as the expanding nation seized California and New Mexico under the banner of "Manifest Destiny" in the Mexican War.[41]

2

THE GOLD LUST

CALIFORNIA GOLD EXCITED the lust of the first great mining rush and the first western mining stock mania. It drew over a hundred thousand gold seekers west and it emptied the pockets of many more thousands who stayed behind. A thirty-seven-year-old carpenter, James Marshall, found the gold that started the frenzy in the tailrace of a sawmill he was building on the American River in the foothills of the Sierra Nevada. He picked up the first flake on January 24, 1848, just weeks before the territory, newly seized in the Mexican War, was officially made part of the United States, while the Senate was still debating the acquisition and that old Whig Daniel Webster was arguing that the whole territory was "not worth a dollar." Word that there was "gold for the gathering" quickly emptied nearby settlements; more gold was found, and tales of instant fortunes began to spread across the seas. But it was nearly a year before the tales were credited in the East.[1]

Finally, on December 5 President James Polk presented to the Congress the electrifying report of Col. Richard Mason, "commander of our military force in California." Mason had visited the placers in July and reported upwards of 4,000 men at work digging out "from $30,000 to $50,000 worth of gold, if not more, daily." Congress ordered the printing of 15,000 extra copies of the report, and the eastern press went wild, filling newspaper columns with even more incredible stories. The Reverend Walter Colton, former navy chaplain and editor of the first newspaper, *The Californian*, at the capital, Monterey, reported that people were running all over the country picking up gold like "a thousand hogs, let loose in a forest, would root up ground nuts." A company of seven from Monterey, employing about fifty Indians to dig for them for seven weeks, got 275 pounds of gold, or nearly $10,000 apiece, a lifetime's wages in the

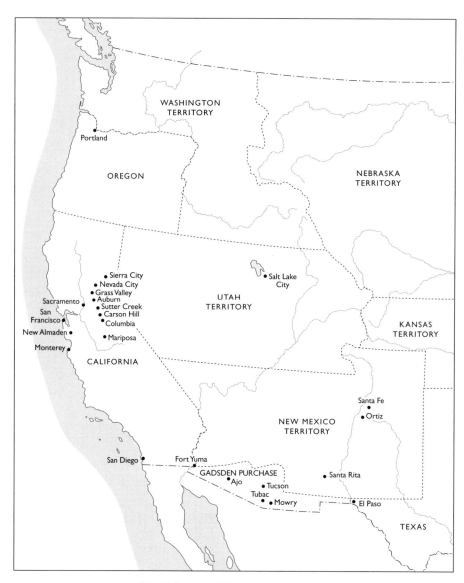

THE MINING WEST IN THE 1850S

East. And one entrepreneur, hiring sixty Indians, was making a profit of "a dollar a minute"! Still, a Philadelphia editor reminded his readers of the great "gold humbugs" of the past in Hispaniola, Jamestown, and the Piedmont, and a Boston preacher, echoing Adam Smith, warned that it was a lottery with "some splendid prizes and many dismal blanks." But the Reverend Colton confidently predicted, "When the wealth in these gold mines is really known and believed, there will not be wagons and steamers enough to bring the emigrants here." Polk's message made believers of most. Within days easterners complained the "gold mania rages, carrying off its victims hourly . . . gold fever runs ahead of the cholera, by tens of thousands . . . a perfect diarrhea of emigration threatens," as the first surge of gold diggers headed west. The world's greatest gold rush was under way.[2]

The first mining companies formed to work the western mines, however, were radically different from those that preceded them and those that followed. These were the emigrant mining companies, organized throughout the eastern states in 1849 and 1850 by the more than 100,000 argonauts who joined the rush to California. Most were inherently cooperative, even communistic, ventures. Each member not only contributed his share of the capital needed to outfit the company, but also promised to share the fruits of his labor. Individual shares were expensive, running about a year's wages, or $300 on the average, but the total capital of the companies was quite modest, at least by later standards. They were typically capitalized at about $15,000 but ranged from barely $3,000 to as much as $50,000. The total depended on several factors—the number of members, the amount of provisions, whether they were going by sea and would buy or lease a ship or were going overland and would buy or furnish their own wagons and teams, and on the intended scale of their operations, if they planned to take trading goods as well as mining equipment.[3]

No matter how small the company, however, they often donned all the trappings of corporate identity: printing elaborate charters, constitutions, and bylaws; electing presidents, vice presidents, treasurers, secretaries, and boards of directors; and issuing stock certificates. They also usually saddled themselves with pretentious, but prosaic, names, such as the Boston & California Joint Stock Mining & Trading Company—although a few, like the Sea Serpent Company, named just for the ship they sailed on, showed refreshing simplicity.[4]

Most of the charters and bylaws placed virtually no restrictions on the companies, allowing them to engage in mining, trading, or any profitable pursuit. But the shareholders were required to devote their full time and energy to the benefit of the company, engaging in no personal business during the lifetime of the venture. They typically expected to operate for at least a year or two and then disband, dividing up their collective earnings among their shareholders. The members were also promised full health care from their comrades without any loss for time they were too sick to work, unless it was due to "immoral conduct," and their full share would be paid as the death benefit to their heirs if they died. But in return many of the companies imposed a number of moral restrictions on their shareholders. Typically, observance of the Sabbath was strictly demanded, with fines of $10 for every violation. In addition, swearing, quarreling, drinking, gambling, and in some companies even "playing cards for amusement" was forbidden, and the penalties ranged from $5 to as much as $100 for such offenses. Persistent offenders ultimately faced banishment from the company, with forfeiture of all claims. Real crimes, such as theft, robbery, and embezzlement, were punishable by much larger fines of $1,000 or more, plus banishment. Violent crimes, like murder, were not discussed, but they were dealt with in a summary fashion when the occasion arose. All charges were to be tried by a jury drawn from the company members, and every shareholder had to agree to abide by their decision. In practice, petty violations were most likely only punished in the breach, for the argonauts found that conditions en route and in the diggings were quite different from Sunday life in the East.[5]

In all there must have been at least a thousand emigrant mining enterprises formed in the eastern states, raising at least $20,000,000 during the frantic gold rush years of 1849–50. There were over 120 in Massachusetts alone, and three-fourths of that state's 6,000 argonauts joined such companies. The success of these ventures was mixed. Most did succeed in providing their members with relatively cheap passage to the gold fields, and those that also brought along speculative cargos for trade made handsome profits. But as mining enterprises they were almost universally failures. Dissension among members, heightened by the hardships of the journey and a divisive feeling that there might not be fortune enough to go around, slowly poisoned the spirit of cooperation that had brought them together. Ill adapted and ill equipped for the realities of life in the

TO CALIFORNIA
ON SHARES.

TO SAIL ON JANUARY 15th.

Entire Expense for Passage and Outfit $160.

A few persons of industrious habits and who can give undoubted references of their good moral character, can join

GORDON'S
CALIFORNIA ASSOCIATION.

This Association consists of **100** members who go in a body well armed, whose object is to go safely, and comfortably fitted out and equipped.

Each member pays **$160**, for which he has passage in a fast sailing vessel, and is found with six months' provisions of the best quality.

The President of the Company, who is a practical geologist, accompanies the expedition, and for the one-fifth share in the profit provides the Association with

CAMP EQUIPMENT,

To every ten men a large tent with oil skin cover.

To every man a hammock, hammock bed, blankets and coverlid, a tin plate, dish, mug, knives and forks, &c.

To every twenty men a Field Cooking Stove, ovens, boilers and kettles. Also,

MACHINERY AND IMPLEMENTS:

Machines for washing in the gold deposits to go by horse power, force pumps, screens, seives, shovels, pick axes, axes, saws, two sets carpenters tools, set of coopers tools, set of blacksmiths tools, including forge, bellows, &c. saddlers and sailmakers tools, tinsmiths tools, &c. saddles and bridles, rifles, fowling pieces, fishing tackle, and agricultural implements.

The Association carries a complete medicine chest and a Physician.

A German assayer with chemical tests and apparatus for trying the value of minerals.

The main body of the Association go by Cape Horn, and a few go across by Mexico as pioneers, to survey locations and prepare the way for the arrival of the rest of the company.

The great object of the Association is, by combining together to make a strong party to insure the *safety* of every member, to have all the assistance that good machinery and science can give in washing for gold, and to have everything comfortable and convenient, so that the trip may be made reasonably pleasant. Good care will be taken of health, and all be brought within a small expense.

The agent of the Association is now in California and doing well.

Apply immediately and if possible personally, at the office of the Association.

A. COCHRANE'S,

143 Walnut St. 2 doors above Sixth, Phila.

gold fields, almost all disbanded soon after reaching California. Nonetheless, the companies clearly helped bring tens of thousands of men and a vast amount of goods to California, accelerating the opening of the mines and the creation of the state.[6]

A few enterprising argonauts, however, organized companies with a definite eye to their own personal advantage. One of the earliest and largest of these private enterprises was Gordon's California Association, formed in Philadelphia in December of 1848 just a couple of weeks after President Polk's electrifying message. It was organized by a sharp and charming, thirty-year-old former London guano broker, George Gordon Cummings, who had dropped his surname and fled to America a few years before, leaving his creditors with a "large quantity of suspicious bills," totaling several thousand pounds. Gordon, announcing himself as a "practical geologist," offered "industrious" gold seekers shares that promised them passage to California, six months' provisions, and a full outfit of camping and mining equipment in exchange for $160 in cash plus a fifth of the profits of their labor in the gold fields. The demand was so great that he easily filled one ship with 127 Pennsylvania shareholders to sail around Cape Horn, and he sold another 113 shares to New Yorkers at $225 to fill a second ship that sailed from the "foot of Wall Street" on a shorter route via Nicaragua. Gordon and his family sailed with the latter group in February of 1849.[7]

Like the emigrant companies, Gordon's association soon disbanded, as unhappy shareholders branded it a "complete humbug." Those who went around the Horn all split as soon as they arrived in San Francisco, and the other group began to break up in Nicaragua after Gordon spent months arranging passage on to California and levied an assessment of a couple thousand dollars. Most finally sailed on an "old tub" that was "slower than justice," while others took a smaller ship that was becalmed off Baja California and some chose to walk the last 200 miles to San Diego. The promised provisions were all consumed en route as the two-month trip stretched into eight and disgruntled shareholders were no longer inclined to share their prospective earnings. Gordon, nonetheless, did very well for himself, using the assessment to buy lumber in Nicaragua that he sold in California for "an almost fabulous price," giving him capital for further speculations.[8]

ALTHOUGH THE EMIGRANT COMPANIES were by far the most numerous of the mining ventures launched during the gold rush, the actual capital invested by the argonauts heading west was soon dwarfed by that poured into the mining maelstrom by all of the vicarious argonauts left behind as the first wave of mining speculation swept the nation and on across the Atlantic. For with each ship from California bringing new tales of quick fortunes and million-dollar cargoes of gold dust, even those unwilling or unable to risk the hazards of going to the diggings eagerly sought ways to share in the wealth to be won there. Even the man who wouldn't lend a dollar to a friend was soon found plunging ten on mining shares.

The flood of gold also lifted both the American and European economies out of the depression of 1847–48 to a new prosperity in the early 1850s that fueled the speculative fever. But the first opportunities open to the armchair argonaut were limited to helping outfit a gold seeker, or to buying shares either in an emigrant company and sending a substitute, or in a company that simply sent a hired crew to California for a share of whatever they might find. Typical of the latter was the Curtis California Mining Company, which sold 2,000 shares at $5 each to send a crew of ten to the diggings. They were to work for two years, paying the company only a sixteenth of their earnings. Expectations of a profit to investors in such a scheme hung on exaggerated reports that miners were making an average of $100 a day. Good earnings in 1849–50 were closer to $10 a day, which after expenses wasn't sufficient to pay back even the original investment, let alone a profit. Still such ventures were for a time the only game in town and those vicarious argonauts anxious to get some sort of stake in the gold rush bought heavily.[9]

The heaviest speculation in these shares was abroad, where news from the gold fields was perhaps the least reliable and the most enticing. In London during the first frenzy in January of 1849, eight California mining companies were formed in ten days to offer the public more than $12,500,000 in shares. In Paris, which soon became the center of the California mania, over eighty companies with shares totaling nearly $70,000,000 were floated to meet the demand of speculation within the first two years of the gold rush. These French companies were more attractive to cautious investors because, like modern corporations, they offered limited liability to shareholders. Unlike British joint-stock companies, which under the law of the time made each shareholder liable for

the full debts of the company, French company law, like that in most of the states in America, limited a shareholder's liability to no more than the par value of the shares held. Thus the French mining shares also sold well in England, and by 1851 one enthusiast was predicting that other British concerns would soon be organizing in France in order to "induce persons of limited income to embark small sums in commerce, which they are now prevented from doing by the liability of the whole of their property in case of failure." But it was the wealthy investors who were much more concerned with liability, because they had much more to lose, while the poorer investors knew that there were much deeper pockets than theirs for creditors to go after. Much of the appeal of the French companies was also to speculators, since the shares were simply assigned to the bearer and were easily sold and transferred without registration in the company's books, as was required for the British companies. But Parliament soon responded by enacting the Limited Liability Act of 1855 to finally shield British investors, aid speculators, and add the word "Limited" to the company names. Even before that the eager promoters of the Anglo-Californian Gold Mining Company simply added the words to their prospectus to catch additional investors, but it didn't fool creditors who later forced them to pay up.[10]

These British and French ventures were much more extravagantly capitalized than their American cousins, averaging around $1,000,000 each with one as high as $9,000,000. They also assumed more grandiose titles, such as the Gold Mines, Lands & Rivers of California Company, the Company of Adventurers for Exploring the Gold Districts of New California, and the all-encompassing Le Nouveau Monde, as well as the more candid La Toison d'Or (Golden Fleece) Compagnie des Placers de la Californie. Others with more mundane names, like the Compagnie Française & Américaine de San-Francisco, flourished much more enticing shares to lure their investors. To inspire public confidence, the promoters also stacked their boards of directors with an impressive-sounding array of nobility and titled gentry, ranging from mere admirals, generals, and members of the Légion d'Honneur to counts, marquises, and even a prince, Louis Lucien Bonaparte, who decorated Le Nouveau Monde. If the excitement failed to reach the fever of the infamous Mississippi Bubble, it was not for want of trying.[11]

But without any property to work, or even a passing knowledge of conditions in the diggings, the declared purposes of these companies were as vague and all encompassing as some of their names. One broadly sought only to satisfy the investors' "*desire* to participate in the advantages to be derived from the wonderful discoveries of gold." Another aimed only a little more narrowly at the "exploitation of the placers of California and the exportation of merchandise." To pursue these nebulous goals, the companies usually sent out some unnamed "gentleman of high respectability" and a crew "accustomed to the washing and extraction of gold in all its forms," to see what they could do. But one company sent its gentleman to Washington, D.C., rather than the gold fields, in an old royalist quest for a special government "charter for the privilege of working or otherwise using the territory of California."[12]

Other companies sent out a variety of marvelous patented gold machines, which were expected to do the work of as many as a hundred men. From such devices one French company, L'Aurifère, unblushingly promised annual dividends of 450 francs on 10-franc shares. As if such profits were not inducement enough, companies also offered free books, chromos, magazine subscriptions, and lottery tickets as bonuses with each share purchased. Shares in these schemes ranged from $1 to $100. But would-be British investors of even the most modest means could reserve a share in some companies for a down payment of as little as 6 pence, and a few companies even took produce in payment for stock.[13]

The press response to these promotions was divided. The esteemed "money market" editor of the London *Times* and criminal phrenologist, Marmaduke Sampson, came down hard on such ventures, arguing that it was preposterous to believe that a remote board of directors in London could guarantee that the men they sent would continue to work for the company after they got to the diggings when the commanders of American troops stationed there couldn't prevent desertions "even by the severest discipline." He flatly warned all who might be "deluded" into putting their money into shares of such companies not to expect even a "shadow of its return." Yet the business manager of the same paper eagerly accepted lengthy advertisements for such promotions. He was, of course, not alone. One paper in Paris, the bourgeois *Journal des Débats*, sold the whole last page to mining company advertisements, while the editor, critical of the business, signed his name at the end of the preceding page and let the devil take the hindmost.[14]

The promoters of some of these companies may have been as naively optimistic as their investors, but others were blatantly fraudulent. A belated investigation by the Paris police found that very little of the money collected from the stock sales had actually been spent on sending men to the mines, or on anything else related to mining. The largest expenditures had been for advertising. But most of the money simply went directly, or indirectly, into the pockets of the promoters, many of whom skipped the country as soon as they unloaded their stock. A handful of disillusioned investors brought suits against some of the companies, but the majority suffered their losses and embarrassment in silence.[15]

How much was lost by investors in these first California mining companies is hard to estimate. The total par value of stock offered the public in the first two years of the excitement was over $80,000,000. This was well in excess of the $50,000,000 in gold actually mined in California during the same years. But the stock was by no means fully subscribed, nor were even those shares taken all fully paid. Thus, the total paid into these first omnibus ventures was probably no more than $10,000,000, but this was only the beginning.[16]

Even if these companies had all been honestly conducted, their chances of success would still have been slight, for during the early years of the gold rush there was very little meaningful basis for outside investment. The shallow placer deposits, already weathered and broken out of the hard rock and concentrated in the bottoms of the streams of California, mostly required no other capital than a shovel, a pan, hard work, and a lot of luck. At best they were "poor men's diggings," where an individual could succeed almost wholly by his own effort, or where a group of miners could temporarily join together to dam a stream or dig a ditch to bring water for the common good. The more ambitious invested considerable time, labor, and savings in erecting wing dams and flumes to divert the rivers and in putting up water wheels, pumps, cranes, and windlasses to work the river bars to bed rock in the hopes of dividing a rich cleanup of gold. But they were usually temporary, rather informal associations, although even the little York Mining Company in Sleepy Hollow with only two hundred shares of $100 each still had them all neatly printed.[17]

Despite wonderful success stories, such as a widely publicized report of a company of three taking out $87,000 in just a few days, most of these ventures were also unprofitable, demonstrating once again Adam Smith's warning on the preponderance of blanks over prizes. An accounting of over a dozen companies working bars on the Tuolumne during 1850 revealed that the partners in only a third of the companies earned better than the going wage of $8 a day. The most profitable paid ten men $20 a day, while the bottom third with over a hundred men found no gold at all and got nothing for their labor. Only some of the smaller associations with modest expenditures had a chance of paying a profit, while the larger companies with more grandiose plans were doomed to fail under their own weight because the placers were simply spread too thin. It was only when prospectors found the rich, deep quartz lodes from which the placer gold had come that a need for outside capital truly arose.[18]

LODE MINING, unlike most placer mining, usually required some capital investment for labor and machinery right from the start. The surface croppings, no matter how promising, were rarely rich enough to pay the costs of exploration needed to try to determine the extent and wealth of the lode. Then, even if a sizable body of paying ore were found, more capital was needed to put the mine into production, hiring a crew of miners and purchasing hoisting and milling machinery needed to extract and work the ore. For not only was it much more costly to dig the gold out of hard rock than the loose sand of a stream bed, it then required costly milling and crushing to break up the ore into fine particles to help free the lode gold, which natural weathering had already done for placer gold. Thus it was, with the beginning of lode mining in the West, that a real basis for mining investment was finally opened to the vicarious argonauts in the East.

These early mining companies sought capital for three basic purposes: to pay for the mining property, to pay the organizers and promoters an upfront profit, and most important from an investor's point of view, to pay through sale of the company's treasury stock for all of the development and operating costs until the mine could be put on a self-supporting and paying basis. If the treasury stock was exhausted before the mine would pay, the directors could either issue additional shares or, as a result of unlimited liability, levy assessments on the existing shares, which were then subject to forfeiture if the assessments were not paid. The chances of success, of course, and indeed the legitimacy of the venture depended directly on the relative apportionment of these three disbursements. But all too often the organizers and promoters would simply take the bulk of the money and run.

Even in an honest venture the risks were still enormous, hanging not only on the skill, economy, and honesty of the management, but on the whims and vagueries of local geology. If bonanza ore was struck, however, the profits could be astronomical, and it was this hope that drove the seemingly endless waves of speculative investments in the western mines.

The first lode mines in California were discovered in Mariposa in the fall of 1849, and within a few years the great Mother Lode of gold-bearing quartz had been traced for roughly a hundred miles along the western foothills of the Sierra. Other rich lodes were found farther up into the range to the north at Grass Valley, Nevada City, and Sierra City, and lesser veins were soon located throughout the country from the edge of Death Valley to southern Oregon.[19]

Much of the prospecting for gold-bearing quartz lodes was seasonal, carried on by small prospecting parties in the dry months when there was not enough water to work the placers. These parties were generally made up of a few miners who contributed their own expenses and agreed to share in their discoveries. If a promising lode was found, they would begin exploratory work, often with the backing of a local merchant, who provided food and supplies in exchange for an interest in the property. Such "grubstake" arrangements rarely lasted for more than a year. If the work failed to develop paying ore, it was abandoned and each was out his time or money. But if the prospect developed well, then a company was formally organized to raise the necessary capital to put the mine on a paying basis.

Such capital was frequently sought from successful merchants or other burgeoning capitalists within the state. A helpful bookseller aided the quest with a pamphlet entitled A "Pile," A Glance at the Wealth of the Monied Men of San Francisco and Sacramento City, listing over six hundred expectant millionaires with assets over $5,000. One of the most prominent local ventures was the Merced Mining Company, formed in March of 1851 by San Francisco port collector and former Georgia Whig congressman Thomas Butler King and private mint owner John L. Moffat. They gave a young Dane, John "Quartz" Johnson, $2,500 cash and $90,900 worth of shares in the $500,000 company for a string of promising claims on his Great Johnson Vein on the Merced River, just below Yosemite north of Mariposa. Touted as "the golden back bone of California," these rich claims, the Pine Tree and Josephine, plus their southern extension the Princeton, eventually produced about $10,000,000, but it would take many years of experimentation and litigation. Most of the local companies, however, were homely endeavors with a more modest capital of $100,000 or less, because, since California laws specifically made corporate stockholders "individually and personally liable" for their share of all company debts without limit, many of the city's monied men were still quite cautious. Moreover, the California courts would rule that even if the shares were forfeited for failure to pay assessments, the shareholder was still liable for a share of the debts.[20]

A few more ambitious men headed east with rich specimens of ore to seek capital in such financial centers as New York, Boston, or Philadelphia, or even abroad. One of the first of these was an exuberant but sly physician, James Delavan. He was the secretary of an unprofitable placer company of forty men who, in the summer of 1849, had invested $40,000 worth of effort at Rocky Bar on the American River and recovered only 107 pounds of gold worth less than $27,000. But they had happily convinced themselves that they had found a great quartz lode that was "the source or matrix of all the gold in the country," and if they just had enough money they could recoup their losses and much more. So in February of 1850 Delavan rented a third-floor office at 54 Wall Street to exhibit the "choicest selection of specimens ever brought from or found" in California, the centerpiece of which was a ten-pound "boulder, covered and filled with gold." Ten days later in the New York *Tribune* he announced the formation of the Rocky Bar Mining Company with a modest capital of $40,000 in 160 shares, of which 120 were offered to the public. At the same time he sent anonymous letters of California news to several newspapers and brought out a pamphlet on the gold fields, signed only "by one who has been there," all well larded with plugs for Rocky Bar. He succeeded in selling all but 37 of the public shares and proudly returned to California only to conclude that the $20,000 he had obtained was far from sufficient to do any serious quartz mining. Thus he and his partners decided to split the stock more than sixty to one and boost the capital to $1,000,000, magically turning the original handful of unsubscribed shares into more than 2,300 new shares at $100 each, which could bring in enough money to do something.[21]

So back Delavan went to Wall Street that fall to issue a classic prospectus to excite investors. Claiming their gold quartz could yield $4 a pound, or an astonishing $8,000 a ton, and that the lode was "literally inexhaustible," he then conservatively calculated that even if it only paid 20 cents a pound, they could put up a stamp mill, crush forty tons a day and produce $4,800,000 a year. This would pay dividends of 400 percent per annum, and as proof he promptly declared a 100 percent dividend. He didn't bother to mention, of course, that this was only on the old capital of $40,000 and would be paid out of sales of new shares! But it worked perfectly. He sold enough shares that the original investors got all their money out and still kept three quarters of the stock, and the company got roughly $100,000 to begin real mining. When that ran out they went to London in 1852 and sold another 1,000 shares for $125,000.[22]

The prospectuses of some of the other companies were even more blatant than Delavan's, and they all became models for those to come. Their pages were packed with tales of astonishing riches that would have made Munchausen blush. No exaggeration seemed too great—"average specimens" of quartz were said to assay $35,000 a ton, and there was enough of it to last fifty years. Fancied profits ran to a dizzying $240,000,000 a year—which the promoters at least conceded would "revolutionize the commercial relations of the world." Even the most conservative companies promised annual dividends of 200 percent. Nothing seemingly could stand in the way of their success; even grossly inefficient mills that saved but a fraction of the gold from the ore were dismissed with "but little embarrassment," for fortunes were so "easily acquired." Geography too was of little import, for every rich strike in California was cited in support of the value of each company's mine, no matter that they might be separated by hundreds of miles. Indeed the whole quartz district was said to be so "auriferous to an extent that defies calculation" that any honest company simply couldn't help but succeed. For final emphasis they also stressed that, unlike any other industry, the product of gold mining was "money itself."[23]

All sorts of easterners also joined in the scramble for a buck in the mining mania. Even New York *Tribune* founder and editor Horace Greeley, who would later advise young men to "go west," instead advised family men, infected with the "California fever, to invest the money it would cost them to reach the gold region in the stock of some quartz-crushing company which they believe will be honestly managed" and stay home. Greeley, of course, had just such a company in mind, the Manhattan Quartz Mining Company, of which he was secretary and treasurer. He enthusiastically promoted the company in both his advertising and editorial columns, as well as in a popular pamphlet on quartz mining, offering 1,500 shares at $100 par to open sixty-four "lots" in Grass Valley with assays as high as $3,464 a ton, and visions of "a golden harvest of large dividends and an increase in the value of stock" with "enormous profits of $38,765 per day! of $11,639,500 a year!"[24]

THE MOST AGGRESSIVE CAMPAIGN to raise money during this flurry, however, was carried out by John Charles Fremont, popularly known as the "Pathfinder" of the West, who claimed the largest tract of mining land in California. Fremont, a pretentious but rather inept opportunist, had acquired both his fame and his land by a very crooked path. Starting as a surveyor in the army topographical engineers, he had secretly courted powerful Democratic Missouri senator Thomas Hart Benton's fifteen-year-old daughter, Jessie, who was eleven years younger. When they eloped, her outraged father promptly had Fremont dispatched out west on the first of a series of surveys of the wagon roads to Oregon and California, officially to encourage western expansion, though obviously in the hope of breaking up their marriage. To her father's chagrin, the romantic and talented Jessie stood by her husband, transformed his prosaic official reports into popular adventures, and transformed the pedestrian path-marker into a celebrity "pathfinder." Arriving in California on the eve of the war with Mexico, however, he became involved in a dispute between the naval and military commanders and was court-martialed. Before he was taken east for trial, Fremont asked the American consul Thomas Larkin to buy him a ranch near San Francisco for $3,000, but he ended up with an unconsummated Mexican land grant to a former governor, Juan Alvarado, on the little creek called Mariposa ("butterfly") more than a hundred miles east at the foot of the Sierra Nevada. By the time he returned to California, however, gold had been discovered on the creek and he had become the accidental claimant to a potential bonanza.[25]

But this was just the start of more trouble, for Fremont's seeming good fortune also proved to be as elusive as a butterfly. The title to the land was faulty, because the original conditions of the grant had never been fulfilled, and even if they had, the grant did not include mineral rights under Mexican law, nor could it be sold! But Senator Benton, with an eye to increasing the family fortune, helped rig approval of the grant by hav-ing his other son-in-law, William Carey Jones, appointed "confidential" government agent to report on the validity of California land grants, while Fremont at the same time named Jones as his Mariposa agent. Not surprisingly, Jones claimed that nearly all of the grants, including the Mariposa, had "perfect titles" and that, contrary to other reports, he could find no legal basis for withholding mineral rights on the grants. Armed with this report, Benton and Fremont, who served a three-week term as one of California's first senators, introduced a bill for a blanket confirmation of all grants, unless there was clear evidence of fraud. But the Senate rejected the bill, accusing Fremont of acting only for "his own private advantage," and the Congress instead created the three-man Land Commission to rule on the validity of each grant. Undaunted, Benton had Jones named secretary of the commission, where he made the Mariposa grant the first case, and then, acting as Fremont's attorney, won its approval by a two to one vote before any standards of validity were established. The U.S. attorney general, however, promptly declared the grant invalid, and the federal judge in San Francisco concurred,

launching a long and costly legal fight with reversals piled upon reversals amid further charges of judicial bribery and corruption.[26]

In the meantime, Fremont eagerly tried to capitalize on his seeming good fortune, despite a contested title. First, he leased the original quartz lode to his San Francisco bankers and heaviest creditors, Palmer, Cook & Company, who formed California's first quartz corporation, the Mariposa Mining Company, in September 1850. Next, aided by gold-studded rocks bought from a dealer, he leased another lode to the Philadelphia and California Mining Company for $250,000, and as its president he helped push $2,000,000 in shares on eager investors. At the same time he dispatched agents to London and Paris offering many more leases to various other gold lodes, both real and imagined. His agents flashed more dazzling ore samples and talked of fabulous assays, averaging as much as $6,500 per ton, and outrageous profits of at least $26,000,000 a year, totaling more than $650,000,000 over 25 years, or a return of 20 percent per day on an investment of $500,000! The agents formed more than two dozen companies and could have floated even more, if Fremont hadn't promised too much to too many, unleashing a public squabble between two of his agents that scared off further investors. When a California newspaper branded his promotions an "outrageous fraud," Fremont sued for libel and went to London and Paris to try to reassure investors, but only made matters worse when he was arrested as a debtor. In disgust at both the crooked promoters and the gullible investors, one British mining expert, Thomas Allsop, seconded Samuel Butler's cynicism that indeed it seemed "the pleasure is as great, of being cheated as to cheat."[27]

The most promising of Fremont's leasers was Le Nouveau Monde, which after failing to make money in the placers eagerly paid $230,000 for seven of his quartz leases for twenty-one years at a one-sixth royalty. But even seven chances weren't enough, for they like all the rest soon found that the claims were bogus and there was really no profitable ground left on the grant. All of the companies threw up their leases in less than a year and most were promptly wrapped up. But Le Nouveau Monde and half a dozen others tried to salvage their investments by buying or leasing real mines elsewhere. Fremont, meanwhile, also floated seemingly endless loans using the property as a collateral cash cow to support his crucial title battle, as well as many pretentions. But in spite of all the wonderful claims for its richness, he let the whole estate go on the auction block to his associates, more than once and for as little as $579, trying to dodge some of his creditors. Fremont's scandalous dealings with the grant and its promotion would become an issue a few years later in his failed campaign as the new Republican Party's first candidate for president of the United States.[28]

THE CAPITALIZATION OF these early quartz mining companies ranged widely from $20,000 to $2,000,000. Their shares at first were pegged at $100 to $500 fully paid, clearly intended only for men and women of means. But some companies, particularly abroad, set their shares as low as $2.50 each, in order to attract a larger flock of investors. That and a promise of limited liability attracted a broad range of investors that even cut across traditional British class lines from laundresses to ladies and laborers to lords. The shareholders of the Anglo-Californian company were fairly typical of these early vicarious argonauts. Even though small investors with holdings of $25 or less made up over 80 percent of the company's 1,060 shareholders, all of their shares combined made up less than 10 percent of the total, so they had little voice in the company's affairs. The majority of the shares were held by the large investors with holdings of $250 or more. Women, two-thirds of whom listed themselves either as "spinsters" or "gentlewomen," accounted for roughly a quarter of both the small and large investors. The rest of the women were mostly shopkeepers and servants. There were only a few widows and no obvious orphans. The largest individual investors among the men were lawyers, doctors, clergymen and gentry, but most of the large investors were merchants, while the smaller investors came from all trades. Such a distribution was most likely representative of mining investors in France and in the eastern states as well.[29]

By late January 1851 the trade in mining shares seemed large enough that the New York Mining and Mineral Exchange tried to open trading at 50 Wall Street, but it was premature. By June of 1853, however, the speculative fever had swelled and the New York Mining Share Board was formed, since most mining stock was unofficially banned from the only marginally more respectable New York Stock Exchange. Abroad, however, California quartz mining shares were traded on the regular exchanges in London, Manchester, and Paris. As the fever grew, the other middlemen also moved in. Three mining journals sprang up in New York in 1853 just to satisfy the "interest of investors," and, of course, cash in on the promotional advertising.[30]

The total offering in quartz mining ventures must have reached over $20,000,000 at the peak of the fever in 1853. The shares actually sold probably ran at least $6,000,000 that year alone, so that the money put into hardrock mining well exceeded the returns from such mines—at most, still just a couple million dollars a year. This pattern was to be repeated in each successive mining boom for more than a century.[31]

The California fever broke that same year, 1853, as the unfulfilled promises of the prospectus writers took their toll on public confidence. The failure of many companies doubtless stemmed from simple ignorance and inexperience of their management. The early years of quartz mining in California were marked by costly experimentation in milling practices—with every imaginable new notion of an ore crusher and gold saver given its chance. By 1852 over a hundred different quartz mills, costing some $6,000,000, had been erected at the mines. But only a few had proved profitable.[32]

All too often, costly mills were set up only to find that there was not even any paying ore in the mines they were to serve. The Rocky Bar company, that first quartz venture formed in the East, offers a illustrative case history except for its tragic ending. When James Delavan returned from New York early in 1851 with both a Chilian roller mill and a stamp mill plus operating capital, he and his partners decided that their quartz lode wasn't quite wonderful enough. So they used some of the money to buy a real mine in Grass Valley from Alonzo Delano, the popular gold rush satirist "Old Block." Delavan then sank the rest of the money into a tunnel to tap the gold lode. But the rock was too hard, the mills were a failure, and he retired from the company at the end of 1852. The directors began levying assessments of $5 a share to carry on the work and ran through a couple of new superintendents. The first was "totally inexperienced," and although the second actually took out a lot of rich ore, he was so "extravagant" that the shareholders still got nothing. After the majority of the stockholders had forfeited their shares rather than pay more assessments, the company reorganized as the Mount Hope in 1856 and finally sent out a conscientious young man, Michael Brennan, who worked the mine more efficiently, and the shareholders got a real dividend of $1 a share. Suddenly heady with success, Brennan mortgaged the mine, put up new mills, and followed an unprofitable stringer of ore down a costly shaft before the creditors seized the property. On Sunday, February 21, 1858, "driven to desperation and despair," he poisoned his wife, three children, and himself with cyanide.[33]

The majority of the mining companies floated, however, were purely fraudulent promotions without even a pretense of mining. As the eminent geologist Josiah Whitney lamented, "a large part of the excitement which sprang up in 1852 and '53 . . . was the mere blowing up of a prodigious bubble." And he concluded, "almost without exception, companies with large fictitious capital, and an enormous number of shares, have been got up for the purpose of swindling the public, and not for *bona fide* mining purposes."[34]

WITH THE FADING OF investor interest in the East and abroad, California miners again turned to local capitalists for financing. The amount of investment capital available in San Francisco and elsewhere in the state had grown substantially by the mid-fifties. But mining companies now faced competition for funds from other ventures, with the result that local investment in the mines was still rather small. More limited capital, however, forced the mine management to seek stricter economy and efficiency in their operations. These efforts, aided by improved milling and recovery processes and by declining labor and supply costs, soon led to profitable operations in quite a number of quartz mines, placing the industry on a sound footing for the first time.[35]

The richest of all of California's lode mines was the Empire in Grass Valley, an enormous, deep bonanza that ultimately produced $90,000,000 in gold, but it didn't come quickly. The Empire was discovered in October of 1850 by a twenty-two-year-old greenhorn, George D. Roberts, just a few months after he stepped off a steamer in San Francisco. Roberts, the son of an Ohio tanner, was energetic and clever, but also impatient. He immediately set to work opening the lode, but he lacked the funds and experience to make it pay enough to meet his expectations. So he sold his bonanza the following year without recognizing its potential. The ore shoots were, in fact, very scattered and the great mass of the ore was deep, so the mine passed through a number of hands before it was finally made to pay well by slow and costly development nearly half a century after its discovery. Roberts, in the meantime, teamed up with a savvy thirty-year-old former farmer and lead miner from Missouri, George Hearst. They prospected, located a number of claims, and opened them just enough for a quick turnover. But in 1857 they found a lode, the Lecompton, that Hearst wanted to work, so Roberts sold his share and they started to go their separate ways. Although they still worked together from time to time, their methods slowly diverged as the two Georges ultimately came to symbolize the extremes in western mining. Hearst would become one of the most successful mine developers and investors in the West and Roberts would become one of the most skillful and notorious swindlers. But while the rough-hewn Hearst usually took a conspicuous role in his ventures, the slier, almost impish Roberts, "badly marked by the flames" in a Nevada City fire, usually preferred to remain in the background, pulling the strings from behind the scenes.[36]

The most spectacular of the early quartz operations was the Carson Hill mine on the great Mother Lode gold belt in Calaveras County. But its surface croppings were so incredibly rich that it was soon plagued by avaricious betrayal and claim jumping that would mire it in litigation for decades. Proclaimed "the richest deposit of quartz gold ever known" the bonanza ores were so rich in gold that the "quartz was scarcely noticed," and the gold could be cut out with a cold chisel! In fact, the largest gold "nugget" ever discovered in North America was pulled from Carson Hill on November 22, 1854, totaling 141 pounds of gold sprinkled with about 20 pounds of quartz. This bonanza was discovered in the fall of 1850 by William Hance and James Finnegan. Within weeks of their discovery Hance talked Finnegan into going east to get machinery to work the mine. As soon as his partner left, Hance, "with the cunning of a fox," sold interests to eight San Franciscans, headed by Capt. Alfred Grey Morgan. They formed the Carsons Creek Consolidated Mining Company and laid claim to 1,700 feet of the lode. Before Finnegan returned the following June, they had taken out some 350 tons of incredibly rich ore, said to have yielded between $1,500,000 and $2,800,000 in gold, or $4,000 to $8,000 a ton.[37]

The returning Finnegan cried foul, sued for a share, and won the first round. But when the ruling was overturned, he rallied some two hundred well-armed and greedy supporters, who drove off Morgan's men and organized the People's Mining Company to hold the property. Morgan sued back and finally regained the ground in December of 1851, after Finnegan and his supporters had taken out the remaining $1,000,000 worth of exposed ore. Still more suits followed, however, that would tie up the mine for another twenty years—for Hance, in the meantime, had gone to London, and doubling the company's capital, offered the British a half for $1,050,000 at $5 a share in February of 1852. With John Sadleir, MP and junior lord of the treasury, as his board chairman, Hance touted the venture as "the richest gold mine in California and probably the richest gold mine in the world," claiming it had produced as much as $150,000 in a single day. Although the Londoners learned of the title dispute, Hance soon reassured them that the title was clear at last, but then in the fall, after they found that the surface ores had been stripped, the shareholders took back half of that money, cutting the shares to $2.50 apiece for 40 percent of the mine. In 1855, after they still hadn't been able to get possession, they finally wrapped it up. Then they found that they had been robbed of $50,000 as a part of massive frauds by their board chairman, Sadleir, who confessed to "diabolical crimes" and drank a bottle of cyanide early the following year. Some were certain, however, that even his "very suicide was a swindle," and that he had fled to New Orleans, followed by his brother. The *Times* called him "a national calamity" and topical Charles Dickens promptly made him the model of a new character, Mr. Merdle, in his serial satire of poverty and riches, *Little Dorrit*. Many years later the mine was finally reopened, in new, deeper ore bodies that produced about $20,000,000.[38]

It was a deeper, more elusive bonanza, however, that made California's first mining millionaire, a thirty-year-old from Vermont, Alvinza Hayward. His spectacular success was touted as a heroic tale of faith and perseverance rewarded. It was told that in 1853 he bought a claim called the Badger on the Mother Lode at Sutter Creek in Amador County, and started sinking a shaft on a small "chimney" of ore. But the ore body was spotty, and after four years he was badly in debt and "his family were next to starving." Yet, certain that he was "on the verge of a great strike," he persuaded an old friend, Oscar Chamberlain, to loan him just $3,000 more, so he could sink the shaft just a little deeper. That was all it took, and he struck a bonanza that started paying over $28,000 a month and made his fortune. Unfortunately, that wasn't the whole story. Hayward had actually worked the mine all those years in partnership with an older fellow New Englander, Thomas Robinson. But in May of 1857 Robinson, like poor Finnegan, temporarily left all the work in his partner's hands for a year, while he made a trip east to visit his family. It was while he was gone that Hayward discovered the bonanza, but he kept very quiet about it. Then, when Robinson returned, Hayward sadly told him that the mine had "failed," and they were "wholly ruined and $36,000 in debt"! Since he had been managing the mine, however, Hayward said he didn't want to panic their creditors by selling out, so he generously offered to give his partner $7,000 for his half. Robinson trustingly agreed and didn't discover how badly he had been deceived for more than a decade. In addition to that, Hayward had also seen that the ore continued into the Eureka, the idle mine adjacent, which had a $30,000 mill, so he had quietly bought up its shares for less than the price of the mill before anyone else became aware of the bonanza. Thus Hayward worked the combined mines privately until 1868, taking a profit of roughly $5,000,000, all the while refusing to reveal any information about the mines. Finally seeing the end of the bonanza, however, he incorporated as the Amador Mining Company, publicly declared profits

of 67 percent on production, and sold out in six months for an added
$1,000,000. By then Hayward had gained a reputation as one of the lead-
ing sharks in the West, and he had made his Eureka mine the deepest in
California, sinking to a depth of 1,230 feet. But he had also worked it as
cheaply as possible, cutting every corner, which also made it the most
deadly mine in California, with a death rate three times that of others,
so his extraordinary profits came at a further cost of nearly sixty lives.[39]

MANY INVESTORS WERE NOT content to wait for years, however, at any
price. The Merced Mining Company shareholders grew anxious after
only seven months, when their superintendent had spent $130,000 and
hadn't yet found a mill that was fast and efficient enough to turn a profit
on their $30-a ton-ore. They were also unhappy that he had run up a debt
of $50,000, or 10 percent, on their capital, even if that was only because
they hadn't yet paid up more than a quarter to a half of what they owed
on their shares. Nonetheless, they sent out an investigating committee
and finally decided in 1852 to lease everything for $35,000 and a third of
all profits to the Nouveau Monde company, which had just thrown up
its worthless leases on Fremont's nearby Mariposa claim. The Nouveau
Monde hired a venerable old British engineering firm, John Taylor &
Sons, who reported opening ore running as much a $135 a ton, but spent
another two years and $650,000 putting up a magnificent works that still
couldn't work the ore profitably. So the shareholders finally called for
an investigation, and some sharp directors sent out a prominent but
pliant mining expert, J. Arthur Phillips, who wrote back that except
for a "few specks" in the Pine Tree and Josephine, he was not able to
"discover any visible gold in any of the company's mines!" Share prices
crashed as the shorts cleaned up, and the stunned shareholders voted
to give up on California entirely and lease a silver mine in Guatemala
that the insiders recommended! The Merced shareholders still squabbled
for another year before they finally decided to focus on getting out the
richest ore and reopened the Josephine, while an exasperated "Quartz"
Johnson began working the Pine Tree illegally on his own account just
to get something more out of his discovery than gilded shares.[40]

When the Great Johnson Vein started turning out gold at last, Fremont
and his creditors suddenly grew covetous and began scheming to grab

even more. He had appealed his case to the Supreme Court, and in the spring of 1855, Benton's old and close ally Chief Justice Roger B. Taney wrote the majority opinion reversing the lower court decision. The Court ruled that the grant to Governor Alvarado, amounting to sixty-nine square miles, was simply a "general gift" for his "patriotic services." It further ruled, however, that the grant did not become a "particular gift" until it was surveyed by a government surveyor, who had the "duty" to insure that the land was indeed "vacant and that the grant would not interfere with the rights of others." Until particular boundaries were surveyed, anyone at any time could obtain a title to land in the area and his title "could not be impaired by any subsequent survey of Alvarado" or his successor, since the earlier title was "the superior and better one." Although an official survey of the grant along Mariposa creek had been made in 1852 from a crude map by Alvarado, the township lines had not yet been extended into the area, so the Court formally called for a resurvey by the surveyor general to connect the grant to them. Since the mines that Fremont had claimed within the grant were nearly worked out, he immediately cut a deal with the California surveyor general, Jack Hays, an old friend of his, and the tenant and obliging crony of his creditors, Palmer, Cook & Company. Under the guise of simply extending the township lines and running a ditch, Hays's surveyor, directed by an agent of Fremont's, secretly staked piratical new boundaries that flipped most of Fremont's grant a dozen miles outside his original survey, flagrantly disregarding Hays's duty not to violate the rights of prior claimants, and fraudulently seizing mines and mills that had been worked all those years by the Merced and Nouveau Monde companies, as well as the homesteads of hundreds of other settlers whose rights had never been questioned. With a little more help from Fremont's father-in-law and his bankers, the rapacious new survey was confirmed without further review in February 1856 by President Franklin Pierce, in a patent which he personally handed to Fremont. The court, however, had still not awarded him the mineral rights.[41]

The eastern press promptly proclaimed Fremont the "Richest Man in the World," and the newly organized Republican Party swiftly made him their first presidential candidate. But Mariposa became a liability in the campaign, when opponents blasted it as "the most stupendous California land jobbing scheme ever known in the history of the country."

Although it remained a minor issue, Fremont still got only a third of all the votes nationally that fall, and he got just 165 votes out of the 2,195 cast in Mariposa County! But he came back to the grant with a vengeance after his election defeat, and started seizing the mines and homesteads. The outraged miners and settlers organized to resist his "fraudulent invasion of the rights of the people," an "outrageous, bare-faced piece of down-right stealing," and they finally took up arms to drive his claim jumpers off the Merced company's mines. But Fremont sent a frantic message to the Democratic governor, John B. Weller, husband of one of Jessie's cousins, who threatened to send troops to enforce the seizure.[42]

To open a legal battle for the mineral rights, Fremont temporarily gave his handyman, Biddle Boggs, a lease to one of the Merced company's mines that he hadn't yet grabbed, and then brought suit in Bogg's name when the company refused to turn over their property to him. Despite popular protest, the compliant county judge ruled for Boggs, but the Merced company appealed to the three-man California Supreme Court and won a reversal early in 1858 by a two-to-one vote. Fremont and his creditors, through their manager, Charles James, made very generous offers to the justices to try to induce them to rehear the case. But only the original dissenter, Stephen J. Field, a younger brother of the prominent New York lawyer David Dudley Field, was willing to consider it. So they waited over a year until one of the resistant judges retired and the other, David Terry, resigned to fight a duel. After Fremont's attorney for the case, Joseph Baldwin, was elected as one of the replacements, the pliable Field, who had become chief justice by default, happily reopened the case for a reported $50,000 in Mariposa bonds. Baldwin had already collected his fee of $100,000 in bonds, but after he and Field leaned on the new third justice, Baldwin modestly abstained when they actually voted to reverse the previous decision and give Fremont the Merced company's mines.[43]

To justify this action, Field turned the world upside down in an obscene distortion of the facts, turning the victims into the villains. He dismissed as "groundless" the claims filed under well-established local mining district laws by miners who had for years worked and developed public land that Fremont had never claimed, until he suddenly decided to grab it in clear violation of their property rights, and instead branded them as greedy squatters and despoilers who had invaded the innocent Fremont's land in violation of his sacred rights. Thus, Field destroyed their property rights while claiming to protect them, and he concluded with savage irony, "There is something shocking to all our ideas of the rights of property that one man may invade the possessions of another!"[44]

Even that old gold rush schemer George Gordon was outraged by such judicial thievery. Although he expressed "admiration for its ingenuity," he cried "it is not justice!" How could it be, he asked, when Field hands the mines to Fremont, saying in effect, "Yes, take them; they are yours. You did not claim the land while these people were building the works; in fact you disclaimed it, but that was a mistake for which you must not be prejudiced; you did not find the vein, or make the tunnels, or build the mills, or level the roads; you have done nothing except manage a survey skillfully; but take the property, the law gives it to you!"[45]

The Merced Mining Company, of course, immediately appealed the decision to the U.S. Supreme Court. But the case would drag on for seven years, until well after Field was further rewarded by appointment to that court in 1863 at the urging of his brother, David Dudley, who was by then both a consultant to President Lincoln and an attorney to Fremont, for an extravagant fee of $200,000 in Mariposa shares. Two years later, on a motion by his brother as counsel for Boggs, the appeal would finally be dismissed for "want of jurisdiction."[46]

THE MOST PROFITABLE mining investments of the fifties, however, were not in the gold mines themselves but in their two essential adjuncts, water and quicksilver. Enormous amounts of water were needed to wash out and separate the placer gold from sand and gravel, and quicksilver, or mercury, was needed to further save the fine gold from both the placers and quartz mills, dissolving and concentrating it into an amalgam from which the mercury could be boiled off and used over and over again. A single company gained a virtual monopoly on quicksilver throughout the West, while water companies, monopolizing local watersheds, were ubiquitous.

The biggest quicksilver mine in all of the Americas was the New Almaden in the Coast Range, just south of San Francisco. For a time it even surpassed its namesake in Spain, which dominated worldwide production, but its promoters and managers turned out to be as slippery as quicksilver itself. The coastal tribes had long worked its brightly colored cinnabar ores for paint and had fought the first of many battles for control of it. Then in November of 1845, over two years before Marshall's discovery of gold, a young army officer from Mexico City, Andrés Castillero, recognized the paint as extremely rich mercury ore that yielded as much as 35 percent metal by weight, and he immediately claimed the ground as the Minas de Santa Clara. It was the only important mining grant in California actually located under Mexican law. Castillero also promptly formed a company in which he kept half of its twenty-four shares and gave the remainder to the local commander, a priest, and two men who would actually work the mine. He then returned to Mexico City in the spring to seek a $5,000 reward that the government had offered for discovering and opening quicksilver mines in Mexico to break the old Spanish monopoly. But with the outbreak of war with the United States in May and its invasion of California, the Mexican government refused to pay him a reward. He then turned to Barron, Forbes & Company, a Scottish trading firm in Tepic on the Gulf of California, one of whose partners, Alexander Forbes, had written the first English history of California in 1839, urging both its acquisition by the British as a foreclosure of Mexican debts and the formation of a company to exploit it. The sixty-nine-year-old Forbes examined the mine in the fall of 1847 and was so impressed with its prospects that he and his partners started buying up the shares. They also renamed it the New Almaden, determined to make it a real competitor to its namesake.[47]

By the end of 1848 Forbes and his partners had bought control of the mine with just over half the shares for less than $20,000. They hired a distant relative, James Forbes, the British vice consul in California, who had also purchased a couple of shares on his own, to manage the mine. Over the next two years they spent roughly $300,000 purchasing some adjacent property, developing the mine, and building furnaces and other works to refine the ore before they finally got the mine into full production in the latter half of 1850. That year they produced nearly 600,000 pounds of mercury, which greatly exceeded the demand in the new gold fields of California, where no one yet worried about the loss of fine gold and the lode mines were just beginning to be explored. But they found a ready market in Mexico and South America and sold the mercury for an average of $1.30 a pound, bringing in nearly $800,000, for a net profit of $600,000, since mining and refining costs were only 24 cents a pound. This was enough to pay back all of their initial investment plus a generous profit of nearly 100 percent. The following year they tripled their production to over 2,000,000 pounds, which they sold for close to $2,000,000 and for a return of over 40 percent a month![48]

But all of the New Almaden shareholders didn't share equally, for such outrageous profits also excited outrageous greed, and as happens over and over again the major shareholders schemed to get it all. Aging Alexander Forbes retired to England in 1850, and a sharp junior partner, William E. Barron, moved to San Francisco to look after the mine. There, with James R. Bolton he quickly cornered most of its profits by setting up a new trading firm, Bolton, Barron & Company, to handle the distribution, and then, as manager of the mining company, he contracted to sell all of its mercury to his trading company at a ridiculously low price of 27 cents a pound, just 3 cents above production costs. Barron naturally was very secretive about the New Almaden company's affairs, but when one of the minority shareholders finally sued for a share of the earnings in 1854, the management claimed that in four years the company had only gotten $1,872,000 for all of their mercury, making a net profit of just $162,000, "less than one per cent a month on the capital employed." They didn't mention, of course, that during that time they had produced just over 7,000,000 pounds of mercury, which Barron and his friends had sold at around 84 cents a pound for a total of $5,900,000 and pocketed the difference of $4,000,000, which didn't show on the books of the New Almaden company![49]

At this point James Forbes, who had fallen from favor and was cut out of the scheme, finally agreed to sell his two shares to the majority holders for a mere $55,000. This was still an enormous return on his own modest investment of less than $4,000, but he had expected much more and he would soon seek revenge. Among Barron's fortunate friends, on the other hand, was a shrewd young Sacramento merchant-banker, Darius Ogden Mills, who became the exclusive agent for mercury sales in the mines, and he got a 15 percent cut of whatever he could sell it for. This became the first of several monopolistic deals from which Mills would eventually extract one of the West's largest mining fortunes. By 1858 the New Almaden mine had produced 16,500,000 pounds of mercury, which Barron and his friends apparently got from the company for

DARIUS OGDEN MILLS
Henry Hall, *America's Successful Men
of Affairs*, 1895

only $4,500,000 and sold for about $10,000,000. Thus the insiders took personal profits of roughly $5,500,000, while the minority stockholders got less than half of the company's official earnings of barely $500,000.[50]

Despite Barron's persistent efforts to make the New Almaden mine look unprofitable, it had excited great envy right from the start. Soon after the discovery, the American consul in California, Thomas Larkin, had joined with others in buying part of the neighboring Capitancillos land grant and forming the Santa Clara Mining Company, early in 1848, to work an extension of the New Almaden deposit. Larkin's consular dispatches, trumpeting the richness of the mercury mines and their great potential, also excited the U.S. secretary of the treasury, Robert J. Walker, an aggressive speculator, casual con man, and consummate political manipulator—and a former Democratic senator from Mississippi and future governor of Kansas. Walker bought a major interest in Larkin's Santa Clara company a couple of years later. Although that company's mine didn't prove to be profitable, Walker joined with another speculator, James Eldredge, who had bought the rest of the Capitancillos grant at a sheriff's sale, and they used it to launch a rapacious scheme to grab control of the entire New Almaden mine by switching the boundaries of the grant, just like Fremont was doing with the Mariposa. The prize was so great that the piratical Walker claimed it was just "too tempting not to be prosecuted with eagerness." It also proved to be a long, bitter, and extremely costly fight, hotly contested at every step by the equally sharp Barron and his allies.[51]

In 1854, Walker and Eldredge reorganized under the all-inclusive name of the Quicksilver Mining Company, with a paper capital of no less than $8,000,000 under a Pennsylvania charter. Walker's protégé, Samuel F. Butterworth, a coldly ambitious lawyer and patronage seeker, was named its president. Then they generously handed out shares in the venture to Democratic president James Buchanan's pliable attorney general, Jeremiah Black, and others in the Congress and the courts, wherever they were needed to influence a decision. It all paid off handsomely, and they finally secured Supreme Court confirmation of a gerrymander Capitancillos grant with a snakelike extension that reached out and swallowed up the New Almaden, despite the passionate protests of dissenting justices who railed at "such a monstrous result."[52]

At the same time, the attorney general also obligingly sent Walker's company lawyer, the pugnacious Edwin Stanton, to California as a special government agent to ferret out fraud in the land grant claims. Stanton first won local acclaim by exposing the forged claims of the notorious José Yves Limantour, who had tried to claim most of San Francisco! Many doubtless hoped that Stanton would next go after the equally fraudulent Santillan claim for another big piece of the city, which the opportunistic firm of Bolton and Barron, together with those schemers Palmer and Cook, had sold to a group of Philadelphia speculators who floated it as a $15,000,000 land company. But the Philadelphians had also hired Stanton as their lawyer. So instead, with the aid of letters that Walker's associates had bought from a vengeful and by then bankrupt James Forbes for $20,000, Stanton surprised them all by proclaiming Barron's New Almaden mine grant to also be a patchwork of "fraudulent fabrications." Even though the New Almaden title had already been confirmed by the federal land board, Stanton got a federal court injunction in the fall of 1858 that would stop all work in the mine for a couple of years. During that time, of course, the resourceful Barron, together with Darius Mills, used the opportunity to raise mercury prices by 50 percent in the mines and double their profits selling the accumulated surplus, which they had previously had to dump, cheaply in China.[53]

The court closure of New Almaden stirred an outcry from gold miners, thinking there was a true scarcity, and their state legislators branded it "an outrageous violation of free government." There was also an outpouring of anonymous pamphlets, mostly put out by Bolton and Barron to indirectly push their case with the public, who still saw them as the promoters of the "Santillan swindle" and may not have been too unhappy to also see them being swindled themselves. One anonymous "Quartz Miner," lambasted the easterners' mine-grabbing scheme with a lively Smart and Cornered, How to Get a Mine Without Finding, Opening or Working One, while another denounced the easterners as a "sweatless class" of "crafty waiters on political providence, whose brains have leisure to hatch plots, because their hands lack the nerve of honest industry." An anxious George Gordon even raised the larger specter of a general seizure of all the mines by such cliques. Thus Barron's forces fiercely fought the charges both in the streets and all the way to the Supreme Court. Although they ultimately lost the battle, they would win the war![54]

THE MINING WATER COMPANIES also paid extraordinary profits, but they were not generally so contentious. Although most of these companies began as working cooperatives, they too soon developed into corporate monopolies that dominated the placer mines. These ventures were formed to dam streams, tap mountain lakes, and build ditches and flumes to bring water year-round to the placer miners. Most of these were started as joint-stock companies among the miners for their mutual benefit, but some of the larger companies also sought outside capital in Marysville, Sacramento, and San Francisco. Despite the shadow of suspicion that had fallen over quartz mining promotions, the water companies found eager investors in merchants and financiers. The aggressive banker and quicksilver broker Darius Mills also became the major shareholder both through purchase and default in two of the largest ventures, the Bear River & Auburn Water & Mining Company and the Tuolumne County Water Company. These two companies were so profitable that they spared no expense, even commissioning exquisite lithographs for their shares. The water companies were, of course, much less speculative,

since their income, like that of the camps' merchants and the mercury sales, came from supplying the miners with a vital commodity, and not from the often elusive profits of mining itself. Nonetheless, their profits were extraordinary. Those of the largest companies ranged from 2 to 12 percent per month in the mid-fifties, while some of the smaller ditches paid even more—one claimed 6 percent per day! Such incredible profits as these from water and quicksilver monopolies gave Mills, still in his late twenties, a heady start, and he would go on to be counted as one of the ten richest men in America, making much of it by controlling the less risky, but essential, adjuncts of western mining.[55]

These outrageous profits, of course, came from squeezing the miners dry. Although any miner who wanted water from the companies often had to buy a share, a majority interest in the companies usually ended up in the hands of just a few investors like Mills. Conflicts inevitably arose between the many miner shareholder-consumers, wanting low water rates, and the few major shareholders, who controlled the management, took most of the profits, and demanded high rates for the water. When yields from the placer mines were high, the miners were happy to get water at any price, but as the placer yields declined, they staged "strikes" or boycotts. One of the earliest confrontations came in March of 1855 at Columbia, the "gem of the southern mines." There, many miners found their daily earnings had dropped to a point where they could barely pay their water bills of $6 a day to the Tuolumne County Water Company, while it still enjoyed monthly dividends of 10 percent. A delegation of miners presented their case to Mills and the other company trustees, requesting a reduction in rates. But Mills and his partners refused, and three thousand angry miners finally decided to revolt. They held a mass meeting in front of the Methodist Church to the sounds of a brass band, church bells, and a cannon to declare "war" on the "controlling cormorants of this monster monopoly." They vowed to pay no more than $4 a day for water; they signed up to build a rival ditch; and as a coup de grâce they threatened to withdraw all their money from Mills's bank. Just a few days later the company capitulated. The Columbia miners' victory inspired similar "strikes" against other "water monopolists" over the next few years that finally brought down water prices throughout the diggings. Nonetheless, the Tuolumne water company still averaged profits of about 40 percent a year, or at least $2,000,000 for the decade, and the Bear River & Auburn company did half as well.[56]

Abundant water also led to the development of the much more efficient, albeit destructive, process of hydraulic mining. This completely transformed the placers as high-pressure jets of water, piped down from the ditches, were put to work, tearing down the hillsides to dig out deeper gold deposits and wash it all through long sluices with slats, or "riffles," to catch the gold. Moving all the dirt by water power rather than man power slashed labor costs, for three men with a hydraulic nozzle could now do the work of fifty. Much poorer dirt and gravel, paying as little as a few cents a ton, could be worked at a profit! Gold in the ancient beds of "dead rivers," from much earlier geologic eras, could also be exploited, and placer mining surged again. It started in 1853, when a thirty-year-old Rhode Island miner, Edward Mattison, used a hose to wash out his claim on American Hill north of Grass Valley. It worked so well that the idea quickly spread throughout the diggings, and in just four years more than half of California's gold, $20,000,000 a year, was said to come from hydraulic mines.[57]

Most of the early hydraulic operations were still done as cooperative ventures, but many of the water companies also went directly into hydraulicking, and even those that didn't still profited greatly from the booming business. As a result, the water companies were so profitable and so successful in getting financial backing locally that by the end of the 1850s some four hundred water companies had raised over $13,000,000 from California miners, merchants, and bankers; built hundreds of dams to hold an estimated 7,600,000,000 cubic feet of water; and dug nearly 6,000 miles of ditches to deliver it. Only a few of the water companies tried to raise capital in the East, where even such promising ventures were rebuffed in the late fifties by investors, who had come to distrust every kind of California gold mining promotion, at least temporarily.[58]

ON THE TAIL of the California mining mania, some of the old Indian and Spanish mines in the Southwest had also begun to attract their first eastern investors. One of the earliest was the New Mexico Mining Company, formed in Washington, D.C., in 1855 with $500,000 in paper by the comptroller of the treasury and former Whig congressman from Ohio, Elisha Whittlesey, and a trio of Democratic politicians in Santa Fe: the territorial Indian agent and former acting governor John Grenier; the territorial attorney general and congressional delegate Miguel Otero; and former North Carolina congressman Abraham Rencher, who soon became the next territorial governor. Two years earlier they had purchased for $1,100 from Francisco Ortiz's widow his Santa Rosalia gold quartz lode in the Old Placers just south of Santa Fe near the mineral-rich Cerrillos, long mined for turquoise by the Pueblos. Ortiz, who died in 1848, had worked the lode only sporadically under an 1833 mine grant of eleven acres from an in-law, the *alcalde* of Santa Fe, who also granted him conditional, and later contentious, use of 69,458 acres of common land surrounding the mine for grazing and farming purposes only, without any mineral rights or "any detriment" to others, as long as he was working the lode. Although the latter grant had expired when Ortiz stopped work in 1842 over a decade earlier, the new company now claimed both grants as permanent and promptly reopened the mine. Assuring investors that the gold was "inexhaustible," they also began unloading as much stock as they could, putting about $20,000 from it into a couple of steam engines and a "crushing apparatus." But the only profit came from selling the stock, not working the rock.[59]

The Southwest's one proven bonanza and biggest profit payer, the great Santa Rita copper mine, was also reopened by private investors after some difficulties. In 1849, soon after the American seizure of New Mexico, Francisco Manuel Elguea's heirs had pleaded with the new acting governor to send troops to take the mine back from the Apaches. But there was a question as to whether the mine was north or south of the new border, which hadn't yet been surveyed, so the heirs tried, unsuccessfully, to sell it for $40,000. The subsequent boundary survey party used Elguea's fort for its headquarters and made sure that the mine was brought into the United States. But then the first Indian agent negotiated a treaty with the Apaches that placed the mine within their reservation, after which eager Texas investors got their Democratic senator, Thomas Jefferson Rusk, to kill the treaty in March of 1857, just a few months before he killed himself in a fit of depression over the death of his wife.[60]

The Texans, San Antonio merchant and soon-to-be-mayor James R. Sweet and his partner, Jean Batiste LaCoste, then promptly joined with their El Paso agent, Simeon Hart, and his father-in-law, Chihuahua merchant Leonardo Siqueiros, who got a seven-year lease on the mine from Elguea's heirs at no cost for the first two years and only $600 a year thereafter. Since the Texans had the contract to haul in the army's supplies to New Mexico, they could haul out the copper on their empty return wagons for little added cost. So, in the summer of 1858, with the help of the Indian agent and the army protecting them from the Apaches, they at last reopened Santa Rita with a force of about a hundred miners from Chihuahua, paid peon wages of only 20 cents a day. By the beginning of 1859 they had sold nearly $50,000 worth of copper to the mint in Chihuahua, and they began shipping over $10,000 worth of copper a month to New York. All of this attracted the Mexican mint's assayer, Sofío Henkel, who with other Texas backers opened a neighboring claim, which he named for his native Hanover, that summer. Within a year the more experienced Henkel was taking out more copper than the Santa Rita, and he already claimed a profit of $100,000. In 1860 the two mines produced 650 tons of copper worth $415,000, to make New Mexico second only to Michigan in American copper production. Santa Rita was again making new fortunes for all of its investors except Siqueiros, who was driven into debt by other ventures and turned over the lease to Sweet and LaCoste.[61]

On to the West, the newly acquired Mexican silver and copper mines in the Gadsden Purchase also attracted speculators and investors. This 29,000,000-acre tract, bought from dictator Antonio López de Santa Anna for a mere $10,000,000 in 1854, shifted the Mexican War boundary south of the Gila River to include the essential southern route to California and more rich mines. At the same time the army established new forts to attempt the conquest of the Apaches and protect new American miners and settlers.

By far the most important of these mines was another great copper bonanza that would eventually rank not far behind the Santa Rita as one of the biggest copper producers in America. In the summer of 1854, immediately after the ratification of the Gadsden Treaty, a thirty-four-year-old San Francisco merchant, Edward Ely Dunbar, formed the Arizona Exploring, Mining & Trading Company, a group of twenty adventurers who set out to find the fabled Arizona silver fields. They eventually found the ground from which the planchas had come, but after several months of exploration they concluded that all the paying silver had been taken out. All was not for naught, however, for en route an Opata Indian had also shown them the rich outcroppings of native copper at a little garlic-scented spring called Ajo, about ninety miles east of Fort Yuma, which had been briefly worked by a small company of Sonorans before they were all killed by the Apaches in 1851. The surface ore, like that at Santa Rita, was exceedingly rich, running as high as 70 percent copper, and on returning from the planchas in the spring of 1855, they staked claim to the mine and reorganized as the Arizona Mining & Trading Company in San Francisco, with a capital of $500,000 in $100 shares. Dunbar began work immediately, bringing in a crew and supplies in June, but by November he had only two men left. One had died on the desert from lack of water, another from poisonous water, a third had had "his brains beaten out while sleeping," and the mounting dangers, hardships, and isolation had driven away the rest. Nonetheless, within a year Dunbar had taken over a hundred tons of the richest ore to San Francisco for smelting at a profit of about $10,000, and he even shipped ten tons to England, where it was hailed as "the best ore ever." Dunbar pushed on for another year, with heady talk of taking out 200 tons a month of $300 a ton ore for a net monthly profit of $40,000, or nearly 100 percent a year on their stock. But although the mine would eventually produce more than 6,000,000,000 pounds of copper, by then worth about $6,000,000,000, the deeper ore was lower grade and just couldn't pay a profit over the high costs of hauling it all the way to California, so in 1858 Dunbar suspended work and headed east to try to raise more money.[62]

Others from Arizona were already ahead of him, however, offering much more alluring silver mines. The first and most aggressive of these ventures was the Sonora Exploring & Mining Company, formed in March of 1856 for a capital of $2,000,000. It was floated in Cincinnati by an eager young lawyer from Kentucky, Charles D. Poston, who scouted the territory for promising old mines right after its purchase and later claimed the title of "father of Arizona," and an older moonlighting, former Fort Yuma commander, Maj. Samuel Heintzelman, who sold the venture to a Pacific railroad–boosting newspaper publisher, William Wrightson, and other Ohio businesmen. Heintzelman became president and Poston the mine manager. Advertised as nothing less than "the most important Mining Company on this Continent," they soon bought with a little cash and a lot of stock no less than eighty old mines scattered all over the hills east and west of Tubac. There they set up their headquarters and launched the territory's first newspaper, the *Arizonian*, to further promote the venture. Their prize, the Cerro Colorado, or Heintzelman, mine on the west, was opened by their energetic young Prussian mining engineer, Frederick Brunchow. It boasted silver-lead ore said to run as high as $9,090 a ton, and Poston modestly claimed that it was "probably the richest in the world-known." Although a shrewd investor like Connecticut gun manufacturer Samuel Colt discounted such claims, he still bought $100,000 worth of stock early in 1858 for $10,000 in cash and an equal amount in guns! And he continued to buy a lot more at a similar discount to become the major stockholder. But the ore, of course, was not nearly as rich as advertised, and it required an expensive smelting works, which quickly put the company in debt. Despite their most conservative predictions of $490,000 a year in profits, in two years they shipped only $15,000 worth of high grade. Undaunted, however, the promoters formed a subsidiary, the Santa Rita Silver Mining Company, around the mines on the east, offering another $1,000,000 in shares, to try to get all they could.[63]

Even the enthusiastic Brunchow had finally given up in disgust and formed a company of his own, with St. Louis backers, to open a promising

mine much farther east of Tubac in what would later become the Tombstone district. But he was apparently a hard taskmaster and was soon
murdered by his Sonoran crew. Colt, nonetheless, was still confident
that the Heintzelman mine could be made to pay with enough "energy
and capital," and "a real class of people." The mine's only problem, he
felt, was the "half-horse" management of midwesterners, who had turned
it into nothing "but a hole to bury money in." So in the spring of 1859,
with the help of San Francisco merchant and vigilante committee leader
William Tell Coleman and others, he took control of the board and the
presidency. When Heintzelman and his Cincinnati friends objected and
called on Poston to help, Colt vowed to "get rid" of them, and when they
refused to sell out for less than 25 cents on the dollar, he refused to buy,
hoping to "let them founder about or wallow for a while in their filth"
until they'd be happy to sell for just pennies. So while they howled about
"sharp financiering," he reincorporated in New York, and through a
dodge with Poston, got a ten-year lease on all the property from the old
company for only $10,000 a year. Then he paid off the debts and sent out
a new mining engineer and $30,000 worth of new equipment and hired
a crew of 140 peons at barely a tenth of that paid Anglo miners. With
such cheap labor Colt's mine manager expected to produce $750,000 a
year at a spectacular profit margin of over 90 percent, or $700,000 a year!
But his manager was also murdered by his peons, and the mine never
produced more than $100,000.[64]

Lastly, another eager officer from Fort Yuma, a young, articulate lieutenant and budding con man, Sylvester Mowry, also got Colt interested
in other mines around Tubac. Mowry, who had gotten himself elected as
a delegate from the prospective territory of Arizona, resigned from the
army in 1858 to go to Washington to lobby both Congress and the president for the creation of the new territory. In a heavy barrage of speeches,
newspaper articles, and pamphlets, he raved over the potential mineral
riches of Arizona, and he soon talked Colt and others into joining with
him to reap those riches. In the summer of 1859 they floated a couple of
ventures, the Sopori and the Arizona Land & Mining companies for
$1,000,000 and $2,000,000, to buy mines close to the Heintzelman. They
sold the stock busily, but they hadn't done much work before Mowry
turned all of his attention to a much more promising mine east of Tubac,
called the Patagonia. Several fellow army officers had bought it for only a

few hundred dollars two years before, and in the spring of 1860 they happily sold it to Mowry for $25,000, for a hundredfold profit. He promptly renamed it for himself and sold Colt a fifth interest, apparently for just enough to pay the purchase price. The ore ran about $200 a ton, and Mowry, thinking it was too good to share with the public, mortgaged his share for $40,000 rather than sell any more, and he and Colt began to develop the mine privately. They put in a dozen smelting furnaces and a large crew, and soon surpassed the Heintzelman, producing as much as $1,000 a day, Mowry claimed.[65]

But the outbreak of the Civil War early in 1861 started a chain of events that soon shut down nearly all the mines in the Southwest. As the war drew near and troops were withdrawn from the Southwest, both the Apache and Sonoran bandits began to raid the mines and settlements. At Santa Rita, Sweet and LaCoste suddenly had to pay an additional 12 cents a pound for hauling out the copper, after they lost the military supply contract that had subsidized it. Then Confederate Texas imposed a 25 percent tariff on all goods shipped north, while copper prices fell to a low of 18 cents a pound and all their profits vanished. Sweet joined the Confederate army as a colonel on the lower Rio Grande, and LaCoste eventually moved across the river to Matamoros to try to ship the copper profitably from Mexico, as the wartime demand began to drive prices up 10 cents a pound. But Confederate forces from El Paso took over New Mexico that summer and confiscated supplies bound for the mine. When Union forces drove the Confederates out the following spring, the Texans finally abandoned the mine and fled just before it was seized by Gen. James H. Carleton's forces from California. Sweet and LaCoste later claimed that they lost 300,000 pounds of copper and $70,000 worth of supplies during the war. In Arizona, Poston abandoned the Heintzelman mine in July of 1861, after the troops were withdrawn and his brother and others in the mines were killed by bandits and Apaches. Colt, Arizona's major investor, died six months later at the age of forty-seven from "an acute attack upon the brain." Mowry, however, fortified his mine, fought off the Apaches, and called for help from Confederate troops when they captured New Mexico. But after Carleton defeated the Confederates in Arizona, he had Mowry and everyone else at the mine arrested for "treasonable complicity with the rebels" and had the property seized in June of 1862. Mowry was imprisoned at Fort Yuma, and although a receiver was appointed to work the mine for him, it was soon shut down because the actual output was barely $100 a day and it couldn't pay expenses. Thus, while gold mining still continued unabated in California mines, silver and copper mining came to a temporary end in the Southwest.[66]

THE GOLD RUSH DECADE had transformed the West and the nation. The news of instant fortunes being made in California drew over 300,000 argonauts from all over the globe and created an instant state. The fabulous amount of gold panned from its streams had reached $600,000,000, nearly three times that previously in circulation, and it instantly yanked the nation out of a depression. But nearly 100,000 hopeful miners had worked those streams all that time, as long as there was water, so on the average they made just $3 a day. Although that was three times eastern wages, the cost of living was twice that, and for every lucky soul who did several times better there were several more who did far worse. Only the rare lucky ones could even afford to share their winnings with backers, if they had and remembered them. So of the hundred thousand, or more, armchair argonauts at home and abroad who'd been enticed to wager over $200,000,000 in hopes of getting a share and had actually put up around $30,000,000, nearly all lost their bets. There simply weren't very many profitable opportunities for investment, or even speculation, in small-scale placer mining. Only a handful of quartz lodes had yet been made to pay real profits and only in the hands of sharp, tight-fisted operators, who like Alvinza Hayward, personally managed and manipulated their mines. The rare big profits for investors came from monopolizing those two basic essentials of gold mining, the water for washing and the quicksilver for amalgamating, and they laid the foundations of several lasting mining fortunes for Darius Mills and a few close friends. But the greedy struggle for control of these essentials, together with fraudulent land-grant grabs of the mines themselves, threatened violence and corrupted even the highest levels of government. The stock sharks, of course, had also done very well feeding shares to all the hungry marks. Yet this was only the start of it, as an even more rapacious new wave of silver mining speculation was already breaking.[67]

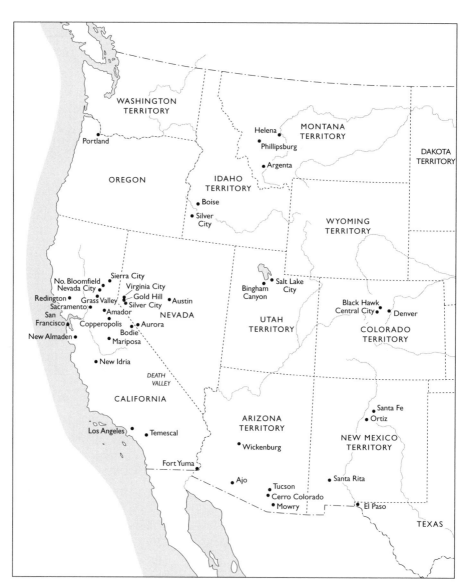

THE MINING WEST IN THE 1860S

3

THE SILVER RAGE

I N THE DEVELOPMENT of western mining, 1859 proved to be as pivotal a year as 1849. For it saw the opening of enormously rich new
lodes in what would eventually become the new states of Nevada,
Colorado, Idaho, and Montana. The exploitation of these lodes together
with deep new California bonanzas shifted the balance from placer to
lode mines as the leading producers in the West. With this shift came a
greater need for capital and a new surge of mining investment. With it
also came a spectacular wave of mining stock promotion and speculation
that engulfed nearly every city, town, or camp in the West as the flood
of bullion began to pour from the new mines and eventually touched
even the most conservative financial centers in the East during the wildly
speculative Civil War years.

Miners, leaving the declining placer camps of California in the late
1850s, had struck out over the Sierra to prospect the Great Basin and
Rocky Mountains. There, in 1859, on the slopes of Sun Mountain above
the Carson River, the dark, heavy rock that had confounded local placer
miners was found to be phenomenally rich in silver. The great Comstock
Lode was opened, and silver camps sprang up throughout the Great
Basin. At the same time farther east, a few former Georgia miners, who
had picked up a bit of gold at the edge of the Rockies on their way to
California, returned in 1858 to trigger an overblown rush to the Pikes
Peak placers. Early the following year other Californians and Georgians
traced the gold to its source in bonanza lodes on Clear Creek, and the
gold camps of the Rockies were born. The Comstock silver excitement
seized the West, and the Rocky Mountain gold excitement captured the
East, as the new mines poured out their riches and sparked a new frenzy
of investment, speculation, and fraud.[1]

The heavy bonanza rock from the Comstock was first brought over the Sierra to Nevada City and Grass Valley from the Ophir claim in the summer of 1859. It assayed an astonishing $1,595 in gold and $3,196 in silver per ton. The news flashed like lightning and the rush was on! George Hearst was among the first. He sold the Lecompton and borrowed $1,000 to buy, with a friend, one-sixth of the fabulous Ophir for $3,500. Others quickly bought up nearly all the rest of the 1,400-foot lode, paying the discoverers just over $70,000 at about $60 a foot. By November the new partners had taken out thirty-eight tons of ore, which they shipped to San Francisco at a cost of $140 a ton, paid another $412 a ton to have it refined, and still cleared a $90,000 profit. That paid Hearst a 400 per-cent return in just three months. News of such instant fortune and so much magnificent rock electrified the gold camps of California. Thousands of fortune seekers rushed over the Sierra to the new El Dorado. Virginia City sprang up astride the lode and quickly mushroomed to become the second largest city west of the Mississippi, surpassed only by San Francisco. Many thousands more, seemingly everyone left in San Francisco, also wanted a share in the mines. The Ophir owners were the first to oblige, incorporating the Ophir Silver Mining Company in April of 1860, offering a generous $5,040,000 in shares at $300 par. Two months later Hearst and his partners also formed the Gould & Curry Silver Mining Company on the Comstock's other, even bigger bonanza mine to offer a more modest $2,400,000 in shares, but at a pricier $500 par. A flood of incorporations followed.[2]

By 1861 the slopes of Sun Mountain were blanketed with over 3,000 claims, stretching for over seven miles along the trend of the lode, and scattered for a mile or more to either side on real and imagined parallel lodes and branches. Most of these claims, however, didn't show even a speck of ore. Indeed, San Francisco preacher Thomas Starr King ridiculed the Comstock as a mess of "Ophir holes, gopher holes and loafer holes." Still, the streets of Virginia City were packed with men who fancied themselves to be fledgling millionaires, yet rarely had even a dollar in their pockets. But they all had shares, commonly traded as "feet" in the running length of the claims, although the rich Ophir allotted one share per inch! Most spent little time working their claims; instead they milled about on A Street, dubbed the "Exchange," swapped shares in each other's dreams, and waited for a mark with a little coin. They

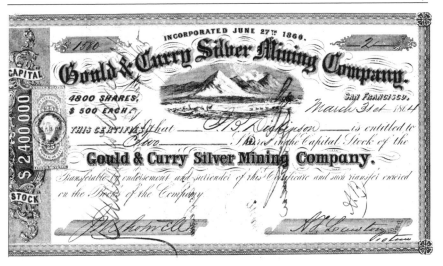

didn't usually have to wait long, for as the bonanza mines, the Ophir and the Gould & Curry, took out more and more ore, their "feet" took off.[3]

The bullion output of the Comstock mines with roughly equal value in silver and gold leapfrogged each year from just over $250,000 in 1859, to $1,000,000 in 1860, $2,500,000 in 1861, $6,000,000 in 1862, and over $12,000,000 in 1863. The value of mining shares jumped even higher, even though few dividends were being paid. Gould & Curry, which had originally sold for as little as $3 a foot in 1859 before it was incorporated, and for nearly $200 right after, bounded up with every bullion shipment until it passed $6,300 a foot in June of 1863—a 2,000-fold jump! Ophir did nearly as well, and the wildcats all leaped far beyond reason. Even San Francisco's most conservative investors joined in, but certain that it all was too good to last, they demanded usurous returns on their money. So even with the steadiest dividend payers, they only bought if the prices were such that the dividends would pay at least 50 percent per annum.[4]

Never before had such fortunes been made in western mining. Is there any wonder that thousands of would-be millionaires happily dreamed of doing the same with a few feet of some future Ophir and frantically grabbed almost anything they could get their hands on? Most bought their shares, however, not as investors anticipating dividends, but purely as speculators looking for a rise in the shares, a sale, and a quick profit. They seemed to care little whether the stock represented a mine or a myth, just as long as it moved upward in the market. Nearly every city, town, or camp on the Pacific Coast was seized by the mining stock mania, and no class, age, or sex was immune to the epidemic. Not only bankers and merchants, but doctors and lawyers, editors and preachers, carpenters and blacksmiths, clerks and laborers, cooks and maids, all owned shares in some silver dream. It was the broadest popular stock mania yet in America.[5]

It was a time when "every man carried 'feet' in his pocket and dividends in his eyes," a time when "one could not throw a stone at a dog without first looking to see if it did not contain a sulphret or chloride of some precious metal," a time when a company was even floated on rumors of a silver lode right in the heart of San Francisco on California Street, and others sold shares in imagined gold mines at Fort Point and Mission Dolores. Wherever the scent of precious metal was found, or even imagined, claims were staked and companies floated. Every foot of the Comstock was soon incorporated, so promoters turned to new silver strikes hundreds of miles to the south in Esmeralda, to the east on Reese River, and in countless other mining districts that suddenly appeared all over Nevada and the surrounding territory. Even many of the old California gold mines were reincorporated to share in the boom.[6]

The total capital solicited from the public in the name of mining soared to dizzying heights, far above the output of the mines themselves. By 1861, when the Comstock had yielded up nearly $4,000,000 in bullion, there were already ninety Comstock companies offering over $60,000,000 in shares. By 1863, at the peak of the boom, when the total bullion had grown to $22,000,000, there were roughly four thousand companies hawking shares totaling close to $1,500,000,000! One observer, noting that the price of shares being offered had "outrun the coined metals of the world," wondered how people were even expected to pay for them. Moreover, since the entire population of California and Nevada was close to 500,000, their full subscription would have required an outlay of $3,000 each by every man, woman, and child, when the average annual wage even in the West was only around $1,000. But the market was so flooded with shares that probably less than a tenth of them were taken up, even at less than a tenth of par on average, for a likely total of no more than $50,000,000, or $100 per capita. So if roughly all 250,000 adult males actually gambled on at least some shares, their average wager would have been only $200, or $70 per year—around 7 percent of wages.[7]

Most of the companies, however, produced nothing but paper—they were mere wildcats, with little more to recommend them than a prospect hole and a promise, conjured up by a legion of mining sharks to entice the marks. They ranged from such staid-sounding ventures as the Webster Gold & Silver Mining Company, through the promising Golden Harvest, Muy Rico and Rich & Rare Consolidated, the inspirational Hope, Confidence and Perseverance, the whimsical Haul-Off & Jump-Up, and the Pride & Big Foot Consolidated, to the downright honest Fool Catcher No. 1, 2 & 3, and the Fly By Night Gold & Silver Mining Company.[8]

A small fraction of these wildcats actually put a little money into the ground, opening a claim on the long shot that they too might strike a pocket of paying rock, or at least enough to stir up a bit of excitement to help sell shares. But the bottom line for most was simply to bring in money for the promoters, if only in the name of a mining claim, either real or imagined. These companies were each usually organized by just a few close friends, who had either located a claim, picked up one for a promised share of the stock, or just pretended they had one. They then took most or all of the stock in exchange for the claim, or their claim of the claim, and dividing up the shares, formed a pool to work up the share price and a market to unload on the public. They occasionally recruited figurehead directors, who were "either lured into it by paid up shares or liberal fees, or they were selected for the facility with which they may be controlled by the schemers." They also placed "puffs" of fabulous prospects in the daily papers, or even published fancy prospectuses with "flaming reports" from highly touted mining experts and college professors, usually bought off with stock options to ensure a strong shared interest in sales. One of the most helpful experts was Yale chemistry professor, and founding member of the National Academy of Sciences, Benjamin Silliman Jr., who traded very profitably on his famous father's name. But he generally worked on a strictly cash basis, reportedly offering a sliding scale of fees "in accordance with the favorable or unfavorable nature of his reports." On one extended trip to the West in 1864, "experting" very favorably on a number of wildcats, he boasted of taking in a shameful total of $200,000 in fees that would cost their trusting investors many, many times that.[9]

Once the promoters had unloaded all the stock they could on the public, that was often the end of it, and nothing remained but the tracks of "wild cat feet" and the smile of its shares. But some promoters occasionally went one step farther, if they could actually point to a hole in the ground or even some enticing pieces of picture rock that they said came from their ground. Then they could start levying assessments on all the shares that they had sold in order to open their potential bonanza, and of course cover additional costs like their salaries and other expenses. If they were skillful enough, with the help of some reliable relative, appointed as superintendent, who would send inspirational reports of new ore strikes and rich mill runs, they could all collect a good income from assessments for years, stringing along their shareholders until they finally gave up hope and forfeited their shares. Then, if the sharks were really adept, they could suddenly find new indications of a great bonanza, sell the shares all over again, and perhaps even start a new round of assessments as well![10]

Most of these wildcats at least showed enough humility to set their asking capital at a modest quarter to half a million each. But some boldly sought just as large a capital as the bonanza mines, and they asked for it in bites of $200 to a $1,000 a share. That, however, just made the shares look like an even bigger bargain, since most wildcat shares rarely fetched more than a fraction of par, most often only pennies on the dollar. Still, a surprising number occasionally jumped almost as high as those of the paying mines, lifted solely by the strength of earnest hopes and flagrant lies. Many shareholders learned, however, that some shares were dear at any price. One all-too-eager mark was delighted when a promoter generously gave him eleven shares with a face value of $1,100 in a little kitten called Mohawk & Montreal. He hopefully held them for nearly three years, dutifully paying every assessment so as not to lose them, before he finally realized that these free shares had cost him $228.50, and he quickly gave them away before they could possibly cost him any more![11]

THE PROMOTERS OF some of the wildcats went for even bigger game. Known as "fighting companies," these were organized around a conflicting claim to part of some more promising mine in the hopes of striking it rich in litigation. Some of the companies, however, claiming separate lodes that paralleled the main lode, raised serious challenges to the wide, single-lode theory proclaimed by the major companies and to their claims of unlimited "extralateral" rights to follow the dip of the lode underground, far beyond the sidelines of their claims, following earlier California laws. The result was an orgy of litigation that carried on for several years as the dozen leading companies engaged in an average of over twenty suits apiece, more than two-thirds of which they initiated. The Ophir alone was involved in thirty-seven suits and spent $1,070,000 in just one losing engagement with the Burning Moscow mine over the single-lode issue. All the litigation cost the companies about $10,000,000 in legal fees and other payments to combative lawyers and blatantly bought witnesses and experts, as well as judges, whose official salaries were a paltry $1,800 a year.[12]

The most aggressive lawyer in these suits was a former Ohio high school math teacher, William M. Stewart, who had gone into law in 1852, at the age of twenty-five, in Grass Valley and Nevada City and come out a decade later as the most arrogant and unscrupulous lawyer on the Comstock. He openly boasted of holding a judge at gunpoint to get his way, and he demanded $200,000 a year for his services, which soon made him a millionaire. One admirer wrote, "Bill Stewart'll worry the witnesses, bullyrag the Judge, and buy up the jury and pay for 'em; and he'll prove things that never existed—hell! What won't he prove!" Even after losing both the Ophir and another major case against the many-ledge theory, he simply bribed and threatened all three justices of the Nevada Supreme Court into resigning, so more favorable judges could be appointed. The ultimate victory of the single lode came with the discovery that all the parallel veins merged at a depth of about five hundred feet. Stewart meanwhile became one of the new state of Nevada's Republican senators and embedded extralateral rights into the new federal mining laws of 1866 and 1872, ensuring a steady livelihood for himself and the rest of his clan.[13]

Yet many of these "fighting" wildcats were basically just blackmailing schemes, filing legal challenges with the sole aim of being bought off with a fat settlement, and they, too, found speculators eager to buy shares in the hopes of getting a piece of the payoff. Typical of these was one of George Roberts's early Comstock ventures, the Grass Valley Silver

Mining Company, which challenged the neighboring Potosí company's Comstock ground, and whose accounts were exposed in a disgruntled stockholder's suit. They raised over $70,000 from assessments of their stockholders and spent less than $6,000 on actual mining, while Stewart, who played both sides of the street, and other lawyers got $13,000 in fees and "loans," their court witnesses were won over with "loans" of $14,000, and other unnamed parties possibly on the bench got "loans" of over $7,000, while, last but not least, Roberts took $9,000 and other officers $6,000 in salaries and "loans," and they mysteriously made off with another $11,000, "nobody knows where or how." This, of course, was on top of whatever they got from the original sale of the stock and the final settlement. A few outside bettors may even have lucked out on such suits. At least it was said that "Johnny Morgan, a common loafer, had gone to sleep in the gutter and waked up worth a hundred thousand dollars, in consequence of the decision in the 'Lady Franklin and Rough and Ready' lawsuit."[14]

By far the biggest and most blatant of these blackmailers, however, was the Grosh Consolidated Gold & Silver Mining Company, which simply laid claim to all the mines from the Ophir to the Gould & Curry. There was no bigger prize, for these were the great bonanzas of the Comstock, the very richest portion that would produce a staggering $120,000,000 in bullion and pay $80,000,000 of that in dividends within the next twenty years! The scheme was wrapped around the tragic fate of the Grosh brothers, who had actually discovered silver on the lode several years before anyone else, and it was led by an avaricious Placerville lawyer and vigilante, Benjamin R. Nickerson, who teamed up with fellow lawyer and San Francisco real estate speculator, S. Clinton Hastings, the first chief justice of the California Supreme Court and later founder of the West's first law school.[15]

Allen and Hosea Grosh, the smart but hapless young sons of a Pennsylvania minister, had rushed to California in 1849 and worked in the placers at Mud Springs near Placerville with little success. So in the summer of 1853 they crossed over the Sierra to try their luck in Gold Canyon, urged by an old Portuguese miner, Frank Antonio, who had found silver there in 1850. Camping near what later became Silver City at the southern end of the great mineral lode, they too spotted traces of silver, four miles from the yet-to-be-discovered bonanzas at the north end. But they had little time to search for its source, since they had to work nearly full time just trying to make a living from the placers. So they headed back to Mud Springs with ore samples the following year and eventually talked about half a dozen friends into forming what they called the Frank Silver Mining Company to grubstake them. Returning with ample grub to Gold Canyon in the summer of 1856, they soon traced out a "perfect monster" of a lode, which they located for the company, and crude tests of the dark "black rock" in a little furnace suggested $50 or more a ton in silver. That winter they eagerly got their father to sign up more than a dozen family members for another company, the Utah Enterprize, so they'd have additional names to stake claims with.[16]

The following summer of 1857 they discovered a new bunch of silver veins on up the canyon at the foot of Sun Mountain and staked those for the family. A "hurried assay" of one small piece of rock even gave $3,500 a ton in silver! With skyrocketing expectations, one of their Frank company partners, Billy Longet, came over to help, while another, George Brown, who ran a station on the overland trail in the summers, offered to put up enough money to open the claims that winter. But suddenly the dream was shattered. First they found that what they thought was their richest silver assay was faulty. Then in mid-August they learned that Brown had been murdered. Just days later Hosea accidentally stuck a pick into his foot, gangrene quickly set in, and he died just two weeks later. His distraught brother finally decided to return to California for the winter and left everything in the care of Longet. Allen, however, got off to a late start, crossing over the Sierra in deep snow late in November, and his feet were badly frozen. Refusing to let them be amputated, he too died within weeks. Without the Grosh brothers to guide him, Longet did nothing more with the claims.[17]

The Gold Canyon miners, however, finally recognized that the great quartz lode was the source of their gold. Early in 1858, just a couple of months after Allen's death, they staked gold quartz claims that may have been on, or close to, the Groshes' claims near the head of the canyon at the southern foot of Sun Mountain. There the camp of Gold Hill sprang up, and the miners organized Nevada's first quartz mining district, but they still hadn't discovered the silver. Not until a year later, in June of 1859, did two Irishmen, Pete O'Riley and Pat McLaughlin, move on over the divide another mile and a half into Six Mile Canyon

on the northeast side of the mountain, and while digging out a spring, accidentally discover the great bonanza ores of the Ophir at the upper end of the lode. Then an opportunistic drifter, Henry T. P. Comstock, quickly insinuated himself. He blatantly claimed title to the spring and all the surrounding ground as a homestead that he had never bothered to record! The badgered discoverers finally gave him and his "homestead" partner equal shares for their fancied claims and water rights. Although they all soon sold out to Hearst and his partners after the ore was finally assayed in Nevada City and Grass Valley and shown to be rich in silver, the boastful Comstock still managed to get his name on the whole lode. He promptly left to claim new bonanzas in Idaho and Montana, but failed and finally put a bullet through his head in 1870.[18]

As soon as the flood of silver began to pour out of the Ophir and Gould & Curry, members of the Frank company started talking about reclaiming part of the lode. But they were quickly outmaneuvered by the fast-footed and faster-talking Benjamin Nickerson. Disregarding the Frank claims, he headed east and began working on the Grosh brothers' aging father. First Nickerson got the trusting old preacher's power of attorney to press a claim, then slowly tricked him, not only into deeding over all his interest in the claims for a mere $5,000, but into signing a receipt for full payment as well, on the promise that the money was on the way—and Nickerson never gave him a cent! In July of 1863, as soon as Nickerson got that deed, he and Hastings formed the Grosh Consolidated company at $5,000,000 in $1,000 shares with Hastings as president, and they recruited half a dozen San Francisco merchants, who put up $20,000 at rock-bottom rates to bankroll the scheme. They then shocked everyone with the startling announcement that the "Grosh boys' claims" weren't in Gold Canyon but were on "the divide" and stretched for 3,750 feet, exactly covering the fabulous Ophir and Gould & Curry, and that just before Allen had left to cross the Sierra, he had entrusted them to Henry Comstock for a quarter interest, and the shifty Comstock had sold not just his own interest, but also the Grosh heirs', to the current companies![19]

With what they proudly called their "absolute deed of conveyance" from the Groshes' father, Nickerson and Hastings promptly filed suits against both major companies and Comstock as well, so that his denials of any connection with the Grosh brothers could be discounted, since he was being accused of fraud. The suits were "bombshells," demanding that the companies be enjoined from paying any further dividends, that the stockholders pay back all of their dividends, and that a receiver be appointed immediately, to start collecting all the money now due the new Grosh Con. Lest anyone think they were overreaching, Nickerson and Hastings also said that the renowned old lawyer Reverdy Johnson, the Democratic senator from Maryland and former U.S. attorney general, had even gone so far as to say that the "brothers, by right of discovery, owned the entire Comstock Lode." At the same time they doubled the shares, which they offered to the public like lottery tickets.[20]

The suits were roundly denounced as "simply preposterous," and the flabbergasted editors of San Francisco and Nevada papers scoffed at the whole business and lambasted the company and its promoters. Like the *Alta California*, most clearly looked upon the suits as "a mere blackmailing operation, intended to compel the owners of the mines to buy off the title." The *Bulletin*, however, suspected that it was also an "effort to bear the stocks," while another simply dismissed it as "one of the most silly moves" of the week. But the playful young reporter on Virginia City's *Territorial Enterprise*, Samuel Clemens, had the most fun, reworking Mother Goose with such ditties as this with its cold close:

> The Ophir, on the Comstock,
>   Was rich as bread and honey,
> The Gould & Curry, further south,
>   Was raking out the money.
>
> The Savage and the others
>   Had machinery all complete,
> When in come the Grosches
>   And nipped all our feet.

And another, spoofing their pretensions:

> Who owns the Comstock?
> We, say the Grosches,
> So loud and ferocious;
> We own the Comstock.
>
> What is your title?
> Shameless audacity,
> Fraud and rapacity;
> That is our title.

Nickerson fired back, ridiculing the "brainless editors" and their "trifling nonsense" and vowing to press the suits "regardless of the threats of resistance and assassination" that he claimed were being "so freely indulged in." But the bonanza companies also had no fear, and they were equally determined to spend whatever it took to kill the case in the courts rather than buy it off.[21]

Thus the suits stretched on for a year and a half, as Nickerson and Hastings kept seeking continuances, while they kept trying to extort a big payoff, but with no success. For although few seemed to doubt that the brothers had found silver in Gold Canyon, hardly any believed that they had discovered the bonanza ores over the divide to the north. Although Nickerson and his partners did produce three questionable witnesses who were willing to swear that they had worked with the brothers on the Ophir ground, their only piece of hard evidence was the alleged contract by Allen Grosh with Comstock that happily floated their claim "beyond the brow of the hill" toward the bonanza mines. That hastily contrived document, however, was clearly fraudulent, since it was clumsily dated December 11, 1857, three weeks after Allen had left Gold Canyon and was already across the Sierra in California at his "wits-end with fever," and writing that he had left everything in Longet's hands, not Comstock's! So the whole scam finally ended quietly in March of 1865, when the court dismissed the claims and ordered Nickerson, Hastings, and their partners to pay costs. Hastings returned to more legitimate pursuits, while Nickerson, unashamed, later ran as the Republican candidate for secretary of state of California, vowing to clean up fraud! Meanwhile, in the summer of 1865 the Republican speaker of the house and next vice president of the United States, Schuyler Colfax, who happened to visit the Comstock, helped dedicate a marble headstone on Hosea's grave, which was the only ground the Groshes had left. [22]

IN SUCH A FEEDING FRENZY of fraud, even some of the sharpest sharks got bitten. In 1863 William Barron, his quicksilver partners, their lawyers, and friendly judges were all taken in by a skillful "salting" of the Santiago at the lower end of the Comstock. The previous year a smooth San Francisco speculator, J. Downes Wilson, had picked up 2,800 feet of ground through a confederate, Isaac Freeborn, for $25,000 and put him to work as superintendent, dressing it up for an eventual kickback of another $40,000 or more. By spring Wilson claimed he had struck a new bonanza, which was such a "big thing" he would let a few friends in on it. Barron and his friends were eager but cautious at first, and before they would completely commit, they prevailed upon Wilson to send fully seven hundred tons of his rock to be worked in a couple of different mills to get an "assay on a grand scale." When the mill runs all came back at a nicely profitable $40 a ton, all caution was thrown to the winds. Barron and his associates pressed Wilson until he generously consented to give them a controlling interest in his treasure trove in exchange for a "stately palace" in the city, a courtly country retreat, and over $100,000 in cash, and they promptly started assessing themselves at a rate of nearly $20,000 a month to buy a mill.[23]

But that summer, despite Wilson's protests, Barron's new engineer started shipping several hundred tons of rock to one of the main Comstock mills, which found only $13 a ton! At the same time it was discovered that, while Wilson was assuring shareholders that the Santiago was "the safest thing in Washoe," he had quietly sold all but just four of his remaining 1,400 treasured feet to eager buyers for another $150,000 or so, and they realized too late that they, too, had been sold. The disillusioned buyers promptly shut down the mine, piled up by then with hundreds of tons of rock "so poor that nobody will haul it away." The stock crashed and they abandoned their shares for assessments. They suspected that Wilson had either gutted the richest rock or secretly "borrowed" a few tons of Ophir ore to add to his mill runs. But when San Francisco *Bulletin* editor and pioneer California historian Franklin Tuthill published a full exposé of the shareholders' claims that they had been "outrageously swindled," Wilson indignantly sued for libel, and to the surprise of many, he cleaned up an additional $7,500 in damages. So in the end he was remembered simply for having "made a good deal of money through judicious management."[24]

Still the most notorious, if not the most lucrative, of these early wildcats was the wonderful Bailey Silver Mining Company, organized around a purely illusory bonanza way out in the midst of Death Valley. Its namesake, Robert Bailey, a crafty miner and prospector, looking for the famous "lost gunsight" silver ledge of the Death Valley forty-niners, startled San Franciscans in January of 1861 with pieces of a strange new ore, "the richest that has ever been found," assaying $16,342 a ton in silver! He claimed at last to have found that long-lost ledge which far surpassed even the Ophir, but he would say only that it was hidden somewhere in the Washington mining district, an area the size of Rhode Island stretching along the east side of the dreaded valley. Bailey, of course, refused to sell such an incredible bonanza, but he did generously consent to sell shares in adjacent claims to a crowd of would-be investors who were eager to buy, if he would just bring a little more rock.[25]

So after giving the slip to those who tried to follow him, he returned in April with four or five bags of rock. Although it assayed *only* $4,200 a ton, an enthusiastic real estate broker and his friends promptly gave Bailey $12,000 "cash on the nail" for a half interest in the "first south extension" of his claim, and together they formed the company with 900 shares at $200 par. Bailey quickly sold his half of the stock at bargain prices of $50 a share or more, and then formed another fleeting company, fittingly named the Comet, around a "second extension" to offer an additional 1,600 shares at $500 par for all of those still wanting to be taken in on his wonderful scheme. The stockholders finally realized that they had been sold in August after Bailey suddenly disappeared, secretly sailing to

Tahiti with his loot of close to $70,000, and a curious geologist, William P. Blake, who discovered a chunk of charcoal in some of Bailey's "ore," proclaimed it all a fake, sloppily concocted in a backyard furnace. Two years later some Comstock imitators ran up the shares of a remote and barren "North Ophir" from $12 to as much as $165 with dazzling lumps of $10,870-a-ton "native" silver, plus a paid puff from Sam Clemens. But their stock also crashed when someone spotted fragments of dates and eagles, as well as the words "ted States of" on one of the lumps.[26]

The proliferation of all these wildcats was undoubtedly encouraged by the ease with which they could be bred. All of the papers, articles of incorporation, ledgers, shares, and prospectuses could be had for only $130, while the title to some stray claim could be picked up for even less. Moreover, the not-so-fastidious promoter, who only wanted the barest of essentials, needed just $25 for a handsome book of stock certificates, printed in color—green was a favorite; embellished with impressive-looking typefaces; and adorned with a couple of inspiring vignettes, of which a miner pushing a carload of ore and a little safe guarded by a bulldog were especially popular. As one admirer noted, "the arts of typography, lignography and lithography were tasked to their local utmost" to fill the demand for such certificates. In fact some, such as those of the Honest Miner Gold and Silver Mining Company were minor works of art, as well as deception, although they didn't match the masterpieces of the great water companies.[27]

NOT EVERY CLAIM HOLDER was a rogue, of course. Many fervently believed that they had at last found a real bonanza and were determined to work it themselves, and naturally enjoy all the profits, without asking for any money or other support from outsiders. But such men were more often pitied than admired, for as Adam Smith had long before warned, most would sink all their time and treasure into what would prove to be only a fruitless quest. As cynics saw it, "it is simply cheating one's self instead of roping in outsiders, and the victim is declared a flat, when he might have been at least a sharp—which is the difference between a rogue and a fool."[28]

In this frenzy of hope and greed, local editors and reporters were also on the take, happily pocketing "complimentary" shares from promoters in exchange for a favorable mention of their latest wildcat. Even that ambitious reporter, Sam Clemens, freely confessed to his mother, "I pick up a foot or two occasionally for lying about somebody's mine." But, he claimed, most didn't really "care a fig what you said about the property," just as long as you said something. Nonetheless it proved to be quite profitable, for he boasted that he had made $5,000 from the sale of such gratis shares and still had a trunk half full of them. He was again careful to point out, however, that not all of them were given by those wanting their claims "noticed." Many, he assured, "looked for nothing more than a simple verbal 'thank you,'" stocking up for a rainy day. He, too, was hooked, however, and plowed most of his illicit profits back into the bigger, more enticing mines.[29]

Some of the investors, of course, also proved to be predators. Two of the most highly publicized and thoroughly exposed were San Francisco jewelers Moses Frank and his partner, Frederick Baum, who stealthily gained control of their prey and made it "bleed profusely" for their benefit. In 1863 Frank bought some stock in the Utah Mining Company, which was steadily producing over $5,000 a month in bullion and was one of the most promising mines in Aurora, a hundred miles south of the Comstock. He took such a "great interest in the welfare of the company" and gave so freely of his time and energy that the other shareholders soon decided that Moses was the ideal man to lead them as their president. He then named Baum, who just happened to run a branch of their business in Aurora, as the perfect and cheapest choice for their superintendent, at a token fee of only $25 a month.[30]

But once Frank and Baum had the company firmly in their grasp, they began to suck it dry. They stuck it with all the petty business frauds still common today. While Frank's official letters were the "very models of business integrity," he secretly directed every step of the looting in a stream of confidential letters to Baum. First he instructed him to take kickbacks on contracts; buy the cheapest supplies and mark up the costs, such as "buy an old horse for $40 and charge $200 for it"; "fix it with the lawyer" to get a cut of legal settlements and report double his fee for an added cut; and, basically, "on all accounts you must see that we make something." Then he started drafting monthly expense reports for Baum to copy "clean" and send back. He also told him to write "flattering" accounts of the mine and report paying twice, or even as many as five times, as many miners as they had at work. So they were soon taking $1,500, or more, a month just in fictitious wages. To cover that, Frank carefully alerted Baum whenever one of the shareholders or others were planning to visit, so the superintendent could temporarily put on extra men to "tally with the payroll."[31]

All the while Frank boasted that the trustees "never ask any questions at all," generously treated him to dinners, and promised him all sorts of gifts the moment he declared the first dividend. They were still so clueless that they reelected him president, to which he consented "as a great favor to the shareholders." He also got Baum a concurrent appointment as superintendent in a neighboring mine for just $50 a month so they could loot it, too, but he vowed "as long as Utah will pay us we stick to it." When one stockholder did get nosy, they bought him off. Next Frank tried to work a penny-ante stock scam, telling Baum to "let nobody in the mine," send an official telegram that he had just struck "very rich" rock, and give the miners an extra "$2 or $3 for drinks to say the strike is good." Frank predicted they would make $3,000, but it didn't happen, so he decided to sell their stock instead, bear it with assessments, and buy it back cheap. At the same time he planned to buy a new mill and other equipment through a friend and split the resale markup to the company.[32]

But after nine months, some of the trustees finally became suspicious and voted to dismiss Baum. With outraged innocence, Frank indignantly called the action "highly contemptible" and threatened "sharp revenge." Like a "spasmodic leech," one critic noted, Frank then frantically tried "sucking the last drop of life blood from the company," instructing Baum to issue over $10,000 in drafts and other charges against the mine and not to surrender it until they were all paid. But when Frank approved one of these drafts and tried to pass it, he was suddenly arrested for forgery. Then, despite his protests of invasion of privacy, the trustees got hold of his letters, which Baum had failed to burn as instructed, and their whole "disgusting and rascally thieving" was exposed. Frank tried to flee to Manila, shamelessly claiming "they are persecuting me!" The police, however, plucked him from the ship just before it sailed out the Golden Gate. He then tried to cut a deal, dropping all efforts to get back the incriminating letters in return for the company dropping all charges against him. The trustees agreed and let Tuthill publish all the larcenous details in the *Bulletin* for the enlightenment of all. But a grand jury then seized upon the letters to indict him again. While still crying "persecution," he bribed jurors in two trials, but that too was exposed, and he was finally sentenced to six years in San Quentin to "take a long sweat for his dastardly villainy."[33]

AS THE FRENZY of speculation swelled and the flood of paper rose, stock-brokers' and mining company secretaries' offices blossomed all along San Francisco's Merchant and Montgomery streets. Soon there were over two hundred new mining stockbrokers, or one for every two hundred adults in the city! First-floor window space was in such high demand that even long-established merchants rented out their front windows to brokers to set up desks and shelves of enticing mineral specimens. Those who couldn't find an office simply made do on the curb with a telltale swatch of black crepe on their hats and their pockets stuffed with rocks and stocks. The mining company secretaries had to settle for second- and third-floor cubicles, which were often made to do service for a dozen companies, or more.[34]

The brokers in turn created stock boards and exchanges to market their wares. The first informal market was San Francisco auctioneer James Olney's Washoe Stock Exchange, which began auctioning Com-stock shares in the fall of 1860. Soon other brokers commenced rival auc-tions, and confusion over prices mounted until the first regular exchange was organized in the fall of 1862. On September 11, thirty-seven brokers, later dubbed "the Forty Thieves," organized the San Francisco Stock and Exchange Board. They held their first trading session two weeks later in a small room in the Montgomery Block. As trading rose steadily from a few thousand dollars a day to hundreds of thousands a day in less than a year, the exchange moved into successively larger and more imposing surroundings. The number of seats on the exchange doubled, and their price escalated from $100 to over $2,000 each. The "mining stock mania" had seized San Francisco, and in the next three years all the frantic shuffling of shares on the exchange totaled $85,000,000, twice the treasure from the mines![35]

The twice-daily sessions of the stock board were a source of mys-tery and amusement to outsiders, who variously likened the affairs to a "Chinook vespers," or the "snarling of a pack of wolves fighting over a carcass." But to lend an air of piety, Sam Clemens offered a playful benediction:

> Our father Mammon who art in the Comstock,
>   bully is thy name; let thy dividends come,
>   and stock go up, in California as in Washoe.
> Give us this day our daily commissions;
>   forgive us our swindles as we hope to get even
>   on those who have swindled us.
> "Lead" us not into temptation of promising wild cat;
>   deliver us from lawsuits;
> For thine is the main Comstock,
>   the black sulphurets and the wire silver,
>   from wall-rock to wall-rock, you bet!

All the while the stock prices surged and swooned, sympathetic to every telegram from the mines, as speculators savored every whisper of a rumor. When part of the Ophir caved in, the shock was felt much more severely in the San Francisco Stock Exchange than in the drifts of the mine itself. Indeed, one enthusiast wrote, "If the Comstock lode should catch the itch, everybody between North Beach and the Mission would go to scratching."[36]

The success of this first exchange quickly prompted imitation. As the fever of speculation raged on, stock exchanges sprang up everywhere "like beer-booths about a military parade." Five rival exchanges were started in San Francisco in little over a year. As the "feet disease" spread, stock boards also sprouted in the inland towns of Sacramento, Stockton, and Marysville, and on up the coast at Portland and even Victoria, B.C. The mining camps themselves were not to be outdone, and any camp with aspirations had to have an exchange of its own. Eight blossomed on the Comstock alone in Virginia City and neighboring Gold Hill, plus one at Carson City, and at least one apiece in the outlying towns of Aurora, Austin, Como, Grass Valley, and Silver Mountain. The most pretentious of these exchanges, the Washoe Stock and Exchange Board in Virginia City, erected a two-story brick hall with a stained-glass dome. But most carried on their dealings in more informal surroundings like the Como Miners' Stock Board, which met evenings in the corner of a saloon.[37]

The hungry pack of instant brokers worked the flock of new investors and speculators for all they were worth and more. Brokering was suddenly the shortcut to fortune. "A young man, without means or credit, can commence the brokerage business easier than any other, and make it more profitable," one enthusiast noted. A seat on some of the lesser exchanges could be had for as little as $25, and it could pay handsomely if you weren't too picky about what you did. Some may have been quite content just to honestly broker sales on demand for a commission of a few pennies on the dollar, which was, after all, ten times the fee of eastern brokers. But like the easterners, many wanted much more, and they eagerly fed the rumor mill, spreading each whisper of a bonanza or borrasca to keep stirring and shuffling the shares. They also aggressively pushed new shares, in which they had taken a block at a low price, and joined a "pool" with the promoters to work it up with "wash sales" of fictitious transactions, reporting ever-increasing sale prices and the illusion of a great rush of excitement, spurred on with hot tips and urgent calls as they brought in the public. Some even staged live performances of feverish sales of favored shares in their offices, where a few shills "begged for the privilege" of buying and others "sighed grievously" at selling, all to excite and entice a particularly promising fat cat.[38]

Even if an investor couldn't afford the price of a share, a generous broker would always offer to sell it to them on margin, for as little as 10 percent down and an interest of at least 5 percent per month on the balance. As long as the market was rising faster, some of these investors turned speculators, leveraged as much as tenfold, could make a real profit if they were lucky enough to sell out before the market broke. Their successes inspired many others who only got to savor paper profits, however fleeting. Even if their customers lost, of course, the brokers still won, and if they were confident that the price would fall, or if they could "bear" it enough themselves to wipe out the margin, they could simply pocket the margin money and never have to even think about buying the stock, unless the marks were willing to put up even more "mud," or money on margin. If customers happened to think a stock was overpriced or had heard some rumor of a imminent decline, helpful brokers would steer them into a "short" sale, where they contracted to sell at a future date shares they didn't yet own at a lower than current price, hoping that by then the price would be even lower, so they could buy them cheaply and "cover the shorts" to fill the contract for a quick profit. It could be a costly gamble. Thus, in the midst of the mania, prudent and conservative

"The Silver Mania at San Francisco"
J. Ross Browne, *Harper's New Monthly*, May 1865

men and women, who had always kept their savings, large or small, in a bank or under a mattress and had never even looked at a stock certificate before, were plunging wildly in over their heads in wildcat shares on margins, shorts, and other devices.[39]

As the mania raged, one exasperated watchdog grumbled, "It is a waste of labor and honest printers' ink to even warn the people of the perils of stock gambling. . . . This whole devilish business is so utterly demoralizing, so entirely bad, so absolutely destitute of one single redeeming feature, that we discuss it with impatience almost amounting to bad temper." He saw the sharp broker as a "stock tarantula" who "with subtle and ingenious deviltry schemes how he may rob and plunder his fellow-citizens," but he also saw the client as a "poor, silly fly" from a "multitude of most egregious fools." For, he confessed, "while we have only contempt for the plundering stock-jobber who systematically robs his victims, we have a feeling nearly akin to it for the victim himself." And, he lamented, "this stock gambling is ruinous to our people, and is laying broad and deep the foundation of a superstructure of crime and disaster to our state and city. It is making a few disreputable and unscrupulous gamblers rich; it is sweeping away the earnings and the accumulations of thousands, and will leave them in poverty, disheartenment and despair. It is destroying the industry of the people; it is sapping the morals of all who engage in it."[40]

YET IN MANY of the mining camps, the fancily printed stock certificates played an added role, frequently passing for legal tender. For it was one of the ironies of western hardrock camps, that although they were surrounded by a wealth of gold- and silver-bearing rock, there was a chronic lack of hard cash. This resulted not so much from any poverty of the itinerant miners as from the steady outflow of money for food and supplies that often surpassed the returns to the camp in miners' wages. This was in sharp contrast to the immediate returns in placer camps, where each miner's daily earnings was some fraction of an ounce in gold dust, readily taken, albeit at a discount, in any store, saloon, or "hurdy house." Thus credit and hope were the lifeblood of most hardrock camps, and mining shares became a medium of exchange, the currency of last resort.[41]

Even claim owners with little means could cheaply incorporate their holdings into a mining company, which, like a wildcat bank with its assets locked up in the earth still untapped, issued shares like banknotes to tender for every bill or debt incurred. Most of these early shares were in fact little larger than banknotes, and to facilitate exchange they were frequently written in small denominations of shares, typically "fives" and "tens," so that it was easy to make change! Though these shares were not taken at anything near par, the rival greenbacks were also locally discounted to as little as 40 cents on the dollar.[42]

Needless to say, not all merchants were willing to accept shares in payment. But many succumbed when it was that or nothing but a promise. Their resistance to shares was also softened by a few well-publicized instances where such shares had suddenly appreciated to reward the holder many times over what was due. Shares in one wildcat company given a Sacramento restaurant keeper in 1863 for a month's board were said to have skyrocketed in value to $15,000. Such stories doubtless grew with every telling and were seemingly retold with every new boom. So forty years later, a story went the rounds that a Spokane saloon keeper, who reluctantly took mining stock for a $16 whiskey bill, subsequently got $25,000 just in dividends, while the stock itself climbed to $300,000! Most merchants undoubtedly got little or nothing from the shares they took in trade. But some of the shares did have temporary value, and the proliferation of local stock exchanges offered merchants the chance of recovering at least something from them.[43]

LATE IN 1863 the mining mania spread east after Gettysburg, when a Northern victory seemed assured. There the fever readily caught hold in the Civil War–inflated economy of the Northern states. Men who had turned quick profits from the inflation through commodity speculation looked to mining speculation for new opportunities. Many less fortunate, who had seen their wages and savings eroded by rising costs and depreciating greenbacks, also caught the mining fever and desperately plunged what they had left into mining stocks in the hopes of quickly recouping what they had slowly lost. By March of 1864 the demand was so large that the dedicated Mining Board was opened in the Gold Room of the New York Stock Exchange, and a rival, the California Stock Board, followed in November.[44]

San Francisco and New York promoters unleashed a rush of Nevada wildcats in the East, and even though eastern investors shied away from the Comstock, they hopefully put their money into the outlying camps. Seemingly the most profitable of these stock "fancies" was the all-encompassing New-York & Nevada Gold & Silver Mill & Mining Company, a pyramid scheme launched with $660,000 in shares in January of 1864 by a trio of sharks—George A. Freeman, John J. Osborn, and William H. Forbes—who lured marks with a show of "large dividends and magnificent pretensions." In March, amid a lavish display of gold bars—quietly bought at the Gold Room, since the company never produced even an ounce—they paid out 5 percent of the take as their first quarterly dividend. This so excited investors that the trio created two subsidiaries, New-York & Washoe and New-York & Reese River, with an additional $1,000,000 and $1,500,000 in shares, to satisfy the demand,

rake in even more profit, and provide a little cash for more dividends
for the New-York & Nevada company. As sales of their subsidiaries
slackened, they repeated the process in January of 1865, launching yet
another subsidiary, the New-York & Santa Fe, to pick up over $700,000,
start quarterly dividends for the two earlier subsidiaries, and, of course,
keep up those of their wonderful New-York & Nevada, which they were
still working on the exchange. This might have gone on forever with
ever-new subsidiaries, but some of the investors finally got wise early in
1866 and brought suit, though not until Freeman, Osborn, and Forbes
had sold them nearly $4,000,000 in shares and generously paid back
$516,000 to keep them hooked.[45]

Yet even when investors actually got into rich mines like those of the
Manhattan Silver Mining Company, it was still no guarantee of profits.
Although these mines had rich ore in sight, the company floundered
for a couple years as its first novice manager squandered any profits on
extravagant surface works and ran it $180,000 into debt by 1866. Then
its clever, twenty-eight-year-old New Jersey bookkeeper, Allen Curtis,
who had become a major creditor, took over as manager. While the
company cut back its $1,000,000 capital to a lean $387,500, Curtis opened
ore averaging $250 a ton, paid off its debts in a year, began annual divi-
dends the next, made it the boss mine of Reese River, and transformed
Austin into a company town.[46]

But Curtis also proved to be a thoroughly modern accountant, for
while the shareholders got only 5 percent a year, he became rich and
even branched out into banking and railroad building. He skillfully
worked every angle of what the New York *Times* editor called the "stick-
ing process," that costly problem in mining where at "every stage of the
process, some of the product sticks to the persons engaged in it and never
reaches the proprietors." So, although the ore was exceptionally rich,
the "skillful" Curtis worked the mine by contract rather than paying
wages, purportedly to trim costs. But the contracts that he made were
suspiciously generous, resulting in outrageously high mining costs of
$98 a ton, nearly ten times the typical costs on the Comstock and most
other Nevada mines, worked by wages at $4 a day! Like any good mine
manager, Curtis also did exploratory work seeking new ore, but again
his contracts were so generous that they added another $50 a ton to the
overall cost. To protect the company against potentially costly litigation,

he also brokered generous deals to pay well over $100,000 for overpriced neighboring claims, until he had expanded the company's holdings from its original four claims of 3,700 feet to an unbelievably excessive 150 claims running to nearly 130,000 feet. He also prudently kept a reserve of more than $100,000, equal to several years' dividends, in his bank for emergencies. But he was generous not only to himself but to the community as well, spending $9,500 on an impressive Gothic church and winning local loyalty. So even though the mine would eventually produce more than $13,000,000 in silver over the next fifteen years, its weary, but still clueless shareholders, tiring of their pitiful 5 percent dividends, finally let go of their shares cheaply, just to try to salvage some fraction of their investment. Curtis eagerly bought them out, named his younger brother as manager, and reincorporated the company in San Francisco at $5,000,000 for added profits on the stock exchange.[47]

DURING THE FLURRY of mining speculation in the East, even Fremont's old Mariposa was raised again, reaching even more scandalous heights. Through political influence and outright bribery, Fremont had gained court-approved title to the land, if not yet the mines, but he had mortgaged it all to buy victory and defaulted. His bankers and partners, Palmer, Cook & Company, had also failed, and their leading creditor, Trenor Park, a "keen-witted" but larcenous forty-year-old Vermont lawyer, took over their half of the Mariposa and in 1859, together with San Francisco banker Mark Brumagim, who had become Fremont's new principal creditor, bought the property at the sheriff's sale for a paltry $1,428! Park worked the mines under a receivership for a fat fee, but he and Brumagim agreed to give "the bewildered Pathfinder" and "foremost debtor in America" one more chance at the title, if he could raise the $1,500,000 needed to pay off the debts. That all came to naught, however, with the onset of the Civil War, when Fremont was briefly given command of the army in St. Louis before Lincoln was forced to dismiss him on a long list of charges ranging from neglect of duty and insubordination to corruption and tyrannical conduct.[48]

Although Fremont had again failed to pay the debt, Park kept him as a lure and succeeded in selling others the chance. In June of 1863 he joined with a trio of New York sharks to offer an obscenely bloated

$12,000,000 in stock and $1,500,000 in bonds on Wall Street in the newly formed Mariposa Company, whose only asset, as one critic put it, was an "overdue pawn-ticket" to redeem his ill-gotten property. They pushed it hard, however, with outrageous praise from pliable mining experts, including even Josiah Whitney, head of the California Geological Survey, who could "hardly see a limit to the amount of gold which the property is capable of producing" and obligingly concluded that it must, "without exaggeration, be termed inexhastible!" As if to prove it, Park began gutting the stolen Pine Tree and Princeton at a rate of as much as $101,000 a month to impress investors before they unloaded, and to grab all he could, he cooked the books and took a cut of the output instead of salary. His partners, New York's Republican mayor and former broker, George Opdyke, and banker Morris Ketchum, who had previously been involved with Fremont in wartime arms profiteering, worked the market end of the scheme with another shark, James Hoy, while a clueless Frederick Olmsted, the designer of Central Park, was set up as manager.[49]

Giddy New Yorkers couldn't get enough of the Mariposa, pushing its shares from their "bargain" opening of $50 on up to $70. Before it

was over, the promoters had raked in about $8,000,000 for "a property
to which they never had a title, and for which they never paid a cent."
Opdyke, Ketchum, and Hoy got away with half of the loot, including the
bond money; Park got out with close to a quarter; and a couple of other
creditors got about a tenth between them. But the inept Fremont got stuck
with all sorts of added "incumbrances," including $200,000 to David
Dudley Field for gaining title to the grant in his brother's Supreme Court,
and he let Ketchum lock up most of his final third of the stock until
the price had broken. Still, a friend claimed, Fremont did get $1,237,000
in greenbacks, worth at least $700,000 in gold, which might have kept
him out of debt for a while. But Jessie was "full of the idea" that they
could turn it into "untold millions" speculating in the gold market, and
they lost it all within six months. It was tricky business, however, and
even those consummate crooks, Jim Fisk and Jay Gould, did little better in their bungled attempt to corner gold. In the meantime, Olmsted
discovered that the rich ore was exhausted; he suspected that Park had
actually salted the mines, and he privately blasted Fremont as "a selfish, treacherous, unmitigated scoundrel." The shareholders also soon
discovered that they owed an additional $960,000 in other mortgages
and liens that Fremont had failed to mention. Amid cries of fraud and
wishful rumors of Fremont's and Park's arrests, the whole rotten business
sank in a morass of recrimination and litigation without ever getting
title. Fremont's heroic image took a further hit when he had to reveal
the embarrassing details of the Mariposa deal in a libel suit Opdyke
brought against the editor of the Albany *Journal*.[50]

Edwin Godkin, editor of *The Nation*, later despaired at both the Mariposa's promoters and investors, arguing "we do not believe that the
popular readiness to be taken in by high-sounding names in the least
diminishes the guilt of those who avail themselves of it. On the contrary,
we think that the frauds perpetrated by men of reputation are amongst
the worst." But, he conceded, "the readiness of people to be swindled
seems to increase every day instead of diminishing, and there seems to
be a fascination about certain methods of swindling, and about certain
names in the history of swindles, that lures not only fresh victims to
destruction, but brings whole troops of former victims with gleeful haste
to a second and more deadly sacrifice. Such a fascination appears to be
exercised upon the public mind by the name of Mariposa, which has

been recently gathering in its fourth or fifth crop of smiling victims." There was indeed a persistent attraction, for like some pied piper the promises of Mariposa were destined to rise again and again with every new mining fever and lure anew its faithful flock of victims.[51]

THE BIGGEST SWINDLE, however, was Robert Walker's final scandalous seizure of the New Almaden quicksilver mine. After the closure of the mine by the injunction granted to Walker's lawyer, Edwin Stanton, in 1858, U.S. attorney general Jeremiah Black brought suit in the name of the government to void William Barron's title on the basis of fraud, so that Walker's Quicksilver Mining Company could take over the mine under their gerrymandered grant. Barron's lawyers, however, led by Mexican War veteran and Mexican law expert Henry Halleck, fought the case very effectively in the federal appeals court. They denied Stanton's charges of "fraud," arguing that his principal piece of evidence, an 1848 letter from Alexander Forbes to James Forbes expressing concern about the validity of their grant under the American occupation, was itself a fraud, since James Forbes claimed the original had been stolen and all he could show was what he said was a copy. Moreover, they claimed that they held title, not only by Castillero's original mine grant approved by the federal land commission but also by the purchase of the adjacent land, and that they had peacefully worked the mine, unchallenged, for more than a decade. The attorneys general, Black and Stanton, who replaced him, both fought back for Walker by trying to ban all testimony by Mexican officials as untrustworthy. But the appellate judges finally concurred that the charges of fraud were false, lifted the injunction, and confirmed Barron's title to the New Almaden early in 1861. Barron promptly reopened the mine with Halleck as director general of the company.[52]

Stanton promptly appealed the case to the Supreme Court, but it stalled with the change of administrations and the start of the Civil War. Still, Walker refused to give up. He simply handed out another round of gratuitous shares to members of Lincoln's incoming Republican administration to ensure that they, too, had a direct interest in the case. Although he was a Democrat, he had become a Unionist with direct ties to Lincoln. He claimed that Lincoln had even offered him a cabinet post for having cut a deal with Horace Greeley to throw the weight of the New York *Tribune* behind the new administration, giving support at the onset of the Civil War that Lincoln vowed was "as helpful to me as an army of one hundred thousand men." Walker built on this connection by giving Lincoln's longtime friend and confidant Leonard Swett at least $10,000 worth of $100 par Quicksilver stock by his own admission, or as much as 3,000 shares, Barron said. Without a mine, of course, the stock sold for as little as $14 a share as a speculation on the New York Stock Exchange. But if they could seize the New Almaden, the price would skyrocket. Walker also apparently extended his largess to both Lincoln's and Swett's close friend and newly appointed Supreme Court justice David Davis, as well as Secretary of the Interior John Usher and Attorney General Edward Bates, who apparently without Lincoln's knowledge would arrange to turn over the mine to Walker's company. Walker may even have gotten to Secretary of State William Seward, who curiously claimed credit for getting Swett into the scheme.[53]

In any event, a cooperative Bates revived the New Almaden case in the Supreme Court, still in the name of the government, but allowed Walker to again hire all the lawyers, led by Black, who had conveniently been appointed Supreme Court reporter, and Stanton, who had become Lincoln's secretary of war. At the same time Black also appeared against the government in the formal appeal of Walker's gerrymandered Capitancillos case, while Barron's lawyers, of course, appeared for the government. The two cases, deciding the title to a mine said to be worth from $25,000,000 to $60,000,000, were hailed as the biggest that had ever been brought before the court. The lawyers' fees of more than $1,000,000 also set new records. Black's share was touted as the "largest ever earned in America," but it may have been mostly in stock, for Bates noted that if Black lost he would be "flat broke."[54]

The New Almaden case was heard early in 1863, and despite both its importance and its sheer volume of over 3,500 pages of earlier testimony, the Supreme Court dispensed with it quickly, reversing the lower court in a five to three decision, with Davis providing the deciding vote; the absent ninth justice said he would have gone with the minority. But the majority justified their decision not on the basis of fraud as Stanton argued but on what seemed to be a mere pretext, a catch-22 technicality that Castillero was not in "strict compliance" with the letter of Mexican law at one point, because he had failed to present his claim to a mining tribunal and substituted the local alcalde instead, since there was no such tribunal in California! The decision was widely denounced in California as a "fraud," and the Washington *Chronicle*, was also outraged that Black had been privately paid $120,000, asking, "Were private parties allowed to contest their rights under the shield of the government and in its name by a man who was also receiving a government salary as Reporter of the Supreme Court?" Even Bates later admitted to being "shamed and disgusted" by how easily in "this highest of all courts . . . the scales of justice rise and sink" with the overpowering pull of political influence and bribery. Comparing this with Walker's rival Capitancillos grant, which was upheld, and Fremont's Mariposa as well, he railed privately at how "one claim was rejected for *lack of registry*—that seeming to be the only ground—and another was confirmed in spite of that objection!" But on Wall Street when the decision was announced on March 10, the shares, which had been "beared" on rumors of a Barron victory down to $14, jumped to $80, as a throng of hungry investors suddenly scrambled to get in, and Walker and his friends eagerly fed them for a quick profit. Walker intimated that he had picked up $500,000 on the rise, and Black, Stanton, Swett, and others also cashed in some of their ill-gotten shares as well. Swett, in fact, "made a great deal of money" from the stock—$100,000, Seward said—but he became so intoxicated with speculation that he stayed a little too long and lost $30,000 of it a year later.[55]

The patriotic but ever avaricious Walker then sailed for London as Lincoln's special agent to raise money for the Union selling government bonds, and for himself as well, pushing a new but unsuccessful suit against Barron, Forbes & Company for an additional $6,000,000 for all the mercury that he claimed they had taken from his land. At the same time he got Bates, Usher, and Stanton to apparently con the war-weary president, who seems not to have actually read the lengthy Supreme Court decision, into ordering the seizure of the mine on the erroneous grounds that the court had decided that the New Almaden grant was a fraud! In May Lincoln quietly signed the order, drafted by Bates, and dispatched Swett to take the mine by "military force" if necessary. Usher and Bates even more quietly then gave Swett instructions to turn over the mine to the Quicksilver Mining Company. Swett arrived at the mine in July, accompanied by the Quicksilver company president, Samuel

Butterworth. But the New Almaden manager, backed by a force of heavily armed men, refused to turn over the property. Swett immediately called for federal troops, but Lincoln was barraged with telegrams to revoke the order and he refused to send them, fearing that the incident might trigger a secessionist rebellion in California. Moreover, when his new general-in-chief, Henry Halleck, the former New Almaden attorney and director, sent the orders to his California commander not to send troops, he further charged that the previous seizure order was "surreptitiously obtained." That thoroughly outraged Bates, who branded Halleck a "liar" and a "viper" and wanted him fired.[56]

The attempted government seizure of the mine ignited a firestorm of protest in the West. Even the pro-Lincoln *Alta California* proclaimed it an "outrage," and excoriating Walker, Stanton, and Black for the whole fraudulent scheme, demanded that those "guilty of this gross attempt at swindling should be exposed, disgraced and punished." But Lincoln excused his corrupt friends and still erroneously insisted that the Supreme Court had declared the New Almaden grant "utterly fraudulent." Much of the fire was extinguished in August, however, after Butterworth and Barron negotiated the purchase of all the New Almaden property by the Quicksilver Mining Company for a relatively modest $1,750,000, with $1,000,000 down in gold and the balance in annual payments of $250,000. The directors eagerly ratified the deal and issued another $2,000,000 in stock, pushing the total to $10,000,000, while the stockholders rejoiced at last in the peaceful settlement of the dispute.[57]

The following year the directors also rewarded Butterworth with a five-year contract as general agent, at an extravagant salary of $25,000 a year including perks, or nearly a hundred times the pay of most of his Mexican miners, making him very much a model of the modern executive. Butterworth immediately pushed production to the limit. But he soon discovered that getting the mines was only a hollow victory because they had no market, since Barron, who'd broken up with Bolton, now wholly controlled all western and Latin American mercury sales. Moreover, just before he sold the mine, Barron had stripped out the last high-grade ore, doubling production to get a large reserve of mercury, and he had opened a rival mine as well. So he refused to even talk about buying any of the Quicksilver Mining Company's mercury for two years until he had sold off his own reserve. The directors, nonetheless, hoping that they could sell their mercury in Europe, bravely declared a "semi-annual" 5 percent dividend at the end of 1864 to drive up stock prices while they sold out at a 100 percent profit. They also hired the popular writer J. Ross Browne, who had been Walker's private secretary, to provide an attractively illus-trated puff piece in *Harper's New Monthly Magazine* to try to shore up the stock again after they failed to pay a second dividend six months later. But Butterworth, meanwhile, had to borrow heavily just to meet oper-ating expenses, and he turned to the new Bank of California, formed by Darius Mills and others, including Barron and his new partner and companion, Thomas Bell. By 1866, when Barron was ready to deal, the company was more than $500,000 in debt and eager to sell their unsold backlog of over 4,000,000 pounds of mercury. So Butterworth, for a quiet share in the take, made a sweetheart deal with Barron and the Bank of California to take both the backlog and future production at 39 cents a pound, and the directors grudgingly agreed.[58]

Thus, with Butterworth's connivance, Barron and Mills kept control of the quicksilver market, fixed the market price at 60 cents a pound, and raked off close to 20 cents a pound profit. This deal continued until 1872, while ore yields dropped sharply from 16 percent to barely 7, and costs rose from 24 to as much as 37 cents a pound. The Quicksilver Mining Company cleared only pennies a pound and couldn't even pay off its debt to the Bank of California, let alone dividends to its stockholders. By 1869 the company was so deeply in debt that they tried to float $1,000,000 worth of 7 percent ten-year bonds to beat Mills's bank's much higher interest rates, but that drove share prices, once up as high as $99, down to $11. When that still didn't bring in enough money, the directors raised a little more from their stockholders by getting many of them to convert a total of 43 percent of the stock into new 7 percent preferred shares, by paying an additional $5 on every share they held. So while Barron, Bell, Butterworth, and Mills were happily sharing a steady income of about $350,000 a year, Walker, his cronies, and all their unsuspecting investors were still paying in and getting nothing out. For though Walker had successfully outmaneuvered Barron politically to grab the mine, he lost all its great profits to Barron, who outmaneuvered the former treasury secretary financially! After all those years of behind-the-scenes schem-ing, Walker died a near pauper in 1869, while Barron grabbed back the profits to become a multimillionaire before he died suddenly, just two years later, at the age of forty-nine. Meanwhile, Butterworth turned over the mine management to his nephew, John Butterworth Randol, and devoted his time to other schemes with the Bank of California crowd, making at least a million of his own before he died in 1875.[59]

THE HIGH PRICES and apparent scarcity of quicksilver during the closure following Stanton's injunction had also spurred the opening of new deposits that produced a couple of profitable rivals. The most successful newcomer was John H. Redington, a New England argonaut who had started the first wholesale drug business on the West Coast. In 1861 he bought a rich quicksilver deposit in the Coast Range north of San Francisco and formed the Redington Quicksilver Company with a very modest capital of $31,500 in 1,260 shares. Within three years he made the mine a major producer, paying dividends of over 100 percent per annum. When Barron and Mills tried to buy him out in 1867, Redington boosted the par value of the shares to $1,000 to dissuade them. He did agree, however, to join with them in a price-fixing combination that gave him about a quarter of the total production at 60 cents a pound and guaranteed a good profit. By 1878 Redington and his shareholders had made just over $1,000,000 in dividends, or earnings of nearly 200 percent a year for seventeen years on their actual investment.[60]

But the biggest new mine was the New Idria, discovered in 1854 by a former New Almaden miner, Henry Pitts, some eighty miles southeast. Like its namesake in Austria, which was the second most productive mine in Europe, the hopefully named New Idria would in fact eventually rank second only to the New Almaden in the Americas. Pitts began working the mine with backing from San Francisco merchants, and early in 1858 he and his backers formed the New Idria Mining Company, also with a very modest capital of $23,000. But that fall they issued another $10,000 in shares in a futile effort to keep Barron and his partners from buying up a controlling interest in the mine. Once he had control, however, Barron put nearly $100,000 into furnaces and other developments, and by 1860 the mine had paid back that investment and was producing over $200,000 a year for an annual return of over 100 percent.[61]

Like the New Almaden, the New Idria also quickly attracted sharks who tried to wrap a Mexican land grant around it. Soon after Pitts opened the mine, the U.S. district attorney for Southern California, Pacificus Ord, a forty-year-old lawyer and political appointee from Washington, D.C., spotted what he thought would be an easy mark. He had handled several Mexican grant claims before the Land Commission, including former Mexican government clerk Vicente Perfecto Gómez's undocumented claim for a roughly five-mile square of ranch land known as Panoche Grande in the valley fifteen miles east of the new mine. Despite Ord's pleading, Gómez's claim had been rejected by the Land Commission in 1855 for lack of any official record of the grant or any evidence that Gómez had ever set foot on the ground. But the following year, on hearing of Pitts's developments, Ord quietly got Gómez to sign over a half interest in the claim for legal services, and then he appealed the decision in his home court in Los Angeles. There, acting as the government attorney, he disavowed the commission's decision and got the claim approved! Next, following Walker's example, Ord moved the ranch up on top of the mountains to the west and claimed title to the New Idria. By that time, however, Gómez wanted out. In December of 1857 he sold his remaining half interest in the "fighting claim" to an opportunistic and tenacious thirty-year-old Irishman, San Francisco liquor dealer William McGarrahan, for only $1,000—plus, he later claimed, a promise of half of whatever McGarrahan got out of it.[62]

Barron and his partners, of course, had also learned their lesson from Walker. This time they got to U.S. attorney general Black first, and he obligingly hired their attorney as his special agent to investigate. Then, in the name of the government, Black shamelessly appealed the case for fraud, exposing Ord's secret interest in the claim, and the district court judge called for a new trial. The retrial ended in August of 1862, however, in a reapproval of the claim, and Black, now openly paid by Barron, appealed the government's case to the Supreme Court, where it would drag on for nearly four years. During that time, McGarrahan the activist moved to the nation's capital to press his case, while Ord, annoyingly, just sat back and waited. At first McGarrahan brashly talked of having that "old thief" Black impeached, but he soon threw all of his effort into a concerted campaign to get an official government survey and patent on his mine-grabbing claim. [63]

To finance it all, he also followed Walker's model, organizing the Panoche Grande Quicksilver Mining Company in 1861 with a generous

$5,000,000 in speculative shares, and when those ran low he just issued another $5,000,000 in worthless paper. He actually succeeded in selling some of the shares for living and lobbying expenses, but he gave away far more to buy favorable decisions and votes. Handing out fistfuls of a thousand or more $100 shares to each prominent recruit, he lined up an impressive following that included Lincoln's postmaster general and political strategist, Francis Blair, who got over 2,600 shares, and more than a dozen Democratic and Republican congressmen on crucial committees, as well as the influential Washington *Chronicle* editor and Senate secretary, John Forney. By 1862 McGarrahan had his official survey in his pocket, and the following year he got the government patent drafted. He was within a pen stroke of getting a presidential signature, when at Black's urging the new attorney general, Edward Bates, persuaded Lincoln to wait for the Supreme Court's decision. There, after fits and starts, rumors and innuendo, Black and Barron finally won the day in 1866, as the court officially rejected the Panoche Grande as a fraud. McGarrahan, undaunted, simply got the Panoche company to deed the discredited claim back to him, while he still peddled its now obviously worthless shares and regrouped for a new attack. Refusing to admit defeat, he would haunt the halls of Congress for more than a third of a century, trying one angle after another to grab the prize.[64]

In the meantime, Walker's protégé, Butterworth, and others schemed with government officials to fraudulently "float" yet another land grant to grab what was said to be one of "the richest tin mines in the world." But, like McGarrahan, they found it could be a long, hard fight. Nestled in the Temescal hills just fifty miles east of Los Angeles, the reddish-brown mineral deposit had long been worked for medicine by the local Cahuillas, but it was not until 1859 that it was recognized as high-grade tin ore, running as much as 60 percent metal and claimed as the Cajalco Mine. News of the tin discovery spread fast and several Los Angeles merchants bought control, but before they could do much work they were stopped by the first of many lawsuits. For since the mine actually lay on the land of Leonardo Serrano's old Temescal ranch, just three miles from his 1818 adobe, another enterprising Angeleno, Abel Stearns, bought the ranch from Serrano's widow for $4,500 and 150 cows. Although title to the ranch had been rejected by the Land Commission for lack of a grant, Stearns successfully appealed, based on the Serranos' "undisputed possession" for over forty years, and in 1863 he won a federal court decree affirming his ownership.[65]

When Stearns shipped several tons of rich tin ore to refiners in San Francisco in 1865, however, he attracted the attention of Butterworth and a couple of friends, pioneer California railroad builder Lester L. Robinson and a skillful land grant lawyer, Horace W. Carpentier, who set out to take it away from him. Conspiring with U.S. surveyor general Lauren Upson, his sharp chief clerk, Edward Conway, the aggressive former attorney general, Jeremiah Black, and that pliable Supreme Court justice, Stephen Field, they bought another grant, the San Jacinto, twenty miles to the east, for $8,000. Then they launched a government suit to scuttle the Temescal title to the mine, while they cut loose a third of the San Jacinto acreage, sailing it west to anchor over the mine to claim it for themselves. Field officially sat out the case, but his Supreme Court brethren saw no merit in the Serranos' long tenure and voided their title in 1867. Although the government lawyers had also attacked the Temescal title for its "elasticity" in covering the mine just a few miles from the ranch house, Upson quickly approved a survey of the San Jacinto offshoot, which had completely snapped free and jumped twenty miles to grab the mine. The following January the conspirators officially floated the San Jacinto Tin Company and divided up its prospective $4,000,000 in stock, with Upson, Conway, Black, and Field taking more than a quarter of the shares. In a new round of activity and publicity they proudly poured what they claimed was "the first bar of tin made on the Pacific Coast," and they presented Secretary of State William Seward with a fancy tin box, while Butterworth and Robinson's brother, Edward, feverishly tried to sell the whole works for $1,250,000, or more. But Stearns fought back, hollering fraud, and even after his death, in 1871, his widow and her new husband carried on the fight, thoroughly tying up the title again, scaring off investors, and leaving the mine in legal limbo for nearly twenty years.[66]

MORE IMMEDIATE PROFITS were to be made from the opening of California's first paying copper mines, which even threatened Michigan's dominance, after rapacious eastern speculators drove the price up from 22 to 50 cents a pound during the Civil War years. Old argonauts had been looking for gold in the Gopher Hills of Calaveras County just west of the Mother Lode near the great Carson Hill mine for over a decade, but all they found was a lot of rock that they tossed aside as "iron rust." One curious prospector, nonetheless, did nearly fill his cabin with specimens of the stuff before his partner made him throw them all out. It was not until the spring of 1860 that the rock was finally discovered to be rich in copper after an old-timer, Hiram Hughes, and his son, Napoleon, returned from the Comstock, started looking around for silver, and actually sent a piece of the stuff to San Francisco for an assay. Hughes and his son promptly staked the Napoleon lode and sold all 2,700 feet to eager investors for over $50,000. Suddenly it was "Copper! Copper! Copper!" as a couple of thousand hopeful miners and speculators blanketed the Gopher Hills with claims, and the new camp of Copperopolis sprang into existence.[67]

By far the biggest and best-paying new copper mine was the Union lode paralleling the Napoleon. It was discovered by another forty-niner turned farmer, William K. Reed, several months after the Napoleon, and although new local mining laws limited each claimant to just 150 feet, he added the names of a dozen friends and relatives to stake a much larger claim. He then sold 600 feet for only $400 to a more experienced miner, Thomas Hardy, who had sunk a prospect hole there in 1852 looking for gold. Since its surface ore was dazzlingly rich, running as much as 56 percent copper, they decided to work it themselves and formed the Union Copper Mining Company as a private venture with just thirteen shares of 150 feet each. Hardy became superintendent, and early in 1861 they began shipping all the ore running more than 20 percent copper to smelters in the East. The shipping ore was worth around $115 a ton at the prewar copper prices of 22 cents a pound, and even though the smelters paid only about $85 a ton for it, that still left a profit of close to $40 a ton after mining costs of $15 and shipping charges, which were under $30 a ton because they were close to a steamboat landing at Stockton. By the summer of 1862 they had shipped over 2,000 tons of ore for a return of about $170,000. But Hardy insisted on putting all their early profits back into developing the mine, so Reed, after waiting two years for little return, tired of the business and sold his half interest in the mine that fall for $61,750, at $9,500 a share, to return to farming.[68]

The buyers were their Stockton shipping agents, Charles Meader and Charles Lolor, who had earlier bought one and a half shares for over $6,000 a share. As soon as they took control, they arranged lower shipping rates and doubled ore shipments, and that winter with rising copper prices they paid themselves a handsome dividend of $11,000 a share—fully repaying their investment with interest in just a few months! The next year they took additional dividends of $20,000 a share from over 5,000 tons of ore as copper climbed to an average of 34 cents a pound, and in 1864, as the average reached 47 cents, they doubled their output again to over 10,000 tons. At that point they refused to say how big a profit they were making, but Hardy was able to sell their Boston partners his four shares at an astonishing price, widely reported at $650,000 or more. Hardy later claimed that he *only* got $375,000, but that still gave him a profit of over $500,000, counting dividends, or a thousandfold return in just four years! The remaining $275,000 apparently was split by Meader and Lolor as their "commission" for placing the shares in Boston. By then California was well on its way to producing as much as a third of the country's copper, seriously challenging Michigan. The Union mine thoroughly dominated California's production as the owners doubled their output again over the next two years. But to do so they had to start shipping lower-grade ore of only 15 percent copper, as the rich ore had begun to run out. Over the same period copper prices slowly declined to 34 cents a pound, and in 1867 they broke to the prewar base of 22 cents while shipping costs rose sharply. Then, with the first-class ore exhausted and even 15 percent copper ore fetching only $48 a ton at the smelters, the company's profits evaporated and they were forced to shut down. Although they had spent $75,000 on a small experimental smelter and talked of spending an additional $250,000 for a much larger one to do all their own refining, the remaining ore just didn't justify the investment.[69]

By 1867, however, the Union had produced roughly 60,000 tons of ore worth around $8,000,000 at a total profit of roughly $1,000,000. That gave Meader and Lolor about $600,000 on their eight shares, not counting their "commission," or a 700 percent return on their $74,000 investment in under five years. Hardy, of course, had made an even bigger personal profit, but he didn't hang onto his. He spent lavishly, winning a couple of terms in the California Legislature, picking up a trophy wife twenty-four years his junior, and moving to San Francisco. But he soon lost most of his fortune in mining and other speculations plus a costly divorce, and he quietly retired to a ranch in Calaveras. The dazzling profits of the Union, however, had sent a new horde of copper prospectors all over

California and Nevada and stirred up a whirlwind of copper shares from a couple of hundred new wildcats, adding further frenzy to the silver mining mania. Even the original Napoleon was finally puffed up in the frenzy, after its first investors, forming a modest company for $270,000 at $100 a foot, had honestly worked it for a few years, and taken out about $500,000 worth of shipping ore. When copper prices began to decline, however, they sold the mine to a couple of New York sharks, who reorganized it as the Napoleon Copper Mining & Smelting Company for $1,200,000 and printed up some dandy stock certificates to work eastern investors. But the mines of Copperopolis and the new prospects opened elsewhere all proved to be much smaller than the great deposits like the Santa Rita, and they were soon worked out.[70]

CALIFORNIA'S BIGGEST and most reliable profits were still being taken from the gold lodes. They were being made by investors who financed deeper exploration either privately or with just a few close partners, as Alvinza Hayward was doing. A number of these ventures were also family affairs. Some of the most profitable gold lodes were those in Grass Valley which had been abandoned in the first rush to the Comstock. Among the most fortunate investors were the Coleman brothers, John and Edward, in their thirties, who after laboring in the mines for several years joined with friends to purchase a forsaken mine and mill, the North Star, for $15,000 early in 1860. Edward began a thorough exploration of the lode and the following year, at a depth of 250 feet, he struck a rich ore body that paid them profits of as much as $12,000 a month. After taking out over $800,000 in gold and clearing close to $600,000, they finally sold out early in 1867 to San Francisco investors for $250,000 cash, which gave the brothers a total return of 5,500 percent on their investment, the equivalent of more than 2 percent a day for over six years! But the new investors didn't find such rich pickings in the mine, which, of course, was why the Colemans sold it. The San Franciscans formed the North Star Gold Mining Company with an optimistic capital of $1,200,000, though they soon cut it back to a modest $300,000, closer to their actual outlay, and they happily took out close to $1,000,000 more in gold. But the ore was lower grade, so they got back only $154,000 in dividends before the ore pinched out, and they paid out $60,000 in assessments to look for more before they finally gave up in the summer of 1874 for a net loss of over $150,000. Others were willing to look longer and deeper, however, and an enormous bonanza was later discovered that yielded a total of $32,000,000 before it was found to merge into that of the great Empire, and the two mines were combined in 1929.[71]

Grass Valley's most prominent mine at this time, however, was the Allison Ranch, worked by John and William Daniel with surface croppings that ran up to $375 a ton. But the ore was pockety, and the mine didn't become a steady producer until 1855, after Michael Corbet and his partners bought in with the brothers and opened a large body of $50-a-ton ore at depth. They worked the mine down over five hundred feet and divided a profit of $1,200,000 among six partners before work stopped in 1866, when the majority balked at the idea of money flowing the other way again and refused to pay an assessment of $3,000 for new exploration. Even the old Rocky Bar briefly sparkled, but it still didn't pay. After a costly legal battle, its creditor, Andre Chavanne, finally gained clear title to the mine in 1863, over five years after Michael Brennan's tragic end. He leased the property to a neighboring company, which started to deepen the shaft the following spring and at the first blast opened ore "literally studded with gold, a few feet below where Brennan had hopelessly abandoned work." Chavanne and his friends eagerly commenced work on a large scale and took out about 12,000 tons of $30-a-ton rock over the next two years. But in spite of wonderfully romantic stories that the ore was so rich that even with costs of $1,000 a day they still cleared over $2,000,000, in the real mine the ore "petered out" before Chavanne even recovered his bad debts and legal expenses.[72]

Up north in Sierra County, Ferdinand Reis and his two younger brothers, Gustave and Christian, also made phenomenal profits. Thrifty German argonauts, they had made enough in the diggings to start a small bank, and in 1856 they bought the Sierra Buttes mine for a "negligible amount," well under $100,000. Even at that the previous owners thought they were getting a good price, because the mine wasn't paying and their hired hands told them the ore was exhausted. But Ferdinand moved up to the mine, and sinking on the lode, luckily hit ever-improving ore that produced over $1,800,000 in bullion and paid them over $1,100,000 in dividends over the next thirteen years. By then the mine was so well known as a steady producer that promoters were bidding to sell it in the

East and abroad. The Reises finally sold out to the British for another $1,125,000 early in 1870 amid the added excitement generated by the discovery of California's second biggest "nugget," a ninety-five-pound mass of almost pure gold, in the neighboring Monumental claim. Thus they must have made well over a 2,000 percent return on their "negligible" cash investment, or more than 150 percent per year.[73]

Down on the Mother Lode, Sacramento merchant, persistent political aspirant, and future railroad baron Leland Stanford laid the foundation of his fortune as a reluctant investor in the Union mine just north of Alvinza Hayward's Amador. It was the first quartz claim in Sutter Creek, located in 1851 and worked for nearly a decade by the Union Quartz Mining Company, organized by a local merchant, Robert C. Downs, and others. They prospered for a time, but falling ore values finally forced both the company and Downs into bankruptcy in the fall of 1859. Stanford was Downs's biggest creditor, and much to his dismay, all he was able to collect was Downs's and his partners' stock in the mine, seventy-six of its ninety-three outstanding shares. He wanted to dump it quickly for $5,000, but Downs argued that there was still good ore in the mine that could pay him much more than that. So Stanford grudgingly gave him a chance to prove it, promising to return a third of the stock if he succeeded. Downs did indeed succeed in opening a new ore shoot that paid a profit of $15,000 within a year. Modest as that was, it was already a big return for Stanford, whose entire mercantile business was worth only $33,000, and he promptly returned shares, just shares short of the promised third so he could keep control of the mine.[74]

Downs's new ore shoot proved far richer than he had hoped for, paying even bigger annual profits and producing a total of over $2,000,000 in gold by the end of 1872. By then the ore had finally begun to peter out, but having shown the mine to be so profitable, he and Stanford, with the help of Mariposa shark Trenor Park, unloaded it on foreign investors for $400,000. In all, Stanford made a profit of over $500,000 from the mine, a magnificent return from a bad debt that he was willing to liquidate for $5,000! The Union mine was pivotal in his career, for its steady profits, together with generous loans from Downs's share, enabled him to join with Collis Huntington, Mark Hopkins, and Charles Crocker in putting up $15,000 each to form the Central Pacific Railroad Company in 1861. It also paid his share of costly legal fees and lobbying for federal funds to build the western half of the first transcontinental railroad, which would pay them enormous profits, both honest and otherwise. His early mining profits further helped Stanford finally make a successful campaign for governor of California in 1862, after two previous defeats. Out of his railroad fortune and other mining speculations, he later paid over $100,000 for a Republican seat in the Senate.[75]

Just over the Sierra to the east, at the fledgling camp of Bodie, however, Stanford let a far bigger, $20,000,000 bonanza slip unknowingly through his fingers. In the fall of 1862 he and others bought up all of the most promising claims in the camp, the Bunker Hill and ten others, promoted by the local justice of the peace, Frederick K. Bechtel. But instead of working them as a private company, like the Union mine, they floated them as the Bodie Bluff Consolidation Mining Company in January of 1863, with Stanford as president and Bechtel as secretary, and offered most of their 11,100 shares to the public at only 10 percent of their $100 par. Encouraged by surface croppings that ran from $50 to $400 a ton in gold, and a prestigious president, who was by then both the governor of California and the head of the new railroad, prices quickly doubled to $20 a share as investors eagerly snapped up the stock. Unfortunately, however, Stanford and his partners opened their vast holdings in the middle of the lode and missed the great bonanza that lay hidden in the Bunker Hill claim at the north end. The rich surface pickings actually amounted to little, and by the fall shares had fallen to $3 and dropped to only half that the following spring, when the first assessment was levied and most forfeited their shares. So in August of 1864 Stanford, Bechtel, and the few remaining shareholders happily sold all their property to that ever-ready resurrector, Trenor Park. He apparently picked it up for a little over $10,000 and promptly, in a fitting encore to Mariposa, floated it in New York as the Empire Gold & Silver Mining Company, with an outrageously "watered" asking capital of $10,000,000! Park immediately hired the talented and accommodating J. Ross Browne to visit Bodie and glowingly extol its prospects in an handsomely illustrated prospectus and in another of his entertaining articles for *Harper's New Monthly Magazine*. As soon as he had unloaded all he could, Park moved on, and the new shareholders unfortunately decided to sink a new shaft farther south, which produced less than $100,000 before they at last gave up in 1867. Thus, they, too, missed the great bonanza, and nearly a decade would pass before it was finally discovered by accident.[76]

In the meantime, other departing miners from California had opened new gold bonanzas all along the Rockies, and since the big gold bonanzas of California were still privately held, it was the raw, new gold ledges that most excited eastern passions. The Pikes Peak rush had brought in tens of thousands of hopeful gold seekers in the depression years of 1858–59 to form the new territory of Colorado, and by the time the national economy was booming again in the speculative frenzy of the Civil War years, it boasted many paying, or at least promising, gold lodes ripe for investors. Leasers working the first lode, discovered by a Georgian, John Gregory, were taking out as much as $1,000 a day, while those on rival bonanzas on the neighboring Bobtail and Gunnell lodes were making $100 a pan from surface rock, and the new little county of Gilpin was being heralded as the Mother Lode of Colorado, and its booming towns of Black Hawk and Central City hoped to rival Virginia City. Even though Colorado's annual gold production was still only a small fraction of California's or Nevada's, many easterners looked upon its mines as virgin opportunities, unlike the Mariposa and Comstock, which they knew were already defiled by the California sharks. Colorado and the Rockies still seemed to promise eastern investors at least a chance of "getting in on the ground floor," and their own Wall Street sharks were all ready to sell them a ticket.[77]

The first Colorado gold mining venture, floated in New York in October of 1863, was nonetheless called the Ophir, in imitation of the pioneer

Comstock company, and others followed in rapid succession. As the fever of speculation swelled, the demand for mining shares seemed almost insatiable. New companies simply couldn't be organized fast enough. Eastern promoters descended upon the Rockies like "a swarm of scheming cormorants," looking for any and every prospect they could wrap a company around. By the spring of 1864 the Gilpin lodes had produced nearly $5,000,000 in gold and roughly six hundred "bubble companies" had been organized in the name of mining in the Colorado Territory. The promoters bought raw claims for a few thousand dollars and stocked them at a premium of as much as a hundredfold. At the end of March when a real bonanza, the highly touted Consolidated Gregory Company, opened their books in New York for subscriptions to 40 percent of their shares, the full $2,000,000 offered were taken within an hour! The latecomers would eagerly have taken $1,000,000 more. So, in a skillful bait and switch, its sharp promoters, Elisha Riggs and Leonard Jerome—who was "one of the boldest, coolest, and most successful manipulators in the Street," part owner of the New York *Times*, and future grandfather of Winston Churchill—sold them another $2,250,000 in shares of a wildcat, the Corydon Mining Company, handled by an enterprising Colorado lawyer and future Republican senator, Henry M. Teller. The excited crowd then rushed the Gunnell Gold Company's office, where its promoter, Anthony Morse, a rising young railroad stock manipulator and broker, having purchased the mine just three weeks before for $300,000, now offered it for $3,000,000. He stirred up such a frenzy that the frantic buyers actually fought one another for shares! Another company had to call in the police to clear their office of angry customers after all of their stock was sold.[78]

The Bobtail, Gilpin's second richest lode, was divided up into several small companies, but it started Colorado's first mining magnate, Jerome B. Chaffee, on the road to fortune. In 1860 Chaffee, a thirty-five-year-old upstate New Yorker, real estate promoter, and ex-banker, had joined with a younger Grass Valley quartz mill man, Eben L. Smith, to set up a twelve-stamp mill to work ore from the various mines. From their profits they bought up claims on the Bobtail to feed their mill, and then sold them in New York for $100,000. Chaffee, reverting to type, also loaned out some of the money to other claim holders for fat profits. The most ravenous was a $600 loan to a neighbor, who wanted to go east for a visit, in exchange for the right to work the man's claim while he was gone. Chaffee then gutted the mine, tearing out the high grade and clearing $35,000 in the six months before the owner got back, a banker's dream of nearly 1,000 percent interest per month! Through schemes like this and worse, Chaffee was branded a "mine butcher," and was also soon a millionaire, acquiring more than a hundred mines, which yielded him a reported $3,000,000. He dominated Colorado mining for the next two decades, and in 1865 he opened the First National Bank of Denver, as both an economic and political tool. Later, boasting he "would just as lief buy a seat in the United States Senate as in a stage coach," he became Colorado's first Republican senator.[79]

IN THE FEVER OF SPECULATION, when they could no longer find enough prospect holes in Colorado to meet the demand, eastern promoters turned to other rich new strikes elsewhere in the Rockies. Early in the sixties, on the heels of the first rushes to the Comstock and Pikes Peak, prospecting parties from California had invaded the Indian treaty lands on the Snake, Salmon, and Boise rivers. There they had found rich new placers in what would become Idaho, and then, crossing over the Rockies, they had struck more pay dirt in Alder and Last Chance gulches in Montana. At the same time gold-seeking soldiers of the California Volunteers stationed at Salt Lake had struck rich ledges in Bingham Canyon and the Wasatch range. Although the placers were "poor man's diggings," offering little chance for capital, the gold and silver lodes that fed them were soon discovered, luring the first investors into mines throughout the Rockies, while the old Spanish mines enticed other speculators into the Southwest. But as always, the results were quite mixed.[80]

The first paying gold lodes in Idaho were found on the upper reaches of the Boise basin in the spring of 1863, and spectacular silver lodes were found that fall at the new "Silver City" in the Owyhee Mountains to the south. The richest lode in the Boise was the Elmore, owned by Henry T. P. Comstock, who after getting his name attached to the great Nevada lode, now boasted that his new bonanza would even surpass his namesake. Wilson Waddingham, a cheery young Canadian, welcomed for his "earnestness" and "indomitable energy," bought the story and much of the lode in 1865 after Comstock had taken out about $30,000.

With Comstock's boast and assays running as high as $7,434 a ton, Waddingham headed east, where he floated the mines as "the celebrated" Ada Elmore Gold and Silver Mine and his own modest Waddingham Gold & Silver Mining Company with $2,000,000 and $600,000 in shares, respectively. Shares sold so briskly that he soon doubled the capital in his namesake and cleared a fat profit, even after squandering nearly $200,000 on a useless mill and other property. He also sold off a couple of new claims for an added profit of over $100,000, and he carried on one or two heated claim fights in the courts. But he only produced about $60,000 in bullion and paid his investors just a token dividend of $7,500. Finally spurned as a "great detriment to Idaho," he sold out the last of his holdings in 1867 for $50,000 and slipped out, leaving debts of nearly $25,000. This was just the beginning of his fortune, however, and he would eventually be hailed as "the largest land holder in the United States," some said the world.[81]

The silver bonanzas of Owyhee did in fact threaten to eclipse the Comstock, and they grabbed new investors in both the East and West. By far the most spectacular was the hotly disputed Poorman claim, whose discoverers ripped out roughly $250,000 worth of incredibly rich high grade in just two weeks in the fall of 1865, working night and day behind armed barricades, before they were stopped by an injunction brought by overlapping claim holders who accused them of claim jumping. The legal battle finally ended when the rival claimants' financial backers, former Portland mayor George Collier Robbins and Columbia River steamboat monopolist Putnam Bradford, agreed to sell out to the New York & Owyhee Gold & Silver Mining Company for $1,050,000 plus lawyers' fees. To entice investors, who soon paid in $1,249,500, the company shipped fifteen tons of their richest ore, assaying $4,000 a ton, all the way to Newark, New Jersey, and in 1867 they sent pieces of a quarter-ton block of ruby silver, running 67 percent metal, all the way to Paris for the International Exposition, where it won a special gold medal. The excitement was so great that promoters pushed an additional $1,000,000 in shares in the slyly named Poorman Gold & Silver Mining Company, formed around a truly poor northern extension of the lode, which had been picked up for $150,000. But in the end even the New York & Owyhee stockholders lost heavily, for the company only got about $600,000 out of the mine, which paid back dividends of just under $100,000, or less than 8 percent of their original investment, while the market value of their stock crashed to $150,000.[82]

At the same time, those old Californians George Roberts and George Hearst floated another flashy venture around the discovery of a startlingly rich silver pocket just south of Silver City. They formed the Rising Star Silver Mining Company, with Roberts as president, to feed the San Francisco market another $3,600,000 in shares. They emptied the pockets of many hopefuls before the rebellious ore pocket was exhausted after barely a year, turning out just $90,000 and taking in $380,000 in assessments. Although the Owyhee mines would eventually produce over $90,000,000, it would be slow in coming, and for more than a decade the managers "treated their investors badly," taking at least $3,500,000 in assessments from them and giving back less than $600,000 in dividends.[83]

James W. Whitlatch offered a very different example. In September of 1864 he discovered Montana's first profitable gold quartz, the Whitlatch Union, the "mother lode" of the rich placers of Last Chance Gulch at Helena. A twenty-one-year-old former Pennsylvania farm boy, whose father had died when he was eight, he had joined the rush to Colorado without learning to read or write, but he proved to be much wiser than most of his far more lettered contemporaries. Instead of opening his claims for a quick speculative sale, he carefully and thoroughly worked the lode for its own riches. Without funds from investors, he took out $150,000 worth of richer surface ores to pay expenses while he opened and blocked out the next $500,000 in deeper ore. Then he simply sold 230 feet of the west extension of his now famous mine for a paltry $15,000 to New York backers of the National Mining & Exploring Company, who eagerly put up a $35,000 stamp mill to more efficiently work his ore. Other companies soon followed, and by 1871, when he finally sold the mine, he had turned out $3,663,000 in gold and was worth well over $1,000,000 at the age of twenty-eight. The New York company investors, including William E. Dodge Jr., a scion of Phelps, Dodge & Co., also made a good profit from milling charges on their very modest investment. Whitlatch soon lost most of his money on the market, but he made another $250,000 in other mining promotions before he finally ended up with a bullet in his head in San Francisco at the age of forty-seven.[84]

Montana's first silver mining investors, however, didn't even get a taste of fortune. They were led down the garden path by Samuel T. Hauser, a sharp, thirty-two-year-old Missouri railroad engineer and Montana argonaut turned banker, who had made a small fortune with loans of 5 percent a month. In May of 1865 he and a few old St. Louis friends, including two former mayors, John How and Luther Kennett, floated the bloated Missouri and Montana Mining Company with a capital offering of $400,000 on a scattering of silver claims they bought for $60,000, mostly in stock. For his part, Hauser got over half the total stock offering and persuaded his friends to unload much of it in St. Louis. To push it, he unblushingly claimed that their biggest bonanza, the Rattlesnake lode in Argenta, a hundred miles south of Helena, was wider and richer than the Comstock. Its marvelous ore, Hauser said, assayed over $2,000 a ton and could be worked for a paltry $45 or less, and he talked outrageously of profits of $38,700,000 a year! In spite of, or because of, such talk, the company name was changed twice in the next six months, the stock increasing each time, until it ended up as the St. Louis and Montana Mining Company with a full $1,000,000 in shares. As with most silver ores, however, the silver was not pure metal but occurred in sulphides, chlorides, and other compounds that required roasting, smelting, and even more sophisticated metallurgical processes for recovery. So in the ensuing year the investors struggled to put up the first smelter in Montana to try to work the marvelous Rattlesnake, but the furnaces proved to be a costly failure and their imagined bonanza was forfeited to creditors.[85]

Still unabashed, Hauser quickly picked up new claims on the Hope lode in Phillipsburg, sixty miles west of Helena, and in 1867 he persuaded some still-hopeful investors to put up another $100,000 for a stamp mill there, buying new preferred stock at $35 a share with a promise of redemption at $80 in a year and a quarter! But the treasurer, John How, was already "wishing I had never heard of Montana." The mill, though heralded as "a model of workmanship," recovered less than 60 percent of the assay value of the ore. On this impressive record Hauser went east, raised more money selling an added $556,000 in shares, and floated another venture, the Montana Silver and Copper Company. By the end of 1868, however, when the ore ran out, the new Hope had produced only $65,000 in bullion. By then the stockholders had bought nearly $1,700,000 in stock; no dividends had been paid; the preferred shares couldn't be redeemed; and the company was $73,000 in debt. All work was stopped, yet the company somehow succeeded in raising an additional $80,000, selling 10 percent bonds at 80 cents on the dollar, to pay off its debts. A few St. Louis investors still had hope for the Hope, and they eventually formed a new company that took out over $2,000,000 by the end of the century and finally paid back $500,000 in dividends. Hauser, meanwhile, became one of the richest men in Montana and a local power in Democratic politics, and in 1885 he was appointed territorial governor by President Grover Cleveland.[86]

Col. Patrick Edward Connor, a Mexican War veteran and former gold seeker who would be remembered as the "father of Utah mining," commanded the California Volunteers stationed on the outskirts of Salt Lake to guard the overland mail and suppress a feared Mormon rebellion during the Civil War. He also vigorously promoted the opening of Utah's mineral wealth to bring in outside miners and investors, arguing that they offered "the only sure means of settling peacefully the Mormon question." While his troops prospected the hills, he bought up interests in nearly two hundred mining claims, organized a few mining companies, built smelting works, and even launched Utah's first daily newspaper, the *Union Vedette*, to try to stir up a rush. By far the biggest find was the great Bingham Canyon copper deposit just twenty miles southwest of Salt Lake City. Its massive, low-grade porphyry ores would eventually produce over $40,000,000,000, to become the most productive and profitable copper mine in America, but that would take some time. A Mormon apostate, George B. Ogilvie, found the first indications while cutting timber in the canyon. He led Connor and some of his men and their wives, including future lawyer and suffragist Catherine Waite, to locate the first mineral ledges and organize Utah's first mining district in September of 1863. At the same time they also organized the first stock venture, the Jordan Silver Mining Company, with just twenty-six shares of two hundred feet each and soon multiplied them twentyfold at $200 par to sell to friends at bargain prices, while they ran exploratory tunnels from monthly assessments.[87]

But Connor's smelting efforts failed, and Salt Lake was still too remote for copper or silver-lead ores to pay shipping east, so after a year and over $100 each in assessments on their shares, most of the investors refused to pay any more and the company folded. Connor's other ventures did no better, so despite "heroic" efforts, the incipient boom died a-borning after he was transferred to Colorado and the troops were discharged in

1865. Little was done except prospecting until the railroad finally came at the close of the decade. The railroad would at last make the silver-lead ores profitable, but much more efficient technologies were needed to make the complex copper ores pay. One enterprising eastern promoter, nonetheless, did rake in some cash in 1867 just ahead of the railroad by offering a generous $5,000,000 in New York & Utah Prospecting and Mining Company shares to explore a potpourri of claims scattered all across Utah and on into Nevada.[88]

At the same time, New Mexico's new territorial governor, Henry Connelly, who replaced the secessionist Abraham Rencher, joined with their congressional delegate, Miguel Otero, and other Santa Fe politicians and speculators, in resurrecting the old New Mexico Mining Company on the Ortiz Mine Grant in 1861. In a salute to Fremont, they also grabbed many surrounding mines by fraudulently claiming them under Ortiz's defunct, nonmineral use of common land covering over a hundred square miles. These they grabbed through a special act of Congress after a generous stock distribution to many of its members, rather than seeking a court to do their bidding. Then, once again proclaiming the mines to be "very rich in quality and inexhaustible in quantity" and a certain "source of immense profit," Connelly and his partners greatly expanded their stock offering, to $2,500,000, and launched it anew in the East. They also teamed up with a middle-aged Missouri physician and incurable schemer, George M. Willing, in floating a subsidiary, the Pecos & Placer Mining & Ditch Company, to bring in water to hydraulic all the old placer ground on their ballooned grant, which they claimed still held an outlandish $350,000,000! After the ditch dried up, Willing moved on to a far grander scheme and Connelly soon died of an opium overdose, but the old quartz lode actually produced modestly for nearly a decade, although the company was by then so overcapitalized that the only profit again came just from selling its shares. Still, like Fremont's Mariposa, the Ortiz Mine would be resurrected again with every new boom, enticing even Darius Mills and that "wizard of Menlo Park," Thomas Edison.[89]

The "largest and richest gold mine" in the Southwest, however, was the spectacular Vulture lode, hailed as the "Comstock of Arizona." It was discovered in the fall of 1863 by an old California prospector, Henry Wickenburg, who named it for a turkey buzzard seen perching on it, but it would attract other birds of prey as well. The first ton of ore that Wickenburg took out paid $360, and he tried eagerly to work the mine on his own. But the return was too slow and erratic, so he adopted the novel practice of selling the rock in the ground for $15 a ton, which was clear profit, leaving the buyer to extract and work it. By the summer of 1865 he had a small army of ore purchasers working it in forty crude *arrastras*, or drag mills, plus a little five-stamp mill. But the swarm of miners, each ripping out high grade without any plan, made chaos of the mine, and it still wasn't paying fast enough to suit him. So that fall Wickenburg sold out for a promised $75,000, with $25,000 down, to a slippery predator, former California land speculator Bethuel Phelps and others from New York. Forming the Vulture Mining Company, Phelps issued $250,000 in bonds, set up his brother-in-law, Thomas B. Sexton, as secretary and manager, hired a large crew, put up a twenty-stamp mill, and took out close to $2,000,000 in gold in six years before he shut down. It must have paid him a handsome private profit, because he kept the company chronically in debt—for over $100,000—and stiffed Wickenburg for his remaining $50,000 as well.[90]

MEANWHILE, Samuel Butterworth of Quicksilver Mining Company fame joined with the late Samuel Colt's company secretary, William M. B. Hartley, to resurrect the old Heintzelman mine and push it on the frenzied new market. In the summer of 1863 they formed the grandiose Arizona Mining Company in New York, with Butterworth as president and an outlandish $10,000,000 in stock for a mine that had barely produced even one percent of that amount and never would produce much more! They gave Heintzelman and the old Cincinnati company only 5 percent of the stock for all of the property, and bought out Colt's and Charles Poston's heirs' interests in the lease for $70,000. At the same time, Butterworth teamed up with Sylvester Mowry and his chief creditors to float the Mowry Silver Mining Company in San Francisco to push another $2,500,000 in shares in the West. For a little extra money, Butterworth also teamed up with New York broker Anthony Morse to form the San Antonio Silver Mining Company of Arizona around a new claim not far from the Mowry mine, offering yet another $3,000,000 in shares.[91]

To push all of this paper, Butterworth took a couple of young mining experts out on an exciting trek to make glowing reports on the mines and launch a promotional frenzy. As the capper, he hired his helpful friend J. Ross Browne to do a few more entertaining and enticingly illustrated puff pieces for Harper's magazine, promoting not only the mines but also Butterworth himself as the daring hero of an Apache attack. Browne touted Heintzelman ores as high as $8,735 a ton and averages above $250; it was so rich, he claimed, that the town of Saric, just below the border, had been built solely on the proceeds of its stolen ore! Even the Mowry ore, averaging only about $60 a ton, was comparable to that of the great Gould & Curry, he said, and a single day's run yielded profits at a rate of $500,000 a year, or 20 percent on its par shares. Mowry had also eagerly brought out a promotional book, Arizona and Sonora, with advance copies of some of the Harper's pieces. For added notoriety, he filed an inflammatory $1,129,000 damage suit against General Carleton for having seized and personally looted his mine, even though he had recovered it for a mere $2,000 at a public auction in Albuquerque. Browne was running low on effusive praise by the time he got around to the San Antonio mine, but Butterworth confidently predicted that his new wonder would pay a profit of over $750,000 a year, an annual return of over 25 percent on its par shares, and if that wasn't enough to lure investors, he generously offered, for a limited time only, the first $2,000,000 in shares at half price, which boosted the predicted profits to over 50 percent a year![92]

But as usual, once all possible stock had been unloaded, little more was heard of the mines. When Browne became the federal mining commissioner after the ventures all crashed, he admitted in a footnote that earlier reports at least of the Mowry mine were "to some extent exaggerated" and that it had in fact "never paid expenses." Still, he didn't bother making any changes when he republished the glowing articles in his book *Adventures in Apache Country*. Mowry tried to cover his tracks by blaming the mine's failure on Mexican seizure of his bullion shipments. Butterworth just sat on his share of the stock sales and made no comment.[93]

Copper, too, was not forgotten. The persistent Edward Dunbar at last succeeded in floating his fledgling Ajo mine in the East as the Arizona Copper Mining Company, for $1,000,000, and he bought some steam wagons to try to haul the ore, but it was still too remote to be able to pay. At the same time, Carleton, smarting from Mowry's libelous charges of corruption, seems to have decided to have the game as well as the name, and in June 1866 he joined with new Republican governor of New Mexico and former Union general Robert Mitchell, Democratic congressional delegate Charles Clever, and others in a bipartisan effort to try to grab the Santa Rita copper mine. As the Santa Rita Mining Association, they filed claims on all of the ground under Senator Stewart's new federal mining law. Two months later Carleton established a new fort, just five miles west of the mine, and under its protection a nominal crew was put in place to do enough work to hold the claims. Although the former operator, James Sweet, tried briefly to reclaim the mine, his lease had expired, and the association confidently applied for patents on the ground. But they were blocked by a Chihuahua lawyer and former Arizona delegate to the Confederate Congress, Marcus MacWillie, who filed a counterclaim on behalf of the Elguea heirs, and the Land Office commissioner scolded the association for daring to claim as a new discovery the most famous mine in New Mexico, known and worked for more than half a century! A jubilant MacWillie then tried to secure the title for the heirs, but they couldn't find a copy of the original grant, so the mine remained in legal limbo for another decade.[94]

SHARES IN MANY of these companies were traded on both the mining and regular exchanges in New York and Boston, where they were bounced about like "speculative foot balls." The wildcat Corydon was kept in play longer than most and gained notoriety after its price was scandalously run up about tenfold and unloaded by a "keen-witted" young Boston shark, John Leighton, before he was arrested on other charges. In all, around $400,000,000 in mining shares were offered easterners in the spring of 1864. Of that the New York *Herald* estimated, perhaps too generously, that about $150,000,000 was actually paid in by frenzied investors and speculators as they plunged headlong into the fray. Only a small fraction of this money, most likely only $10,000,000 to $20,000,000, actually went into the purchase and development of mines in the West, for the great bulk went straight into the pockets of promoters in the East.[95]

Even the money that did go into the mines was too often foolishly spent and gave little or no returns. As in California a decade before, many of the Colorado companies sent out incompetent managers who exhausted their working capital on an expensive, but unworkable, array of ponderous mills and reduction works, before they ever bothered to try to develop a paying ore body. Even those with ore fared no better, for once they got below the weathered surface croppings they found that Gilpin County's gold turned "rebellious"; not only was it mixed with silver and copper sulphides that resisted simple milling and amalgamation, but much experimentation would also be required to find suitable roasting and smelting processes for effective recovery. The Consolidated Gregory squandered its capital on what proved to be utterly useless smelting works, built by its quixotic promoter and manager, James E. Lyon, who succeeded in producing only $350,000 in gold before the company failed. The Gunnell Gold Company turned its management over to a pompous ex-general, Fitz-John Porter, just drummed out of the army, who drew a fancy salary, surrounded himself with a "full staff of high-priced clerks and subalterns," and wasted $150,000 on a monumental stone mill, dubbed "Fitz John's Folly," crammed with machinery that was never set in motion. He shipped only $109,000 in gold and left the company tangled in litigation and heavily in debt. Although a handful of companies did succeed in paying dividends, they had been so overcapitalized that they were unable to pay back more than a pittance of what had been put into their stock before the prices plummeted to barely a penny on the dollar. Like most eastern investors, popular writer Bayard Taylor found Colorado to be the "scene of exorbitant hopes and equally extravagant disappointments." Its entire lode-gold production during the boom was only $10,000,000. Yet it was the beginning of an era of hardrock mining in the Rockies that would eventually produce many thousands of times more. And, of course, lure many more thousands of investors.[96]

As the bubble began to burst, a few persistent promoters rushed their wares to London in 1864 and attempted to work up some excitement there. But uncertainty over the outcome of the Civil War damped enthusiasm, and the Americans barely stirred up a ripple, floating just over a dozen ventures for a total offering of under $10,000,000. The promoters still made a little, but investors lost every penny, despite such assurances as those of the grandiose Washoe United Consolidated Gold & Silver Mining Company, which claimed that in Washoe "almost every investment is yielding enormous returns" and "fearlessly affirmed that where capital is judiciously employed failure is impossible as the sources of wealth themselves are inexhaustible!" The company's managers, however, were not judicious, squandering as much as $100,000 on a single twenty-stamp mill, and soon burying the company in debt.[97]

THE EXPLOSIVE DEMAND for news of the mines and markets had also prompted the establishment of several mining and stock journals. Two, the *Mining & Scientific Press*, founded in 1860 in San Francisco, and the *Engineering & Mining Journal*, started in 1866 in New York as the *American Journal of Mining* and renamed in 1869, became the leading mining magazines in the country. Throughout the 1860s the western paper was edited by a veteran mining camp newspaperman, Warren B. Ewer, who had published the pioneer papers in Nevada City and Grass Valley in 1851 and had started California's first mining magazine several years later. He had also briefly been a partner of George Roberts in the Grass Valley paper, but in spite of that, or perhaps because of it, he became a determined crusader for legitimate mining interests, exposing countless mining stock and management schemes, including those of his one-time partner. The eastern rival was ably edited by an incorruptible mining engineer, Rossiter W. Raymond, who carried on a campaign against scams and frauds from the late 1860s through the 1880s. Together they tried to provide a solid base of facts in a treacherous and ever-shifting sea of hype and deception. The *E&MJ* eventually absorbed the *M&SP* in 1922, and it still dominates the field.[98]

The facts, of course, were not universally appreciated, either among the sharks or the marks, then or now. "Men seemed to yearn for misinformation and misrepresentation," one editor lamented, "and regarded with disfavor those who sought to open their eyes to the facts." And, of course, "the manipulators were ready with calumny to assail those who exposed their deception. . . . No attempt was made to disprove accusations; a rejoinder from the accused that the accusing editor had been 'stung' sufficed." For as the editor concluded, "It did not occur to a community obsessed with the desire for gain to reflect that experience is excellently adapted to qualify a person to give advice."[99]

The mining stock mania again got mixed reviews from the press at large. Many papers profited from the speculative boom with increased circulation, advertising, and job printing. The metropolitan dailies devoted a column or two to mining stock quotations and exciting news of rich strikes and enticing rumors of "indications," which attracted a throng of eager readers following every fluctuation. But it was the burgeoning columns of advertisements of company meetings, assessments, and delinquent stock sales that brought in the real revenue.

Still, editorially the press both assailed and praised the madness. The San Francisco *Call* decried "the insane traffic in all sorts of mining stock" as a "loathsome disease" and denounced Montgomery Street as "worse than the hall of an insane asylum when the doors of all the cells have been opened, and the whole crowd of furious maniacs let out. God save us," begged the editor, "from the society of those afflicted with feet on the brain!" Ewer of the *Mining & Scientific Press* further railed at the abysmal moral and legal corruption of a community and an industry where men, who openly "salt claims and cook dividends," are "allowed to quietly enjoy their ill gotten gains, still retaining the respect, if not the confidence of the community." Yet, in the East, the Boston *Commercial Bulletin* saw the mania as the very fountainhead of progress, exclaiming "the entire country springs forward an age at once under the influence of a mining mania, which leaves such an era of advancement that the inevitable revulsion and crash does no more than breathe upon the mirror of its reflection, the solid silvering is yet behind and untouched. In that light we view mining manias as especial blessings—sorrowing, it is true, for the sufferers by bogus and unadvisably formed companies, the victims whose ruin marks, as the bones of dead camels do, the track where the rich caravan has passed."[100]

THE MINING BUBBLE BURST in the summer of 1864. The exhaustion of the Ophir bonanza, followed by rumors that the Gould & Curry, and perhaps the whole Comstock, was worked out, punctured the speculative market in San Francisco. Gould & Curry shares collapsed from $4,550 to $900 a foot in little over three months, while Ophir fell from $1,580 to $425 and wildcats like Pride of the West crashed from $95 to under $2. Mills's new Bank of California and other lenders quickly foreclosed on margined shares and brokers' loans, hastening the collapse. By the end of 1865 nearly $40,000,000 in "market value" had vanished. At the same time in the East, the failure of promised dividends and tales of mismanagement broke the market for Colorado and other Rocky Mountain stock. Most speculators and investors alike were left with only a hatful of paper to show for their lustful dreams. The profits, of course, nearly all went to the promoters who sold out early, and any lucky speculators who happened to get out before the crash.[101]

The enthusiastic Samuel Clemens, happily enjoying his paper fortune, went down with all the rest. "What a gambling carnival it was!" he later wrote in *Roughing It*. "And then—all of a sudden, out went the bottom and everything and everybody went to ruin and destruction! The wreck was complete. The bubble scarcely left a microscopic moisture behind it. I was an early beggar and a thorough one. My hoarded stocks were not worth the paper they were printed on. I threw them away. I, the cheerful idiot that had been squandering money like water, and thought myself beyond the reach of misfortune, had not now as much as fifty dollars when I gathered together my various debts and paid them." So once again he "took a reporter's berth and went to work."[102]

The amount of money thrown into mining maelstrom in those fevered years from 1860 to 1864 greatly exceeded that put into the California gold mania in the 1850s. There were nearly 4,000 mining companies pushing stock in both the East and the West at a total par capital of around

$2,000,000,000, split about three to one between the silver wildcats of Nevada and the gold bubbles of Colorado. Since only a fraction of this was subscribed and much of that was greatly discounted, the total paid out in the name of western mining was probably close to $200,000,000. Unlike the offering, the subscription was likely split nearly two to one in favor of greater greed and gullibility in the East, even if we discount the New York *Herald*'s estimate of $150,000,000 squandered on the gold mining bubble versus no more than $70,000,000 on the silver wildcats, of which about $20,000,000 was paid in assessments. The majority of these ventures were again out-and-out swindles, and most of the remainder were at best foolish exuberances. One commercial editor concluded that not one in twenty actually tried to work a mine. In the greed of the moment, even the owners of solid, producing mines had succumbed to the temptation to grossly overcapitalize them and take their money out of the investors instead of the ground. It was, after all, so much quicker and more certain than actually working the mines.[103]

Only a small fraction of all the eager investors came out ahead. Even the fabulous Ophir, which had been grossly overcapitalized at $5,040,000, or $3,600 a foot, barely matched that amount in total production. Most of its stockholders, who had to pay anywhere from $1,425 to $3,400 a foot for their shares, recovered only $996 a foot, or less, in dividends. Selling out at the postboom price of $425, the best that any of them could have done was to just about recover their investment, but most undoubtedly lost half or more. Yet once hooked on the game, many would stay in and bet all they'd gotten back and much more on assessments continuing on into the twentieth century in the hopes of finding another bonanza. Many stockholders in the Gould & Curry, on the other hand, did very well indeed. Its capitalization was less than half and its dividends were three times that of the Ophir, or $3,155 a foot. During the boom it sold for anywhere from a bargain $190 a foot, while early assessments of $145 a foot were being levied, all the way up to $6,300 a foot, and after the boom it still sold for at least $900. So anyone who bought before the opening of the San Francisco Stock Exchange in the fall of 1862, while it was still selling for less than $2,000 a foot, could have at least doubled their money, and those who bought at the earlier bargain prices could have made more than a 800 percent profit! Even though Gould & Curry stockholders too would eventually put more than $5,600 a foot back into

WILLIAM SHARON
Hubert Bancroft,
Chronicle of the Builders, 1891

the mine in assessments, it was the prospect of fabulous profits like this that drew them and many more back into the market again and again.[104]

In the wake of the mania, one California editor, reassuring his readers that "a child once burnt avoids the fire ever after," happily proclaimed the era of wildcat stocks a thing of the past and confidently looked forward to a new era of economical mining and prudent investment. But it was a vain hope. Though many brokers closed up shop and all but one of the mining stock exchanges disbanded, a small flock of dedicated plungers still stayed in the game. And their numbers slowly grew again as new mining excitements lured more back to the fold to be shorn again.[105]

IN THE FINANCIAL PANIC that followed the bursting of the mining bubble, the vultures descended upon the Comstock. They hungered not for any single bonanza, but for control of them all. A grand scheme was devised by William Sharon, a dapper but coldly calculating forty-three–year-old former lawyer and real estate speculator who had come west from Ohio in 1849 and would become for a time the effective "King of the Comstock." In September of 1864, just months after the crash, Sharon convinced another, slightly younger Ohioan, William Chapman Ralston, cashier and manager of Darius Mills's newly formed Bank of California, to open a Comstock branch under his management. Sharon argued that there were still enormous profits to be made from the mines, and he laid out a plan to get them. He proposed to undercut the high local interest rates of 5 percent per month and secure the indebtedness of the major mines and mills. By taking stock as security for both mining companies' and investors' loans, the bank put the shares in its name or its clerk's, and voting the "pawned stock," it could elect the directors and control the mines. Then it could systematically withhold ore from its debtor mills, force them further into debt, and foreclose, gaining control of the mills and through them reap all the profits from the mines. Although Ralston was a sucker for almost any acquisition, Sharon's plan for control of the mills was particularly appealing to Darius Mills, who had, after all, built his fortune on control of such mining essentials as water and mercury, rather than the mines themselves.[106]

Sharon did it all with icy efficiency. In less than three years, what came to be known as the "Bank Ring" controlled nearly all of the big mills and held a virtual monopoly on the extraction of bullion from the Comstock ores. In June of 1867 the Ring transferred the mills to their privately held Union Mill and Mining Company, of which Sharon was president. Then, through a variety of crooked practices, they grabbed the lion's share of the profits from Comstock ores for nearly a decade. Carefully manipulating the stock of most of the working mines through assessments and other bearing actions, they got control just long enough each year to keep stacking the boards of directors, so that the mining companies were worked for the benefit of their mills. Their directors signed milling contracts that not only paid high fees of $12 per ton of ore, more than twice the cost, but allowed the mills to keep an additional 20 percent of the bullion and required them to return only 65 percent of the assay value of the ore, keeping the remaining 35 percent in the tailings for later reworking. The Ring further bribed the mining superintendents of the rich mines to mix waste rock with ore to inflate the tonnage milling charges, while the mills mill the ore inefficiently on the first round to leave richer tailings. They also obtained tax laws that penalized high-grade ore to further encourage mixing in lower-grade rock. In the poor mines the puppet directors levied assessments to pay for the shipping of unprofitable rock to further feed the mills, and finally the

Ring milked the public of millions by manipulating share prices at will, "bulling" them with dividends and rumors of rich strikes and "bearing" them with assessments and rumors of worked-out ore. Thus, under Sharon's rapacious management the Ring worked every angle to steal untold millions in private profits at the expense of the mining stockholders. The profits from the milling schemes alone totaled $10,000,000 by 1875, with Sharon and Ralston each taking $4,000,000 and Mills $2,000,000. Their dividends and insider stock manipulations netted even more.[107]

At the same time Sharon further extended their control over the Comstock to include all of the ore haulage from the mines down to their mills on the Carson River, and virtually all of the other shipping in and out of the district. They formed the Virginia & Truckee Rail Road Company in March of 1868 and the following year built what came to be known both geographically and financially as "the crookedest railway in the United States." Sharon extracted subsidies of over $800,000 from the mining companies that they controlled and $575,000 from county bond issues in a still-all-too-familiar pattern of promising big tax revenues that they later dodged. With this money Sharon completed about half the work on the railroad and then mortgaged that to finish the task. He later boasted, "I built that road without its costing me a dollar." The railroad cut shipping costs by roughly half and allowed the companies to ship much lower grade ore to the Ring's mills, further swelling their personal profits.[108]

Their avaricious practices soon brought forth an outcry from the press. Although guardedly hiding under the pseudonym of "Mary Jane Simpson" as a correspondent of the San Francisco *Chronicle*, Comstock lawyer James J. Robbins first spoke out in 1870, accusing the Ring of having already stolen more than $12,000,000 from stockholders through the "most shameful and infamous robberies that have ever been perpetrated by any who were charged with the administration of other people's affairs." Soon even the San Francisco stock market champion, the *Stock Report & California Street Journal*, and that great Comstock defender, the *Territorial Enterprise*, openly joined in the protest against the Ring's looting of the lode. Perhaps the most trenchant critic, however, was Ewer of the *Mining & Scientific Press*, who decried how "a few unscrupulous men had completely monopolized the control of mining interest on this coast, and how, for years, by virtue of their positions as trustees of nearly all our prominent mining companies, they had continued their nefarious practices, plundering without the slightest sense of shame, or shadow of remorse, everyone outside their own clique." But despite such protests, reform would be slow in coming, for "the weight of the great moneyed power" of the Ring was so heavy and so widely felt that for many years they could smugly quash all redress. In the meantime, their obsessive skimming and manipulating were enough to make cynics of all, and one wary editor saw the shadows of lurking fraud in every aspect of mining, warning "the diamond drill, the mill where ores are crushed, the timber, the tailings, the chemicals, the water, the machinery, the bullion, the assays, the books, the superintendent, the secretary, the assessments, the dividends—everything above ground and under the ground—tunnel and incline, shafts and drifts, cross cuts and air-holes—all are means of stealing."[109]

The flood of wealth that Sharon drained from the Comstock made the fortunes of Mills, Ralston, and himself plus several friends that they let into the Ring. It also made their bank California's richest, but Ralston would nearly bring the bank down—for unlike the other two, with their cool calculation, he proved to be an impulsive and profligate plunger, consumed by a seemingly insatiable lust for possession and power.

A RIVAL, BUT initially more constructive, scheme to rejuvenate and ultimately control the Comstock mines was also begun right after the panic. As the mines went deeper, flooding ground water had become an ever-increasing problem, requiring expensive pumps to keep the works dry. A drainage adit, or tunnel, seemed like the best solution, but the first tunnel company, formed in 1863, had barely begun when the panic hit. Early in 1865 a cocky and aggressive thirty-five-year-old speculator and former San Francisco tobacconist, Adolph Sutro, revived the idea. With the energetic backing of Senator Stewart plus the endorsement of Sharon and Ralston, Sutro worked special bills through the Nevada Legislature and the U.S. Congress the following year. These Sutro Tunnel acts gave him a swath of land and exclusive rights for fifty years to run a tunnel almost four miles into Sun Mountain to tap the Comstock Lode at a depth of nearly a third of a mile below Virginia City. To carry out this grand scheme, he formed the Sutro Tunnel Company in July of 1865 with an extravagant capitalization of $12,000,000. Stewart was president, while Sutro served, generously, as general superintendent without salary, but with a free draft on the treasury for any and all expenses. To make the shares more attractive to investors, they were priced at only $10 each and not assessable. They were also lavishly illustrated with a heroic figure of himself posing as the "honest miner." He kept about half of the shares and put the remainder into the company treasury to fund the work. On the condition that he would raise at least $3,000,000 by August of 1867 and begin work, Sutro negotiated contracts with all of the major companies that would benefit from the tunnel, to pay a royalty of $2 a ton on all ore taken out after the tunnel's completion.[110]

Sutro was confident that he could complete the tunnel within just a few years at a cost of around $5,000,000, after which he predicted the royalties and other income could total as much as $6,000,000 a year for the lifetime of the lode. So the tunnel appeared to be a magnificent investment, and in the fall of 1866 New York investors agreed to raise the initial $3,000,000 and more, if he could raise just $500,000 in the West to show local "indorsement." By the following spring Sutro had secured pledges of $600,000 from the major mining companies, and hoping for an additional $1,000,000 from San Francisco capitalists, he looked forward to a triumphant return to New York to wrap up the funding before the August deadline. But by June the Bank Ring had taken over most of the mills and mines, and now recognizing the tunnel as a clear threat to their control of the Comstock, they suddenly declared war on Sutro. One by one the puppet directors of each of their mining companies repudiated their tunnel subscriptions, and Sutro's support, both eastern and western, evaporated.[111]

On Sutro's failure to raise the money and start work, the Bank Ring's mines declared the royalty contracts void, which left the tunnel without even prospective income. Finally Senator Stewart, the tunnel's previously enthusiastic president, blasted it as a "humbug" and charged Sutro with "tunneling the Congress and not the mines." A bitter fight raged on for years, as Sutro, struggling to win sympathy and support, fired off a barrage of pamphlets charging the Ring with corruption and conspiracy. But he alienated nearly everyone on the Comstock by boasting that once the tunnel was completed, only "owls would roost" in Virginia City, as all would have to move to his tunnel's mouth. There, of course, he had

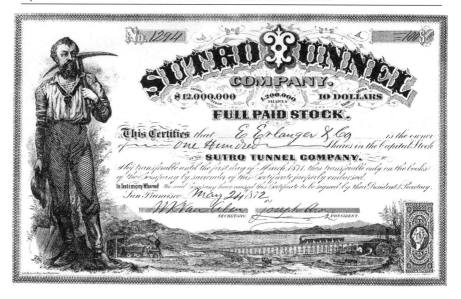

laid out the new town of Sutro, expecting to make an extra $3,000,000 just selling lots. With that the Comstock press branded the tunnel "one of the most infamous and bare faced swindles ever put forth in Nevada." Sutro next turned to Congress for a $5,000,000 loan, and even the New York *Times* joined the fray, lambasting the "impudent" ideas as an "audacious attempt to plunder" the Treasury. In October of 1869, however, calling upon the "laboring men of Nevada" to "shake off the yoke of slavery and assert your manhood," Sutro finally gained some support from the Virginia City Miners' Union, which subscribed for $50,000 in stock. That at last gave him enough money to begin work, though even then he hired only fifteen miners and paid a quarter of their wages in stock. With such limited funding progress was excruciatingly slow, while Stewart taunted that "with ordinary progress" the work would take thirty years, "and with Sutro's progress, one hundred and fifty years!" He eventually secured enough money, however, and "only" thirteen years after incorporation the tunnel would finally be completed. But by then the Comstock would be changed dramatically.[112]

WHILE SHARON AND SUTRO were still fighting for control of the Comstock mines, California's great gold lodes were at last being more fully opened as the Coleman brothers and others explored deeper yet into the neglected mines of Grass Valley and the Mother Lode for slower but much more impressive profits. The Colemans had gone partners in another seemingly unpromising mine, the Idaho, following their spectacular success in the North Star. But after spending over $19,000 in exploration, some of their partners, like those in the Allison Ranch, balked at paying their share of the expenses and all work was stopped. Dumping the deadbeats, the Colemans finally took over in the fall of 1867, forming the Idaho Quartz Mining Company with a modest nominal capital of $310,000 at $100 a share. They subsequently collected another $19,000 in assessments, while Edward sank a shaft over 500 feet before he struck another bonanza. Then in January of 1869 the mine started paying monthly dividends, and two years later it had paid back the entire investment. By 1873 it had returned fourfold and the shares had climbed to $750. As the dividends continued month after month, the brothers' fortunes were made, and they, too, turned to railroad building and other ventures. Their company paid roughly $5,000,000 in dividends before they sold out in 1891, or over 60 percent per annum for twenty-four years for its early investors! The Idaho was eventually worked to a depth of 3,300 feet and became California's second-richest mine, producing over $64,000,000 in gold before it was shut down in 1956.[113]

Fortunate investors also made extraordinary profits from the Eureka mine in Grass Valley during the 1870s. Young George Roberts had tried unsuccessfully to develop it, too, shortly after its discovery in 1851. But it was finally made to pay by local investors, led by a lucky Frenchman, Jules Fricot, dubbed "the Emperor," and a couple of practical Scots, William and Robert Watt. In the fall of 1863 they sank an exploratory shaft that tapped $400-a-ton ore at a depth of only a hundred feet, and they were soon taking out profits of as much as $60,000 a month and touting it as the "richest mine in the world." After two years, apparently fearing that the ore would soon run out, they decided to share their good fortune by selling the mine and adjacent claims for $500,000 to the Eureka Quartz Mining Company, which they incorporated with a small group of San Franciscans and New Yorkers. The company was held in just twenty shares of $25,000 each, among the priciest shares in the western mines.

But the bonanza was far bigger than expected, and even at $500,000 the investment still proved to be exceedingly profitable. In the first year over $500,000 in bullion was produced, at a profit of $360,000, and one share sold for $43,000! Before the pay lead was exhausted at a depth of four hundred feet in 1873, the Eureka had produced some $4,500,000 and paid $2,054,000 in dividends, a profit of over 300 percent in eight years, for an average annual interest of nearly 40 percent.[114]

Other deep gold lodes were also paying good profits, in Amador County on the Mother Lode to the south and in Sierra County to the north. By far the most profitable was the Keystone in Amador City, just north of Alvinza Hayward's old mine. It was a merger of a couple of claims that a local operator, Albert Rose, had gotten by foreclosure in 1857. But it was a losing proposition until the fall of 1865, when, in a flurry of excitement after a nearby mine struck a small bonanza, Rose succeeded in unloading it on San Francisco speculators for $120,000. At first Rose couldn't believe his good fortune, joking to a friend that "no child born would live to see the mine pay for itself!" Within just a couple of months, however, he was eating those words after the San Franciscans started a crosscut from the paltry lode that he had tried to work and discovered a massive bonanza parallelling it just a few yards away. It produced $40,000 within a month and went on to repay itself many times over. The San Franciscans, headed by a seasoned forty-niner from Kentucky, James Monroe McDonald, promptly formed the Keystone Consolidated Mining Company with six hundred shares at $1,000 par and soon began taking regular dividends of as much as $550 a month per share. That was a phenomenal earning even on their par value, and it blew away their actual cost of $200 a share. That was also more than Rose could handle. Three years later he desperately tried to get the mine back by bribing a government surveyor to mislocate the mine on the maps. Then, through a dummy, he bought a government patent on the actual ground as farmland for $400 and brought suit. But the fraud was exposed, and despite heavy lobbying, the secretary of the interior ruled against him in 1873. Rose still tried other legal schemes, which finally collapsed in 1880, and he was bankrupt a few years later. McDonald, in the meantime, had bought out the other stockholders and dug out a fortune to rival Hayward's, steadily working the mine and drawing dividends until his death in 1907 at eighty-two. By then the Keystone had produced about $17,000,000 in gold and paid out more than half of that in dividends, paying for itself fully seventy-five-fold in forty-three years at an average of 175 percent a year for all those years![115]

Most of California's gold, however, was still coming from the hundreds of smaller hydraulic placer mines whose combined output surpassed even the Comstock during the 1860s. These mines could be quite profitable, since a hydraulic outfit only cost about $10,000 and the average mine produced about $100,000. That left a good margin of profit, but the deeper mines often required expensive drain tunnels as well. The best paying was the Blue Gravel Mining Company at Smartsville, working what became for a time "the richest placer mine in the state." The Blue Gravel was the first to open the deep deposits in the ancient bed of the Yuba River, which held a lot of the placer gold washed out of the rich lodes around Grass Valley. It paid spectacular profits, but they were slow in coming. The company was organized in 1855 with a capital of $20,000, and by the end of 1859 it was $60,000 in debt driving a 1,700-foot tunnel to tap the deep gravels. The cash flow finally turned around, but before the mine began to turn a profit, most of the investors had lost faith and sold out. The market value of the whole mine fell to $11,000 in 1862, when most investors were rushing to the Comstock shares. Still four resolute stockholders, led by a former sea captain, Stephen W. Lee, held on to finally win the prize in March of 1864, when they finished the tunnel and struck pay dirt. Over the next four years they washed out nearly $900,000 in gold, giving a total production of over $1,200,000 from their 1,100-foot run of the old Yuba channel, a yield of more than $1,000 a foot. In all, they divided $564,500 in dividends for a magnificent profit of 2,700 percent on their initial $20,000 investment, or an average return of over 200 percent a year over the full thirteen years. Moreover, the mine paid a 5,000 percent profit on any depressed shares bought in 1862, which yielded an incredible 1,000 percent a year for five years! There was still even more gold to be had, since they hadn't yet reached bedrock, but that would require a lower tunnel, so they left that to others and sold out.[116]

Such fantastic profits from the deep gravels of the old Yuba channel excited a rush of investors for hydraulic mines and inspired that ever-acquisitive plunger, William Ralston, to try to monopolize hydraulic mining on the old Yuba in partnership with half a dozen friends, most notably Barron, Bell, and Butterworth from the quicksilver monopoly, but with the much shrewder Sharon and Mills notably missing. In 1866 they privately formed the North Bloomfield Gravel Mining Company and bought up the claims and mines until they held over seven miles of the old river channel, over thirty times that of the Blue Gravel company. To work this bonanza, which they believed to hold no less than $45,000,000 in gold, they also bought up all of the available ditches and dams, and built more, until they eventually had over 13,000,000,000 gallons of water in their reservoirs and 160 miles of ditches to deliver it, and they started a two-and-a-half-mile drainage tunnel to flush out all the gold and gravel. They also brought in all the latest technology, connecting all parts of the mine by telephone lines and installing electric floodlights so they could carry on work all through the night. But all of these acquisitions and developments took a lot of time and money. Slower even than the Blue Gravel, it would be eleven years before the North Bloomfield would finally pay its first dividend. Typical of Ralston's obsessive spending, even though they started modestly enough with a working capital of only $400,000, by the time they were through they would spend just under $5,000,000, drained from forty-four assessments and 9 percent mortgage bonds. They proved that even the profitable hydraulic mining business could be overdone.[117]

Captain Lee and his Blue Gravel partners, in the meantime, decided to push their luck and put their winnings into George Roberts's original Empire at Grass Valley, which still hadn't turned a steady profit more than a decade after its discovery, even though it would eventually become California's biggest gold producer. In May of 1852 a dozen local investors had picked up the mine at a sheriff's auction for about $20,000. They formed the Empire Mining Company, and during the next dozen years they took out just over $1,000,000. By 1864, however, they were barely making expenses, so they finally sold out to Lee and his partners. But by then their luck had run out, too. They confidently hired a crew of more than a hundred men and started sinking a couple of hundred feet and taking out ore yielding $175,000 a year. The wages alone, however, ran $120,000 a year, and Lee spent $100,000 putting up the most magnificent mill in the state and another $50,000 sinking a shaft in the "wrong place." After putting another $50,000 into a new shaft, they also decided to start cutting their losses.[118]

In 1867 they sold half their interest to a group of San Franciscans for $125,000 and the remainder two years later to another San Franciscan, insurance broker William Bourn, who had already lost a lot in other local ventures, including the all-inclusive Nevada Gold, Silver and Copper Mining Company. In the Empire he and his partners fared even worse. Although they, too, managed to take out another $1,000,000 in gold over the next seven years, they ended up in debt. For the mill and surface works burned to the ground the following year for a loss of $140,000; mining was halted for months while they rebuilt; and the ore shoots thinned and expenses rose with depth. As the mine continued to run at a loss, assessments were levied totaling $300,000, and finally, in 1874, a deeply depressed Bourn, whose additional mining investments, as well as his insurance business, had all suffered punishing losses, put a bullet through his heart. Ironically the ore shoots then improved briefly and paid modest dividends for a few years. But they disappeared at a depth of about six hundred feet, and the mine shut down in the fall of 1878. So, after another dozen years George Roberts's promising but seemingly cursed Empire had produced less than $3,000,000 and still paid only $110,000 in dividends, much less than its assessments, for a total loss of at least $500,000. That was in sharp and painful contrast to the Eureka's $2,000,000 return on $500,000, and Idaho's payout by then of $2,500,000 on only $310,000.[119]

By the end of the 1860s the mining frontier had surged out from California in wave after wave of prospecting and mining rushes, opening the intermountain West and laying the foundation for at least eight new states and territories. A myriad of small placer mines still poured out most of the gold, with nearly 100,000 miners taking out $400,000,000 during the decade. But quartz mining was growing rapidly. Thousands of paying lodes had also been opened, with nearly 10,000 hardrock miners delving in the depths to take out $200,000,000 in bullion at the same time as both mining and milling technology rapidly developed. The Comstock and other Nevada mines had led the way, producing $135,000,000, and California provided most of the rest, which was more than double what wild-eyed westerners had bet on all their wildcats. But Colorado's lodes had trailed badly, producing only a disastrously disappointing $16,000,000, which was little over a tenth of what credulous easterners had blown in all their bubbles. Even its placers did no better, turning the Pikes Peak rush into a roundly denounced "humbug." Yet, at least a couple hundred thousand giddy investors and speculators, intoxicated by all the bonanza talk, had gone on the biggest stock binge yet in America. They had swooned over the outpouring of $2,000,000,000 in paper promises from thousands of companies, suddenly fermented in the name of mining, and they sank about $200,000,000 of their savings into the flood. Out of all that, only a handful of California and Nevada bonanzas had actually paid dividends, but these totaled a handsome $20,000,000, which at least gave many of their investors real profits and allowed sharp insiders to grab real fortunes. Such heady returns would also soon entice many westerners to plunge back in again, but the calamitous losses in the Colorado bubbles would frighten away most easterners for a decade.[120]

With the advent of corporate mining had also come the inevitable separation of the shareholder ownership of the many from the management control of a few and the ever-increasing temptations of fraud. Indeed, in looking back on the period a few decades later, pioneer California lawyer and historian Theodore Hittell clearly saw that

> even at that early day, many of the methods of fraud in the management of mines and mining companies, that afterwards played so important a part in the sociological history of the state, were started and developed. It was then or soon afterwards learned, how to water stock and sell the new issue for about the same price as the original shares; how to pay large dividends in worthless mines for the purpose of working off the stock; how to get control of good mines and do the reducing at one's own mills and prices and thus absorb all the profits; how with the aid of diamond drills and otherwise to find out the real value of a mine in advance of others and buy or sell in violation of trust and in fraud of the rights of others; how to conceal discoveries of rich deposits; how to combine and conspire; how to expand or depress or "bull" or "bear" the market, and in fine, how to cheat, steal, rob and commit other felonies in the neatest, smoothest and safest manner up to that time invented.[121]

4

THE DIAMOND DREAM

A NEW WAVE of prosperity came in 1869 in the wake of the first
transcontinental railroad, which brought cheaper goods and
quicker access to the West. The coming of the railroad also
stimulated new mining excitements on either side of the tracks, opening rich new silver-lead mines all across the West, spurring the rise of a
western smelting industry, and inspiring deeper exploration that revived
the Comstock. The outpouring of new wealth from the mines created
a self-sustaining boom in the West that lasted a decade and excited yet
another frenzy of speculation, stockjobbery, and spectacular frauds,
culminating in that grandest of all speculative dreams, the "Great Diamond Swindle." Although most investors in the East were still recovering
from their "bubble mine" madness and the subsequent depression, the
new mining mania also raged furiously among British investors, since
their resistance was down. They had all but forgotten the failures of the
gold rush fever twenty years before, and the Civil War had spared them
from much of the Washoe and Pikes Peak frenzy, so they were now all
too eager to rush in again.

In the depression of the late 1860s, a horde of silver seekers made rich
new strikes throughout the Great Basin that exploded with the coming
of the railroad. The first boom was in the White Pine district in eastern
Nevada. In January of 1868 a Shoshone, Napias Jim, had led prospectors
to an extraordinarily rich silver-lead deposit they named the Hidden
Treasure. Soon after, even richer ore that reached $27,000 a ton was
discovered on the nearby Eberhardt claim, and the rush was on, drawing at least 7,000 stampeders to the windy slopes of Treasure Hill and
bringing many thousands more back to the brokers on Montgomery
Street. Numerologists, certain that 1869 was a year of destiny just like

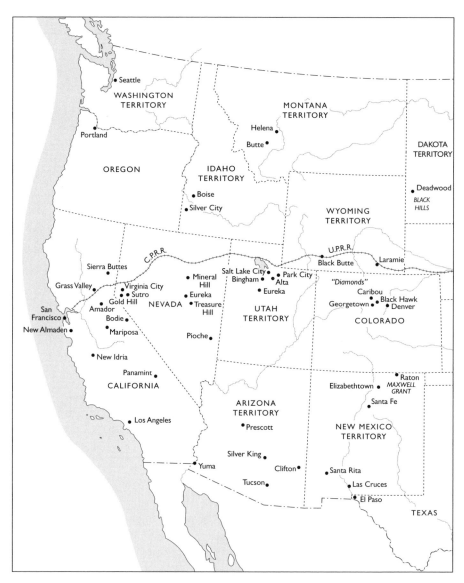

The Mining West in the 1870s

1859 and 1849, proclaimed "White Pine the Wonder of the World" and confidently predicted that Treasure Hill would be the "new Comstock."[1]

George Roberts and his friends bought the original claim in January of 1869 for $200,000 and promptly formed the Original Hidden Treasure Mining Company with 21,333 shares at $100 par. The Eberhardt's lucky owners, led by an old-time miner, Frank Drake, who held a controlling interest, reportedly turned down a $4,000,000 offer from San Francisco speculators. Instead they incorporated their own company with an extravagant paper capital of $12,000,000—but, after all, it was currently being hailed as the richest silver mine in the world. There was one other enticing bonanza in the neighboring Aurora claims. Drake and his partners quickly bought the North Aurora, and transcontinental railroad millionaire Leland Stanford and his brother, Phil, got the adjoining South Aurora. All this made a big splash that raised yet another wave of greed and fraud. Nearly two hundred new companies, mostly wildcats with such solid-sounding names as the Whang Doodle, were formed within just four months to feed the demand for shares, and three new stock exchanges were formed to help handle them. The total offering in White Pine mines was over $250,000,000, twenty-five times the entire bullion production of the district to the end of the century. In the end only two of these companies would pay even token dividends, for the extraordinary White Pine bonanzas turned out to be only shallow, highly enriched surface deposits, but it took several years for that to be fully realized.[2]

In less than a year Roberts ripped out most of the exposed bonanza from the Original Hidden Treasure to recover the purchase price and a quick profit. Then he unloaded the stock on the clueless but conscientious William Bourn and others at bargain prices around $20 a share. They got back only one dividend of $31,999 before the last of the ore ran out at a depth of a hundred feet. But they bravely soldiered on, paying out an additional $330,000 in assessments to search for deeper ore, before they gave up in 1874 for a total loss of over $700,000. Bourn suffered heavily from that loss too and finally gave up all hope.[3]

Drake and his partners, in the meantime, also took out most of the bonanza rock from their celebrated Eberhardt and the North Aurora. That ore produced $1,500,000 in bullion, gave them a spectacular profit, and excited British investors. The British eagerly bought both mines in

1870 for $750,000 cash and an equal amount in fully paid shares of the merged Eberhardt & Aurora Mining Company. The entire capital of $2,500,000 was frantically snapped up in less than two days, and shares soon sold at a 300 percent premium for a grossly inflated market value of $10,000,000. The company was managed by an extravagant amateur, Thomas Phillpotts, the brother of one of the largest shareholders. He commenced work in grand style, putting up a giant mill and a two-mile-long aerial tramway to carry the ore, the longest in America at that time. They actually took out $2,125,000 worth of ore before it was worked out, but they recovered only $192,695 in dividends, an appallingly low profit margin of less than 10 percent on such rich ore. Both Drake and the shareholders blamed the manager, not the mine, so Drake, who still held his vendor shares, which were frozen until the mine paid a profit, was put in charge in 1873 to try to prove its worth. Financed by 10 percent debentures, he sank a 1,400-foot shaft and ran a mile-long tunnel looking for more ore. All he found were a few small pockets, but their discoveries were carefully leaked by the directors to bull stock prices for a quick profit. In 1878 they all finally gave up, with a net loss of over $1,500,000.[4]

In the heady excitement of 1870, clamorous British investors also grabbed up Stanford's South Aurora, after he and his brother had cleaned out most of its rich surface croppings and made it a big producer to rival the Eberhardt. With expert predictions of dividends as high as 75 percent per year and expert help from the Seligman brothers, bankers and brokers of New York and London, Phil succeeded in selling their shell of a mine for a full $1,000,000 in cash and $500,000 more in unrestricted shares, which they quickly shed to hungry buyers at double par, to get away with a total of $2,000,000! The British dispatched a dentist, Dr. Goodfellow, as their mine manager, much to the amusement of old-time miners. But after he exhausted the remaining ore to pay a paltry $182,500 in dividends, he followed his trade and explored for ore using a method that was much quicker and cheaper than shafts and tunnels—with diamond drills. So the British learned the truth sooner than the rest and wrapped up the company in 1872 at a loss of more than $1,800,000. Some of the stockholders hoped to recoup something by reorganizing as the South Aurora Consolidated and naively tried to merge with the failing Eberhardt & Aurora, but were curtly told there was "not a ghost of a chance."[5]

When White Pine crashed, it took with it "millions of dollars with-drawn from savings-banks, etc.," lamented an old miner, Henry DeGroot, "the gatherings of long and laborious years; farms, the support of families; homesteads, the only shelter of wives and children; the servant girl's earnings, and the widow's mite, with the millionaire's surplus, and the miser's hoard—all swept forever away into this bottomless gulf of min-ing speculation!" Yet he hoped "our experience in that district" would serve in the future "to restrain reckless investment, crush the wild spirit of speculation, and cure all classes of blind confidence in undeveloped mines."[6]

INSTEAD, THE WHITE PINE mania actually launched a whole new and expanded phase of western mining and speculation, introducing both prospectors and promoters to the existence of rich silver-lead–bearing limestone, previously ignored in their narrow search for quartz lodes. These rich deposits, precipitated from hot underground waters into old, near-surface sediments, formed large, nearly horizontal beds or blankets of ore, rather than the often steeply dipping quartz lodes or veins of ore, precipitated in deeper fractures. So, in the backwash of the White Pine rush, new bonanzas were suddenly recognized in the surrounding camps of Nevada and farther east in Utah. William Stewart's new hardrock mining laws, however, based on his earlier experiences with quartz lodes, gave "extralateral" rights to the owner of the lode's highest outcrop, or "apex," to follow the lode beyond the boundary lines of his claim. So the opening of these new ore bodies soon generated seemingly endless and costly new legal battles over the nature of mineral deposits and the notion of extralateral rights.

Eureka, just west of White Pine, proved to be the biggest bonanza, pro-ducing over $5,000,000 a year by the late 1870s for a total of $40,000,000 by 1885, when the ore values finally began to decline. During that time, it reigned as the biggest mining camp in Nevada outside of the Comstock. It also proved to be an investor's dream. George Hearst, always on the lookout for another potential bonanza, was lured to the new strikes in the spring of 1870, and in May, in partnership with George Roberts and others, he bought the best claims for a total of $1,000,000, with $400,000 cash and the rest in a mortgage. Two months later Hearst and his partners formed the Eureka Consolidated Mining Company with $5,000,000 in stock at $100 par, but they sold just enough stock at close to cost of $20 a share to cover expenses until the mine began to pay its own way. They hired a clever mining engineer, William S. Keyes, from White Pine, who erected several large smelting furnaces and made the mine a steady producer. It ultimately yielded over $19,000,000 in bullion, and in its first decade it paid back over $100 a share, or a total of $5,000,000 in dividends, giving a handsome profit to its organizers and any investors lucky enough to get shares early.[7]

Eureka's other bonanza was the Richmond Consolidated, floated in London in 1871 with a modest capital of $1,350,000 and managed by a pious but sharp Anglican minister, Edward Probert. After sparring with "quivering eloquence" against charges of graft and incompetence, he ultimately produced over $28,000,000 in bullion and paid the mine's investors $4,400,000 in dividends through 1895, or a return of over 300 percent on the original investment, which made it the most profitable nineteenth-century English mining investment in America. Still, the Eureka Con charged that fully $2,500,000 of the Richmond's production came from ore that its overly zealous reverend had taken from their property under the banner of "extralateral rights." After a long and bitter battle on both the ground and in the courts, the Eureka Con won a landmark decision from that old shark Stephen Field, voiding such rights in these deposits and getting $100,000 cash and a "piece of ground" from the British company. Other British investors, however, wouldn't be nearly so fortunate, for they exuberantly bought up at about half par most of the $65,000,000 shares offered them in about a hundred companies claiming mines in the West, and they would lose fully 90 percent of their money![8]

Although Pioche, south of White Pine, produced only about half as much bullion as Eureka, it made such a spectacular show that the westerners saved it all for themselves. Some of the claims were so rich that the companies built forts, hired gunmen, and fought a couple of deadly battles to maintain possession, which only heightened the interest of many speculators. William Raymond and John Ely owned the richest mine, and in December of 1870 George Roberts and others bought out Ely's half for about $500,000 and together with Raymond formed the Raymond & Ely Mining Company to offer the waiting public $3,000,000 in

$100 par stock. They hired another White Pine mining engineer, Charles Lightner, from the Eberhardt & Aurora, who began taking out ore as fast as possible, and in two months they started paying dividends. In the frenzy that followed they cornered as much as a third of the business on the San Francisco Stock Exchange, as they feverishly worked the shares up from $9 to $180. Raymond & Ely even surged ahead of the Comstock mines to become the biggest dividend payer in the West in 1872 and 1873, paying out just over $100 a share, or a total of $3,000,000 in two and a half years. That gave early investors, who had bought in at $9 a share, a profit of as much as 1,000 percent! But speculators could have done more than twice as well, selling at the ballooning prices after taking early dividends. By contrast, the camp's second biggest producer and pioneer mine, the Meadow Valley Mining Company, was a loser to both investors and speculators, and its distraught and overextended principal shareholder, San Francisco banker François L. A. Pioche, finally put a bullet through his head. He had capitalized it at twice that of the Raymond & Ely with twice the number of shares, and he started them on the market at three times the price of his rival, but he paid dividends at only a third the rate. So his investors never recovered more than half of their money, and since the share prices fell rather than rose, the speculators did no better.[9]

But the frenzy of Raymond & Ely drove the boom, and its manager, Lightner, also lusted for a piece of the action. His chance came in August of 1871, after striking a new ore body at the edge of the company's claims. He quickly conspired with the company's attorney, Duncan Perley, to secretly acquire the adjacent claims for themselves, rather than for their employers and shareholders. Financed by George Hearst, they organized as the Hermes Silver Mining Company with a capital of $3,000,000 and offered to sell it all to their employers for a bargain $600,000. But the Raymond & Ely directors refused to deal and filed suit for the property, charging Lightner and Perley with a breach of trust. Although the jury was "confined like prisoners," they were treated to liquor and cigars by order of the judge, and the trial was said to have ended in a bidding war for jurors with bootfuls of coin dangled outside the window of the jury room. The jury "weighed the evidence" and came down on the side of the Hermes, but the judge, charging jury misconduct, threw out the verdict. Hearst then sold out his interests to the Raymond & Ely crowd for a boasted profit of $250,000.[10]

To the north of White Pine, spectacularly rich ore from the Mineral Hill mines also stirred up an excitement and attracted Roberts and Hearst, together with William Ralston and another cagey shark, William M. Lent, disarmingly known as "Uncle Billy," who with the connivance of Roberts had "put up his house in a lottery and drew the prize himself"! They bought up forty claims for $400,000 in the fall of 1870, but they soon found that the ore was only in a shallow surface deposit less than fifty feet thick, so Hearst sold to his partners and they decided to let the British in. They quickly gutted nearly all of the high grade, paying back most of their investment and at the same time making the mine an impressive producer. But they left just enough rich ore showing everywhere to nicely dress up the mine, and they carefully filled in any workings that had gone into barren rock. Then, with the help of a crafty young confederate, Asbury Harpending, they sold it the following April for $1,225,000 cash to a notorious English shark and Liberal-Conservative member of Parliament, Baron Albert Grant, born Abraham Gottheimer. He promptly arranged a fat profit for himself, mostly in stock, selling the mine for $2,425,000 to the Mineral Hill Silver Mines Company, Ltd., which he formed. With the blessings of Marmaduke Sampson of the London *Times*, Grant sold 15 percent mortgage debentures for $1,500,000 to raise the cash for the Americans, and for his part he took the remaining cash for a commission, plus 80 percent of its $1,500,000 in stock to work on the market. This left the company hopelessly overcapitalized and heavily in debt, with only $300,000 in treasury stock to pay for a new mill.[11]

The mine was basically left to pay its own way, but its vast riches were vouched for by an enthusiastic British mining engineer, John Taylor Jr., youngest son in the venerable old firm John Taylor & Sons, who had inspected the mine, predicted profits of over $1,000,000 a year, and accepted a position as its manager. Taylor did manage to produce $680,000 in bullion and pay Grant dividends of $36,000 on his shares. But in the summer of 1872, before Grant could unload much stock, the ore suddenly ran out. By then Taylor had discovered the filled-in workings, but the embarrassed management covered it up and beseeched Ralston and his friends in vain to return part of the purchase money. After an investigative whitewash cleared all parties, they wrapped up the company in January of 1873. A peevish Harpending, in the meantime, feeling he had been cheated out of additional profits from the debentures, sued Grant, and although he won a modest settlement, he held a deep grudge.[12]

BY THEN, HOWEVER, a far bigger swindle had blossomed out of some dazzlingly rich silver-lead outcroppings in the Emma mine, far to the east of White Pine at Alta, just twenty miles southeast of Salt Lake City. It was a devious operation from the start. The discoverers, James Woodman and Robert Chisholm, who first staked the claim as the Monitor in 1868, were desperate for cash and for a few hundred dollars gave a one-third interest to James Lyon, late of the Consolidated Gregory. But when Woodman and Chisholm's son, William, discovered bonanza ore a year later, they tried to cut Lyon out by relocating the claim as the Emma without him. Then they sold nearly half their interest for $140,000 to Salt Lake bankers Warren Hussey and the Walker brothers, Sharp, Robinson, Frederick, and Matthew. When Lyon brought suit, hiring Nevada's powerhouse lawyer and senator, William Stewart, to press the case, however, Hussey and the others decided it was time to sell. They first offered the mine for $400,000 to Wilson Waddingham, who turned it down. Lent, Roberts, and Ralston were interested at that price and sent Harpending to look at the mine, but by then the owners had upped the price to $1,200,000 and they too declined.[13]

Finally, another sharp San Francisco speculator, Erwin Davis, trying to recoup his fortune after a bankrupting fight with Ralston, took an option on the mine for $1,500,000. He showed its potential to Trenor Park of Mariposa notoriety, and in April of 1871 Park and an old Vermont friend, Gen. Henry Baxter, ex-president of the New York Central Railway, each bought a quarter interest in the mine for just $187,500 apiece, on the condition that they settle Lyon's claim. Park got Stewart to talk Lyon into settling for $200,000, but they had to pay Stewart and others $275,000 for cutting the deal! Park then formed the Emma Silver Mining Company of New York at $5,000,000 in 50,000 shares of $100 par and moved to take complete control. Buying additional stock at $30 a share, he secured 30 percent outright and arranged with Baxter and the other shareholders to buy all of the remaining stock for $50 a share. Once he had control, Park launched a mining blitz that surpassed even his rape of the Mariposa. With Hussey as manager, he nearly gutted the bonanza in just four months, taking out $1,500,000 at a profit of $1,200,000. That suddenly made the Emma the most fabulous mine outside of the Comstock.[14]

After paving the way by shipping the high-grade ore to England for smelting, Park and Stewart headed for London to sell what was left of the bonanza in the thriving British market. Joined by Albert Grant, they formed the Emma Silver Mining Company Ltd. in November to take the property off Park's hands for a full $5,000,000, half in cash and half in stock. To help justify the price, they assured investors that the mine would pay a profit of $3,500,000 a year, based on its production blitz. Park also bought an expert appraisal of the mine by that pliable Yale professor, Benjamin Silliman Jr., for $5,000 down plus $10,000 to $20,000 more, to be decided after the report was made. Silliman rose to the occasion and collected the maximum, praising the Emma as a "deep-seated" vein with ores running up to $2,200 a ton and ranking it "beyond all reasonable doubt" among "the great mines of the world." Park also published a carefully edited comment by the eminent geologist William P. Blake that the "rapidity and ease" with which the ore was extracted made it "unique in the history of mining in the United States," but Park didn't bother to include Blake's conclusion that "at the present rate of extraction, the mine will soon be stripped." To further lure British investors, Park quietly bought several prominent names for directors and trustees, including London banker and member of Parliament John H. Puleston, for a princely $125,000, and the American ambassador and former Republican congressman, Robert Schenck, with a "loan" of $50,000 to buy stock. The solid old London *Mining Journal* heartily recommended the new company to its readers, declaring the Emma to be "one of the largest bodies of ore ever seen," and envisioning "train after train" laden with its ores headed for England. Headlined "How an Honest Silver Mine Was Sold in London," the editor also seconded the claims of an earlier *Wall-Street Journal*, that the Emma mine was "unquestionably the best ever offered to English capitalists" and that "its success will go far to counterbalance the outrageous swindles which have been thrust on unsuspecting purchasers."[15]

In a final flourish, Park announced in the prospectus that monthly dividends at a rate of 18 percent per annum would begin on December 1, just three weeks after the offering. The initial $2,500,000 in shares were gobbled up immediately, and Park promptly collected the cash half of the

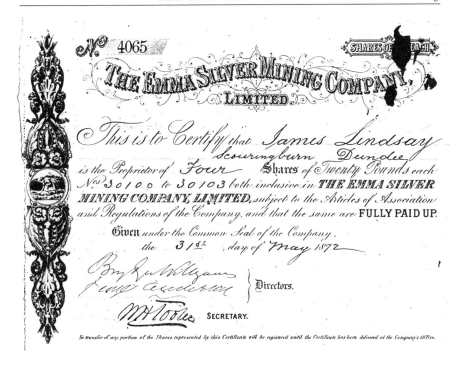

Nº 4065

SHARES OF EACH

THE EMMA SILVER MINING COMPANY, LIMITED

This is to Certify that James Lindsay Scouringburn Dundee is the Proprietor of Four Shares of Twenty Pounds each Nº 30100 to 30103 both inclusive in THE EMMA SILVER MINING COMPANY, LIMITED, subject to the Articles of Association and Regulations of the Company, and that the same are FULLY PAID UP.

Given under the Common Seal of the Company. the 31st day of May 1872

Directors.

SECRETARY.

No transfer of any portion of the Shares represented by this Certificate will be registered until the Certificate has been delivered at the Company's Office.

loot and paid Grant $500,000 for his services. They had, in fact, worked up such a market that the shares were greatly oversubscribed. But so as not to look too anxious, Park waited several months to sell his half of the shares, which required approval of the directors once they were confident the Emma had not been misrepresented. So, like clockwork, the Emma paid thirteen monthly dividends for a total of $968,000. Most of the money came from the last of the bonanza ore, already mined but still being shipped to England, while the remainder came from loans of $170,000 by Park. The steady stream of dividends kept share prices nicely above their $100 par, despite such annoyances as the international political flap that forced Ambassador Schenck's resignation as a director, which further publicized the mine; the first rumors from the *White Pine News* that the Emma had been "gutted," which was indignantly denied; the subsequent cave-in and flooding of the gutted ore chamber, which helped to cover up the evidence; the brief seizure of the mine by a rival company, which actually made it look like the mine was still worth having; and a barrage of pamphlets and broadsides, such as the playbill for the Anglo-American comedy "Bubble and Squeak, or The Emma-Sculated Mine of Utah" at the Bull and Bear Theater. All were dismissed as the growls of a "bear conspiracy."[16]

To quiet any investor concerns, Park sent Silliman back to the Emma for a reaffirmation of its worth and paid one of the directors, Edward Brydges-Willyams, a member of Parliament from Cornwall, $25,000 to also go look at the mine, with the added incentive of a stock option that could pay him as much as $100,000 on a favorable report. Needless to say, Silliman reported more than 8,000 tons of new ore with "sample assays" of $2,000 a ton. The Cornish MP was also so impressed that he picked up the stock option after happily assuring investors that the Emma would still be paying dividends after they were all "dead and gone." Thus reassured, on April Fool's Day the directors approved the sale of Park's shares. Although Park protested that he was selling only at the urging of the directors, he and Grant feverishly worked up the market again with the help of a widely quoted new Emma puff piece entitled "The History of a Great Investment," an elaborate deception planted in leading journals as excerpts from a nonexistent book, and extravagant endorsements from the prestigious London *Mining Journal*, bought for a modest $2,500 cash plus stock options. In the ensuing frenzy, Park unloaded all the remaining stock at well over par for more than $3,000,000, paid Grant his 20 percent commission, and headed home in May of 1872 to pay off the original shareholders at half par. In all, Park must have cleared close to $3,000,000 from the Emma, nearly ten times his actual cash outlay. It was his biggest and most infamous swindle.[17]

The sad awakening of British investors came at last in January of 1873, when the directors announced the end of monthly dividends and a new mine foreman found the Emma in a "most dilapidated and miserable state." The government geologist Clarence King soon confirmed that the Emma was indeed gutted. Then the directors revealed that the actual profits from the mine had been only $150,000 and that Park had loaned the money for the last dividends. A diligent shareholder dug up even more fraudulent business, and all hell soon broke loose. London papers called it "one of the boldest and most impudent swindles attempted in modern times." The stockholders finally revolted, electing new directors and suing Park, Stewart, and Baxter for "false and fraudulent representations." But the unrepentant Park promptly seized the mine for nonpayment of his loans for the illegal dividends! Lyon and others also brought suits, and the litigation dragged on for years.[18]

In the churning wake of the Emma, strained diplomatic relations also forced Schenck to resign his ambassadorship in disgrace and led to a voluminous congressional investigation in 1876 that exposed more damning details of the swindle. Edwin Godkin of *The Nation*, leery of libel, succinctly summarized "only undisputed evidence" to show how Park "swindled shareholders." But he concluded, "Fortunately for Mr. Park, criminal breaches of trust in this country . . . are not punishable by fine and imprisonment." Park only had to face the civil cases, where despite the overwhelming evidence, he convinced New York jurors that the British purchase of the Emma "was made solely on the report of Professor Silliman," and that his own actions were just common business practices. But the British finally won a victory at home. Grant not only lost his seat in Parliament after the Emma scandal, the courts ruled that his failure to disclose his promotional profits amounted to fraud, and the ensuing flood of shareholders' claims forced him into bankruptcy in 1879. Park, suddenly seized by a "desire to do right," at last signed over the mine to a reorganized New Emma Silver Mining Company in 1882, in exchange for payment of his loans and an end to further litigation. The only satisfaction that the Emma shareholders could take was that Park didn't get to enjoy his loot for long; he died of apoplexy before the end of the year. Grant, on the other hand, tried for the next twenty years to regain a fortune but failed, although he is said to have gained a dubious immortality as the model for the villainous financier Melmotte in Trollope's 1875 novel, *The Way We Live Now*. Diehard Emma shareholders also carried on for another twenty years of fruitless reorganizations, spending over $750,000 searching for ore before they too gave up.[19]

An eager Erwin Davis had also followed Park to London in 1871 to play on the Emma boom, floating three ventures of his own and gaining himself a new fortune at the expense of British investors. With a clique of Londoners headed by a distinguished old diplomat, Sir Alexander Malet, Davis packaged a rich extension of the Emma, which he'd optioned at $500,000, as the Flagstaff Silver Mining Company of Utah, Ltd., for $1,500,000, taking $500,000 in cash and $1,000,000 in shares at $50 par. Mimicking Park, he appealed to the most impatient and avaricious investors, offering immediate dividends of 36 percent per annum, and he quickly "picked the eyes out of the mine" to provide them. He paid out over $10 a share to push the price to as much as $85 while he unloaded his. Even after the Emma crashed in 1873, his board chairman reassured shareholders that the Flagstaff's bonanza "would outlast all their times and be handed down with undiminished success to their children." Davis meanwhile loaned the company $290,000 on "future ores" to pay another round of $10-a-share dividends that surpassed the Emma's yield, while he manipulated the shares for added profits. This scam paid so well that he repeated it with two other Utah mines, the Last Chance and the Tecoma, for an additional $2,000,000, taking roughly half in cash and the rest in shares. After the Flagstaff finally fell, Davis moved safely out of reach to a hotel suite in Paris, to claim that the company still owed him $700,000! While the outraged shareholders damned him as "Mephistopheles," he too, like Park, brought suit for the money he had loaned to pay the illegal dividends. That, however, was too much for a Utah court, which quashed the suit. But like their Emma soul mates, some Flagstaff hopefuls hung on faithfully, only to be conned still again by their directors and managers through several reorganizations to the end of the century.[20]

IN THE MIDST of all this feverish prospecting and fraudulent promotion, however, two great bonanzas were also opened in what would become Utah's two biggest silver-lead districts, the Eureka Hill in the Tintic district south of Salt Lake and the Ontario in Parley's Park to the east. Over the next hundred years these two districts would produce about $1,000,000,000 in bullion, and although their total output was nearly equal, their development was very different.

The bold surface croppings of the Eureka Hill bonanza were discovered early in 1870, and the richest claims were bonded by a penniless amateur promoter, John N. Whitney, for $100,000. He got a thirty-three-year-old former St. Louis steamboat captain, Joab Lawrence, later dubbed "one of the shrewdest and brightest men" in Salt Lake City, to put up $50,000 cash and give notes for the rest to be paid from profits. They formed the Eureka Mining Company in the spring of 1871 with a nominal $500,000 in $100 par shares, Lawrence as president, and Whitney as vice president

and manager. Lawrence took 40 percent of the stock, Whitney 30 percent, and they soon sold the remainder at only a fraction of par to a
prosperous gold rush merchant, John Q. Packard, in order to finance
development and fight off neighboring claimants. But a year later, after
seeing the quick profits that the Emma promoters had grabbed, Lawrence
and Whitney also decided to simply unload for all they could get, and
they scrambled to find a buyer. Although they had opened good ore
and easily gathered honest appraisals from prominent mining experts
testifying to good profits, they didn't stop there. With ever-escalating
hype, they soon spread about 500 tons of ore over a small knoll and
claimed it was an ore dump of 10,000 tons with "frequent assays as
high as \$5,000 and \$10,000 a ton." Next, after failing to entice British
investors, they promised extravagant dividends of 60 percent a year to
Eber B. Ward, an avid spiritualist and overbearing, but gullible, heir of
his uncle's Detroit steamboat business, if he would buy half the stock
at par. Then, when Ward agreed to send out his own experts, Whitney
secretly had their sample bags "salted" with high-grade rock, while he
entertained his visitors at a nearby saloon![21]

After the experts gushingly reported that the ore was "almost inexhaustible" and predicted even more extravagant profits of 150 percent
a year on par, Ward bit and bought in October of 1872, giving over
\$260,000 in cash, stock, and Detroit real estate for the controlling share
and proudly electing himself president. But after a full year without
being able to pay any dividend at all, he finally realized that he had been
conned, and a detective he sent to the mine not only got a confession
from the "salter," but incriminating letters of instruction from Whitney
on exactly how to "doctor their samples." Armed with that evidence,
Ward angrily demanded his money back, and when Lawrence refused,
he was slapped with fraud charges and arrested in New York on an
extradition warrant from the governor of Michigan. Lawrence claimed
that it was all a "blackmailing scheme" to get money from him, and he
suggested that Ward had actually bought the stock on the advice of a
"wonder-working gypsy, who saw big Eureka dividends in the bottom
of a teacup." Coming on top of the Emma scandal, defensive Utah miners initially rallied behind Lawrence, denouncing the "salting" charges
as "damnably absurd," until Ward published his evidence. Then they
watched in silence, while Lawrence remained under house arrest in a

posh hotel for four months until his lawyers finally got the extradition warrant rescinded. Right after that Lawrence suckered Ward again, by foreclosing on a mortgage for $43,500 in salaries and other debts he claimed against the company. He had carefully written the mortgage to himself as its lame-duck president a month after the sale, but before Ward formally took control, and fraudulently predated it weeks before the sale! With that, Ward swore they were all "robbers and thieves" and vowed to fight them to the end, but he died of apoplexy just four months later. Although his heirs carried on the fight, charging that it was all a "fraudulent devise" to concoct a "fictitious debt" and defraud the stockholders, they lost. So Lawrence took over the entire property from the sheriff in July of 1876 for the cost of the debt plus 20 percent interest.[22]

By then, however, the tables had also been dramatically turned on Lawrence. For the cagey Packard had cut a deal with Wells Fargo & Company, who also held an unpaid debt from the Eureka mine, and foreclosed on it. Backed by intimidating financial clout, Packard had forced Lawrence to settle for only a third interest in exchange for not opposing his foreclosure. Thus they all reorganized the company as the Eureka Hill, and doubled its stock to 10,000 shares in a deal that gave Packard a controlling 5,001 and Lawrence 3,500, while the remainder was split between Packard's well-connected lawyer and his wife's brother-in-law, Supreme Court justice Stephen Field. Packard then finally put the mine into profitable production, and over the next twenty years he paid out dividends of $1,450,000 and collected his half. With a bit of that money he also bought up most of the neighboring claims on the sprawling bonanza that blanketed the hill and cleared millions more. By the turn of the century he was heralded as the "Mining King of Utah," and its richest man, before he returned to California to retire in Santa Cruz. There he died in 1908, at the ripe old age of eighty-six, leaving an estate estimated at $20,000,000. Lawrence, on the other hand, for all his scheming, collected only a quarter of the dividends on his shares before he was suddenly stricken with an "affliction of the brain" in 1885 and died three years later, at just fifty-one. He had never shared the new stock with Whitney, who then finally sued his estate for a share. But since Whitney never brought suit while Lawrence was able to defend himself, the judge ruled that he had simply waited too long to try to get any more.[23]

BY COMPARISON, the opening of Utah's other great silver-lead bonanza was simplicity itself. George Hearst had also been attracted by the Emma excitement. But the Emma's ex-foreman, former Comstock miner Marcus Daly, steered him on to a new prospect called the Ontario, over the crest of the Wasatch from the Emma in what would become Park City. Although the Ontario would reign as the richest silver-lead mine in Utah for the rest of the century, it was still just a "little hole," barely shoulder deep, when Hearst arrived in July of 1872 and bought it for $30,000. But it looked promising, so he opened it patiently and thoroughly. That was also costly, so he formed a new, but lasting, partnership with a sharp San Francisco lawyer, speculator, and Turkish doctor's grandson, James Ben Ali Haggin, who with his law partner and brother-in-law, Wells Fargo president Lloyd Tevis, had built a real estate empire worth millions. Haggin got a half interest and put over $600,000 into the mine and a big mill over the next five years before it finally began to pay. It was an excellent investment, however, for over the next twenty-five years the Ontario produced about $35,000,000 and paid 40 percent of that in dividends. The Ontario was the foundation of Hearst's real fortune, and he later claimed "from that $30,000 everything else came." It was also one of those rare profitable investments for outsiders. For once the mine was ready to pay, Hearst and Haggin incorporated the Ontario Silver Mining Company in December of 1876 for $10,000,000, listed it on the New York Stock Exchange, and sold a third of the stock to the public at about a third of its $100 par, recovering their costs plus 100 percent profit. At the same time they began paying some two hundred dividends, which totaled $14,932,500 before they stopped in 1902, and anyone who bought shares during the first decade got their money back with good interest of 10 percent a year or more. Hearst and Haggin did far better, with dividends of 16,000 percent on their investment, or 50 percent a year for thirty years![24]

Hearst also bought into yet another enticing Owyhee bonanza, the Golden Chariot, with fabulously rich ore "all sparkling with gold and streaked with silver." But it turned out to be just a few rich pockets, so he and his friends worked it mostly in San Francisco with all the classic manipulations. It had been discovered in the fall of 1867 not far from the Poorman, and its ore, assaying as much as $2,000 a ton, was so rich that rival claims were staked on it. The Golden Chariot was held by Idaho

stage-line king Hill Beachy and the infamous Robert Bailey of Death Valley, who had fled to Idaho after his Tahitian retreat, while the rival Ida Elmore claim was held by Owyhee pioneers David H. Fogus and J. Marion More, formerly John N. Moore, who had fled to Idaho from Mariposa and changed his name after an "embarrassing fracas." At first the contending parties had agreed on a "neutral ground" between their works while they each feverishly ripped out high grade during the winter. But as spring approached, the Golden Charioteers broke into the no-man's-land, and both sides quickly hired gunmen, fortified their underground works, and started taking potshots at one another. Then the gunmen from the Golden Chariot stormed the Ida Elmore, two men were killed and several wounded outright, and a drunken More was killed a few days later, on April Fool's Day. After the governor called in the U.S. Cavalry, both sides finally agreed to give up the fight, divide the ground, and go back to mining.[25]

The deadly war and the fabulous ore had attracted wide attention, and after the shooting stopped Hearst and a few friends, in November of 1868, joined with Beachy and Bailey, who had gotten the better part of the ground, to form the Golden Chariot Mining Company divided into 10,000 shares at $100 par. They promptly cleaned out the exposed ore and paid themselves four monthly dividends totaling $10.50 a share, while unloading a reported $300,000 of it on the San Francisco exchange at $45 to $50 a share. Without further dividends, however, the stock declined to $10 by the fall, and Hearst and his partners slowly bought it back as they explored for, and blocked out, more rich rock. A year later they opened another rich pocket and began an even grander spate of dividends, paying $22.50 a share over six months and driving the stock up to $90, while they unloaded again. The prices fell once more after the dividends stopped and further collapsed to $5 in November of 1871 with punishing assessments of $8 a share, while the partners picked up more on short sales. Depressed investors forfeited over 40 percent of the shares in the delinquent sale, while their trustees quickly loaded up yet again for one last quick deal. Just a week later they announced the discovery of a new bonanza, and in just forty-eight hours the stock jumped from $5 to $21 and Hearst and his friends unloaded for the last time. Most of the clueless investors finally concluded that there wasn't even enough ore for another dividend, but some kept hoping there was still a big bonanza hiding down there somewhere. Enticed by one more pocket of fabulous ore that briefly paid back $80,000 in dividends, they hung in for seven more years, pouring a total of $869,500 in assessments into deep exploration before they finally abandoned all hope at a depth of 1,400 feet. During all the excitement, others had put the Ida Elmore and the old Poorman on the market too, but their shareholders gave up sooner.[26]

FARTHER EAST, Colorado's first major silver lodes were being worked in different ways at Georgetown and Caribou. The rugged slopes of the Rockies surrounding Georgetown at the head of Clear Creek west of Denver were ribbed with silver quartz lodes that made it Colorado's leading silver producer through most of the 1870s, eventually producing over $50,000,000, as silver production pushed ahead of gold in Colorado for the last quarter of the century. The Terrible Lode, discovered in 1867 three miles west of Georgetown, was the first to pay. It was named just as a joke by its discoverer, but for its investors it proved to be a very bad joke indeed. A local speculator, Frederick Clark, and a partner picked it up for a mere $500 and took out $86,000 in 1869. They then joined with a skillful publicist from England, Robert Orchard Old, who operated as the self-styled British and Colorado Mining Bureau, to sell it for $500,000, as the first major Colorado mine in England. First they attracted British attention by shipping their richest ores to Swansea and Liverpool for smelting. Next, rounding up a bevy of titled directors for the "lord-loving public," they formed the Colorado Terrible Lode Mining Company, Ltd., with offices backing on the Bank of England. Then early in 1870, with a prospectus predicting lavish profits of $500 a ton, they sold eager investors $525,000 in shares at $25 each, nearly all of which, plus most of the remaining $100,000 in stock, was pocketed by the promoters for the purchase of the mine. Old served as superintendent until the end of the following year, taking out roughly $300,000 in silver and paying out four dividends totaling $42,000, or about 5 percent per annum, to the investors. But as soon the company began to pay, that opportunistic shark Jerome Chaffee descended. Through his bank he foreclosed on the adjacent Brown mine, which had proved as productive as the Terrible but had been extravagantly managed. In the fall of 1871 Chaffee sent his thirty-two-year-old bank cashier and eventual partner, David H. Moffat, to London to try to sell the adjacent lode to the British shareholders for another $500,000. Moffat not only claimed it was just as rich as the Terrible, but most likely also raised the possible threat of overlapping claims and costly litigation. Moffat was still a novice, however, and even though he cut the asking price to $300,000, the shareholders—looking for returns, not further outlays—rejected the offer and called his bluff.[27]

Chaffee pulled back for a while to mount a new attack. He began financing an aggressive young English speculator, William A. Hamill, who had previously challenged his control of the Brown mine and had now gained control of the Silver Ore claim just below the Terrible as another "fighting claim." Early in 1875 they broke into the paying lower levels of the British mine and then got an injuction against the British that stopped all work in the Terrible. Soon after that Chaffee went to London, offering to sell both the Brown and Silver Ore mines for $400,000 to settle the litigation. Outraged British shareholders denounced the "blackmailers," but they finally caved. Reorganizing as the Colorado United Mining Company and more than doubling the capital stock to $1,625,000, the British gave the entire added $1,000,000 in stock to Chaffee, Moffat, and Hamill. This also gave them control of the mine, and they named Hamill as superintendent. With glowing reports from the mine, Chaffee and Moffat unloaded their stock at premium prices for $400,000 in the summer of 1878, but Hamill held his to keep his hand in the till. Under his management the mine turned out about $150,000 a year in bullion, but he cornered it all in local expenses, and all that he ever gave the hapless stockholders was a single dividend of a shilling a share, until they finally bought him off in 1885.[28]

Georgetown's most spectacular lode, however, was the Pelican-Dives, which produced over $2,500,000 in the early 1870s. But these quick riches only led to a scandalous and deadly struggle that once again reached all the way into the White House. The Pelican claim was located in the fall of 1868 by two seasoned prospectors, Eli Streeter and Thomas McCunniff, and the Dives, which crossed it at a slight angle, was located the following spring by Thomas Burr. Neither claim attracted much attention until late December of 1870, when bonanza ore, running as much as $1,300 a ton, was struck in the Pelican. Its two owners were besieged by would-be investors, but they were determined to go it alone and instead bought up a few adjacent claims. Georgetown's aggressive young city attorney, John H. McMurdy, the twenty-three-year-old, hot-headed son of a "distinguished divine," beat them to the Dives, however, quickly buying out Burr and his partners for $28,500. McMurdy immediately tried to move onto some of the Pelican ground, got into a brawl with a leaser, and took a few blows to the head that laid him up for a week. So he backed off for a while, hoping to find an extension of the bonanza, and formed the Perdue Gold & Silver Mining & Ore Reducing Company in Indiana in May of 1871 to bankroll the exploration. But after more than two years' work with little return, he decided to go after the Pelican again, both in the courts and underground.[29]

In November of 1873 McMurdy pushed the underground workings of the Dives into the Pelican bonanza, suddenly claiming it was part of the Dives, and a flood of litigation and injunctions followed. As the legal fees began to climb toward $500,000, Streeter and McCunniff finally sold out a controlling interest for about $150,000 to two old California miners, Edward Naylor and Jacob Snider, who had also turned banker and bought the largest share. McMurdy also sought help from Hamill and

Chaffee, who was by then the Republican territorial delegate to Congress. So when two territorial judges ruled against the Dives, granting an injunction to restrain him from trespassing, McMurdy arrogantly boasted that he'd have them replaced, and he reached much higher. In December of 1874 he gave, or sold, depending on the source, a quarter interest in the Dives to two of President Ulysses Grant's close friends, John A. Logan, the Republican senator from Illinois and former general, who had served under Grant in the Civil War and had nominated Grant for the presidency, and Henry H. Honore, a wealthy Chicago realtor, whose daughter had just married Grant's eldest son, Fred. Two months later Grant summarily removed both judges and replaced them with two Dives-friendly jurists, and the injunction was promptly dissolved. The New York *Sun* denounced the action as "a shameless piece of scoundrelism," claiming that Grant's son also shared in the spoils, and the New York *Herald* charged that the Republican presidency had been "prostituted" to make a "profit for the family of the President, through the corruption of the courts." At the same time Grant also removed the territorial governor, Edward McCook, a political foe of Chaffee. Indeed, it was rumored that Grant acted after a poker game with Chaffee, whose daughter would also marry one of Grant's sons, Ulysses Jr.[30]

But this still didn't give McMurdy control of the Pelican bonanza. Even during the injunction, however, the preacher's son had continued looting high grade underground from the "dispute ground," secretly hauling out as much as $65,000 at a time under cover of darkness at midnight on the Sabbath! He had also recruited "a small army of fighting men," giving some of the most belligerent added incentive as leasers, and the Pelican owners had armed their miners in response. Tensions reached a breaking point in May after another judge ruled against the Dives claims. This restored most of the disputed ground to the Pelican owners, who quickly brought in armed guards to hold it, even after one of Grant's new judges tried to reverse the judgment, issuing a writ turning over the ground to the Dives. The Pelican attorney and soon-to-be Republican senator, Henry Teller, telegraphed the owners to hold out and warned the undersheriff, a friend of McMurdy's, that if he set foot on the Pelican "they will shoot you and your men like dogs." In defiance two days later, with a posse of McMurdy's hired gunmen, the undersheriff kidnapped and held Pelican owner Ed Naylor hostage in an unsuccessful attempt to force the surrender of the ground. When that failed, Jack Bishop, one of McMurdy's "hired assassins" and leasers, promptly chased down and murdered the other owner, Jake Snider, in cold blood on the street in Georgetown, while the undersheriff stood by and let him escape. A "reign of terror" followed for several days as the Dives gunmen threatened to kill Snider's son and other Pelican men. Finally, yielding to the general outrage, Teller reached a settlement with the Dives, agreeing to a line that divided the disputed ground and ended hostilities.[31]

It was all over for McMurdy as well. Just before Snider's murder, McMurdy and his wife had briefly fled to California for his "health," and while returning early in June, he died suddenly, at age twenty-eight, in Carson City. Some suggested his death resulted either from an "attack of brain disease," caused by his earlier head injury in the claim brawl, or that it was a remorseful suicide for his complicity in the murder. The Nevada doctor attributed it to peritonitis, but the insurance company fought with his widow for years over the cause of death. There was little else to fight over, for the Pelican bonanza had been nearly gutted by then. The Pelican owners had taken out over $1,500,000, and McMurdy and his friends about $1,000,000, but all the profits had been lost to hired guns both on the ground and in the courts, and the two biggest investors had lost their lives as well. William Hamill picked up what was left of both mines for under $100,000 and sold them in 1880 for $600,000 to a group of eastern telegraph and telephone executives led by Norvin Green of Western Union. Playing on the notoriety of the mines, they floated the Pelican and Dives Mining Company at $5,000,000 for a total loss to yet another round of investors, but at least they all got out alive.[32]

The other spectacular bonanza at the time was the Caribou, discovered in the fall of 1869 by a prospector who happened to see some Comstock ore on a train in Wyoming and remembered similar-looking rock in a great quartz "lode blossom" just up the canyon west of Boulder. The rock assayed as much as $16,500 a ton, and a year later the discoverer sold out for $50,000 to a sharp old millionaire casket maker from Cincinnati, Abel Breed, who would eventually be accused of exhibiting "some of the meanest characteristics of the human race." But at first Breed was widely praised for bringing in capital, opening the lode, putting up a mill, and turning out nearly $400,000 in bullion in the next three years. But Breed was looking for quicker and bigger profits, and he had teamed up with the ever-ready shark Chaffee to try to sell it in England. Chaffee also entrusted that effort to David Moffat, when he went to London in the fall of 1871, but he failed with it, too. So Breed brought in a clever publicist, Moses Anker, to drum up more press for the Caribou with such spectacular displays as paving a walkway with its silver bricks for President Grant to tread when he visited Colorado. Then he sent a new agent, that "keen-witted" shark John Leighton, fresh from jail, to England armed with more picture rock to try to find buyers again. Leighton soon lined up an offer of $500,000 from the British, then double that, but Breed wanted even more. So Leighton and Anker went on to the Netherlands, where they finally got the price up to an outrageous $3,000,000 in the spring of 1873 and Breed at last agreed.[33]

By then Breed was taking out $5,000 a day in $200-a-ton ore, and he claimed to have 34,000 tons in sight! So the Dutch investors, led by former Minister of Finance Pieter Philip van Bosse, confidently chartered the Myncompagnie Nederland and gave Breed $1,500,000 in cash and the other half in stock, which he gave to Anker and apparently Leighton as their share. But when the company took over the mine, they discovered that the double-dealing Breed had just gutted the richest ore reserves and saddled them with an additional $100,000 for machinery. Anker still argued that there was plenty of ore left, but over the next three years they managed to take out only $300,000 and still ended up nearly $200,000 in debt. With that, the Dutch finally declared the company bankrupt and a total loss, the Emma of Holland. The miners, meanwhile, had taken over the mine at the end of 1875 to work it for back wages. Anker and other creditors also attached it for $140,000, and the ever-present Chaffee, who had been loudly bad-mouthing the mine, grabbed another $50,000 in debts and convinced Anker, who was also in debt to him, to let him handle all their claims together. Then, in a new round of double-dealing, Chaffee dropped Anker, and he and Moffat went to the sheriff's sale and bought all the property for $70,000, to fix it up for the next boom.[34]

AT THE SAME TIME, Chaffee pushed a new land-grant scheme in northern New Mexico that was destined to become an even more infamous fraud than Fremont's Mariposa, and he unloaded it on the British and Dutch as well. It started modestly as an old grant on the Cimarron, which one of Fremont's guides, Lucien Bonaparte Maxwell, had bought for $38,745. In 1869, soon after rich gold strikes had been made at Elizabethtown over the hills west of the grant, Chaffee optioned Maxwell's grant for $600,000 in partnership with that budding land baron Wilson Waddingham and a corrupt clique of Republican politicians known as the "Santa Fe Ring," led by the aggressive young U.S. district attorney for New Mexico, Stephen Benton Elkins, nicknamed "Smooth Steve," who also acted as Maxwell's lawyer. Then, following Fremont's lead, they conspired with Maxwell to run a survey of the grant that reached out and grabbed not only the new gold fields but also the vast Raton coal fields, which extended into Colorado, as well as countless ranches and homesteads. When the secretary of the interior stopped the government surveyor from going beyond the original grant limits, Chaffee hired him privately to continue, and under a new secretary the Ring finally got Elkins's brother, John, appointed to run the official survey of the expanded grant. Before they were through, Maxwell's grant had magically mushroomed from the already very generous limit of 22 square leagues, or 96,000 acres, of the original grant to an obscenely fraudulent 1,714,764 acres, nearly 2,700 square miles! "Larger than the State of Connecticut and some other states," a carried-away Chaffee claimed.[35]

Even though the secretary of the interior had declared that the grant could not exceed the original limit, the scofflaw banker and soon-to-be Republican senator Chaffee simply ignored the order and bought letters affirming clear title by former attorneys general of both the United States and the Confederacy for $5,000 and $10,000, respectively. Then with a glowing prospectus, touting its vast ranch lands, bonanza gold mines, and enough coal to run a railroad "for hundreds of years," the entire grant was sold in April 1870 to a sharp British promoter, John Collinson, and a couple of partners for $1,350,000. Paying Maxwell a fifteenfold profit on his investment, Chaffee and his confederates took a comparable profit of $750,000 on their expenses.[36]

Collinson, who was welcomed as "a gentleman of shrewd business capacities," promptly formed the Maxwell Land Grant & Railway Company with $5,000,000 in stock. He took over 70 percent of it for himself, giving Waddingham nearly 20 percent for his interest in the grant and Elkins, who had become the company's attorney, a much smaller share for his previous services. Chaffee, of course, only wanted cash. In a new flurry of promotion, Collinson and Waddingham, in collaboration with a clever young Dutch banker, Gideon Maria Boissevain, pushed the scheme in London and Amsterdam. With the added help of a Dutch mining "expert," who extravagantly appraised the grant at more than $90,000,000, they successfully unloaded most of their shares at half of par, together with an additional $3,500,000 in 7 percent first mortgage bonds at a 28 percent discount. They claimed the bond money was needed to help fill the grant with colonists, but it nearly all went just to fill their pockets. In all they got away with close to $3,500,000, after paying off Chaffee; Collinson most likely took about half of it, while Waddingham and Boissevain split the remainder. They were so successful, in fact, that they soon sold another $1,375,000 in 7 percent second mortgage bonds, at a discount of less than 25 percent, to make payments on the first mortgage bonds![37]

But when the British and Dutch investors arrived to claim the land, they were shocked to discover that they didn't actually have title to it. Worse yet, they were met with armed resistance from the miners and settlers already on the ground, who refused to recognize the "fraudulently enlarged" boundaries and were determined to fight a bloody war for their claims and homesteads that would drag on for twenty years. The grant was widely denounced, not only as "the largest and most atrocious . . . steal of the public domain" ever attempted, but also as a "grand swindle" of "gullible European investors." The Dutch finally forced Collinson out in June of 1873, but by then the company was so deeply in debt that another $5,000,000 in preferred stock had to be issued to try to keep it afloat. Even that wasn't enough, however, and the Dutch defaulted on their mortgage bonds the following year.[38]

The Santa Fe Ring, meanwhile, got Elkins elected as the territory's delegate to Congress for two terms starting in 1874, despite charges of massive voting fraud, including flagrant "official" recounts and other devises by Elkins's law partner, Thomas Benton Catron. In Washington Elkins lobbied desperately to circumvent the government order disallowing their ballooned grant, trying both to push through a patent to confirm the grant and legislation to support it. At the same time he and Collinson schemed to grab control of it again. Since foreign companies could not legally hold land in the United States, the company had been formed in New Mexico, and when Collinson went to Europe to unload, he had named Elkins to replace him as president with a generous salary

of $10,000 a year. Collinson and Elkins first tried to get the Dutch bond-holders of the still-profitless venture to organize a new holding company, issuing an additional $11,750,000 in new bonds and shares for fat personal profits, but the bondholders wisely refused. Then, without warning the Dutch, Collinson had Elkins, who still managed local affairs, let the property go into default by not paying the property taxes, and in 1877 their agent quietly bought it at a sheriff's sale for a mere $16,479.46, less than a penny an acre, deeding it to Elkins's partner Catron for $20,000. Collinson and Elkins hoped that the bondholders wouldn't learn of the scheme before the time expired to redeem the property. But the Dutch, getting wind of the sale, quickly raised an assessment to buy it back for just $20,961.85 and regain control. To grab the property back again, Collinson and Elkins placed the company in bankruptcy for mortgage defaults, naming Catron's other partner as receiver. All the while the Maxwell paper had fallen to just pennies on the dollar as the internal and external struggles dragged on and the title remained in limbo. Like the Mariposa, the Maxwell grant became a serial swindle, resurrected in every new mania to shear a new flock of sheep.[39]

Elkins, Catron, and Waddingham in the meantime tried to repeat the ballooning scheme on a few other land grants, including the old

Ortiz grant in 1880. But as soon as they launched it on the New York market, protesting New Mexicans sank it in a sea of litigation. The trio nonetheless grabbed a lot of other land that helped make Waddingham "the largest land holder in the United States," but that didn't profit them much, either. So after helping to loot the federal treasury through the fraudulent mail contracts of the notorious "Star Route" rings, Elkins turned to Colorado mining ventures, which finally made him a fortune. Then, after marrying the daughter of a prominent West Virginia coal mine owner and Democratic senator, Henry Davis, Elkins too turned to eastern coal and railroad ventures. He founded the town of Elkins, West Virginia, and after serving as secretary of war under President Harrison, he bought himself a Republican seat in the Senate, which he held from 1895 till his death in 1911, reigning for many years as its richest member. Catron stayed in New Mexico and in partnership with Waddingham acquired more than 250,000 acres of his own. After Waddingham's death in 1899, he was said to have controlled as many as 4,000,000 acres. When New Mexico was finally admitted into the Union in 1912, Catron too became one of its first senators, along with fellow Republican Albert Fall, who was later convicted of bribery in the Teapot Dome scandal.[40]

MEANWHILE, soon after Chaffee had unloaded the Maxwell grant, he joined with Elkins and others in the Santa Fe Ring in a scheme to take over New Mexico's real bonanza, the old Santa Rita copper mine. They planned to cash in on the artificially high copper prices, from 18 to 45 cents a pound, that Michigan producers had worked up in the spring of 1872 with the aid of a 5-cent-a-pound "protective" tariff. First, the Ring helped block the efforts of the Elguea heirs' lawyer, Marcus MacWillie, to secure title to the mine. Then they talked the heirs into hiring their own man, former chief justice of the territorial supreme court John S. Watt, to pursue the case, and he soon persuaded the heirs to sell their claim to the Santa Rita to Chaffee's agent, Martin Hayes, for just $15,000. In the summer of 1874 Chaffee and his partners reopened the mine and put up a new smelter to begin producing crude copper melt, or "matte." But they still had to haul it by wagon, at a cost of up to 12 cents a pound, over seven hundred miles to the railhead in Trindad, Colorado, for shipping on to refineries in the East. By then, however, copper prices had fallen back to 22 cents a pound, and with all the additional costs of mining, smelting, rail shipping, and refining, there was no longer anything left for a profit. So after producing about five hundred tons of copper at a loss, Chaffee finally shut down the mine again in 1876, while he waited for the coming of the railroad to make it marketable, and, of course, the government to finally clear its title.[41]

The high copper prices had also inspired a few investors to open what would prove to be not only the biggest copper bonanza in Arizona but the third largest in America, an enormous low-grade porphyry copper deposit, heralded as a "solid mountain of copper," just seventy miles west of Santa Rita at what would become Clifton and Morenci. Although it would eventually surpass Santa Rita, it too took time. Its richer surface croppings were discovered in 1865 by a twenty-five-year-old prospector, Henry Clifton, who soon turned to farming, and it was rediscovered in 1870 by a seventeen-year-old, Joseph M. Yankie, and others. Yankie soon sold his claims for about $8,000 to a local mining engineer, Edwin M. Pearce, and he brought in that wealthy Detroit steamboat man and hapless Tintic investor Eber Ward, who formed the Detroit Copper Company in the summer of 1872. Ward put up an additional $30,000 and sent one of his steamboat captains out to open the mine and set up what became the camp of Morenci. But all work came to a halt after copper prices fell and Ward died of apoplexy, at age sixty-four, the day after New Year's in 1875. The mine remained tied up for years as Ward's adult children and his young trophy wife fought a bitter battle over his estate.[42]

But one hopeful investor in the neighboring Longfellow claim was more determined, and for nearly a decade he doggedly struggled to make it pay, plumbing the very depths of mining economy. He was an unsuccessful Australian gold seeker from Poland, Henry Lesinsky, who in his late thirties had opened a small general store with his brother, Charles, in Las Cruces, east of Santa Rita. Yet he still longed to make a fortune. So late in 1872, with his brother, his uncle, and a few friends, he bought a controlling interest in the Longfellow for $10,000 from its locator, Robert Metcalf, and formed the Gila Copper Mining Company. But after continuing disagreements with Metcalf over how to develop the mine, Henry and the others finally bought him out for $5,200 and reorganized as the Longfellow Mining Company in 1874. Henry then took over the management of the mine and put $40,000 into a fancy smelter that collapsed as soon as it was fired up. Undaunted, he put up simple but inefficient adobe furnaces that managed to produce 200 tons of copper matte, which were hauled over eight hundred miles to the railhead in Colorado in 1875.[43]

But by then copper prices had crashed, and the mine had still failed to make a profit. At that point most of the partners had lost enough and wanted out, so in 1876 Henry and his brother and uncle gave them $20,000 for their shares and then sank another $20,000 into one more fancy furnace that also cracked and crumbled. This time, however, Henry tried patching it with the crude copper matte, and this worked so well that he rebuilt the furnace in copper. With a smelter that finally worked he started turning out five hundred tons a year of matte that was 90 percent copper. He also cut costs by hiring mostly Mexican miners at only $2 a day. Although this was much more than the peon wages paid a decade before, it was still less than half that paid to Anglos elsewhere. Much of it was only an illusion too, because he paid them in red cloth scrip, or *boletas*, good only at his Clifton store and saloon, where he charged gouging prices for everything, including watered whisky. He further cut shipping costs by using his own teams to haul out the copper and bring back goods for the stores.[44]

By late 1877 Henry had slashed mining and smelting costs to 8 cents a pound of 90 percent matte and shipping costs to 4 cents by wagon to Colorado and 1 cent more by rail to Baltimore, where he paid 2 cents a pound for refining. Thus he had cut costs to 15 cents a pound for matte, or 17 cents a pound of copper metal, which sold for 18 cents. But just as he was at last starting to make a profit, the Michigan mine owners, fearing competition from new western mines, cut copper prices to just 16 cents a pound. While other western mines temporarily shut down, Henry still fought back in the summer of 1878 by contracting for four hundred Chinese "coolies" at just over $1 a day, which brought down condemnation from miners and merchants alike. But that cut costs by a couple more cents a pound, just enough to make a thin profit again, and he boosted shipments to about forty tons a month. Over the next two years, as copper prices also crept back up a few cents, he shipped nearly a thousand tons of copper and cleared about $100,000, which all went to paying off the company's debts. It was only after the Southern Pacific Railroad moved across eastern Arizona in 1881 that he finally turned a real profit. The following year he and his relatives sold the business for $1,200,000, and he retired to New York with a fortune at last, but it was hard won.[45]

THE OPENING throughout the West of the much more profitable, but "rebellious," new silver-lead ores, requiring more expensive smelting, in turn opened additional opportunities for true investors in that rapidly burgeoning industry. Smelting was essential to the silver-lead mines as mercury and water were to the gold mines, so it too provided a much more stable base for investors that did not depend on the fortunes of a single mine or even a single mining district, and it became the source of several new fortunes. The advent of the transcontinental railroad, which made possible the economical shipping of rich ores, first spurred the establishment of giant smelters at both ends of the tracks in San Francisco and Omaha. Then rivals appeared at Salt Lake, and successful experiments in treating Gilpin County's refractory gold ores led to the establishment of the first profitable smelter in Colorado.

Thomas H. Selby, an enterprising argonaut who had become San Francisco's leading lead importer, led the way in 1867. He spent about $100,000 putting up the Selby Lead and Silver Smelting Works out on North Beach, where the sulfurous smoke from its tall stacks would usually blow out of the city and into the bay. He started by bringing in lead ore by steamer from mines along the Colorado River, and once the railroad began bringing in ore from Nevada and Utah, he had dozens of furnaces roaring night and day. His smelter was soon hailed as "the largest of the kind in the United States," and he now became the West's leading lead exporter.[46]

Rather than charging a fixed fee for smelting, Selby took a percentage of the value of the bullion, over 40 percent on the lead, half that on the silver, and 10 percent on the gold. Since the miners shipped only their high-grade ores, it was a very profitable business. He was soon producing roughly 8,000 tons of lead a year, paying $100 a ton to the miners and selling it for $180 in New York through his partner, Peter Naylor. The lead alone paid him over $500,000 a year, at least half of which must have been profit, and he made a comparable amount from silver and gold. In little over two years he had accumulated $600,000 and was elected

mayor. By 1875 he was worth well over $1,000,000 and was being talked
of as the Republican candidate for governor, but he died suddenly of
pneumonia at the age of fifty-five. His thirty-year-old son, Prentiss, who
was superintendent of the works, incorporated as the Selby Smelting and
Lead Company with a capital of $200,000 and expanded the operation to
forty furnaces with a peak production of over $20,000,000 a year, mostly
in gold and silver, at a profit of about $1,500,000 a year. It remained the
leading works on the Pacific Coast for the rest of the century, paying a
total profit of around $20,000,000, a magnificent hundredfold return, or
nearly 1 percent a day! But Prentiss, like his father, would die prematurely,
just twenty years later at the age of fifty, a deadly testament, perhaps, to
the hazards of the industry.[47]

In the East the leading gold and silver refiner, a noted old German
metallurgist, Edward S. Balbach of Newark, New Jersey, also wanted
a share in the new flood of silver-lead ores. In 1870 he sent two of his
sons, Charles and Leopold, to the Union Pacific terminus at Omaha to
establish a smelter much closer to the source. Together with local investors, they formed the Omaha Smelting and Refining Company with a
very modest capital of $60,000 and free land from the railroad, eager for
the ore hauling business. They started up with over a dozen furnaces
early in 1871 and business boomed. Within a year they were producing
$1,000,000 annually, and in five years they were up to $5,000,000 a year.
Although they originally charged fixed fees for smelting, they quickly
saw the advantages of Selby's percentage charges, and by 1899 they must
have cleared a grand profit of around $10,000,000.[48]

But the success of the new smelting works soon spurred competition
in Chicago, Kansas City, St. Louis, and Salt Lake, and the fierce fight
for ores threatened to reduce the percentages to "dangerously small
margins" of profit. Much of that competition came from the Germania
Separating and Refining Works at Salt Lake in the heart of the silver-
lead mines. Two other practical German metallurgists, Gustav Billing
and Henry Sieger, who were soon joined by a third, the more talented
Anton Eilers, erected the works at a cost of $58,000 in 1872 at the height
of the Emma excitement. Since most of the smelters put up by the mining
companies produced only a mixed silver-lead bullion that still required
further costly refining to separate the silver from the lead, Billing and
his partners were able to concentrate on refining bullion rather than

ores. Because of the added shipping costs of $25 a ton to their rivals in San Francisco and Omaha, they could also charge more per ton and make larger profits, while their customers still paid less per ton overall. Thus, they prospered from the start and remained the largest bullion processor in Utah for nearly thirty years, with an output of around $2,000,000 a year and handsome long-term returns of about $5,000,000, or nearly ninetyfold on their investment.[49]

At the same time, Colorado mining investors were still struggling with the rebellious gold, silver, and copper sulphide ores of Gilpin County. Among the most persistent experimenters was an ambitious young chemistry professor from Brown University in Rhode Island, Nathaniel P. Hill. He had given up teaching in 1864 in the hopes of making a quick fortune in a venture on the Bobtail with other Providence investors but, like the rest, had been frustrated by the resistant ores. Having burned his academic bridges behind him, he was determined to find a way to work the ore. After taking part in James Lyon's famously unsuccessful experiments, he headed for England with seventy tons of very rich ore he had bought for $7,000 to see what could be done there. The smelters of Swansea successfully recovered $19,000 from the ore, but with shipping costs of nearly $5,000 and hefty smelting charges of about $8,000, that still wasn't quite enough to repay the cost of the ore, let alone make a profit. Nonetheless, the experiment showed that with the Swansea process of recovering the gold and silver in the copper matte, the ore could be worked effectively, and it also showed him clearly that the smelters made the surest and biggest profits.[50]

So in May of 1867 Hill persuaded four Bostonians, led by James W. Converse, the son of a shoe manufacturer, to form a new private venture, the Boston and Colorado Smelting Company with a nominal capital of $200,000, and put up a Swansea-process smelter in Black Hawk, just below the Bobtail. Hill probably got an equal share for managing the operations, and he doubtless increased it when one of the partners soon sold out. Starting with a single furnace, he smelted his first ore successfully a year later, and although he filled the canyon with "villainous vapors," he was hailed for reviving the mines of Gilpin at last. But his smelting process was costly, as well as offensive, and mine owners soon complained of his "cupidity" in charging so much for smelting that they could only afford to work their richest ores. Hill's charges were indeed high, for he had also adopted, with boosted rates, the complex Swansea system of charges where the mine owner's fractional return on the gold decreased with the value of the ore, and any copper values were curiously deducted from any silver with a large penalty, leaving little or no return from either. Thus, even though he boasted a recovery of 95 percent of the assay value of the ore, he gave the mine owners only 31 percent of that, even on their rich ores yielding $120 a ton. That gave him the largest profit, a very generous 40 percent after deducting his costs of under $30 a ton, while the mine owners still had to deduct all of the costs of mining! Yet despite the howls of gouging for "immense profits," Hill took in over $1,000,000 worth of business by the summer of 1870. That paid him and his three partners roughly $400,000, or about 65 percent a year on their investment, and it made Hill, who claimed by then to be worth $100,000, the richest man in Black Hawk.[51]

To bring in even more business, he hired a very talented and experienced smelting man from Swansea, Richard Pearce, in 1872 to thoroughly overhaul the works. Leaving him in charge of the works, Hill moved with his family out of the sulfur smoke–choked canyon to Denver to oversee affairs from a more comfortable distance. While the partners increased their nominal capital to $500,000, Pearce not only remodeled but expanded the works, greatly increasing the efficiency and reducing the costs, and within just a few years they were producing $2,000,000 a year in bullion. Only part of the savings were passed on to the mining companies however, since Hill could still charge much more than Balbach and other more distant smelters, because the miners didn't have to pay the added transportation costs. Thus, even though Hill now charged only a flat smelting fee of $35 a ton, which amounted to 29 percent on the same average ore of $120 a ton, he also took discounts of 15 percent on gold and silver, and 80 percent on the copper, which still boosted his total charge to 56 percent of the bullion value. With reduced costs, his profit must have been at least 30 percent, nearly doubling his annual profits to about $700,000, an annual return of 140 percent even on their increased capital. Part of Hill's profits came from cutting down all the trees for miles around to feed his furnaces, and in 1877 the federal government brought a damage suit for $100,000 for timber illegally taken from public lands. Although his lawyer got a favorable jury verdict, he clearly needed a new fuel supply, so in 1878 he finally moved the operation to

Denver, where he could get cheap coal by rail. There he built a giant new smelting works at what he dubbed "Argo," just north of town. By then he had turned out over $14,000,000 in bullion for a total profit of at least $5,000,000, and he and his three partners had become millionaires. In January of 1879 Hill spent a bit of his fortune securing enough votes in the state assembly on the fourth ballot to succeed Chaffee as Colorado's next Republican senator.[52]

ALL THE WHILE, California's great deep mines, the Idaho and Eureka in Grass Valley and the Keystone in Amador, were steadily returning magnificent profits to their hands-on investors, and in 1871 Alvinza Hayward picked up another Amador gold mine just north of the Keystone for only $10,000. He called it the Empire, and he soon opened $100 ore at a depth of 500 feet to add another million to his fortune in hundredfold profits. He also saw that the ore shoot, like that in his great Badger, ran underground into the neighboring mine, the Pacific. But this time he simply stole all he could secretly, while he worked out the Empire. Then, seeing that the ore still went deeper into the Pacific, he sold his stock in the Empire and quietly bought up the Pacific from its unsuspecting stockholders. Once in control, the conscienceless Hayward immediately brought suit in the name of the Pacific company against the Empire for recovery of the $250,000 in ore he had stolen. By showing that the Pacific had been robbed, he won damages, and then took the Empire back by foreclosure for the damages from its impoverished shareholders. Combining the two companies as the Plymouth Consolidated, he cut in a young nephew, Walter Scott Hobart, and they worked the ore to a depth of 1,400 feet, collecting $2,300,000 more in dividends before a fire gutted the mine in 1888.[53]

The marvelous profits of these private ventures naturally excited public envy and inspired an obliging array of tempting opportunities. Among the most seductive, of course, was Alvinza Hayward's original bonanza. In 1868 Hayward had successfully unloaded the stock in his famous Amador Mining Company on a handful of San Francisco speculators, led by a "closed-fisted miser," Michael Reese, at over $300 a share, or a market value of over $1,000,000, on the strength of his own spectacular profits and expert predictions of $1,250,000 more just from ore in sight.

But after nearly exhausting the remaining ore to get back just $200 a share, the new owners added some claims and reorganized as the Amador Consolidated in 1872 to let the public in on a grossly inflated $3,000,000 in new shares. By then the mine was too well known to float in San Francisco, so Hayward generously helped entice "Witch of Wall Street" Hetty Green's not-so-successful husband, Edward, into buying a major share and getting several British friends to take the rest for a bargain price of $2,000,000 plus. The new owners, however, got back barely $350,000 in dividends over the next decade, taking out all the ore they could find. Green tried to keep looking, running on assessments, but all the other investors forfeited their shares, he ran out of money, and his wife would have nothing to do with it. After conferring once again with helpful Hayward, he finally gave up in 1886.[54]

Other British investors, headed by a wealthy London merchant, Robert McCalmont, were steered by his cousin George Coulter into forming a company to buy the famous North Star in 1869. At an offering price of $1,000,000, only one quarter in cash and the rest in stock, it could have been an exceedingly profitable investment, since it would subsequently produce over $30,000,000 more in gold. But the investors sought the advice of a rising, if suspect, young mining engineer, Henry Janin, who was cultivating a reputation of "condemning most every new scheme he was called to report on." He confidently told them that the mine was exhausted and recommended against buying it, so they turned down the best opportunity they would ever get and wound up the company. But McCalmont still wanted a gold mine, so Coulter next arranged for the purchase of the Reis brothers' equally famous Sierra Buttes for $1,125,000, half cash, half stock. Janin did give it his cautious approval, predicting a return of at least $500,000 on ore "in sight," so they bought it in 1870. It eventually produced twenty times that and paid out $1,705,000, returning their purchase money plus a 50 percent profit, but that return was spread over thirty years. Still, it was very profitable at first, paying back half the investment in the first couple of years, so they also bought the nearby Plumas-Eureka mine for $1,400,000, and it produced another $9,000,000, with $2,700,000 in dividends for a 90 percent profit over the same period. Such rates of return amounted to only 2 to 3 percent per annum, however, which was less than bank rates and at much higher risk. Still they did get some profit, which was far more than most mining

investors, who didn't even get their money back. But in their enthusiasm
for even more profits, they pushed their luck too far and ended up losing
nearly all their future profits.[55]

For Janin now seduced them with what he claimed was "the most
splendid combination ever offered to the English public." It was, in fact,
an unproven little mine enticingly called the Original Amador, which
the Keystone's superintendent, John Faull, together with George Hearst
and others had picked up for a trifle. Janin, however, heaped lavish praise
on it and predicted "without any doubt or hesitation" annual profits of
25 to 35 percent, which would place it right alongside Hayward's real
Amador. So despite pointed criticism by Marmaduke Sampson of the
London *Times*, in 1872 McCalmont and his friends, still heady over their
early dividends from Sierra Buttes, eagerly floated yet another venture,
the London & California Mining Company, to buy this new wonder
for an outrageous $1,400,000, all cash! But by the end of the year they
found that the mine was running at a big loss, not big profits. The big
ore reserves that Janin had assured them of were also nonexistent, and
the hoodwinked investors angrily denounced all of his claims as a "tis-
sue of falsehoods." Janin tried to excuse himself by claiming that he had
only taken Faull's word for it, but the investors could only conclude that
he had taken something else from Faull as well. They were sure that a
"more disgraceful swindle was never concocted," but then, Emma had
not yet come to fruition. Still, they kept looking for more ore for another
eight years and did find a small pocket, but all they ever got back was
2 cents on the dollar before the company went into debt for a total loss
of $1,600,000.[56]

THE SUCCESS OF deep mining had, of course, also inspired Mariposa
creditor Mark Brumagim, in 1873, to resurrect that company once again
to get his money back and much more. But, as a cynical San Francisco
editor noted, it had so far produced little more than "private rascality,
Wall-Street jobbery and hope long deferred." Brumagim carried on the
tradition in an uneasy partnership with a crafty, sixty-seven-year-old
speculator and millionaire banker, Eugene Kelly, a former San Francisco
gold rush merchant and William Ralston's early backer and banking partner, who later moved to New York to handle their business in the East.

They magically metamorphosed that ugly, ravaged, and debt-ridden old hulk of a caterpillar into a beautifully alluring new butterfly of promise. Consolidating all of its outstanding debts for $800,000, they formed the Mariposa Land and Mining Company to push another round of bonds and an outrageous offering of $15,000,000 in shares on a whole new generation of eastern innocents! With glowing visions of deep new bonanzas from that ever-helpful Yale professor, Benjamin Silliman Jr., fresh from the Emma mine, Brumagin unveiled a grand plan, à la Sutro, to drive a three-mile-long tunnel straight from the mill to tap the riches of a great, deep "Mother Lode" in the old Pine Tree and Josephine mines. Using newly introduced dynamite, power drills, and Chinese labor, Brumagin further planned to extract all this prospective ore cheaply, to pay dividends at last. Some $5,000,000 in new shares were unloaded at about 20 percent of par for a quick profit of $1,000,000, while another $1,000,000 in shares were exchanged for those of the old company which were selling at only a few per penny. [57]

The company, however, was left without funds to push the great tunnel. So in 1874 Brumagin reincorporated in California in order to levy stock assessments, not allowed under the New York charter. He then dutifully drove his tunnel on assessments for several years, but he never struck the envisioned paying rock. Kelly, who still held a third of the stock, tried to stop the levies, but failing that, he seized the mines in 1878, foreclosing on a $250,000 mortgage that Brumagim claimed was fraudulent. They battled in the courts for years over title to the Mariposa until Kelly finally won out in 1882. All the while the stock had drifted on at just a couple percent of par, as the faithful shareholders paid out its market value several times over in assessments totaling $1,400,000 and waited in vain for their promised dividends.[58]

The only other western mining venture on the New York Stock Exchange at that time, however, the equally infamous Quicksilver Mining Company, was finally on the verge of actually paying a dividend. But that created even worse problems. The quicksilver monopoly was temporarily broken in 1873, when market prices rose to over $1 a pound and John Redington decided to sell his own production. At the same time Eugene Kelly, who was also the Quicksilver company's treasurer, and a clique of the directors seized the opportunity to break briefly into the sales scheme. They sold themselves the next year's production for a much more generous 65 cents a pound, and they gave the company $1.10 a pound in 1874, when the market price reached $1.55. So, even though ore values had dropped and production costs had climbed to 78 cents a pound, the company was actually earning enough to be able to pay a dividend on its preferred stock for the first time. But when the word leaked out, the greedy descended. An old Wall Street sharp, "Uncle" Daniel Drew, and a few friends "beared" the stock, bought control, and elected themselves directors and Drew president. Then, over the protest of the minority shareholders, they proposed to convert their own common stock to preferred for the same $5 a share that the faithful had paid in the dark days of 1870 and promptly get back a dividend of $7 a share! That was just too transparent, even for the otherwise clueless Quicksilver shareholders, and a couple of them got injunctions and brought suits that dragged on for nearly a decade, halting all possible dividends.[59]

Once again the higher quicksilver prices spurred the opening of new mines, and this time it also excited a brief flurry of stock speculation as well. Over fifty new quicksilver mining companies were organized within a year after the prices went well over $1 a pound in the spring of 1874. Every bit of red dirt anywhere in California became a prospective cinnabar bonanza, and every mining promoter from George Roberts to George Hearst floated a company, or two, to capitalize on the excitement, offering more than $250,000,000 in slippery shares to "quicksilver crazy" investors. But a few of the companies actually had some workable ore and went into heavy production. By the spring of 1875 the flood of quicksilver from the upstarts had taken over half the production. But Darius Mills and Thomas Bell, who had inherited the interests of his late partner, William Barron, were anxious to regain control of the sales and production. So they broke the market, trying to put both Kelly and the rival producers out of business by halting purchases at their Comstock mills and slashing quicksilver prices from $1.55 to 65 cents a pound in just two months. That price was close to cost for the Quicksilver company, wiping out Kelly's margin, so he gave up the sales. Then Samuel Butterworth's nephew, John Randol, obligingly gave the business back to Mills and Bell at 58 cents a pound and found ways to trim the costs just enough. The stockholders, of course, couldn't expect profits anyway until the litigation ended. Mills also bought control of some of the more productive rivals, so despite declining ore in the New

Almaden the Bank Ring maintained control of the quicksilver market, but the monopoly profits were a thing of the past, and they never got the price above 65 cents again.[60]

Bell and Mills also enjoyed the profits of the New Idria, which had grown to $250,000 or more a year. But those profits were also an irresistible spur to the irrepressible William McGarrahan. Rounding up more recruits with a fresh $5,000,000 batch of shares in his newly formed California Quicksilver Company, he launched another campaign in 1869 to push the Panoche Grande onto the New Idria. This time he claimed that Congress should give him title on the wonderfully specious argument that since he had bought Vicente Gómez's title in good faith, he was entitled to it, irrespective of any fraud on Pacificus Ord's part, and seemingly irrespective of whether Gómez had any title to sell! When that failed, McGarrahan cast himself as a poor victim of trickery, cheated out of his rightful dues by agents of the powerful Bank Ring who had stopped Lincoln from signing over the title patent to him. He actually succeeded in getting a District of Columbia judge to order the secretary of the interior to issue the patent to him. But the secretary, Jacob Cox, instead wanted to give title to the New Idria company. Cox finally resigned in protest after President Grant ordered him to do neither until Congress also had its say in the matter. McGarrahan rejoiced that "Cox is dead and damned," but the New York *World* damned Grant instead for using "presidential influence in abetting a notorious fraud." McGarrahan's forces then lashed out furiously in the House, claiming that for the last thirteen years the government had been the "cats-paw" of the New Idria company, whose "aiders and abettors" were "blistered all over with corruption." They carried the House with a resolution demanding that McGarrahan be given the title, but Bell's and Mills's men blocked it in the Senate.[61]

McGarrahan, now thoroughly transformed into a poor folk hero "trampled on by wealth and power," soldiered on year after year, Congress after Congress. In 1878 the Senate Committee on Public Lands held a lengthy hearing, before which McGarrahan wailed of "corrupt combinations forming on every side to destroy me and break down my title to the property," and laid it all on a vast conspiracy of perjury and deception by the New Idria owners and a long list of corrupted officials. Bell shot back in kind. Once again imitating Robert Walker, Bell had bought the confidential letters and records of the Panoche Grande company from the estate of its deceased secretary. They were a "bombshell" of self-incrimination by McGarrahan, showing in his own words his own conspiracy of perjury and deception. For maximum impact Bell also published over four hundred pages of them as *The History of the McGarrahan Claim, As Written by Himself*, exposing McGarrahan and his own clique of prominent politicians who had taken stock bribes for their support. But even that couldn't stop McGarrahan. He and Bell would go on hurling charges of fraud and corruption at one another until they died. Many would soon concur with an amused congressman who concluded, "They both accuse each other of fraud, and I believe them both." As the battle dragged on and on, another amused observer suggested that Charles Dickens would never have chosen *Jardyce vs. Jardyce* for his epic of legal corruption if he had only known of the McGarrahan claim. And long before it ended, Bret Harte took up the challenge, elevating the case to pure fiction in his *Story of a Mine*.[62]

THE COMSTOCK FEVER had also come raging back in the spring of 1871 after the discovery of a dazzling new bonanza deep in the Crown Point mine near the center of the lode at Gold Hill. As the great wealth of the ore body became apparent, shares in the mine skyrocketed from a low of $2 in November of 1870 to $310 the following May, a spectacular rise of more than 150-fold in just six months, and they eventually rose nearly a thousandfold to $1,900 a year later, before the fever finally broke! Shares in the adjacent Belcher mine, into which the bonanza ore extended, staged an equally spectacular gain, while the shares in other mines along the lode were carried up a factor of three or so in the excitement. During the next half a dozen years, the bonanza produced $58,000,000 in bullion, out of which the Crown Point paid dividends of nearly $12,000,000 and the Belcher over $15,000,000.[63]

But the trusting stockholders of the Crown Point, who had paid patiently and dearly for the search for ore through stock assessments amounting to over $50 a share and totaling $630,000, got very little of the bounty in dividends, which totaled nearly $900 a share. For their genial but slippery superintendent, John Percival Jones, betrayed them. Jones, a forty-year-old, out-of-work politician, had been appointed superintendent of the Crown Point through the influence of his brother-in-law, Alvinza

Hayward, a major depositor in Darius Mills's powerful new bank and one of the few outside investors in the bank's Mill Ring. When he found the first indications of the bonanza in the fall of 1870, Jones was seized with greed and saw a chance to grab a fortune. The Crown Point was nearly out of money, and Mills, Ralston, and Sharon were ready to suspend work, so Jones, concealing the magnitude of the discovery from the shareholders and the directors, went to San Francisco and gave them just enough encouraging news to get them to levy one more assessment of $3.50 a share. The assessment allowed him to continue work and at the same time helped drive share prices down to a rock-bottom $2. Then he conspired with his brother-in-law to quietly buy up a controlling interest in the depressed stock before the news of its richness was made public. To further help depress the stock and to cut out brokers who had been carrying speculative shares for him, Jones also sold those shares, telling the brokers he only did so because his young son was sick in the East. This deception also worked perfectly, for the brokers, deciding that "Jones' sick child" was actually the Crown Point, started selling shares short and lost heavily as Hayward's quiet but heavy purchases caused the price to rise.[64]

Hayward succeeded in buying more than half the 12,000 shares for less than $30,000 and got control before Jones finally announced the "discovery" in March. Despite his duplicity, Jones was actually seen as a hero by many for having cheated the hated Mill Ring. But Mills, Ralston, and Sharon personally were not that badly cheated, for Hayward had cut a deal with Sharon, effectively swapping the Ring's shares in the Crown Point for his shares in the Belcher, which actually gave them control of what turned out to be the larger part of the bonanza, and the Ring made millions from it. Jones and Hayward, however, did hurt the Ring by contracting with themselves to work all of the Crown Point ore through their own new Nevada Mill and Mining Company, formed in imitation of the Ring to rake in even greater profits. They also wrested control of another old Comstock mine, the Savage, away from the Ring and ran it to help feed their own mills for personal profit just like the Ring, crushing its rock at a loss to shareholders of nearly $10 a ton in assessments, which totaled $640,000 in just two years. In all they pocketed an extra $2,000,000 from their milling scam.[65]

Still, even this was not enough. Jones and Hayward didn't want to just wait for dividends and milling fees when the market showed such promise for much quicker gains from manipulation. So they worked the market up to let the public back in and added some more bait. The Savage mine had held the bigger part of the early bonanza in the great Gould & Curry and in its prime had actually paid $4,500,000 in dividends. Many, then, thought it might still hold other hidden treasures and in February of 1872 Jones and Hayward exploited that belief. Cutting into a small pocket of rich ore, they temporarily confined the miners underground at bonus pay so no details could leak out and started booming the stock on rumors that they had struck another huge bonanza that would rival the Crown Point. Within three months they had worked the Savage stock from $62 a share to $725 and Crown Point up to its high of $1,900, putting market "values" of $11,600,000 and $22,800,000 on the mines. Savage "like a whirlpool drew almost everybody into the vortex of speculation." Then in early May the bubble burst after Sharon, still stung by their betrayal in the Crown Point, "beared" the market by unloading $5,000,000 in shares for a hefty profit and then unleashing charges that Jones had previously started a deadly fire in another mine just to "bear" the stock. Savage shares plummeted to $110 and Crown Point to $830, as over $20,000,000, or two-thirds of their market value, evaporated, and all the other Comstocks came tumbling after them for a total loss of $70,000,000! But Jones and Hayward had still made another killing, unloading their Savage shares plus some Crown Point at a profit of close to $6,000,000.[66]

After the crash, a *Chronicle* reporter sought out the principals. Jones refused to talk, but a haughty Hayward simply blamed it all on "the immense volume of worthless wildcat stocks that flooded the market," and admonished, "It will teach the people to keep clear of such trash." A pious Sharon, on the other hand, blasted Jones and Hayward for advising everyone to buy into a mine "they knew all the time to be worthless." And he concluded, "The whole thing required a prodigious amount of lying; but it succeeded, and now everybody is ruined. Well, it may be all right, but I don't make my money in that way. I don't rob poor people of their little savings by lying and tricking."[67]

The outraged editor of the *Alta California* zeroed in to complain that "one of the greatest evils in mine management is the almost unlimited power of the superintendent to make or unmake the market value of a mine. Where there is a disposition to work the market rather than the mine, a collusion between trustees here and a superintendent on the spot causes the profits or expenses of the mine to rise and fall exactly in accordance with the stock account of the operators." And he called upon the California Legislature to make "false statements and malfeasance on the part of those officers punishable as felony." But one cynic, reviewing the scandalous history of Comstock speculation, later concluded that

> to insist on the faithful performance of duty by mine trustees and superintendents would be the death-knell to stock deals, and it is fair to assume that the majority of shareholders in the Comstock mines did not wish this conclusion. . . . A well-managed stock 'deal' was as acceptable to most holders as an actual development of ore. Stock deals were naturally easier to produce than ore-bodies, so that the gambling public was commonly given chaff instead of wheat. If buyers and sellers were willing to deal in counterfeit coin they had only themselves to blame when their riches turned into ashes in their hands.

Thus, Jones was punished for his crimes the following year by using over $500,000 of the loot to buy enough votes in the Nevada Legislature to get thirty years as a Republican senator. Two years later Sharon bought Nevada's other Republican seat for a reported $800,000, replacing an embarrassed William Stewart.[68]

At the height of the fever, a whole rash of new swindles burst forth, culminating in the most outrageous of all, the "Great Diamond Swindle," which far surpassed the "Savage Deal" and became the most widely known and least understood of western mining frauds. It seems to have began simply as a joke, as an impromptu sideshow of a wildcat promotion, the Pyramid Range Silver Mountain Company, that George Roberts, together with Asbury Harpending, was trying to sell in London. During the summer of 1870 they had worked up an excitement around a "mountain of silver" in the Burro Mountains in the southwest corner of New Mexico, modestly touted as "the greatest treasure ever discovered on the continent!" Before leaving for London that fall, Harpending took a visiting British "mining expert," Henry Morgan, out to inspect their wonderful "treasure" and pick up some "picture rock" to show potential investors. Accompanying them were a couple of employees—a colorful old miner, Philip Arnold, and a token director, James B. Cooper, who was also the bookkeeper for the newly formed Diamond Drill Company in San Francisco, which had just begun promoting those great new probes for hidden bonanzas in the West. For amusement Cooper had brought along a few small industrial diamonds used in the drills, which he and Arnold delighted in flashing before Morgan and the local rubes, claiming they had just picked them up nearby. Harpending, who must also have gotten a quiet laugh, apparently encouraged them, hoping to create a little extra excitement for the sale in London, where the newly discovered South African diamond mines were just starting to create a stir. But when he and Morgan got to London, they bungled the sale, putting out too fanciful a prospectus and failing to make connections with the powerful Albert Grant, so the scheme was roundly condemned by Grant's friend, the influential Marmaduke Sampson of the *Times*.[69]

In the meantime Arnold and Cooper had returned to San Francisco when Harpending and Morgan headed on to London, and they also flashed their diamonds to an unsuspecting Roberts, who was fully taken in by the hoax. After quietly confirming that the stones were real, an exuberant Roberts made grand plans to have the placer mining law amended to include "precious stones," so that a company could locate 160-acre claims at only $2.50 an acre and they could form as many companies as they pleased. But when Roberts started pressing for a secret expedition to stake the claims, Cooper decided that he wanted out. So

Arnold told Roberts that the diamonds had really been discovered by his cousin, John Slack, Roberts's Pyramid Range superintendent, and that only he knew where they came from. When Harpending returned from London in March, however, Roberts learned that it was just a joke. But it was too late. Roberts, now in his early forties, frequently overextended, in debt and desperately trying everything, even swamp land reclamation, saw in the diamonds a chance to finally make it big. So despite the risk, he decided to push the scheme for all it was worth as a gigantic stock swindle. He really couldn't lose, for in the end he could honestly say that he too had been fooled—if only briefly.[70]

First, to help finance the deal, Roberts took in his Mineral Hill associate, William Lent, and his cohort, Gen. George S. Dodge, selling them "Slack's interest," a quarter share, for $50,000 each in June. He and Harpending also conned into the deal that slippery, but still gullible, shark and their biggest creditor, William Ralston. Then Roberts upgraded the bait, secretly dispatching Arnold and Slack to London to buy $19,440 worth of uncut South African diamonds. When they returned with the stones, Roberts sent Harpending, Lent, and Dodge to New York, to show a bag of diamonds to the shrewd lawyer and "railway wrecker," Samuel Barlow, jeweler Charles Tiffany, and others. Tiffany all too hastily appraised the stones at $150,000 and they agreed to form a $10,000,000 venture, subsequently incorporated as the San Francisco and New York Mining and Commercial Company. The New Yorkers would get an option on a quarter of the stock at half of par, which Roberts and his friends would "buy" from Arnold. Lent and Dodge would keep their quarter, and Roberts, Harpending, and Ralston would split the remaining half. No one seems to have asked how Roberts and his friends got a half interest to start with. But this was only the opener, for as Roberts had planned they would then spin off eighteen additional companies, each with a 160-acre claim and $10,000,000 in shares, plus a British subsidiary, for a total stock offering of $200,000,000! Out of all this a paltry $100,000 in stock was later "paid for securing the passage of the necessary legislation in Washington" after California Republican Congressman Aaron Sargent amended the new Mining Law of May 10, 1872, to allow placer claims on "all valuable minerals." Barlow also got $500,000 in shares for his "services."[71]

Before closing the deal, however, the New Yorkers wanted to send an expert to confirm the richness of the diamond fields. They got Henry

Janin, relying on his hard-nosed reputation. But he was sold even before he started. Impressed with Tiffany's appraisal of the stones and Harpending's and Lent's large investments, Janin eagerly accepted not just a fee of $2,500 for his report but stock options on 3,000 shares and a future job as the company's superintendent. Late in May of 1872, right after the new mining law went into effect, Janin—together with Harpending, his old friend Alfred Rubery, and Dodge—went to St. Louis. There they waited a couple of weeks for Arnold and Slack to arrive and lead the way to the secret site that they had just salted with diamonds and rubies in the desolate northwest corner of Colorado, right across the line south of Black Butte Station on the Union Pacific Railroad in Wyoming. Janin was easily conned, spending only two days examining the ground and foolishly letting Arnold be his assistant and dig out most of the diamonds. Then they hurriedly staked out 3,000 acres of placer claims and headed back to New York. Janin was "wildly enthusiastic," claiming the ground was worth the "Crown Point and Belcher combined" and boasting that with just twenty-five men he could "take out gems worth at least $1,000,000 a month." On August first Roberts held a press conference in San Francisco to announce the discovery of the "New Golconda" and release Janin's glowing report. The diamond bubble took off instantly and grew steadily throughout the summer of 1872, amid wild speculation over where the fabulous but mysterious diamond fields might be located.[72]

The San Francisco and New York company was incorporated at the end of July with Lent as president and a board of directors decorated with agents of Baron Lionel Rothschild and other British capitalists, plus California Democratic governor and senator Milton S. Latham and presidential candidate and Civil War general George B. McClellan. Ralston was the treasurer, and still seemingly clueless, he planned for this company, which would handle all of the diamond sales, to operate like the Mill Ring, controlling and manipulating all of its subsidiary mining companies for its benefit. So he insisted on keeping control of just over half of the 100,000 shares, and he offered twenty-five of his biggest depositors up to 2,000 shares each at a bargain $40 a share. The depositors quickly picked up nearly 34,000 shares, and the remaining 16,000 shares, offered to the public, were snapped up in a day. John Milton Hay, New York *Tribune* night editor and later President McKinley's secretary of state, begged Barlow to sell him a hundred shares, claiming that Janin had convinced him to "sell my shirt and pawn my Bible for stock," and like a true speculator he added "If I lose it all, I will be grateful to you. If I make $6,000, I will sue you for the other $9,000." Even the editor of the staid old *Alta California* giddily wrote that the "public was heaved as if a financial earthquake had taken hold of the pillars of adventure and opened an Aladdin's cave for the daring."[73]

But there were still doubters in San Francisco who voiced suspicions about Roberts, Harpending, and Lent, suggesting that it just might be "the biggest job that was ever put up." The New York *Times* editor in fact concluded that the "whole thing was a hoax" put up by a "few reckless and unprincipled speculators." In response to such charges Roberts personally took a new party of "experts" to the diamond fields to "confirm" Janin's report and officially survey the claims. With Alfred Rubery as a "guide," Roberts led the party around the badlands for over a month to throw off trackers, confuse the participants, and kill time till winter drew near before he finally reached the spot, picked up some more diamonds, and officially surveyed their 3,000 acres. He returned triumphant early in October, and the San Francisco *Chronicle* spread headlines proclaiming, "The Tales of the Arabian Nights Surpassed," telling of "Ant Hills Built of Precious Stones!" Crowds thronged the Bank of California to see the new diamonds, and the California Academy of Sciences met to examine bottles of the diamond-bearing earth. With the demand for shares swelling again and the original company fully subscribed, outside sharps also joined in with half a dozen new diamond companies to get a bit of the action in the New Golconda, and Roberts began forming the subsidiaries, starting with one appropriately called the First Choice Diamond Mining Company, on November 6. Thus the final play began. Roberts had set up the perfect stall, announcing that because of the severity of the winter, no further work could be done at the diamond mines until spring. It all seemed perfect. That gave them at least four months to unload $200,000,000 in shares on the public before anyone could get back in to discover that the diamonds were only "salt." Then, of course, Roberts and Harpending could innocently cry that they were just victims like all the rest and point the finger at Arnold and Slack, who had collected their final payoff late in October and could disappear with roughly $500,000.[74]

But suddenly, on the evening of November 10, Clarence King, who was in charge of the U.S. Geological Survey's work on the Fortieth Parallel in

northern Colorado, arrived in San Francisco and spent the night conferring with Henry Janin. In the morning King told stunned directors of
the diamond company that the diamond fields were a "fraud." Some of
King's men, finishing their work for the season, having just happened
to return from Wyoming on the same train with members of Roberts's
party, had concluded that the diamond fields must be in their area. When
they told King, he couldn't believe that he had missed "so marvelous a
deposit" and promptly headed back to investigate. He found the survey
stakes, and despite the snow and temperatures of fifteen below, he soon
discovered that the ground had been salted. The directors secretly sent
a committee together with King and Janin to confirm the fraud. Roberts
and Harpending immediately sent Rubery to quietly grab back $200,000
from Arnold. When the committee returned two weeks later, the company finally revealed the location of their diamond fields and publicly
denounced them as a swindle! The *Chronicle* headlines blazed "Salted!
The Diamond Bubble Bursts. The Money Kings of California Taken In
and Done For." Janin also declared himself the victim of an "ingenious
and infamous fraud." Investors were shaken again, and to protect himself
and his bank, Ralston returned all of the big depositors' investments,
since most were also his own biggest creditors. Early in December Lent
wrote the report of the company's investigative committee, appointed
to "ferret out and punish" the swindlers, and laid the blame solely on
Arnold and Slack. Lent then sued Arnold, who had returned to his
home in Kentucky, for $350,000, but settled out of court for $150,000 with
a promise of no further prosecution.[75]

Edwin Godkin of *The Nation* declared the fraud "the most ingenious
and masterly ever perpetrated among men of intelligence and wealth,"
and he concluded that King's "timely exposure" had saved the entire
country from a "diamond fever . . . which can only be likened to the Mississippi Scheme and the South Sea Bubble." The San Francisco *Chronicle*
fully concurred: "We have escaped, thanks to God and Clarence King,
a great financial calamity. A hundred and eight millions of diamond
stock would have been put upon the market, and before Spring would
have been sold in vast quantities to our people."[76]

As soon as the scheme was exposed, Roberts and Harpending, of
course, claimed that they too were innocent victims and had suffered
great losses like the others, although many doubted them. Most outsiders

nonetheless smiled at the thought of San Francisco's "smartest operators" from Ralston on down being taken in by a backwoods hoaxer like Arnold. But historian Hubert Howe Bancroft, who knew most of the participants, was much less generous. He suggested that only one or two of the "original capitalists—Harpending, Roberts, Lent, Dodge, etc." may have been innocent victims, and he dismissed Arnold and Slack as the "willing tools of the swindlers." Harpending, however, seems to have gotten the last laugh many years later. After all of his cohorts and critics were dead, he published his "inside" story of "The Great Diamond Hoax," which once again made Arnold and Slack the sole perpetrators and became the popularly accepted version.[77]

Ironically there was one other—belated and quite unexpected—victim of the diamond swindle, Marmaduke Sampson of the London *Times*. Before the swindle was officially exposed, Sampson had come down hard on Roberts's and Harpending's roles in the scheme and Rubery's association with them, and he brought the scheme full circle by reminding his readers that they had concocted the Pyramid Range company. Seeking double revenge, Harpending convinced the hapless Rubery to sue not only Sampson for libel but also Albert Grant, accusing him of bribing Sampson to do so. At the trial in the winter of 1874–75, Sampson's attorneys brought out a lot of damaging new evidence on the diamond swindle, but they failed to prove that Rubery knew it was a fraud. Rubery's attorneys also failed to show that Grant had bribed Sampson to libel Rubery, but they shocked the jury with Grant's own testimony that the trusted Sampson had actually taken nearly $40,000 from him on other occasions to "recoup him for losses" in some of Grant's schemes. The indignant jury exonerated Grant in this case but convicted Sampson of libel. Although Rubery was awarded only $2,500, Sampson was forced to resign in disgrace after thirty years with the *Times*, and he died a year later.[78]

WHILE THE MINING MANIA RAGED, shares of mining companies throughout the West had surged upward nearly tenfold in just a year and a half. The total market price of the thirteen leading Comstock companies alone jumped from $4,746,000 in late 1870 to $85,340,000 in 1872. Shares, in fact, soared to such heights that the Crown Point had to split its stock by more than eight to one, and the Belcher ten to one, in order not to price most of the speculative crowd out of the market. Not only had shares "multiplied like leaves in autumn," so too did new mining companies. Some 1,200 new California and Nevada companies were floated during 1869–73, with a lofty paper capitalization totaling around $3,000,000,000. The volume of trade on the San Francisco Stock Exchange swelled to over $35,000,000 a month at the peak in April 1872, twice that for the entire boom year of 1863! A rival, the California Stock Exchange, opened in February to let more brokers in on the action, and the total trades on both exchanges during the three boom years topped $450,000,000.[79]

Once again the market was so glutted with shares that even the ravenous western investors and speculators must have been stuffed after gobbling down no more than $150,000,000 in cut-rate wildcats and thin-margin speculations in the frenzy from 1869 to 1873. Their gluttonous British cousins also consumed another $25,000,000 out of a $65,000,000 offering in only seemingly more savory ventures. For the westerners, still mostly Californians and Nevadans now swelled to about 700,000 souls, that amounted to an average diet of roughly $50 a year in mining stocks for every man, woman, and child over the four years. That was about five times the later mania average for easterners, but western wages were also three times higher, so just divided among about 400,000 adults it was still less than 7 percent of typical incomes. For British investors, mostly Londoners who numbered just over 3,000,000, their per capita investment was much more modest, averaging under $3 per year. Although outside investors in the dozen or so real bonanzas actually made a profit from dividends of around $30,000,000, most recovered little or nothing from their stock and lost probably 90 percent of the $175,000,000 they had ventured. Most insiders, of course, did quite well from both the profits of the mines and the losses of outsiders. About a fifth of that money was raked in just by the major Comstock sharks of the hour—Hayward, Jones, Sharon and his Mill Ring—from their stock deals and milling scams. Roughly as much was doubtless also divided among a hundred or more other sharks, particularly by Park, Roberts, and Chaffee from the British. Hundreds of brokers and at least a thousand lesser wildcat breeders probably took a like share, and finally about 10,000 hardrock miners and surface hands, about a third of them all, undoubtedly got

about $30,000,000 in wages paid out of treasury stock sales and assess-
ments in those mines that at least tried to find ore.[80]

Such dismal showings for investors in these and many other western
mines, of course, lay not in the mines themselves, but in their outrageous
overcapitalization by rapacious promoters and brokers. "Plenty of good
mines have been floated," Warren Ewer noted, which "if worked economically by private individuals, would have proved good investments. But
when these properties are first bonded for a small sum to parties who
put them on the market through other individuals, all of whom want to
make a large sum out of them, by the time stockholders buy their shares
they have to pay about ten times as much as they are worth." And at
that price there was no way they could ever pay a profit! But even as
Ewer wrote, a deep new shaft had just tapped into the Comstock's biggest bonanza, and the mania raged anew as capitalizations and market
prices soared higher yet.[81]

5

THE BIG BONANZA BOOM

THE ENORMOUS OUTPOURING of silver from the Comstock and other western mines enriched San Francisco and the nation, but it also caused a steady decline in silver prices. Silver had fallen from a Civil War high of nearly $3 an ounce to less than half that by the beginning of 1873, greatly cutting potential mining profits. Then Congress further spurred the decline by demonetizing silver in February. This in turn further accelerated the decline in silver prices, cutting profits in western mines. The great financial panic in the fall of that year also made eastern investors more conservative and less inclined toward mining ventures. Yet ironically, as eastern markets sank in depression, the western markets rose to new heights. For just a couple of weeks after the demonetization, a vast new ore body was discovered on the Comstock that proved to be its biggest bonanza of all, flooding San Francisco with its wealth and launching the biggest mining boom yet. Even the falling silver prices had little effect on the Comstock, because the deeper new ore bodies were increasingly richer in gold.

The "Big Bonanza," as it quickly came to be known, was for a time "the richest known mine in the world." It was also the greatest triumph of the assessment system, and its memory would inspire hopeful investors in countless other companies to keep on paying every new assessment for years to come in the hope that lightning just might strike again. The first glimpse of the Big Bonanza came on March 1, 1873, when paying rock was struck on the 1,200-foot level of the Consolidated Virginia mine, just south of the great Ophir in what was considered barren ground, though it was still diligently explored by the faithful on assessments. In contrast to the secrecy and duplicity that had surrounded the earlier Crown Point discovery, the Con Virginia discoverers announced their strike immediately, because they already held about three-quarters of the stock. The stock briefly doubled from $40 to $80 a share before it fell back under two more assessments, while further exploration slowly revealed the enormity of the bonanza. Their initial openness in revealing their discovery did allow others, often at their urging, to buy in low and make modest fortunes on the rise of the market, and it won them praise from an otherwise cynical press. "This is a bright spot of good honest charity in a desert of fraud and subtle iniquity," the San Francisco *Chronicle* cooed.[1]

The Big Bonanza paid its first dividend in May of 1874, and dividends continued every month with only a brief lapse until the summer of 1878. By then it had produced a magnificent $104,000,000 in bullion and had paid out an unprecedented $70,000,000 in dividends, a record not surpassed until the twentieth century. Most of this fortune plus fat insider profits went to the Comstock's newly crowned "bonanza kings," who would be both hailed for their generosity and assailed for their larceny. They were four Irish immigrants in their forties who had come to California in the gold rush, and they had first joined together in 1868 to work a much smaller ore body in the Hale & Norcross mine, but they were otherwise quite disparate men. Two were veteran miners, the modest and popular, but shrewd, John W. Mackay, who held the largest interest, a three-eighths share, and was their workhorse in developing the bonanza, and the gregarious, ever-calculating poser James G. Fair, known as "Slippery Jim," who held a quarter interest and claimed credit for discovering the bonanza. The others were a pair of San Francisco saloon keepers turned speculators, who split the remaining three-eighths, the forceful, seemingly heartless shark James C. Flood, who handled their stock deals and became their financier and lightning rod for public wrath, and his quiet and modest partner, William S. O'Brien, who simply stayed in the background and soon died, in 1878.[2]

Unlike John Percival Jones and his brother-in-law, these four, soon known as "the Firm," had bought up a majority share of the Con Virginia well over a year before they struck bonanza ore. They had personally paid a major part of the assessments they spent in the search for that ore, and they were repaid beyond their wildest expectations. The Consolidated Virginia Mining Company had been formed in 1867 by the quicksilver kings, William Barron and Thomas Bell, from a merger of small claims totaling 1,160 feet, issuing shares at ten per foot with an optimistic par value

JOHN W. MACKAY
Henry Hall, *America's Successful
Men of Affairs*, 1895

of $300. But by the end of 1871 they had expended over $150,000 in assessments without success, and the shares were selling for a paltry $6 to $10.
At that time the Firm bought up a majority interest for under $50,000 and
took control in January of 1872. Mackay and Fair then began a vigorous
exploration, and after levying another $176,000 in assessments, they first
hit bonanza ore a year later. They levied two more assessments totaling
$142,000 and picked up all the additional stock they cheaply could before
they tapped into the bonanza with a new shaft in October of 1873. Then
they began taking out ore at a rate of $250,000 a month, and they eventu-
ally reached a phenomenal $3,600,000 a month in March of 1876! But even
this was no longer enough, as the full flush of greed overtook them too.
In August of 1874 they too formed their own mill ring, the Pacific Mill &
Mining Company, which would rake in additional personal profits of at
least $9,000,000; they also formed their own bank, the Nevada Bank, to
buy the bullion at added discount; and they soon succumbed to most of
the fraudulent practices and manipulations as the old Ring. As a final
indictment, Mackay even got so far into the role that he considered buying
Emma conspirator Baron Albert Grant's mansion in London.[3]

As the shares kept climbing out of reach of many eager speculators, the Firm obligingly split their 11,600 shares again and again and again, reaching nearly a hundred to one just after the price topped out. In the meantime they had also picked up an additional 150 feet in a couple of adjacent claims to the north, which as trustees they sold to the company for large personal profits in stock. Then they split off the northern 600 feet as the California Mining Company in December 1873, increasing the total bonanza shares to 216,000 and distributing 63,000 shares of the California as a stock dividend to Con Virginia shareholders at seven and a half to one. Even all these diluted shares still skyrocketed to a peak of $750 for Con Virginia and $800 for California in January of 1875, so they soon split them again for a total of 1,080,000 shares at $100 par. At that fleeting peak the market price was equivalent to over $11,000 apiece for their original Con Virginia shares, counting the value of their California dividend shares, an incredible rise of 320-fold from their $35 a share cost plus assessments in just three years! As the market value of the two mines rocketed to $167,000,000, the Firm eagerly sold as much of their stock as the market would bear for a quick profit of easily $20,000,000, possibly much more.[4]

Even as they unloaded, however, like the Bank Ring before them, they still kept control of the mines by listing all their shares in the "street name" of James Coffin, one of the young clerks in their bank, who simply signed them off to unnamed bearers, while the Firm still held his proxy. Then they encouraged buyers to also keep the shares in Coffin's name rather than transferring them to their own on the books, by making it very easy to collect dividends on shares listed in his name from their bank and its agents across the country, and very inconvenient to collect on shares in any other name, which would be paid only at their bank in San Francisco. Thus they sold the shares without voting rights, and the resulting anonymity in sales also made it impossible for anyone to find out just how many shares they still held—shares that they unloaded heavily over the next four years as the diamond drill cores secretly revealed the end of the bonanza ore. So by the time the Big Bonanza finally neared exhaustion in 1879, the Firm had probably taken over half of the $71,000,000 in dividends from the two mines, more than $10,000,000 in milling profits and over $20,000,000 from stock sales, giving them a total of at least $70,000,000, all from an initial investment of about $200,000. That was a truly fabulous profit of 35,000 percent in seven years, or an average of 13 percent a day. And that, of course, was just their legitimate profits! Their total wealth was eventually assessed at over $140,000,000. Mackay's fortune alone amounted to about $65,000,000, Fair's $42,000,000, and Flood's $37,000,000. Each surpassed all the other bonanza barons, as well as Stanford and the western railroad kings, and they were ranked just behind Jay Gould among the richest men in America.[5]

The Bonanza Kings, however, were not alone in their good fortune, for many outsiders also made undreamed-of profits. Indeed, those who had paid anywhere from $50 to $100, including assessments, during the exploratory period from March to October of 1873, could have sold their combined Con Virginia and California shares at the peak in 1875 for an equivalent of nearly $5,600 apiece on their original shares, a profit of anywhere from about 6,000 to 12,000 percent in just over a year to a year and a half! With such magnificent profits as these, just "one share of stock," it was said, "was sufficient to secure a poor family against future want; ten shares placed a man of mediocre circumstances in a position of ease; and any greater number of shares enabled the holder to roll in wealth"! It was, of course, hard to lose money on these shares during the rise, since anyone could have at least recouped their investment if they sold at the peak.[6]

By December of 1874, just before the peak, the San Francisco *Chronicle* counted at least sixty outside of the Bonanza Kings who had also made their fortunes on the Con Virginia rise. Perhaps the most successful of these was fellow Irishman Archie Borland. He had come to America at seventeen in 1849 and worked three years on a farm in upstate New York for just $10 a month, before he scraped together enough to buy passage to California. There he found heaven in the mines at $4 a day and soon saved enough to buy claims of his own. He had many up and downs, but by 1866 he had accumulated $25,000, and he started buying stocks. He claimed that he had amassed $400,000 before he was laid low in Jones and Hayward's Savage deal in 1872. But his biggest score was soon to come, for early in 1873 he put $25,000 of what he had left into five hundred shares of Con Virginia, plus $5,000 into five hundred shares of the adjacent Central, which was merged into the California. He held on when others sold and paid an additional $3,000 in assessments before the great strike was made and the stock took off. It was a heady ride. One Saturday late in 1874 when the price of his shares, already multiplied severalfold by stock splits, had reached $350, he just about sold, but on Monday they jumped to $500 and he held out till they went up to $750. Then he cashed in all his stock for close to $3,000,000. That was a profit of nearly 9,000 percent in just over two years; no other outsider even claimed to do as well! He had scored like most outsiders could only dream of doing, and he was soon to be counted as an insider. Borland did talk fancifully of going back to North Ireland to buy a seat in the British Parliament, but he was too thoroughly hooked on mining speculation.[7]

AND THAT WAS just the beginning, for the lode's leading mining engineer, Philipp Deidesheimer, whose innovations in timbering had made deep mining possible, extravagantly declared that $1,500,000,000 was already in sight that would pay dividends of $5,000 a share! With the hope of a share in fabulous profits like that, tens of thousands flocked back into the market. Encouraged by eager sharks, the Big Bonanza created the longest and most intense mining mania that the West had yet seen. It raged for five years, from 1874 through 1878, in the bullion-driven prosperity of San Francisco, while the rest of the nation suffered through one of its longest depressions. Throughout it all the stock exchange was the very "heart of San Francisco," as Robert Louis Stevenson dryly observed, constantly pumping up "the savings of the lower quarter into the pockets of millionaires upon the hill." The daily business of the exchange again drew crowds of spectators, who could watch the show for a $5 ticket, and a special section was set aside for ladies. The sessions began with a commanding call of "Order! Order!" as the anxious brokers all went "scowling to their seats." Then the chairman began calling the names of each company listed, and as soon as some exciting stock was called the whole board came alive. "They spring from their chairs and rush furiously into the 'cockpit' or open space in front of the caller's stand," one amused onlooker reported.

> There is no order. All cry out at once. They shout their offers to buy or sell. They jostle and push each other about like frightened animals before a stampede. They rush from one place to another, wildly gesticulating, stamping and chafing as if infuriate. They froth at the mouth from excessive screaming. They yell and scream until their voices grow husky. A midnight serenade from the howling coyote is not more confusing. Bedlam let loose would scarce rival the scene. Yet, amid this Babel of voices, the quick ear of the secretary seldom fails to catch the sales that are made.[8]

The rival California Stock Exchange was joined by two more contenders, the Pacific Stock Exchange, opened in June of 1875, and the Nevada Stock Exchange Board, in January of 1877, clamoring for a piece of the action. The old "Big Board" vainly tried to fight back, banning its members from dealing with brokers from the rivals. All the while hundreds of "curbstone brokers" also crowded the street, where they screamed "like vultures at a carcass, whether rain or shine." Shares on the San Francisco boards sported a paper value of as much as $350,000,000, and a new onslaught of wildcats scrambled for over $3,000,000,000 more! Even bought on 10 percent margin at a heavy discount, there was nowhere near enough money in all the West to support such speculation, so despite the severe depression that still gripped most of the nation, New York brokers organized four successive mining exchanges from 1875 to 1877 in an effort to stir up a speculative market in Comstock shares. But the volume of trade there never amounted to more than a fraction of that on the San Francisco Stock and Exchange Board alone, which averaged around $200,000,000 a year and totaled over $1,000,000,000 for the boom. San Francisco soon suffered nearly six hundred brokers in mining stocks, triple the number from the previous mania, which magically matched the old peak ratio of roughly one broker for every two hundred adults.

SAN FRANCISCO STOCK & EXCHANGE BOARD
Frank Leslie's Illustrated Newspaper, November 24, 1877

Seats on the exchanges sold for as much as $40,000, for broker's commissions on mining shares ran 1 percent, generating $2,000,000 a year just for handling the paper, and of course far, far more from manipulating it. As the money poured in, the original San Francisco exchange built grandiose new quarters on Pine Street, looking to one cynic "more like the banqueting hall of some great potentate than a place of business." But the din within was still frantic as the brokers cried out "like the unseemly harpies of Dante's hell" with "every cry carrying the Comstock higher."[9]

Once again it seemed everyone was caught up in the frenzy of speculation. One visitor was certain that not one in ten abstained. Everyone he met—his bootblack, his barber, his waiter, his friends, even strangers on the street—talked nothing but mining stocks. He too succumbed. It was simply "irresistible." He couldn't "go into Wall Street with $25 and become a stock-gambler," he explained, but "here that amount is fully enough to get ten or a dozen shares of a very promising stock 'carried' for you 'on a margin.' It is the opportunity to speculate with such small amounts that drags so many into the whirlpool." Yet the fate of his friends should have been enough to discourage him. One bought three hundred shares in a $100 par wildcat at a mere 75 cents a share, having heard of another that jumped up to $10 or more and thinking that this one too could only go up, but instead it just "dropped out of sight altogether." Another went for a "comparatively high-priced and seemingly respectable stock," looking for steady dividends, but got instead only steady assessments. Despite occasional good luck stories to the contrary, mining stocks were just too badly manipulated to be profitable for most investors or speculators.[10]

SOME WOULD-BE INVESTORS were lured not only by envy and greed into speculation but to larceny as well by its temptations. Among the most publicized was the "young Monte Cristo," Charles Kuchel, the seventeen-year-old son of a prominent San Francisco artist, who was helping support his widowed mother and sister as a $75-a-month stockbroker's clerk. Among his duties was keeping track of the shares that the broker was holding for security on margins, and while the market was rising to record heights in January of 1875, he could resist the temptation no longer. He secretly started leaving some of the shares he was looking after with other brokers as collateral for a margin of his own, and went plunging in. Within two weeks he had shuffled shares back and forth to the tune of nearly $160,000; within six months his dealings totaled at least $2,000,000. He made money for a time, and once was as much as $70,000 ahead, a real fortune on his salary! A typical teenager, he bought a buggy and a couple of fast teams that could out race his boss. He also became a dandy dresser, spread money among friends and strangers, and sent his mother and sister on a European tour. Suddenly the "beardless boy-speculator" was the talk of the street with fabulous rumors that the "young Monte Cristo" had made a profit of as much as $5,000,000 in just two weeks! Then stock prices broke and it was "all swept away." He scrambled trying to recover, selling short for a while and using more and more of his boss's stock. Finally that July Kuchel's secret came out, and he was arrested, tried, and convicted of embezzling a thousand shares of California stock worth $63,824. He was sentenced to five years in San Quentin, despite pleas for leniency from William Ralston's lawyer, but through his heroism in fighting a prison fire he got out in six months. Nonetheless for a few more years, still not fully reformed, he fancied himself a "capitalist," before he settled down at last as a bookkeeper. The *Chronicle* editor blamed the corruption of "enterprising youth" on the "fatal fascination" of stock gambling. And he hailed the defrauded broker's warning, "If you have a boy that you care anything about, set him at anything but the stock trade. Put him into a gambling den if you like, set him to dealing faro, but don't make a broker's clerk of him!"[11]

But those who were led into larceny by a greed for quick riches reached far beyond reckless teenagers in San Francisco to such conservative Boston bankers as Royal B. Conant. He was a trusted pillar of the community, "as honest and true as any man in town." He had worked his way up from bookkeeper to become the cashier of the Eliot National Bank, "one of the safest and most conservative banks in the city," and in 1875 he had joined with sixteen fellow bankers to form the American Bankers Association. The banks' directors had also rewarded him with "an unusually good salary—$4,500 a year," fifteen times the average wage, on the belief "that men who hold such responsible positions as that of Cashier of a bank should receive salaries sufficiently high to guard them against temptation." In 1877, however, just after he turned forty, with the East still suffering under a depression, the directors had trimmed his pay by $300 a year, so he turned to mining stocks to recoup the loss. Having helped some of the directors with Comstock speculations, and confident that he "knew enough about finance to act with good judgement and to avoid being ensnared," Conant cautiously put some of his savings into "stock of the two most conservative" companies—Con Virginia and California, which had dropped to what he thought were bargain prices of $25 to $30 and $40, respectively. But he bought on margin, and as the stock fell still lower, cutting into his margin again and again, he finally exhausted his savings putting up more and started tapping the bank's till for one "temporary . . . secret loan" after another. By the end of July 1878, after the stock was down to $8 and $9, he realized that all was lost and stunned the directors by confessing that he'd embezzled $70,000. Then he tendered his resignation, broke down, and "wept like a child."[12]

Despite "the most strenuous efforts of the bank officials to keep the matter secret," the shocking news soon got out, and the U.S. attorney promptly charged him with embezzlement. A less repentant Conant then pleaded not guilty, and only compounded his problems by insisting that since "there was nothing mean or dirty about my transactions," he couldn't understand why everybody was being so hard on "a sad-hearted and penitent man." But one hard-hearted editor could see little difference between stealing from the bank's depositors and picking their pockets, except that the pickpocket was "more courageous" and if detected was "sure to be punished more severely than the Cashier, because he steals a great deal less." He argued that if Conant is "allowed to escape because of his high-flown notions of morality, then there is no good reason why other thieves should be punished." Then, before the case came to trial the Con Virginia shares rose to $17 and the directors happily imagined that they could recover their losses. So they suddenly urged that the criminal case be suspended and announced that the matter should be "settled out of court," as if it were only a civil case, counting his victims as mere "creditors"! The New York *Times*'s editor blasted the directors for being "perfectly willing the criminal shall go free, provided they recover the money," apparently believing that "what made him a criminal was not committing a crime, but being found out"! Similarly, the editor of the Washington *Post* concluded that in banking "It is not theft but unsuccessful theft that is punishable." Indeed, once the stock turned down again, the sudden compassion vanished, and early in 1880, as the price fell below $4, Conant was finally brought to trial, found guilty, and sentenced to seven years in Denham Jail. After only two years, however, his politically connected family friends got him a pardon from Republican president Chester Arthur by claiming that further imprisonment would "result in serious injury of his mental facilities." Since all his old banker friends wanted nothing more to do with him, he had to take a job as a traveling salesman for an oil distributor. Just a few years later, rushing for a train, he fell under the wheels and lost his right leg. He ended his days in the Holy Ghost Hospital for Incurables.[13]

At the same time, a promising thirty-five-year-old Presbyterian minister, Thomas M. Dawson, also fell from grace to be scorned as a "clerical stock gambler" and worse. Soon after arriving from New York in 1872, he had stayed with the Reverend Daniel Poor, a professor at the San Francisco Theological Seminary, who revealed to him the earthly treasures of Raymond & Ely dividends, and the neophyte was hooked. But the dividends ended in 1873, and Dawson turned to speculation when the market began to rise again. He did very well by his telling, making as much as $100,000. He did so well, in fact, that his envious brethren and parishioners pressed more than $10,000 of their savings upon him to "use" for them in his name. He was confident that Con Virginia would go to $2,000, and all went well until the market broke in the spring of 1875. Then "his little speculative bark was swept away out into the stormy sea and finally engulfed among the waves." They all lost heavily, and his fellow divines began "squealing pathetically" for their money back. Things got still worse. Dawson, scrambling for cash, sold Poor's sacred shares of Raymond & Ely, which he held as security for a $84 loan and assessment payments of $297. He got about $1,300 from the sale and poured nearly all of it into briefly covering his own margins, giving poor Poor only $100. Furious, Poor attached all of Dawson's stock, still worth $25,000, and then let the "margins sink away" until there was only $136 left for everyone! Poor and his reverend brothers all the while damned Dawson for everything from embezzlement to unwanted flirtations, and tried to bring him to trial before the local Presbytery, vowing if guilty they would "blast his character, ruin his reputation, and deprive him of the means of support." Dawson promptly resigned and fled to Virginia City until his lawyer got a judge to enjoin them. The crusading *Chronicle* denounced the whole shameful affair as one more example of the evils of stock speculation, the "direct outgrowth of a wicked, carnal desire on the part of certain worldly clergymen to get rich by dabbling in stocks." Dawson, banned from the ministry forever, became a newspaper correspondent and three years later was appointed the American consul in Samoa by Republican president Rutherford B. Hayes.[14]

While some were driven to crime by their speculative losses, others were driven to the ultimate despair. In the frenzy of exhilarating gains and devastating losses, the suicide rate in San Francisco nearly doubled. The coroner reported that mining stock losses were the cause of at least 40 percent of the suicides of known cause. Hardly a week passed without one or more tragedies, and the editor of the San Francisco *Call* decried this "horrible nightmare" and lamented the fate of all these unfortunates "driven to their graves by the terrible system which organized capital has established in our midst!" Still, not all were sympathetic to the victims, even as they raged against the "demoralizing tendencies of stock gambling." For when George Taylor "blew his brains out," leaving only a terse note that "Stocks have brought me to this; California owes me a coffin, no more," the editor of the *Post* protested that the good people of California owed him nothing at all, arguing "If he had made a fortunate speculation, he would never have troubled himself where the money came from, how many wretches were driven to desperation to find the coin he raked in, how many pleasant homes he made desolate. What he wanted was money, which he did not earn, and when he died impecunious, he claimed burial at the expense of others whom he would have despoiled of their wealth with the utmost alacrity." Yet the editor concluded, "George Taylor was neither better nor worse than the vast majority of stock operators. He simply brings into striking prominence the utterly selfish and reckless side of this whole business."[15]

VIRGINIA CITY TOO was swept up in the frenzy as most of the Comstock miners themselves were drawn inextricably into the maelstrom. "Here," one boasted, "everybody is a speculator." They knew the mines far better than any San Franciscans, and they were sure they'd hear of the latest strikes first. But they would find again and again that it wasn't the simple vagaries of the mines they were bucking, it was the calculated manipulations of the mine managers. Unfortunately, most never seemed to learn that lesson. One such unfortunate was James Galloway, a forty-year-old miner who, in 1875, had come with his wife and three children to work in the Con Virginia. Within months he was hooked, after he bought ten shares of Overman at $66 a share and sold them just one month later at $75, making an effortless $90, which was almost as much as he had earned that same month working! After that he put every dollar they had saved or could borrow into stocks. A year later he had sunk $1,095 into stock and lost $800 of it. "I am financially bankrupt after twenty five years of hard work," he wrote in his diary. "Life is almost a blank for me. Stocks have ruined many." His wife died six months later, and he plunged back in on margin. Luck and a rising, but rigged, market were with him for a bit, and a year later he held $5,900 in Gould & Curry and other old favorites. Then the market broke and in two weeks he lost $5,000. Still he couldn't quit and kept on feeding the tiger, falling for every speculative rumor to the day he died, five years later, in a mining accident at the age of forty-nine.[16]

But the miners weren't alone. Even that seasoned mining engineer and pioneering superintendent of the great Ophir, Philipp Deidesheimer, became a frantic speculator, honestly believing his own extravagant estimate of the boundless riches of the Big Bonanza, and he paid dearly. "To him," one San Francisco editor chided,

> the crown of the bonanza was but the apex of a body of solid gold and silver, extending to the center of the earth! . . . He bought and bought and bought. Then he begged and borrowed, and almost stole, and bought more. Then he got some more carried for him. Then the profits on that he hypothecated and bought more. Deidesheimer was a duplex millionaire—on paper. He lived in great style. Champagne soaked his Axminister carpets. The finest strains of blood occupied his stables and the most brilliant of gems lighted up his exultant person. Then one day there came a tumble. Con. Virginia suddenly broke and Deidesheimer got jammed in the door.

He was bankrupt for a staggering total of $534,620! Still, the editor concluded, "nobody can say that Deidesheimer hasn't died game," and he conceded, "he takes his place among those who have done the thing respectably at least." But he was in debt mainly to that old shark Alvinza Hayward, and thereafter he became involved in several scams.[17]

Some, nonetheless, like seamstress Mary Mathews, were successful at least for a while. A thirty-five-year-old, college-educated Civil War widow, she had come to the Comstock with her nine-year-old son in the fall of 1869 to look into the affairs of a murdered brother, and she decided to stay. Doing washing, teaching, and sewing, she had saved over $300 by the summer of 1871, and after watching Crown Point take off, she bought shares in Chollar-Potosí at $52 when it was paying dividends of $2 a month. Early the following year the dividends stopped and the stock fell, but she held it, and soon it began to rise again in the Savage deal. When it reached $89 she sold, only to buy it back at $99 and finally sell out at $320 in April just before it peaked! She promptly put her profits into a boarding house with a store underneath, which paid her steady rental income. She still speculated, of course, with tips from boarders and friends, but she failed to collect on a couple of good tips when her broker steered her into losing shares of his own. After that she put more faith in dream trips to the mines than in brokers' tips. She did well as the market boomed with the Big Bonanza, accumulating a "handsome fortune" in California and Con Virginia. So early in 1878 she returned home to New York, planning to live quite comfortably on an income of $200 a month in dividends and rents. But all of that would end when the bonanza dividends finally ceased. The stocks crashed, taking Comstock real estate down with them, and she was left out like all the rest.[18]

Still, even those like Alf Doten, editor and publisher of one of the leading dailies on the Comstock, who confidently thought he was close to the insiders, found out the hard way that he wasn't always in. Doten, a seasoned forty-niner, had started as a local reporter on the Gold Hill *News* in 1867, worked his way up to editor, and bought the paper when the owner died five years later, borrowing the needed $10,000 from William Sharon and doing his bidding. He also speculated heavily, and in just three years, with inside help, he had cleared his debt and was worth $20,000 or more. Suddenly overconfident, he decided to be independent, and though he plunged on in the market on margins, his insider tips began to fail him. He again appealed to Sharon, then Jones and Mackay, for help, but they gave him little. In four years he had lost everything and was over $5,000 in debt.[19]

Indeed, the majority of the speculators, particularly those of limited means, bought on margin still for as little as 10 percent, and most of their losses resulted from that. Not only San Francisco reformers, but solid mining advocates like Judge Charles C. Goodwin, editor of Sharon's

Virginia City *Territorial Enterprise*, vowed that "men are little better than crazy to purchase stocks on margin. . . . There are something over one thousand men in San Francisco who subsist on the fluctuations in stocks," he pointed out. "They command a vast amount of money, and some of them are restrained by no sort of conscientious scruple. When they, by a preconcerted arrangement, pounce upon any stock, no matter how good, they are bound to break it. They do not scruple at circulating any rumor which will answer their purpose, and the mass of people on the outside will, despite the experience of ten years, be influenced by those rumors." Thus, he warned, "it is simply putting one's head into a lion's mouth to buy stocks on margins. The break often is only intended to last for a day or two, but a day or two is sufficient for the purpose of the men who inaugurate a movement intended simply to rob people who have purchased stocks without paying for them. So long as they continue to place themselves exactly in a position to excite the cupidity of thieves, they are going to suffer continued loss and continued heart-aches." Goodwin's motives, of course, weren't altruistic either; his boss, Sharon, simply wanted dedicated shareholders who paid assessments to feed his mills, not margined speculators who forfeited them.[20]

STILL, ANXIOUS SHAREHOLDERS hungered for every new rumor, and the brokers and operators fed them well, both for churning their clients' stock for added commissions, or much more profitably working every manner of manipulation and "deal" in their own pet shares. The skillful operators knew well that the quieter the whisper and the stricter the confidence in which they shrouded their tips, the farther they would go and the more compelling they would be. One wealthy shark, wanting to unload some shares, gave a strictly confidential tip to his minister. Then after he had cleaned up and the price had collapsed, he offered his deepest regrets that the tip had proved false and generously made good the divine's loss, while all the deacons and others in the congregation who had secretly shared the pointer suffered their losses in silence.[21]

Sharp brokers also worked countless other swindles on their customers. One very common scam was simply pocketing the customer's profit, if he happened to make one. Suppose a customer deposited money with his broker to buy a hundred shares of a stock at a limited price, say $50 a share, and his broker bought it at that. Then if the price rose to say $55 before the customer returned, his broker would quickly sell it, taking the $500 profit for himself. When his customer came back, the broker simply returned the money he had used to such good advantage and told his trusting victim that he was sorry but he wasn't able to buy the stock at the limit set! Or, perhaps just as often, the broker might just "put up" the price, if he were able to buy the stock for, say, $2 less than his customer's limit and it then rose to that or a little more, he would simply tell his customer that he had to pay the limit for it, and pocket the $200 difference—in addition, of course, to his commission for the service! Little wonder, then, that one prominent broker could boast that he made far more off other people's money than bankers did, and with far less investment of his own. But it was the brokers' very success, so proudly displayed in palatial mansions and luxurious yachts, that raised a red flag for some and inspired that now famous question, "Where are the customers' yachts?"[22]

Prices seesawed continuously, and one observer blasted the entire market as "a dishonest, dishonorable business," claiming "there is not a single operator, inside or out, who is not impelled by greed and avarice to engage in a pursuit which his conscience and his judgement alike condemn." Swings of a factor of ten or more, up and down, year after year, were not uncommon, as the knowing ones reeled in their catch and then cast out their bait again, and many brokers systematically gutted and cleaned their margin clients in "bucket shop" scams. Such looting became so commonplace that even the editor of the conservative old *Alta California* began to complain that

> enormous losses have been sustained by the public through operations on the part of brokers, which will not bear the light. If the long list of stock buyers, including widows and single women, who have been compelled to put up margins as long as they could raise a dollar, only to be closed out finally, when the last dollar was gone, could search the transactions, a large proportion of them would be surprised to find that they had been the victims of a 'hogging game.' They would discover that the broker they trusted was, in fact, no broker at all, but merely a gambler who professed brokerage only to ensnare deposits on the pretense of buying stocks for the depositor. That if he really bought the stock, he had sold it out again immediately on his own account, and then by rumors, aided by unscrupulous confederates, caused a fall in price, which he made the pretense to call for more deposits.

Sadly, he concluded, "It is somewhat curious to observe a crowd of people anxiously watching the market in expectation of a rise, under the impression that they have an interest in the stocks, when the fact is, that the gambling brokers who have got their money, have sold out all their stock and are operating together in the common object of draining them of their last dollar."[23]

Early in 1874, the persistent press crusade against the "evils of stock gambling" finally spurred the state legislature into attempting reforms that struck at the very heart of stock speculation. The "most revolutionary" bill was one to prohibit both margin sales and wash sales, and another would abolish short sales. Each offender could be fined up to $1,000 and whistle-blowers could get half of the fine. But the startled brokers quickly rallied enough votes with cash and stock tips to block any "such monstrous follies." Only a much more modest bill, aimed instead at secrecy by the mining company officers, succeeded in passing. It at least offered stockholders better information by guaranteeing the rights of all to inspect the company books, and for those holding shares of at least $500 par the right to inspect the mine itself, and it imposed a fine of $200 on any officer who refused to comply. This was only the first step, however, for the press crusade for much stronger reforms was growing more insistent every year. But public support flagged for a while, since just a couple of months after the first reform the Con Virginia paid its first dividend and the cries for reform were drowned in the frenzy.[24]

So STRONG WAS the new fury in mining stocks that it blew away several crises that would have quenched a lesser furor. Fanning the first frenzy to a fever pitch was Ralston's frantic fight for control of the old Ophir, right next to the California and Con Virginia. Spurred by the plunder of the Belcher bonanza, Ralston had become ever more extravagant in his obsessive speculations. As Sharon noted, "if he got into anything there was no end to it." When he simply tried to build a house, he ended up ten times over budget! By the summer of 1874 he was mired in the construction of a lavish hotel of oriental opulence, casually paying an outrageous sum for a ranch just to get oak planks that he never used, and plunging on with abandon. Sharon, who also held a half interest in the hotel, finally asked in exasperation, "If you are going to buy a foundry for a nail, a ranch for a plank, and a manufactory to build furniture, where is this thing going to end?" The answer of course was bankruptcy.[25]

By then Ralston had saddled the bank with bad debts in his profitless speculations running to over $3,500,000, or 70 percent of its capital, and he had illegally overissued bank stock by more than 25 percent to cover other debts. Desperate for more plunder from the Comstock, he saw the Ophir as his one chance for a piece of the new bonanza, for he was certain that the vast ore body must extend into it. To control the mine he and Sharon needed control of a majority of the stock by the annual board election in December, so they arranged with a charming and skillful operator and soon-to-be president of the stock exchange, James R. Keene, to quietly begin buying up shares at $22 in September. But within a month their plan became clear to all, and share prices escalated as sellers held out for more and more. Before it was over the "knightly" Keene had made well over $1,000,000 just through his purchases for Ralston. But the biggest profits went to Elias J. "Lucky" Baldwin, a shrewd speculator who had held the largest block of stock, fully half of the 108,000 shares, and won $5,000,000—and his nickname—after he sold his last 20,000 shares for $2,700,000 to give Ralston and Sharon control. Baldwin shortly became president of the rival Pacific Stock Exchange, but, in his late forties, he soon took his fortune to Southern California to lead a scandalously flamboyant life, speculating in real estate and raising horses, which he raced at his Rancho Santa Anita. He always disputed that he was lucky, claiming instead that his success came just from knowing "when to go into a deal and when to go out, and don't waste any time doing either."[26]

The scramble for Ophir had run its shares up to $212 by December, up nearly 900 percent in just two months, and the frenzy carried the rest of the market with it. But even after they got control, Sharon kept the excitement alive for another month, pushing Ophir on up to $315, while he unloaded some of his own shares and then let it break to clean up on the shorts. During the frenzy, Sharon had also had to pump $800,000 into the Nevada legislature to buy the Republican majority, who shook him down for all they could get, for his election to the Senate that January. However, he generously rewarded his greedy electors with a tip to put the money into Ophir for a further rise, and he was said to have gotten back all of his money and more from them at $200 to $250 a share before the price broke. But while Sharon got out ahead, Ophir proved to be Ralston's undoing. It sank him more than $3,000,000 deeper into debt, and his longed-for bonanza turned out to be only a worthless and devastating illusion. As the Ophir-driven market bubble also collapsed, much of his paper assets evaporated too, and his creditors pushed even harder. Without a fresh infusion from the Comstock, Ralston "kited" checks to try to keep afloat after James Flood and the other new Bonanza Kings took their money out of his bank to start their own, the Nevada Bank, and cash generally became scarce in August of 1875. On the twenty-sixth, as Ralston still scrambled for loans, a run started on the bank, and it closed its doors at 2:35 that afternoon. The next day the directors demanded his resignation as president. An hour later his body was floating in the bay.[27]

The old San Francisco Stock Exchange suspended the same day, but the rival California exchange continued uninterrupted. Despite hopeful predictions that this would at last bring an end to stock gambling, the impact proved to be surprisingly light. Sharon, who was also one of Ralston's heaviest creditors, took over his assets and settled his debts of nearly $10,000,000, some at as little as 50 cents on the dollar. He and Mills then reopened the bank in October. The old exchange reopened at the same time and the market quickly recovered, with Con Virginia topping $400 again. Ralston's failure, nonetheless, marked the end of the old Mill Ring's reign on the Comstock and the crowning of the Bonanza Kings as its new unchallenged monarchs.[28]

Sharon, the former "King of the Comstock," proudly retired into Ralston's extravagant estate of Belmont, south of San Francisco, and he basked in his new title of senator from Nevada. However, he refused to spend much time in Washington, and after being widely criticized for his chronic absence, he lost that too to one of his usurpers, James Fair, who bought enough Democratic legislators to take the seat in 1881. After that Sharon's tightly wound life began to unravel. Two years later he was briefly arrested for "adultery" on charges by Sarah Althea Hill, one of the girls of the notorious black madam and entrepreneur Mary Ellen "Mammy" Pleasant. Sharon had kept Hill as a mistress at $500 a month after his wife died. But she and Mammy wanted more, and suddenly producing a marriage contract and letters, later ruled forgeries, she filed for divorce and demanded alimony. The first round of several sensational suits and

RALSTON PANIC AT THE BANK OF CALIFORNIA, AUGUST 26, 1875
Frank Leslie's Illustrated Newspaper, September 11, 1875

countersuits ended on Christmas eve of 1884 with a surprise victory
for Hill. Sharon promptly appealed, but broken in health, he died the
following year at the age of sixty-four. Hill then sued for a share of the
estate, hiring and marrying the former California Supreme Court justice
David Terry. His old adversary, Stephen Field, now on the U.S. Supreme
Court, finally ruled against Hill in 1889, and when the hotheaded Terry
later slapped Field, he was shot dead by Field's trigger-happy bodyguard
and "gun fighter," Dave Neagle. Soon after that Hill was declared "hopelessly insane." Sharon's estate was managed by his son-in-law, lawyer
and Democrat Francis G. Newlands, who later got the Senate seat back
in the family.[29]

MEANWHILE, the seemingly endless flood of Big Bonanza bullion fueled the speculative euphoria. Even a disastrous fire in Virginia City couldn't quell the exuberance. Neither could a "bear raid" by Keene the following spring that savaged margin holders of the bonanza stocks. Keene spread rumors that the Big Bonanza was exhausted after Fair had issued an exaggerated report. When Mackay increased production to over $3,600,000 a month in March of 1876 to try to counter that claim, Keene, aided by George Roberts and even Fair, claimed that Mackay had gutted it. In five months Keene succeeded in wearing down bonanza stock prices by almost half and clearing a couple million on shorts and margins. He then headed east, where he "fell into the New York market as easily as any man generally falls among thieves," and the *Wall Street Journal* would later proclaim him the "Napoleon of traders." The Bonanza Firm, meanwhile, had responded by commencing regular monthly dividends from the California as well and splitting all their stock yet again, five to one, to a total of 1,080,000 shares so that everyone in San Francisco could afford them.[30]

As the clamor for shares raged on, an average of four hundred new companies were incorporated in California each year throughout the boom. Many were merely resurrections of claims left idle from the 1863 and 1872 booms. But there were several new excitements throughout the West that hoped to rival the Comstock. The opening of the enormous silver-copper deposits of Butte, Montana, and the frantic new gold rush that overran the Sioux reservation in the Black Hills of the Dakotas would, in fact, not only rival but ultimately surpass the Comstock. These new strikes offered some incredibly profitable investments and, of course, many, many more wildcat speculations.

For those who just couldn't wait for the next big bonanza on the Comstock, Nevada senators Jones and Stewart promised something "even bigger than the Comstock!" It was a bonanza modestly named the Wonder of the World, located in the spring of 1873 by a few California miners and a band of Pioche stage robbers high in the Panamint range on the very brink of Death Valley. The ore ran as high as $900 a ton and was said to average nearly $100. But most important, it was so enticingly displayed. It was a promoter's dream. Its surface outcroppings rimmed both sides of Surprise Canyon for more than a mile, standing almost vertically, and ore just as rich was found six hundred feet below in the bottoms of the steep side canyons that cut through it. For all the world it appeared to be a truly magnificent lode with at least $100,000,000 worth of $100 ore already fully exposed! It was just that appearance that Jones and Stewart would work. Never mind that they knew the ore at the bottom had simply fallen there and the underlying rock was barren.[31]

Joined by fellow robber baron Trenor Park, they bought out the petty stage robbers in the summer of 1874 to try to stage a much grander robbery. They paid around $250,000, it was reported, though most was doubtless in future shares. Then, as a measure of the grandeur of their new wonder, they formed nine separate companies, led by Stewart's Wonder, with a total capital offering of $50,000,000 for the public. In case anyone doubted their sincerity, they also formed a private Surprise Valley Mill and Water Company to actually work the ore for added profit. As a further show of faith, they even planned a railroad from Los Angeles to the mines. Hoping to unload the new wonder in London, they immediately began shipping more than $150,000 worth of high grade to Swansea to excite British interest and get quick cash as well. But calling attention to their scheme in England backfired, for some of the press noticed "the shapes of the Emma and Mariposa speculators flitting about in connection with Panamint" and warned of "a new Emma looming up." So they soon shifted their aim to San Francisco investors, merging their mines into just two companies, the Wonder Consolidated on one side of the canyon and the Wyoming Consolidated on the other, all at a new bargain capitalization of only $12,000,000. All the while Jones and Park extravagantly predicted that the great Wonder would truly surpass the Comstock, producing as much $50,000,000 a year! As a final inducement, when they put the first 20,000 shares on the exchange in January of 1875, they offered a rock-bottom price of only $15 a share on a $100 par. By then the excitement was so great that frantic investors snapped up over 10,000 shares in three hours, and the rest were gone in two weeks. One latecomer offered $10,000 for another 500 shares, but he was only allotted 100. They were saving the remaining 100,000 shares for an even bigger push at much higher prices.[32]

With this money they erected an impressive mill with an aerial tramway to deliver the ore, and by the fall of 1875 they were turning out bullion at a rate of $1,300 a day, or $500,000 a year. Although they were still far short of their gargantuan predictions, they had turned the property into a very solid and prosperous-looking venture, and they proudly boasted that this "puts the blush on all Panamint croakers and its enemies gener-ally." By the beginning of 1876, after several months of steady produc-tion, they were finally poised to unload the bulk of the stock in a grand finale when the ghost of Emma returned again to haunt them. Until then the scandal had been confined to the British courts and British press, but in February of 1876 the U.S. Congress began its investigation of the involvement of the American ambassador in the scam. Suddenly the American papers were filled with accounts of Parks and Stewart's crooked dealings, and the Panamint bubble popped. The congressional hearings ended on May 4, and two days later, despite months of glowing reports from Panamint, Jones abruptly announced that the mines had pinched out and the great mill shut down. The deal was dead. The trio publicly claimed personal losses of more than a million to blunt lawsuits from those early investors who had already paid in over $300,000, but the mines had also produced nearly $500,000, so their actual losses, if any, were likely no more than $100,000 at most. Their real losses were all those millions that they had hoped to grab when they unloaded, and the real winners were all the prospective investors who would have eagerly paid out those millions. That was the real Wonder of the World.[33]

But even before Panamint collapsed, the profitable real bonanzas were already being opened. The flashiest was the Silver King, discovered in central Arizona, and it actually proved profitable to investors. It was a massive quartz vein packed with silver-lead, first noticed in 1865 by a soldier crossing over it on a military road. He picked up a piece of almost pure silver that he showed to local ranchers, but then he disappeared. The ranchers hunted for the ledge for a decade, and one was killed by Apaches before the others finally located the lode in March of 1875. Two of them, Benjamin Reagan and Charles Mason, bought out the others for $80,000, which they took out in six months, and the following year they began shipping more than $30,000 a month in high grade to San Francisco for smelting. Much of the ore was shipped through Yuma merchant James M. Barney, a very sharp and enterprising, forty-year-old New York Quaker, who had come to the territory in 1865 and soon cor-nered all of the freighting contracts for the army. But when the Southern Pacific Railroad reached Yuma in May of 1877, breaking his monopoly, Barney decided to go into mining, and he bought Mason's half interest for $260,000, mostly in notes to be paid off out of his profits from the mine. He and Reagan then formed the Silver King Mining Company with a paper capital of $10,000,000 in 100,000 shares, and he promptly bought Reagan's half of the shares for $350,000, also in notes.[34]

Thus, Barney had full control of the mine bought almost entirely on credit, but he angled for even more. For he quietly leased the mine from his company for the next two and a half years for a flat fee of $457,000, paying just $7,000 down to set up a company office in San Francisco and the rest in installments out of earnings. That allowed him to sell shares and still keep most of the profits. He began work on a large scale, concentrating the lower-grade ores for shipping, and by the fall of 1878 he had taken out $1,900,000 in silver. Out of this he paid off his $600,000 in notes, and cleared more for himself. For he also paid off early the remaining $450,000 on his lease, most of which he immediately took back as dividends totaling $4.50 a share—out of one pocket, into the other, and back again! That, of course, made the Silver King a solid dividend payer, attracting eager investors. Indeed, as soon as Barney paid himself his first dividend, he listed the stock on the exchange, where it started at $12 a share and climbed to $20 by the fall of 1878, while he sold at least 40,000 shares for an added profit of over $600,000. Then the dividends stopped, and Barney directly pocketed all of the remaining profits until the lease expired at the end of 1879. By then the shares had fallen to barely $4 a share. Barney doubtless bought back whatever he could from disillusioned investors, while he quietly exposed new ore to a depth of seven hundred feet. On the promise of that bonanza, he offered the mine to a California speculator for $20 a share, but when the sale fell through, he finally resumed dividend payments in the fall of 1880 and the stock climbed to $25 a share in six months, while he undoubtedly shed more shares again.[35]

Despite all of Barney's manipulations, however, many of his investors also did well, for the dividends continued for the next seven years, ultimately totaling $19.50 a share, or nearly $2,000,000, while the mine produced $6,500,000 before shutting down in 1888. The earliest investors, who bought in at $12, made more than 6 percent per annum for a decade just on dividends. Those who happened to buy, rather than sell, at $4 during the suspension of dividends could have done far better, making nearly 300 percent in seven years. A handful did even better than that. For right at the start, Barney had spread some of the shares just as cheaply among his army cronies and their friends to give them a direct interest in seeing that the mine was protected from attacks by the Apaches, who hadn't yet fully conceded their title. Not the least of these was former General James Garfield, who was by then a Republican congressman and soon to be president. Garfield bought a thousand shares at a bargain price of $4 apiece, while they were selling for $12 or more, and by the time he was assassinated in 1881 he had already collected $7,500 in dividends. The stock was worth $20,000 if his widow had sold then, or she would have at least earned another $12,000 if she had held it, for a total profit of 400 percent. None, of course, did nearly as well as Barney, who cleared around $2,000,000 from an actual cash outlay of no more than $20,000, for the Silver King paid its own way after that.[36]

At the same time another very sharp and enterprising man, a former professor of rhetoric at the Kalamazoo Theological Seminary and future owner of the Cleveland *Plain Dealer*, Liberty E. Holden, made his fortune opening Utah's first big lead producer, the Old Telegraph mine at the head of Bingham Canyon. But his partners would all sue him for fraud. Holden had turned from a modest teacher in Kalamazoo to a prosperous real estate speculator in Cleveland after the Civil War, but the collapse of the market in 1873 left him with over $200,000 in mortgages. So at the age of forty he decided to try his luck in mining and got a job managing a Lake Superior iron mine. Then he was lured by western silver, and late in 1874, borrowing $30,000 for his share, he talked wealthy old friends and neighbors in Cleveland and Kalamazoo into forming a company to buy the Utah silver-lead mine for a highly inflated $240,000. The mine, only thirty miles southwest of Salt Lake City, was one of many located by General Conner's soldiers a decade before, but it had not paid well, and the owners were doubtless delighted to turn it over at such a price to a bunch of easterners. It was immediately profitable, however, at least for Holden, who became both vice president and general manager, with a very generous salary of $500 a month plus expenses for moving to Salt Lake City to personally direct development. Still, he had no better luck in the mine than his predecessors until January of 1876. Then his superintendent, George Doane, changed the direction of the tunnel and suddenly discovered a massive ore deposit of glistening galena just three feet beyond the side wall! This bonanza, dubbed the Black Stope, averaged nearly 50 percent lead by weight plus about $25 a ton in silver, and it was so big that it could even "excite a Comstocker, for a moment."[37]

Doane quickly and secretly telegraphed Holden, who was in Cleveland at the time, "Buy up the stock, the thing was immense." Holden immediately set about doing just that, with a share for Doane. One by one, he quietly confided to the three largest owners, his trusting fellow church members, that the mine was a costly failure and their shares, which had cost them a total of $86,000, were a total loss. Out of Christian charity, however, he kindly offered to salvage something for them by swapping their shares for an old mare, a secondhand buggy, and a lot of other truck, plus $2,200 in cash. In two weeks he held just over half of the stock, and he rushed back to the mine. Doane had opened much more ore, and he had already turned out enough bullion for Holden to quietly pay off his $30,000 loan out of the company till. Finally, at the end of April Holden called three of the leading Kalamazoo shareholders out to Salt Lake City to give them the bad news too, that the mine was in dire straits. He showed them all the costly improvements he had made plus disappointing assays of low-grade ore that was too poor to work, but not the new bonanza, of course, and then he told them that they had run out of money and the next payroll was due in less than two weeks. All the shareholders would have to start paying assessments, and he had already billed them for $1,500. Within a couple of weeks the shocked stockholders had eagerly sold him most of the remaining shares for $40,000, only a fraction of their cost. Holden paid them as well with the proceeds of the mine, rightfully their own money, and he became virtually the sole owner with 95 percent of the shares. By the end of 1877 the Old Telegraph had become "the largest ore-producing mine in the territory," having turned out close to $1,000,000 in lead bullion, roughly a seventh of the entire U.S. production. Out of that Holden had cleared about $600,000, but his badly deceived partners had also finally gotten wise.[38]

Late in 1877 the former stockholders got hold of dozens of Holden's secret letters and telegrams that clearly proved he had lied to and defrauded them. They promptly filed fifteen separate suits to rescind the sales and enjoin Holden from further exploitation of the mine. The injunction hearing came up first, and under cross-examination Holden blandly admitted to withholding word of the rich strike just to get the shares as cheaply as possible, and he casually claimed that nearly $500,000 was owed him for "his services." Although the Salt Lake City judge refused to grant an injunction, after the rest of the cases had dragged on for a year with costs reaching $75,000, Holden finally decided he had to settle with his former friends. By then he was also ready to sell the mine, for he had taken out most of the high-grade ore. So he optioned the property for $1,000,000 to former San Francisco operator Erwin Davis, who had successfully placed the infamous Emma with Park and sold other Utah mines in London, and Davis successfully unloaded this one in Paris in February of 1879. As part of the deal Holden also arranged for his defrauded partners to get an additional $200,000 for dropping their claims. That barely returned their original investment, but by then they were undoubtedly happy to even get that. Holden, of course, took all the profits and got away with at least $1,600,000, his fortune made. He later returned triumphantly to Cleveland, bought the daily *Plain Dealer*, which he turned into the city's leading paper, and became a pillar of the community. His son Albert would also go into mining, but he would make a much better name for himself. For his part, Davis teamed up with a flamboyant Parisian "bubble-blower," Simon Philippart, to share a fat profit floating the mine for $3,400,000 as La Société des Mines d'Argent et Fonderies de Bingham, with extravagant estimates of over 3,000,000 tons of ore in sight and enticing promises of 100 percent a year in profits! Two years later it was abandoned as worthless.[39]

THE MOST IMPORTANT new silver lodes opened at this time, however, were those on what would come to be known as "the Richest Hill on Earth," the garantuan ore deposits of Butte, Montana, the biggest producer in America for nearly a century. Although their silver yield would soon pale in comparison with that of copper, the low price of copper in the late 1870s discouraged attempts to develop them, and it was silver that first attracted investors to the mines. Butte had, in fact, begun as a placer gold camp in 1864, and early hardrock miners had staked claims all along the great black quartz lodes that ribbed the hill above the camp. The ore proved to be "rebellious," however, and too low in gold to pay, so most of the miners eventually moved on. But one of them, a persistent thirty-five-year-old miner from Pennsylvania, William L. Farlin, returned after he discovered that his souvenir ore samples were rich in silver, and on New Year's Day in 1875 he relocated his old claim and a dozen others. He soon put up a ten-stamp mill with a roasting furnace that successfully worked the ore, and he started a silver rush to Butte. In doing so, however, he ended up $73,500 in debt to a local banker and a merchant. So early in 1876 he named the banker, another thirty-five-year-old fellow Pennsylvanian, William Andrews Clark, as trustee to work his mill and mines for eighteen months to take out enough ore to pay off his debts. But Clark had other plans.[40]

Clark was a humorless, tight-fisted, driven little man, who had turned from teaching to placer mining at the age of twenty-three, gotten a stake, and moved into merchandising and banking. He made money mostly by buying gold dust at a discount for resale in New York, and he spent the winter of 1872–73 there studying assaying at the Columbia University School of Mines to more tightly gauge the value of the dust, since he was handling over $1,000,000 worth a year. He had also soon recognized the promise of Butte's ores and began acquiring an interest, one way or another, in many promising claims, including the Frank Moulton, which was rich in silver. But he needed a mill to work them, so he eagerly ran Farlin's mill mostly on ore from the Moulton mine and turned out a steady and profitable flow of as much as $4,000 a week in bullion for both the mill and himself. He didn't touch much of Farlin's rich ore, however, and he put all of the mill earnings back into further improvements. So when the time was up, Farlin's debts were still unpaid, and Clark foreclosed. He personally auctioned off the mill and most of the mines for under $25,000 to his brother, Joseph, who served as his agent, and then took back half for his share of the debts. Clark thus became the sharpest operator in Butte, and he would later play a lead role in opening its vast copper bonanzas. Clark's milling success excited Butte's other banker, Andrew Jackson Davis, the partner of Helena banker and mine owner Samuel Hauser and the much older and more prosperous brother of the notorious Erwin. Davis quickly refitted an old mill, which he had also picked up by foreclosure, and opened the Lexington claim, which he had likewise taken for a $50 debt, or as legend would later have it, a bobtailed sorrel horse. The Lexington proved to be even richer than the Moulton, and Davis was soon turning out as much as $8,000 a week.[41]

The sudden flow of silver bullion from Butte in the summer of 1876 attracted the attention of Salt Lake City bankers and former Emma mine owners the Walker brothers, principally Rob and Sharp, and they sent their expert, former foreman Marcus Daly, to Butte to look for a promising mine. Daly was a genial, generous, ambitious Irishman who had gained a thorough education in the mines. He was two years younger than Clark and his very antithesis. But he too saw great promise in Butte, and he would stay to even outdo Clark by forming what was for a time the biggest copper producer not only in America but the world, the great Anaconda! Daly was still looking for silver then, however, and he recommended the Alice mine, adjacent to the Moulton, and the Walkers bought it that fall from Clark and his partners for a bargain $6,500. The Walkers made Daly the superintendent and gave him a small share in the mine as an added incentive to make it productive. They put an additional $160,000 into it before they got anything back, as Daly opened the lode and put up an efficient twenty-stamp mill. Then in November of 1877 the bullion began to flow, and in the next two and a half years the Alice produced $1,600,000 at a profit of about $1,000,000, or 500 percent on the Walkers' investment, making it Butte's first big producer. By then Daly was flush enough from his share of the profits to buy into mines on his own, so he resigned as superintendent and sold out his interest in the Alice for $100,000.[42]

On the strength of this performance, the Walkers went public in May of 1880, as the Alice Gold & Silver Mining Company, with Rob as president and Sharp as vice president. They took all the shares, $10,000,000 at $25 par, in exchange just for the Alice mine and listed them on the New York Stock Exchange. With a giant new sixty-stamp mill they turned out another $1,600,000 in bullion and paid out $1 a share, or $400,000 in dividends, even though the company secretly owed the brothers a like amount for the mills, which Rob boasted were given free! But the dividends, of course, kept the stock up at around $8 a share, while they unloaded at least 40 percent of their stock for $1,300,000, or more. Then the dividends stopped in December of 1881, and the stock crashed to under $2, while they may have picked up a little more on short sales. Yet the bullion production continued undiminished at $1,000,000 a year, while the perplexed shareholders howled for dividends and a public accounting.[43]

Not until early 1884 did investors finally discover how their trustees, the Walkers, had managed to soak up all of the potential profits, to the amount of nearly $900,000. First the brothers had taken in nearly

$500,000 right off the top for their bank discounts on the purchase of all the silver bullion, five times what they had to give out on its sale in the East. Then they had taken an additional $200,000 in payment for two mining claims that they had bought for less than $10,000, and finally they had taken another $200,000 in interest and payments on the undisclosed debt for the mills, on which $225,000 was still due! Rob and Sharp truly lived up to their names, clearing at least $3,800,000 from the Alice in eight years for a profit of over 2,000 percent of their investment, while the trusting shareholders had lost over 80 percent of theirs. Although the mine continued to produce for another dozen years, tripling its production to an impressive $12,000,000 and finally paying more dividends for a total of $1,075,000, it never paid a profit to anyone but the Walkers. The final dividends still amounted to only $2.69 a share, leaving a 70 percent loss for early shareholders, and even those who might have bought in at just under $2 a share after the crash only got back $1.50.[44]

The Walker brothers' successful float of the Alice in New York in 1880 soon inspired Clark and Davis to take in investors as well. Clark incorporated the Moulton Mining Company on Christmas Eve to push $10,000,000 in $25 par shares in the East. Although the mine hadn't made a splash on its own, he played on the Alice's early popularity, and with the help of his New York banking partner, Robert W. Donnell, he was able to privately place about half his stock at as much as $4 a share for at least $500,000, as he modestly reduced par to $5. Clark then steadily worked the mine to feed his mill, producing over $500,000 a year in bullion, which paid him well in both milling fees and sales discounts, but all the shareholders ever got back was $1.37 a share in dividends spread over nearly twenty years.[45]

Meanwhile, Davis, with the help of brother Erwin, succeeded in selling the Lexington to the French in 1881. By then he had taken out $1,500,000 in silver from the top eighty feet for a profit of at least $300,000, and the Alice had shown that good ore could be found much deeper. So the struggling French investors on whom Erwin had unloaded Holden's Old Telegraph in Bingham were offered a chance to recover their losses in Butte, and they bit. Reorganizing as the Société Anonyme des Mines de Lexington with a fully paid capital of $4,000,000, they bought the mine from Davis for $1,000,000 in cash, $250,000 in a short-term mortgage, and 15 percent of the stock, which he soon unloaded for at least $500,000.

Together with his earlier earnings, that gave Davis a total profit of over $2,000,000 from his $50 outlay and made him for a while "the richest man in Montana." The new French stockholders, of course, weren't so fortunate. Their energetic manager, Frank Medhurst, put up a new fifty-stamp mill and rivaled the Alice. He pulled out over $1,000,000 a year in silver and paid out $545,000 in dividends, until he fled to London in 1884 to escape gambling debts of $75,000. The company only paid one more dividend, of $64,000, in the next fifteen years. Yet there was an enormous bonanza of copper beneath all of these mines, and Clark, Daly, and others were already beginning to tap it, so even these despondent shareholders could still make money if they just held on.[46]

WITH SILVER PRICES on the decline, many investors were also attracted once again to new gold strikes, both large and small. A rich new strike was even made in the old Rocky Bar that finally let it pay a handsome profit, although this was shared by only a handful of investors. William Watt and friends had bought the mine from Andre Chavanne in 1870 but couldn't make it pay either. So in the fall of 1877 they finally sold nine of its eleven shares for a bargain price of $18,000 to a new enthusiast, David B. Kelly, and others still willing to try. They paid a 10 percent assessment, put up new machinery, and opened some more $50 ore that roughly paid expenses for the next couple of years. Then in the summer of 1880 they struck a bonanza of dazzling rock, which sold to jewelers in San Francisco for $43,000 a ton—the richest ever found in Grass Valley! In all they took out over $3,000,000 for a undisclosed profit that probably ran to a million, amply rewarding their faith by about fiftyfold.[47]

But a much bigger gold bonanza was also finally found at Bodie, which would produce over $20,000,000 in gold and become one of California's biggest mines. After Trenor Park's Empire failed in 1867, the Bunker Hill was foreclosed by their black caretaker, "Uncle Billy" O'Hara, and relocated as the Standard. In 1874 he offered it to two Scandinavians, Louis Lockberg and Peter Eshington, for $8,000 to be paid from any ore they found. They sank a shaft 120 feet but found nothing that would pay, and they were about to give up when their poorly timbered shaft suddenly caved in, exposing bonanza ore. In the next two years they paid off the mine, took out an additional $35,000, and exposed enough remaining ore to sell the mine for $67,500, more than a tenfold profit on their investment of three years' labor. The buyers were an old argonaut, Seth Cook, who with his younger brother, Dan, had finally made some money in Comstock speculations. Together with that seasoned shark "Uncle Billy" Lent and a younger but rising sharp, Johnny Skae, they organized the Standard Gold Mining Company in April of 1877 with a nominal capital of $5,000,000. But they held the shares for two years as a close corporation, while they began full-scale development with all the trappings of an aerial tramway and a stamp mill. During that time they took out over $1,000,000 a year in bullion, opened massive ore reserves, and paid themselves $600,000 a year in steady dividends, a nearly tenfold return on their investment each year.[48]

They suddenly made Bodie the new wonder of the world, and 8,000 eager new argonauts rushed in to make it, briefly, the biggest mining camp outside of the Comstock. Investors all across the country wanted a piece of its instant riches, and over seventy new companies were floated to feed over $500,000,000 in shares to the hungry crowd. Lent led the pack with over half a dozen companies nearly surrounding the Standard to offer $54,000,000 in paper. The best of these was his Bodie Mining Company, holding the claim immediately south of the Standard and offering $5,000,000. Lent and a few friends had to pay about $50,000 for it as a barely opened prospect hole, and that didn't include a reputed $15,000 for three gunmen sent out in an unsuccessful prior attempt just to run off its owners. But there were no clear surface croppings on the claim, so its $100 par shares fetched as little as 50 cents on pure speculation until May of 1878. Then his men suddenly tapped into the extension of the Standard bonanza, and the stock climbed steadily from 50 cents to nearly $20 by late July, when they hit an extraordinarily rich pocket. The San Francisco Daily Stock Report, alluding to the part "Uncle Billy" had previously played in such "malodorous deals" as the "Great Diamond Swindle," cautioned investors against Bodie as just another scam, and the price stalled just over $20. But the paper quickly turned bullish in mid-August, just as Lent suddenly launched a spectacular blitz of dividends, paying out $11 a share, or $550,000, in just one month, mostly to himself and his friends. That nearly gutted the bonanza, but it sent the price up to $52, as he and his friends unloaded over 20,000 shares for another $500,000, or more. By then the cynical editor of the Bodie

Standard suggested that "Uncle Billy" himself had hired the *Stock Report* to help hold down the price until he was ready to run it up and unload.[49]

Lent wasn't the only one to clean up on the Bodie deal, however, for some outside investors were also said to have made surprising profits. Right after Lent's first dividend the New York *Times* happily told of one penniless young speculator who pleaded with everyone he knew until someone loaned him $200 to buy four hundred shares of Bodie on margin at $1 a share, and, it was said, just two weeks later his shares were up to $10,000, with a $1,200 dividend on top of that. But the *National Police Gazette* topped that with a frantic story of an accidental beneficiary, a Comstock saloon keeper who reluctantly took a hundred shares of Bodie in payment for a $50 whiskey bill, when it was said to have been selling for only 40 cents a share. A while later he noticed that it was up to $3, but he couldn't find the certificate to sell. Each day the price climbed another dollar or two as he and his wife looked frantically everywhere, but it was still nowhere to be found. By the time it reached $35 they were "nearly crazy" for want of the stock. Then he suddenly noticed his little boy about to paste some "pretty pictures" on his kite. He grabbed the stock and sold it the same day, for a magnificent profit of 6,900 percent! That was such a good story that it lived on in several variations, doubtless hooking countless new gudgeons into the further manipulations of "Uncle Billy," for he still wanted more.[50]

In August of 1878, just before Lent unloaded the Bodie stock, he split off the poorer half of the claim as the Mono Gold Mining Company to get an additional 50,000 shares to work. He promptly started sinking a new shaft to look for a continuation of the Bodie bonanza and soon began sending out "confidential" telegrams "hither and thither," spreading tales of "wonderful developments" in the Mono. That worked the shares up to $10, while he unloaded them for yet another $500,000 or so, before it became apparent that there really wasn't any ore after all. In the meantime he and his partners had been buying back Bodie shares, which had fallen from $50 to barely $5 after the suspension of dividends and the commencement of assessments. The following spring they launched another "deal" to work the Bodie once again, this time on an eager new crowd of eastern innocents. Striking a new pocket of ore running as much as $1,000 a ton, they proclaimed it a "new bonanza" and resumed monthly dividends, which totaled $8 a share and drove the price up over $40 for several months while they unloaded once again. Then the dividends ceased, and the price collapsed. *Engineering & Mining Journal* editor Rossiter Raymond lamented that "the public was more success-fully duped in this operation than in any other," suffering losses of over $1,000,000, and he sadly concluded, "the Bodie rings are becoming worse than the Comstock." The Mono shareholders, meanwhile, in searching for its bonanza, paid out another $560,000 in twenty-two assessments before they finally struck a stringer in 1886 and were rewarded with a single paltry dividend totaling just $12,500![51]

The Cooks too, growing impatient for quicker personal profits, finally gave the public a chance at their private bonanza. Early in 1879 they reorganized as the Standard Consolidated, doubling the capital for a total of 100,000 shares at $100 par to place in New York. George Hearst's partner, James Haggin, made the market, selling the first 20,000 shares privately at $21, and then put it on the New York Stock Exchange, where it opened at $35. But even these watered shares still turned out to be a good investment for many, since over the next five years investors got back $35 a share, or $3,500,000, in dividends, so all who bought at those early prices at least got their money back, and many nearly doubled their money, before the dividends stopped for a while and the bottom fell out in 1884 as stock prices crashed to 20 cents a share. The Standard bonanza eventually produced over $20,000,000, paid out an added $800,000 over the next thirty years, and spread its wealth farther than most. Dan Cook showed off his new riches by joining the New York Yacht Club and buy-ing their fastest sloop, the *Tidal Wave*, while Seth plowed his money back into other ventures. Even old Frederick Bechtel, who had hung onto an adjacent claim, made a small fortune and retired to San Francisco at age fifty-four to marry a trophy bride thirty years younger. All of these daz-zling profits also inspired two former Empire investors, New York lawyer Samuel Barlow and banker Butler Duncan, to bring suits against several of the Bodie companies, claiming all of their mines still belonged to the defunct Empire, but the western press howled "blackmail!"[52]

By far the biggest new bonanza, however, was the great Homestake Lode in the Black Hills of the Dakotas, that surpassed even the Comstock. It would become "America's greatest gold mine," the biggest producer and

the longest dividend payer in the West, paying faithfully for over 120
years! It would also be the real source of George Hearst's great fortune.
But Hearst wasn't among the first on the ground this time, and he had
a rough start.[53]

The Sioux reservation in the Black Hills, granted in perpetuity by a
treaty in 1868, was one of the last treasure troves in the West from which
prospectors had been excluded. But rumors of gold in the Hills had circu-
lated for many years, from the coming of the first missionary, Father De
Smet, to a few furtive intrusions of prospectors. Pressure to renegotiate
the Sioux treaty had slowly grown, and Gen. George Custer, a part-time
mining speculator, was finally, in the summer of 1874 in violation of the
treaty, sent into the Hills with ten companies of cavalry, a battery of
Gatling guns, and a small battalion of prospectors and newspaper cor-
respondents to determine whether there really was enough gold to fight
over. They quickly fanned the flames with fabulous reports of gold "in
quantities so great that with pick and pan a single miner may take out
$100 per day"! The Sioux chiefs were soon summoned to Washington,
and as New York *Times* editor John Foord sarcastically noted under
the head of "Preposterous Pagans," that "they were offered $25,000 for
a region which was supposed to contain an enormous quantity of gold
and which was large enough to make a good-sized state." Yet, "only those
who knew the deep depravity of Indian character could have foreseen
that this magnificent offer would be refused, but refused it was." Thus,
they "drove the Government to leave them to their fate." When gold
seekers rushed in, no one was surprised when the Sioux struck back
and drove them out. So General Custer set out in the summer of 1876
to slaughter them into submission and was slaughtered himself. Some
called it "a fiendish massacre," but Foord countered, if Custer had actually succeeded in slaughtering the Sioux instead, that would surely have
been called "a glorious victory." The Sioux couldn't stop the "Black Hills
fever," however, for it was by then totally out of control.[54]

By the summer of 1876 nearly ten thousand had swarmed into the
Black Hills, wallpapering them with overlapping mining claims. Most
worked the placers, but the wiser ones started looking for the quartz
lodes that fed them. Among these were the Manuel brothers, Moses
and Fred, who had left Quebec in 1867 and wandered all over the West,
coming into the Hills late in 1875. They soon acquired an interest in a

couple of claims, the Father De Smet and Golden Terra, on what became the Comstock Lode of the Black Hills, an extraordinarily broad lead that stretched for over a mile south from Deadwood Gulch, just a few miles above the booming new camp. The following April, with two partners, they located two claims of their own, the Homestake and the Old Abe, on what turned out to be the richest ground at the south end of the lead. That winter they took out $5,000 from the Homestake and proudly proclaimed it "the mine of the Hills." In the spring of 1877 they sold the Golden Terra for $35,000 to an enterprising Denver speculator and former Nevadan, John W. Bailey, and they optioned the Homestake for $40,000 to California speculators, led by one of George Roberts's old cronies, John W. Gashwiler. A Missouri preacher's son, he sported the title of "General," but he was popularly known as "Gash," and he failed to come up with the cash.[55]

George Hearst, in the meantime, had been getting offers for claims in the Hills but held off because of the Sioux until the summer of 1877, when he finally sent an old miner, L. D. Kellogg, to go look over the mines. Kellogg picked the Homestake as the best buy, and got an option on it for Hearst for $70,000, payable in thirty days. Interested, Hearst asked his Ontario mine partner, James Haggin, to go in with him, but Haggin declined, claiming, "It is too far off, I don't want it." So Hearst offered Gashwiler a second chance at it, which he accepted, but then Gash got drunk and failed again to get the money. As time drew short, Hearst headed to the Hills to see the ground for himself and pay down his half of the money. He was quickly convinced that it was worth exploring, and by the end of October he had finally convinced Haggin and his brother-in-law, Lloyd Tevis, to put up the other half. They ended up paying fully $105,000 for the Homestake, $50,000 to the Manuel brothers, $45,000 to their co-locators, and an additional $10,000 to a Deadwood merchant who had taken a hundred feet for supplies and a small stamp mill. In November they formed Homestake Mining Company with a nominal capital of $10,000,000 and added quicksilver king Thomas Bell and a couple of other nominal trustees. Within six months Hearst clearly recognized that "while the quartz is not very rich, the amount of quartz that will pay a profit is truly enormous," and he prophetically told Haggin, "You nor your children will never live to see the end of the time when this property will not be worked for a profit!" But, he added,

GEORGE HEARST
Memorial Addresses on . . . George Hearst, 1894

George Hearst

"as you know, it will take time, patience and money," for "we have all kind of people to contend with and all on the make." But foremost they needed enormous and efficient milling power to work all that quartz, so they promptly assessed themselves an additional $100,000 to put up a giant eighty-stamp mill. Hearst also hired an aggressive and penny-wise superintendent, Samuel McMaster, to push the work.[56]

Hearst soon realized that the Homestake claim itself covered only a fraction of the great Homestake Lode and he wanted to control it all, so they started buying up all of the claims along its swath. But he had already let some crucial claims get away, and with many speculators now on the make, he would have to pay dearly to get them. Hearst particularly wanted the Father De Smet, which he at first thought was "the greatest gold mine yet discovered," but the owners wanted $700,000 and he turned them down. Other Californians, led by Archie Borland and San Francisco Stock Exchange president Coll Deane, moved in right after that and bought it for $400,000 in December of 1877. They too promptly put up an eighty-stamp mill, and they fought with Hearst for several years for control not only of the mines but of the water. Hearst did get a controlling interest in the intervening Golden Terra from John Bailey for $90,000, mostly in Homestake shares. But Gashwiler and his friends in the meantime finally scraped up some money and bought the Old Abe, right next to the Homestake, for $250,000. Hearst had passed it up at $45,000 as overpriced and was still waiting to get it cheaper, but he ended up buying out Gash at top prices. That blew the top off, and every claim holder felt the sky was the limit. The owners of the Golden Terra Extension wanted even more, so Hearst challenged their title in the courts, and when that failed he and McMaster simply seized the ground. They were promptly arrested for claim jumping, while the *Black Hills Times* lamented that the "California company is determined to possess all the good mines on the belt, and if unable to buy a mine, why then they steal it, and hold it at all hazard." Borland then further fanned the flames by buying that ground too for an exorbitant $200,000.[57]

The situation took a deadly turn after Hearst and his partners assessed themselves another $100,000 to build a second and even grander 120-stamp mill in the fall of 1878. They started buying up ground for the new mill site on the flat just south of the Homestake, at the east edge of what was becoming the company town of Lead. They paid anywhere from $100 to $10,000 for town lots and speculative mining claims. But one aggressive young operator, Abe Cohen, and his partners wanted much more for their Pride of the West, and Hearst balked. In January when McMaster began grading the rest of the site around them, Cohen started sinking a shaft and fortified it. In response McMaster hired some Deadwood gunmen to seize the shaft house and tear it down. When Cohen and his men returned to rebuild it, McMaster's gunmen opened fire with a shotgun from a nearby cabin, killing one of Cohen's men, Alex Frankenberg. The gunmen escaped, leaving behind only the shotgun, said to belong to McMaster.[58]

There was immediate talk of a vigilance committee to mete out justice to the assassins, who the *Black Hills Times* correspondent charged were "only too willing to do the dirty work of our large mining companies," vowing that the citizens had "stood the insolence and abuse of these companies long enough." But several of the gunmen were quickly arrested, along with McMaster, who was soon released. Hearst hurried back to Deadwood and wrote Haggin, "I found all of our troubles at fever heat, and most of them at a culminating point," but, he reassured, "We have all the boys out of jail except three, and they are there for murder and will be tried." The Homestake's lawyers defended them, and in a situation reminiscent of Hearst's notorious Hermes trial in Pioche, despite all evidence the jury declared them "not guilty." The outraged judge, Gideon Moody, denounced the jurors, charging the "Homestake Company have bought you outright," and he declared he'd "sooner have a jury of Pagan Indians!" McMasters, however, drank a toast to the jury, and the three gunmen quietly left the Hills "splendidly mounted and equipped" with "rifles, revolvers and whisky in abundance." Hearst finally defused that dispute by shifting the mill site slightly.[59]

By the spring of 1879 Hearst and his partners had paid out at least $500,000 for additional claims. They bought some of the claims with Homestake stock, and they recovered much of the remaining costs by temporarily incorporating the claims in several new, well-watered companies, the Deadwood, Golden Terra, Highland, and Old Abe, in which they needed to sell barely 10 percent of the stock to pay back their purchase price. But they still had many more claims yet to buy, and the Homestake was also mired in nearly forty civil and criminal cases in the courts. A frustrated Hearst traced their root problems back to Gash. As he wrote Haggin in March, "I think the advent of Gashwiler into this camp has been the cause of nearly all our troubles." Buying him out of the Old Abe at too high a price "has given outsiders confidence in thinking that they could at any time force us to pay out money." Gash would not be forgiven. Two years later Thomas Bell foreclosed on him for debts of $200,000 and forced him into bankruptcy. He died two years after that at the age of fifty-one, unable to make a comeback.[60]

Despite all these troubles the Homestake advanced rapidly. The second mill started up in August, and the mine was soon employing five hundred men and producing about $150,000 a month in gold. On January 15, 1879, it paid its first dividend, and Hearst and Haggin put it on the New York Stock Exchange the next day. It opened at $21, and with steady monthly dividends of 30 cents a share it rose to around $40 by summer, giving it a market value of around $4,000,000 and a solid return of about 10 percent per annum. Though "bitter feeling towards the Californians" still existed among some, the practical editor of the *Black Hills Times* praised the great progress in the mines and mills and asked, "What difference does it make whose money puts up such great enterprises—whether it

is Geo. Hearst's or the Angel Gabriel's—only so we get the big mills and the wealth of our country is thus developed? Then why this insane cry against the Californians who have made us what we are? They are our benefactors, and instead of speaking of Mr. Hearst as that 'd–d old Hearst,' he should be honored . . . by everyone expecting to live and prosper in the Hills. Our ore is low grade thus far, and the salvation of the country depends upon big mills." After Hearst made it clear that he was in the Hills "to stay," he wrote, "Many have come to me and fairly thrown up the sponge and signified a willingness to take anything I will give them. I think they are getting tired of fighting." Judge Moody too came around, becoming chief counsel for the Homestake and soon after that Republican senator from South Dakota.[61]

Archie Borland also won local praise for his development of the Father De Smet, but he worked it for short-term personal profit and soon won the condemnation of investors. He and his partners had incorporated as the Father De Smet Consolidated Gold Mining Company with what was by then the common California value of $10,000,000, although they worked it for the first two years as a close corporation, repaying costs and taking private profits from the richest ore before going public. Then on Christmas Eve of 1879 they too commenced regular monthly dividends of 30 cents a share, and just before Easter they put 45 percent of the stock on the New York Stock Exchange at $20 for an initial offering of $900,000. With the stock thus paying at a rate of 18 percent a year and Borland confidently predicting profits of $60 a share just from ore reserves, the shares were gobbled up as he unloaded those and more. But soon everything fell apart. The company's superintendent, an ethical mining engineer, Augustus Bowie, refused to swear to such an exaggerated ore estimate and resigned in protest, with many others quitting in sympathy. Then, after paying out only $2.40 a share, the dividends stopped and the shares crashed to $5, as the stunned investors howled. Unashamed, Borland then grabbed an additional $400,000 by selling the company a "worthless" water system, and the stung investors howled even louder. Hearst eagerly took advantage of the depressed price to finally, on New Year's Day of 1881, buy up a majority interest in the mine. Dividends were resumed under his management, and the shares jumped back to nearly $20. He paid out over $9 a share over the next five years and finally made it a very good investment for any who had bought in at bargain prices around $5 a share.[62]

With his purchase of the Father De Smet, Hearst at last gained full control of the entire Homestake Lode. Nonetheless, he continued to pick up additional property until he held fully 250 claims covering every inch of the great lode for over four square miles. Hearst also steadily increased the milling capacity, running six hundred stamps by 1889. But unlike so many other bonanza kings, Hearst didn't work the mills separately for his own private profit; he played it straight with his shareholders and gave them honest management. He also, of course, gave them good, steady annual dividends of 10 percent or more. The Homestake had paid out over $15,000,000 by 1900, and over $54,000,000 in its first half century, making it a marvelous investment for its early stockholders, who got an average return of 35 percent a year for the first twenty years and an average of 50 percent a year for the first fifty years! For the great lode was an enormous ore body that averaged only a fifth of an ounce of gold per ton, just $4 a ton, but it could be steadily worked at a profit of close to 25 percent, so it continued to pay dividends month after month for more than a century, as it was worked deeper and deeper. It ultimately reached a depth of 8,000 feet, the deepest mine in America, before it finally closed in 2001, having produced roughly 40,000,000 ounces of gold, then worth over $15,000,000,000.[63]

THE BLACK HILLS RUSH and a sympathetic excitement in the Big Horn Mountains just to the west also attracted countless cons. The most innovative was James Monroe Pattee, who freely offered mining stocks as lottery prizes and pioneered several other new promotional gimmicks. The son of a New Hampshire farmer, Pattee, in his early forties, had in 1864 turned from being a writing teacher to a wildcat promoter, recruiting postmasters to offer instant fortunes to their patrons with two-hundred-fold annual returns on shares of his People's Gold & Silver Mining Company from a mythical mine "on the borders" of the Comstock! A couple of years later he went to Nevada City as manager for another wildcat. While there, he helped raise money for a local school and suddenly discovered the even quicker and richer possibilities of lotteries. He promptly headed east again to launch a variety of charity lotteries in Nebraska and Wyoming. He advertised the tickets all over the country, and vowed, "I intend to get one dollar out of every man, woman and child; in ten years I will be the richest man in America." But once he had the cash in hand he was increasingly reluctant to hand out even part of it to either the charities or the lottery prize winners. Since Pattee soon dispensed with public drawings, it was easy enough for him to keep the large prizes, which went to less than one percent of the "winners," by assigning them to friends or simply to fictitious names. It was the great mass of small prize winners, who were promised $1 or 50 cents to keep up their hopes and keep them in the game, that amounted to his biggest outlay.[64]

So in August of 1876 Pattee created the Bullion Gold & Silver Mining Company, with 500,000 attractively illustrated certificates of $10 par, and sent out one to each of at least 100,000 small prize "winners" of his latest Wyoming Lottery, instead of their paltry cash prizes. He also included a circular extolling how its rich quartz veins, which assayed as much as $47,000 a ton, glistened like "golden ribbons" way up in the Big Horn Mountains, how its great Bullion mine held $50,000,000 in gold, never mind that was only a prospect hole a couple of hundred miles to the south, and how its shares would soon sell for $100 and pay monthly dividends. But there was a catch. He couldn't start paying dividends until he sold enough stock to buy machinery, and he had only succeeded in buying this amazing bonanza by assuring its discoverer that those who received a free certificate would surely buy ten or even a hundred more at a bargain price of $2 to raise the needed money. The lucky "winners," nonetheless, could still get a free ride and win yet another free certificate for every five shares that they sold to a friend![65]

After the New York *Times* exposed the Bullion as "a new swindling device" and the U.S. Postal authorities cracked down on the Wyoming Lottery, the ever-versatile Pattee reorganized in January of 1877 as the Silver Mountain Mining Company with 100,000 even more attractive shares at a much pricier $100 par. He promoted these shares with a new scheme that he called the "California plan," sending out half of the stock "free of cost, to reliable business men" in small towns throughout the country, on the sole condition that they agree to pay $1 assessments when called upon to develop the mine. Even before the first assessment was exhausted, he speculated, they might strike a bonanza and the shares could jump to $2,000. Even better yet, he offered to give the more enterprising businessmen one free nonassessable share for every five assessable shares they could get their friends to take. "Such an opportunity to become suddenly rich, in a bona fide legitimate transaction," he claimed, "was never before offered to any one east of the Rocky Mountains."[66]

When this too was exposed not only by the *Times*, but also by the later notorious Anthony Comstock of the New York Society for the Prevention of Vice, the irrepressible Pattee pushed the shares with yet another new scam in the spring of 1878, to recycle all of his victims again. Posing as New York claims adjusters, under a variety of names, he clipped the signatures off old letters from previous victims and attached them to letters from the "adjusters," advising them that, if that was their signature, they were entitled to five Silver Mountain shares with face values of $500, which the adjusters would register for them in the company books. All that the helpful adjusters asked was a nominal service fee of $2, and the money again poured in. Soon after the arrival of the shares, of course, came the added assessments, and out of all the money taken in, the company's superintendent later admitted, none was ever spent on the mines. After Comstock and the postal agents finally drove Pattee's operations out of New York, he simply moved them to Canada, where he continued until 1885. Throughout it all Pattee readily explained, to any who asked, that "the people wanted to be humbugged, and it was his business to do it." Still, he always stayed just far enough ahead of the law to keep out of prison, and he finally retired to St. Louis, where he died in 1888.[67]

Many others also worked mining scams on much smaller scales in small towns all across the country. Often they were simply the work of returning prodigal sons who seemed to take a special delight in exciting the envy and then picking the pockets of their not so fondly remembered old classmates and neighbors. Perhaps typical was the widely told tale of Alex Wilson, who in the summer of 1875 returned to his old home in Hickman, Kentucky, after six years in California. He threw a little money around and told everyone of his fabulous good fortune, finally striking it rich in the Raccoon ledge in the wilds of Inyo County. The local farmers were so impressed that they finally persuaded him to sell them a quarter interest, and they elected one of their own, William Garrison, as a trustee to return with Wilson to California. There Wilson took him to the "company's office" in San Francisco, where the "superintendent" informed them that they had arrived just in time, for before they could start taking out ore, water had started flooding into the mine and they needed to buy pumps. An assessment was quickly voted, and the clueless Garrison helped persuade the folks in Hickman to send money for the emergency, assuring them that the rich ore was there waiting and "the money would all be paid back with interest" in just a couple of months. As soon as the cash came, Wilson took it east to buy the equipment, the superintendent said he was heading back to Inyo to ready the mine, and Garrison was left in charge of the office. Garrison became suspicious after three weeks, when the rent came due and he had heard nothing from either man. He hurried to Inyo, but found that no one had ever heard of either the Raccoon ledge or Wilson, and he finally realized that he and his friends were out $20,000. An Inyo editor, however, questioned the story itself, claiming that he couldn't find anyone who had seen a man named Garrison, either.[68]

STILL, IT WAS the supposedly more sophisticated urban investors who seem to have suffered the greatest losses through the manipulated bonanzas and countless wildcats worked in San Francisco. Even the wildcats wanted $10,000,000 apiece, so by the end of the Big Bonanza boom in 1879, an outlandish $24,000,000,000 in new mining shares had been offered the public, equal to the total wealth of the nation! That was also more than ten times the total western production of gold and silver up to that time. With the rest of the nation sunk in a long depression, the potential market for nearly all that paper was also in California and Nevada, still buoyed up by that accumulated flood of over $2,000,000,000 from the mines. But their combined population was still under 800,000, so it mattered little how much stock the sharks pushed into the market, there was still a limit to how much they could stuff into the public. All the shares of the leading Comstock mines ran at a market value of around $200,000,000 through much of the boom, so even if they were mostly held on margin they still amounted to at least $20,000,000. Moreover, their sales on the San Francisco Stock Exchange topped $1,500,000,000 during those frantic years, shuffled back and forth ten times over, as sharks savaged the margins again and again with dire rumors and assessments. Those assessments too sucked over $50,000,000 out of the speculating public during the decade. So, after several turnovers on the exchanges, the marks could have actually paid out around $100,000,000, and counting assessments and another $50,000,000 they may have dropped on all those other untraded wildcats, would run the total for the Big Bonanza boom years to about $200,000,000. Adding in a likely $150,000,000 from the earlier boom years of 1869–73, the total taken from frenzied investors and speculators was probably about $350,000,000 over the whole decade of the 1870s. Again, dividing this by the roughly 500,000 adults in San Francisco and the mines, where it was said that nearly everyone held some shares, would require that they gambled on average about $70 a year apiece on wildcats and margins, or just 7 percent of typical miner's wages. Many abstained, of course, and plungers squandered far more.[69]

Indeed, just to stay in the game, speculators often had to pay as much, or more, than the margined purchase price of their shares in subsequent assessments on them. But the undying hope of still more bonanzas drove the seemingly endless willingness of many to keep putting up "more mud" for assessment after assessment on the Comstock mines. The biggest sinkhole was the Savage, which drained back $4,964,000 in "Irish dividends" in the 1870s, but at least it had actually paid out $4,460,000 in real dividends in the 1860s. Much more shameless were the Sierra Nevada, the Bullion, the Justice, and the Overman, which each sucked out over $3,000,000 in assessments from their faithful followers. Only the Sierra Nevada, which took in $3,850,000, ever paid any dividend and that was only $102,500! The abuses of assessments on the Comstock became so notorious that many companies, especially in the East, proudly publicized that their shares were nonassessable.[70]

Countless investors were sucked dry by the drain of Comstock assessments until they could pay no more and had to forfeit their shares at the company's auctions. The daily newspapers carried a steady stream of such auction notices, the death knells of shareholders' dreams, with seemingly endless lists of the names of all the "delinquents" whose shares were to be lost. Some were bankrupted and driven to desperation. One frantically wrote Sharon,

> Your time and mine is short. I have hoped that the assessment would be rescinded and save me from ruin. To come to poverty and misery, No, No, I would die first, you force me to it. The assessment is unnecessary—you and your party are now trying to freeze out the weak ones, for 18 months I have held on but now alas the worst is upon me & the Day of Grace is a short one and you shall have it. So help me God. PS You will not Believe this till you see the muzzle of my Revolver and then it will be too late.[71]

Investors, of course, were led to believe that their assessments were paying for exploration work in the mines, sinking shafts and running drifts and crosscuts searching for new bonanzas. But, in fact, the larcenous management of many of the Comstock's great "assessment bonanzas" systematically stole nearly all of it through one scam or another, and the investors were lucky if even 10 cents out of a dollar actually went into the mine. Moreover, as the shareholders of the ironically named Justice Mining Company learned all too painfully, things could get even worse if they did happen to strike ore. The Justice on the lower Comstock had been cursed with corrupt and lawless management for years, and in October of 1874 a dispute between rival factions of the directors, when the newly appointed superintendent of one faction tried to take possession from the other, culminated in a bloody shoot-out in which five men were killed. Then hopes suddenly brightened the following year when, after nearly $800,000 in assessments were paid, rich ore was at last discovered. The discovery only attracted new sharks, however, in the neighboring Woodville company, which had been taken over by George Roberts and John W. Pearson, a speculator and sometimes partner of George Hearst, branded by the company's former superintendent as "one of the most consummate liars, deliberate swindlers, and damnedest scoundrels that I have ever known!" They had been working their shareholders through both the market and assessments, when they saw a much bigger opportunity with the Justice's new bonanza. First, they reorganized as the Woodville Consolidated, doubling the number of shares and issuing all 60,000 new shares to themselves in payment for pretended rival claims. Then in December, pressing title claims against the Justice, they cut a deal with its directors to buy their controlling 60,001 shares of Woodville at $8, more than twice the market price, to preempt any legal challenges to the new bonanza. With this scam they grabbed a quick $480,000, while the Justice shareholders, of course, had to pay it off in added assessments.[72]

Outraged shareholders protested loudly and demanded an investigation. Although they didn't succeed in getting their money back, they did succeed in exposing the frauds, dumping the old management, and electing new directors. The Justice stockholders, however, only went from bad to worse. In May of 1876 they brought in a commanding forty-year-old Prussian, George Schultz, a longtime manipulator of the Bullion, one of the great "assessment bonanzas" that had already drained $1,800,000 from its shareholders in forty-five levies. Nonetheless, the Justice hopefuls seemed willing to suspend belief when he promised conservative management and big dividends. He promptly appointed one of his brothers as superintendent and aggressively worked their bonanza ores. In a year and half he had cleaned out most of the ore and more, producing an impressive $3,000,000 in bullion, which made it briefly the biggest producer on the Comstock outside of the Con Virginia and California. But instead of getting their promised dividends at last, the long-suffering investors instead had to pay $1,050,000 more in assessments![73]

For their new champion Schultz had learned well from the tricks of the Bank Ring and the Bonanza Firm. He had stacked the board of directors with his liquor partner and another brother, and they had given him a sweetheart contract to work all the ore in mills that he leased. Then, with the help of his other brother, the superintendent, who mixed in waste rock until the ore just paid milling charges, Schultz the reformer had stolen nearly every dollar of profit out of the mine. At the same time his brother had also used all the standard business scams of inflated and fictitious charges to further loot the treasury. But even that wasn't enough, and as the stockholders began to cry foul, Schultz secretly took another $250,000 from another neighboring company, the Alta, for signing a quit claim to that part of the Justice ore body which extended into their ground. Finally, in December of 1877, the irate shareholders launched

another investigation, headed by Lucky Baldwin, which exposed the whole scandalous record of "criminal mismanagement, fraud and villainy," and threw out Schultz and his clique. They also sued to stop the Alta from working their ore, but the dispute escalated as the companies hired crews of gunmen, fortified their works underground, and built an armored, "subterranean" juggernaut for a bloody showdown. Only the prompt intervention of the miners' union stopped bloodshed and forced a peaceful settlement. But there was no justice for its "woefully fleeced" stockholders. Although Schultz was threatened at gunpoint by one angry victim, sued by all the rest, arrested by the police, and indicted by the grand jury for embezzlement, a judge to the great surprise of most dismissed the charges and let him off on a seemingly irrelevant technicality. If he wasn't simply bribed, perhaps the judge was just repulsed at the hypocrisy of only prosecuting a small barracuda, when all the great white sharks were still free to roam all the way to the Senate.[74]

BUT ONE STUNG STOCKHOLDER, Squire P. Dewey, would soon correct that inequality with a spectacular class action suit against the Bonanza Kings, charging them with "gross frauds" in their management to the extent of $35,000,000! This was by far the biggest fraud suit yet in the western mines, and it was self-inflicted. It all grew out of escalating rancor, verging on a vendetta, between James Flood and Dewey. But the suit soon took on a life of its own, exposing flagrantly illegal dealings and costing the Firm well over $1,000,000 in judgments, legal fees, and a final settlement, plus far more in public opinion. Dewey was a humorless, rather self-righteous sixty-year-old speculator who had come to San Francisco from New York in 1849 and made a modest fortune in real estate speculation before turning to mining stocks as one of the original trustees of the great Ophir, and like others he also did well on the rise of the bonanza stocks. After the crash of the Bank of California in August of 1875, however, he had been among those who turned against the "Bonanza Gang," denouncing Flood for "hounding" poor Ralston "to his death." Flood was furious when he heard of the attack from one he claimed he had "always been kind and friendly towards," and he swore to his friends, "Some day I may give him something that will entitle him to feel bitter, and . . . make him howl in earnest."[75]

Flood's chance came just a few months later after the disastrous fire in October of 1875 that leveled half of Virginia City as well as the hoisting works of the bonanza mines, temporarily shutting the mines down. Fear that the Con Virginia wouldn't be able to pay its next monthly dividend of $1,080,000 caused stock prices to start down. An anxious Dewey asked the company secretary how much cash was on hand and was referred to Flood, who told him it was between $400,000 and $500,000. Dewey, concluding that no dividend could be paid, sold his shares before the price dropped any further, expecting to buy them back when the stock bottomed out. But, as he later learned, the actual cash balance had been $2,300,000, so the full dividend was easily paid, the stock fell no further, and Dewey lost $50,000 buying back his shares. Flood later claimed that the added money was actually in bullion, not cash, and that Dewey simply hadn't asked the right question. But Dewey claimed that the company checkbook clearly listed it as cash and that the day after the dividend was paid an additional $700,000 was transferred to Flood, so that only then was there just $500,000 left in cash.[76]

Dewey, now furious at being "deceived and defrauded" by Flood's "falsehood," suddenly turned reformer, and after looking into some of the Firm's operations, he showed up at the annual stockholders' meeting in January of 1877. There he succeeded in blocking the customary blanket endorsement of the trustees' actions and offered two modest resolutions, asking to have monthly financial statements and up-to-date maps of the underground works made available to the stockholders. These could have spared him from his own stock losses. In response, Flood blustered that Dewey was an "enemy of mine" who had only come to the meeting "expressly to insult me," and when a prominent British shareholder dared support Dewey, a "hotheaded" Mackay falsely accused him of complicity in the Emma swindle and contemptuously charged that "there was not an honest man in London." Nonetheless, after the company's lawyer assured that the resolutions were only advisory, since the stockholders had already turned over all of their power to the trustees, the Firm let them pass, hoping to wrap up business.[77]

Dewey, however, still had another, far more pointed resolution, calling for the company to buy its own mills and end the illegal milling contract that the Firm, as trustees of the mining company, had made with themselves as mill owners, cheating the rest of the shareholders out of millions. Over the protests of the trustees, Dewey pointed out that the Firm had already taken $5,320,000 from the company for working the ore at $13 a ton, which was twice the actual cost. Moreover, he bitterly claimed that they only recovered for the company 70 percent of the assay value of the ore, which they, not the company, determined, and that they then kept the tailings and slimes that held the remaining 30 percent, which they reworked for their personal profit. By then the Con Virginia had produced $39,000,000 in bullion, which he claimed left a staggering $16,700,000 in the tailings, which could have paid the stockholders another $10,000,000. Thus, in an impassioned plea Dewey called on the stockholders "to rid themselves at once of this vampire which is sucking out the very life blood of the mine." But the Firm quickly killed that resolution, and the meeting was closed. Even though Dewey was confused about the recovery, which was actually just over 80 percent, the tailings still held $8,000,000 and at least half was privately recovered. So if the company had indeed milled their own ore like most others outside the Comstock, the shareholders could actually have gotten an additional $7,000,000, or 25 percent more, in dividends. And that still wasn't the half of it.[78]

Such heated proceedings naturally found their way into the San Francisco and New York papers, where Dewey was generally praised for "his manly defence of justice and right," and Flood and Mackay were condemned for their "indecent and outrageous, bad temper" and "dirty purse pride." Yet even these modest resolutions were without force, for the Firm still tightly held the proxies for a majority of the shares. Dewey's only effective recourse was to bring suit. When he went to his lawyer, however, he found that Flood already had him on retainer. At this point he offered to drop everything if Flood would simply split his losses for $30,000, but Flood flippantly refused. So Dewey finally turned to an energetic young English lawyer and former ballad singer, John Trehane, to launch the giant stockholders' suit for fraud. For a nominal plaintiff, Trehane recruited a young *Chronicle* reporter, John H. Burke, and Dewey gave him a hundred shares of Con Virginia to make him a shareholder just two days before they filed the suit, on May 18, 1878.[79]

The "Bonanza Suit," *Burke et al. vs. Flood et al.*, in behalf of the Con Virginia stockholders charged the Firm with fraud, not only in their private milling contracts with the company, but also in their private lumber, water, and land sale contracts with the company and in the company's large discount bullion sales to their bank. Although each of these charges soon became a separate suit, the basis of all the suits was that any contract between the Firm as private parties and the company of which they were trustees was explicitly outlawed by the California Civil Code. If trustees were found guilty of making such illegal contracts, the stockholders could demand repayment of all potential profits at their highest market value. When it was all added up, the suits called for the Firm to repay rampant misappropriations totaling more than $35,000,000, plus about $10,000,000 in interest![80]

The Firm first fired back with an anonymous and "libelous" pamphlet, blasting Dewey as a contemptible "blackmailer" unworthy of consideration, and the *Territorial Enterprise*, which Fair had acquired, eagerly urged the public to "speedily consign S. P. Dewey to the Penitentiary." But Dewey indignantly denounced the blackmail charges and fired back a barrage of details, further exposing the "gross frauds" of the Bonanza Kings, in illustrated pamphlets of his own. The Firm made no public answer to the charges against them and in still haughtier indignation they even refused to testify until they were threatened with contempt of court. When they did finally talk. they displayed such an amazing ignorance of their business affairs that newspapers across the country ran a lengthy column of "What Mr. Flood Does Not Know!" Although Trehane had thoroughly documented the Firm's actions from the company's books, their lawyers bravely argued that their contracts with themselves were nonetheless legal, since they had been ratified by the majority of the stockholders, when the trustees voted the proxies of the very stockholders they were cheating, all still held in the name of their clerk, James Coffin. But the judge rejected the notion that trustees had the power to exonerate themselves of wrongdoing. Thus, the Firm was finally forced to resort to the statute of limitations.[81]

The first case to finally come to judgment was one charging the Firm with defrauding the stockholders of $10,000,000 by the illegal sale of a mining claim to the company. Their lawyers confidently claimed, however, that even if the sale had been illegal, the Firm could no longer be prosecuted because the sale was recorded in the company books, where a diligent stockholder could have found it and brought suit before the statute of limitations ran out. Burke had just waited too long. Trehane showed, however, that one part of the claim, which had actually been sold in half a dozen separate pieces, was in fact never recorded, so the stockholders could not have previously discovered it. Thus they were still entitled to sue for the fraud in that piece, and the judge concurred. Trehane then showed that this piece, just 12½ feet out of the total 1,160 feet in the mine, had been bought by trustee Flood for only $1,250 in 1872 and soon sold to the company for 125 shares at the par rate of 10 shares per foot, even though the shares were already selling for more than ten times the purchase price. Moreover, by the time of the suit in 1878, they had exploded with stock splits into 6,125 shares of Con Virginia and 3,573 shares of California, with a market value of about $200,000, plus dividends of $704,380. The Firm did not deny the sale. Thus the judge found them guilty of fraud and ordered them to repay the stockholders profits totaling roughly $1,000,000 including interest![82]

The brokers and financiers of both Pine Street and Wall Street were shocked by the decision, which could threaten countless other corporate boardrooms if the shareholders decided to revolt. But despite the triumph of stockholder rights over the managers of the richest corporation in the West, only a handful of additional stockholders signed up for their share of the payout. By then, of course, the Firm's Nevada Bank was the largest in the West, so it is not surprising that most weren't willing to face their displeasure for what amounted to barely $2 a share. Although they still had much more promising and profitable suits coming up against the milling and other frauds in the Con Virginia, and Dewey had filed another set of suits for frauds in the California, he and his associates began to see that it could be a hollow victory. Since they still weren't able to oust the Firm from control of the company, all of the money that they had just recovered, plus whatever else they might recover, was still effectively in the hands of the Firm, and there was no other penalty for breach of trust other than returning the funds. With the prospect of further defeats, however, the Firm was also tiring of the game, having lost irredeemable influence in the market, not for the charges of fraud themselves but for their failure to quash them. Moreover, as New York *Times* editor John Foord observed, the suits had also held them up to national scorn as egregious examples of "the arrogance of wealth and the shameless greed and trickery with which it carries out its plans." So in April of 1881, just a couple of weeks after the court decision, the Firm eagerly paid Dewey and the others a settlement to end all further litigation. The amount was rumored to be $180,000; Dewey crusaded no more, Trehane retired comfortably to England, even young Burke moved to the country, and the managers breathed easy again. The rest of the stockholders, who had refused to help fund the suits, of course got nothing. The biggest impact of the suits, however, had already been felt in their added impetus for the corporate and market reforms enacted in California's new constitution of 1879.[83]

MANAGEMENT FRAUDS HAD in fact become so flagrant that Warren Ewer of the *Mining & Scientific Press* joked that even

> a mine which turned out twenty-dollar pieces would never pay any one of its stockholders, except the immediate managers themselves. They would have caves, fires, overflows and accidents of all kinds to prevent the mine being worked, and would at the same time be taking out the "twenties" privately from an "old shaft" or "deserted tunnel." When all the twenties were gone, as they soon would be, they would sell out themselves and wouldn't even pay the bills, but leave them to be paid by assessments. Then with an air of injured innocence, they would deny everything on general principles, or with Tweedish effrontery ask: "What are you going to do about it?"

"That is just the difficulty," Ewer grumbled further. "Of all the barefaced swindles and salted mines of which we have heard, we have yet to learn that any punishment has been meted out to those connected with them. The larger the swindle, the less danger of trouble. A poor devil of a miner who steals a bucketful of rock from a 'specimen mine' will be driven from camp, while a 'promoter' who hangs around honest men, bonding mines, salting mines, selling holes in the ground at exorbitant prices and swindling right and left, is looked up to as a smart chap, and his society cultivated." The best that Ewer could hope for was a new vigilance committee that would round up all those "who have practiced these infamous operations, put them down in the deepest shaft on the coast, and cover them up with the worthless shares they have sold and the salted rock they have exhibited!"[84]

For most stockholders, however, assessments were the biggest and most obvious drain, and they alone had become so heavy that investors seriously began to ask, "Does mining pay?" Boosters eagerly tried to prove that it was indeed profitable by simply showing that for the 103 Comstock companies traded on the San Francisco exchanges, dividends collectively exceeded assessments $116,000,000 to $62,000,000 by mid-1880 for a net profit of $54,000,000. What they didn't bother to mention, however, was that individually dividends topped assessments in only six of those companies and that the vast majority, fully 94 percent, were all losers. Moreover, the simple difference between dividends and assessments in this tally was still more deceptive, because it completely ignored the initial cost of all that stock, which greatly exceeded its potential sale value. Indeed, the nominal capitalization of those shares topped $1,000,000,000, and the total paid in capital was at least $200,000,000, so in truth the total cost, including assessments, was more than $262,000,000, while the total return in dividends, plus potential sales at the 1880 market value of less than $20,000,000, was only $136,000,000, leaving a collective loss of over $126,000,000! And it was really much worse than that, because the list didn't include hundreds of other Comstock companies not listed on the exchanges that also levied assessments and never paid a dividend. So, as Adam Smith had observed a century before, mining investment was "perhaps the most disadvantageous lottery in the world."[85]

Lack of dividends from wildcats, of course, was to be expected. But what finally cooled the speculative fever was first the skipping and then the suspension of regular monthly dividends in the Big Bonanza mines, as the bonanza ores were finally exhausted. Their stock prices had already begun to reflect concern over their longevity. In the initial excitement

THE CURSE OF CALIFORNIA
The Wasp, July 14, 1877

the Bonanza stocks had surged up until their dividends paid barely
7 percent a year, but after that the public, looking for new bargains,
hadn't pushed their prices on up as fast as their dividends, so their
annual return had reached 12 percent at the peak in 1875. Then when
the Con Virginia began skipping dividends in 1877, the price fell off until
the annual returns reached a cautious 40 percent. But even that wasn't
enough to make it a profitable buy, since the Con Virginia dividends
finally stopped entirely the following June, and California did the same
two months later. Even the long-awaited completion of the Sutro Tunnel
couldn't reverse the decline.[86]

DESPITE THE DETERMINED EFFORTS of the Bank Ring with a $200,000
"corruption fund" for lobbying Congress, Adolph Sutro had defeated
further congressional attempts to repeal his franchise, doubtless through
his own generous distribution of shares. He had at last obtained financing abroad through the dedicated efforts and close family connections
of two friends, Joseph Aron and George Coulter. Aron, who had personally supported Sutro's wife and children for several cash-strapped
years, was an in-law of the Parisian bankers the Lazard brothers, and
Coulter, who had promoted the highly profitable Sierra Buttes mine in
England, was a cousin of London bankers Robert and Hugh McCalmont.
After the close of the Franco-Prussian War, his friends persuaded their
relations to take $1,350,000 in shares; others soon joined in for an addi-
tional $750,000; and Robert McCalmont eventually took a mortgage on
the property for another $1,000,000, while the company ballooned the
shares to $20,000,000. That finally gave Sutro enough money to really
begin effective work in December of 1871, putting on a crew of over two
hundred men and bringing in machine drills. Once again Sutro talked
exuberantly of completing the tunnel in just a couple of years, which
would have brought it in at the peak of the bonanza years. But it was not
until just before midnight on July 8, 1878, that the tunnelers broke into
the Savage mine at the 1,650-foot level, and a triumphant but exhausted
Sutro climbed through the hole. The 20,500-foot tunnel was finished at
last, at an actual cost of just over $2,000,000, though Sutro spent another
$1,000,000 on management and promotion, as well as a 6,000-square-foot
mansion and other improvements at the tunnel mouth.[87]

But it was a hollow triumph. For the Big Bonanza ores were nearly worked out, and the Bonanza Kings and most of the other mining companies still refused to pay Sutro a royalty on ore, claiming the tunnel was of no use to them. Only the Savage and two others, not taking out any ore, agreed in principle, which cost them nothing. Some, however, did talk of pumping water out of their mines into the Savage to drain out through the tunnel at no cost as well, and a frustrated Sutro built a watertight bulkhead to block the flow and had started a drain to return the water to the Savage when his men were arrested. After sparring for a year, Sutro cut his royalty demands in half and reached a deal, with most of the companies agreeing to pay $1 a ton for all ore averaging less than $40 a ton. During the bonanza years the mines had produced over 500,000 tons a year running over $55 a ton, which would have paid the tunnel over $1,000,000 a year, but those days were past. In 1879 the companies took out less than 100,000 tons, averaging $39 a ton or less and paying the tunnel less than $100,000 a year, and they held it to less than that thereafter.[88]

Nonetheless, at the end of June 1879, as the first water finally began to drain from the Comstock, Sutro proudly celebrated, with cannons and bonfires, the end of the royalty dispute as a final triumph. But privately he had come to realize that the royalties would not pay a profit on the shares, of which he was still by far the largest holder, with roughly 600,000. His only chance now to make the fortune that he had so long dreamed of and fought for was to dump his stock. So after listing the tunnel shares on the New York Stock Exchange in December of 1878, Sutro boomed the tunnel all across the country, triumphantly proclaiming that it gave the Comstock "a new lease on life." It opened a new, "glorious era" of low-grade ores "worth hundreds of millions of dollars" that he promised would pay tunnel profits of millions. But at the same time he secretly contracted with a New York broker to slowly unload his shares at anywhere from $5 to as little as $2 a share. That all went so quietly and smoothly that he also appropriated an additional 300,000 shares out of the treasury, or so Aron later claimed, and privately unloaded them on many of his old friends for added cash. In all, his total haul must have been at least $3,000,000, some claimed as much as $5,000,000.[89]

Then he surprised everyone with the announcement that he was quitting as superintendent, and when it was learned that he had dumped all of his stock, a furious Aron turned against him for his fraudulent practices and Coulter denounced him as an "arch fiend." He was also dumped by his wife, after she caught him with a mistress he had picked up in Washington. After a failed bid to buy a Nevada Senate seat, he retired to San Francisco, put his money into real estate, and settled for a term as mayor. The tunnel investors never received any earnings, and the shares they paid as much as $5 for could be had for a penny. Robert McCalmont, who suffered a paralytic stroke, was the biggest loser; out $650,000 in stock and nearly $1,600,000 on his mortgage with interest, he recovered only $800,000 when his brother eventually foreclosed in 1887.[90]

AFTER NEARLY TWENTY YEARS of consecutive bonanzas, most of the Comstock's stock addicts just couldn't believe that it was over. One cheerleading broker, San Francisco Stock Exchange president Coll Deane, still tried to keep hope alive and shares trading with fantastic tales of three secret, hidden bonanzas. But the inevitable decline was halted only fleetingly that fall, when Johnny Skae's "Sierra Nevada Deal" brought stocks soaring back for a deadly slaughter on the promise of a fabulous new bonanza.[91]

Johnny Skae, a clever Canadian in his late thirties, had come to the Comstock as a telegraph operator just before the discovery of the Big Bonanza. He was rumored to have secretly cracked the coded communiques from Fair to Flood, cleaning up, some said, as much as $3,000,000, on the market. But after losing access to telegraphic tips, he turned to various long shots. Among them was that old assessment eater, the Sierra Nevada, which he bought into with Senator Jones in 1875. Skae became its president, and in a quest for a long-dreamed-of bonanza at the north end of the lode, he levied heavy new assessments on shares—already smothered by stamps of old levies—to sink a new shaft on ground disputed with the adjacent Union Consolidated. Early in the summer of 1878 some indications of rich ore were seen on the 2,100-foot level of the mine, and Skae ordered his broker to start buying up Sierra Nevada stock, then selling for less than $5 a share. Buying on margin, Skae eventually was said to have cornered between 24,000 and 50,000 of its 100,000 shares before the price reached $8 in late July. Then in a theatrical play, while en route to New York, Skae got an urgent telegram, he said, telling of the discovery of bonanza ore. He immediately ordered all work stopped and, in a grand flourish, chartered a special train to rush back to the Comstock. He even paid a local bard $100 to pen an epic poem about the charade. All this, plus rumors that the "whole face of the drift assays an average of $500 a ton," ignited a stock frenzy that drove its shares up to $80 by the end of August.[92]

After much lobbying, Skae also got the federal land commissioner to declare the Union Con's title invalid. And in celebration, the "Little Napoleon," as he had come to be known, threw a spectacular Sunday brunch of trout, wine, and song for about 150 of his friends. But the Bonanza Kings had picked up the Union Con just in case there really was a new bonanza, and although they fought the title grab, they eagerly helped the play proceed, as Flood made a show of paying his friend Robert Sherwood $1,000,000 for 5,000 shares of Union Con. The deal, of course, also greatly helped them unload again their own Con Virginia and California shares, which rose from $5 to $50 in all the excitement. The rest of the market followed, rising by more than $100,000,000 from May to September. Flood, however, milked the market a little too heavily, causing a temporary reversal. That in turn caused Skae to almost blow the whole deal by threatening, horror of horrors, that if "any further attempt be made to break the stock, the work of extracting ore will be commenced!" Instead, Skae staged a stall to further raise anxiety and prices by starting a new incline down to the 2,200-foot level and then a crosscut to test the depth of his great bonanza. And despite his claims of a vast treasure vault of ore all ready for the working, he also cheekily levied another $4 a share in assessments. Most speculators only chalked that up to a "bear movement" by the management trying to get their stock and cheerfully paid it, hanging on all the tighter as the shares rose ever higher. Sierra Nevada topped $261 in late September as the incline inched downward. Then the price settled back to around $200 as Skae, Jones, and other insiders slowly began to unload, while, trying not to break the price, they earnestly assured outsiders that the shares would go to $500 when the incline at last struck the bonanza. Most faithfully held on for the great day, though one critic wondered if any had ever considered just how big a bonanza they would need to pay a profit on an investment of $50,000,000 in the two mines.[93]

Finally, on November 18, just as the crosscut reached the lode and Skae arranged a grand celebration with the biggest band ever known on the Comstock, the stock crashed, plummeting from $180 to $60 in just three days. Stockholders were slaughtered as their margins evaporated and their stock was "swept away in the grand whirlwind of ruin that was leveling rich and poor alike." Archie Borland, who again held his shares, claimed he lost $1,400,000 in ten days. An angry mob, "ruined by the panic," surged through the streets of Virginia City, threatening to seize the mine and hang Skae high! He hastily let a few outsiders into the mine, while Jones, who had already sold out his 5,000 shares for close to $1,000,000, still assured that "it never looked or promised better," but it was too late, "the bubble had been pricked!" Flood, one jump ahead of Skae, had dumped his Union Con stock, and to wipe out Skae's remaining stock value, spread rumors that the vein was barren. The extravagant "Little Napoleon" had gotten himself over $1,000,000 in debt to Flood's Nevada Bank and others, and Flood effectively foreclosed. Although he would remain for a while as a figurehead, a week later Skae quietly signed over control of the Sierra Nevada to Flood, who appointed a new superintendent and deeded most of the disputed "bonanza" ground to the Union Con. Thereupon Skae's "immense bonanza" was finally revealed to be no more than a small, stray pocket artfully opened by a mere "coyote hole." Three years later, with a sense of retribution, it was widely reported that Skae was "penniless," after he was arrested for drunkenness and was unable to pay $5 for bail, and just a few years after that, at the age of forty-four, he was said to have died "in poverty and obscurity." But, to spoil a good story, he was in fact far better off than his followers, because, like any modern con man, he had put more than $150,000 plus valuable real estate in his wife's name before the crash, and the two lived quite well on that.[94]

The Sierra Nevada Deal had a devastating impact. It had inflated the market for three months with the illusion of another magnificent bonanza. It bloated Sierra Nevada shares by $25,000,000, Union Con by $19,000,000, and the rest of the leading stocks by over $60,000,000. At the same time other stock sharps unleashed a pack of more than 150 new wildcats to push an additional $1,000,000,000 on eager speculators. Tens of thousands were sucked into the frenzy to be ruined in the crash. So great were the swirling paper profits that they had even excited the envy and greed of speculators as far away as that "staid old Hub" of Boston. There a syndicate, led by bankers Stone & Donner, sank about $5,000,000 into the maelstrom, while an excitable bank cashier embezzled $64,000 to lose in the crash. The deal drew in tens of thousands, and in the "fearful carnage" of its crash nearly all suffered and countless were ruined.[95]

YET NOT ALL had sympathy for the "victims of Sierra Nevada." "It has been a notorious bubble from the first," the Sacramento *Record-Union* editor argued, the speculators "had no one to blame but their own credulity and gullibility. . . . No man buys stocks because he believes the mines are worth what they stand for in the market, but because he thinks that some other people will be bigger fools than himself and he will win their money. Those who have been cheated this time went in hoping to cheat others . . . none of them have any right to howl when the trap they have laid falls on themselves." But Comstock editor Judge Goodwin of the *Territorial Enterprise* again blamed the crash on margins, claiming that "all had squeezed and stretched their last cent in buying" and couldn't raise another cent to save their shares when the crash wiped out their margins. "Had everybody owned his own stock the panic would have been stopped . . . without margins such a panic would have been impossible," he argued. "Margins are the misery of both brokers and their customers. . . . They should be abolished!" The following spring, with the wounds of the "Sierra Nevada Deal" still fresh, California voters would do just that, adopting a new state constitution that came down hard on stock speculation.[96]

The action was the culmination of a great groundswell. Public and press criticism of the "corroding ulcer" of mining stock speculation had grown ever louder as every new stock deal took its toll on the fortunes and futures of the credulous. The San Francisco *Chronicle* had raged for years against the "stock-jobbing juggernaut," and the "army of robbers that fatten at the Stock Board." Ministers warned that the "darkest winged angel of hell and destruction had established the seat of his empire on Pine Street." Even faro dealers claimed that stock gambling was "more vicious than bunk." The San Francisco *Call* ran a long, black-bordered column with a skull and crossed bones, listing scores of obituaries, headed the "Roll of Death: Suicides Assigned to Stock Gambling." Still, a few defenders saw "a brighter side" to mining speculation, warning that the great "dense cloud that mantles the manufacturing district of the city would drift far out to sea, and in its stead would droop upon the foundry walls the gloom of quick decay, were this source of speculation cut off." Others argued more mildly that the speculation simply "scatters coin, but nothing is lost to the community." But Ewer of the *Mining & Scientific Press*, always a solid supporter of honest mining, countered that in fact the coin was instead just gathered "too often from the one who can little afford it, into the coffers of the rich and unscrupulous, who do the manipulating." And he concluded, the "stock gambling mania is the grandest scheme yet devised to make the rich richer and the poor poorer, and much of it may be termed legitimate robbery." Other old mining men, such as Almarin Paul, also joined the attack with an impassioned condemnation of this "idle, non-productive, blood-sucking, parasite of mining," and called for efforts to "destroy this wild stock gambling spirit."[97]

The call was first taken up in December of 1877 by the Democratic governor, William Irwin, who called upon the legislature to protect mining investors from "the mismanagement and rascality of the directors." He happily suggested a tax of half a percent on all mining stock sales as the best way to curb the rampant speculation that had wreaked "such terrible and widespread evils." Such an act, the Act to Encourage Mining and Suppress Stock Gambling, was promptly introduced in the Assembly that included not only the tax, but a grand state mining czar to whom all brokers had to report monthly and all companies weekly, with fines of up to $100,000 for failure to comply! Despite anguished outcries from brokers and much of the press against the threatened "strangulation of the stock business," the bill passed the Assembly by a resounding 52 to 15. But a reported cash infusion of $28,000 from brokers killed it in the Senate.[98]

Next the banner was taken up by the radical Workingmen's Party, which won control of the convention drafting the new Constitution of the State of California, adopted in May of 1879. There they too argued that "if there is any one evil above all other evils that is affecting the people of California today, it is the evil of stock gambling," which impoverished both the people and the mining industry alike, taking "the very capital that ought to invest in legitimate mining" and flushing "it into the worst form of gambling." Although they failed in an over-the-top bid to abolish all stock exchanges in the state, they did deal a crippling blow to mining speculation by finally adding a section declaring all margin and future sales of stock void and allowing buyers to sue for lost margins! With a direct eye to the Bonanza suits, the subsequent legislature also

passed the Act for the Further Protection of Stockholders in Mining
Companies, but it lacked teeth. The act attacked the proxy control of
the management by requiring that all stock be listed in the names of the
"real owners" and that only they could vote their shares, but the Firm's
officers, meeting in Virginia City, defiantly refused to comply, since there
was no penalty attached. The legislators also went after insider deals,
requiring that at least two-thirds of the stockholders approve any sales
and purchases of claims, but through management control of the proxies it had no effect.[99]

The margin ban, however, was stunningly effective. Supporters argued
that it did not interfere with "any honest sale of stock" and would only
"prevent those swindling sales of stock which the seller has not got and
the purchaser does not want, which is simply betting upon the rise or
fall of that stock." It did indeed stop such sales, and trading volume
plummeted, as the brokers angrily decried the "iniquitous law" and hotly
denounced its authors as "communists." Foord of the New York *Times*
also joined the chorus of outraged wealth, damning the whole constitution as "the first practical illustration of 'Communism' . . . designed to
bleed the rich and fill the pockets of the poor," and direly predicting that
"the Communists will soon find that they have killed the goose which
laid the golden egg." Most expected that the margin ban would promptly
be repealed, but it was upheld by Oliver Wendell Holmes in the U.S.
Supreme Court, and it remained on the books until 1908. Ultimately,
regulation of the margin rate to control stock speculation would also
become a very effective part of the first federal market regulations, the
Securities Exchange Act of 1934.[100]

The passage of the new constitution further hastened the decline of
the market. By the beginning of 1880 the principal Comstock mines had
a market value of only $7,000,000, compared to almost $400,000,000 at
the peak of the Big Bonanza excitement five years before. After nearly
two decades the Comstock's dominance of western mining finally came
to an end with the virtual exhaustion of the Big Bonanza. With that
also came the end of San Francisco's dominance of the mining stock
market, and together with the great backlash of demoralized and impoverished investors and speculators at the polls, that in turn sent most of
the sharks east.[101]

STILL, THE GREAT mining stock orgy had raged nearly a decade, sparked by the completion of the transcontinental railroad and fed by the discovery of the Comstock's biggest bonanzas, quickly gutted for instant fortunes and an irresistible show of dividends. A host of other new bonanzas throughout the West also fed the frenzy, driving western gold and silver production to a new high of $730,000,000 for the decade, sending mining dividends up tenfold to over $130,000,000, and inspiring such infamous schemes as the Emma and diamond swindles. Yet, ironically, the biggest discovery was the great Homestake bonanza, which in sharp contrast to the Bonanza Kings, George Hearst would patiently work to ultimately surpass even the Comstock. But it was all those flashy fortunes and dazzling dividends that had excited the avarice of investors, and the ever-ready sharks plied them with over $24,000,000,000 in gilded paper. Investors had wildly wagered roughly $350,000,000 and lost nearly all of it. For even in the Big Bonanza mines, the biggest mining dividend payers of the century, most investors lost because the market prices rose so quickly that they got in too late. By December of 1874, only six months after Con Virginia started paying dividends, its market shares shot above the record-breaking $42,000,000 that it would eventually pay back in dividends. Although the price fell off after it peaked at $80,000,000, it dipped only fleetingly below what it would eventually return, before the dividends finally stopped and its price crashed to barely $1,000,000. So even those who bet on the richest mine of all once the boom started lost most or at least some of their money. Worse yet, those who bet on the runner-up California even before it started paying out its grand $31,000,000 also lost heavily, because its market price shot above $50,000,000 just in anticipation and never fell to less than its future dividends. Finally, since the prices peaked early and fell slowly, there wasn't even much chance for outside speculators. Thus it was all these sobering losses, not just in the wildcats but in the biggest dividend payers as well, that finally drove most western investors and speculators out of mining shares for nearly twenty years. But all the while a whole new flock of lambs had grown up in the East, and George Roberts and his friends were ready to shear them.[102]

THE MINING WEST IN THE 1880s

6

THE SILVER SHARKS

The year 1879 marked another watershed in western mining, and the numerologists played it up like 1869, 1859, and 1849. It marked the fall of Virginia City and the rise of Leadville, Colorado, as the new boss camp. It also marked a great shift in the vortex of mining stock speculation from the West to the East. For soon after the passage of California's new constitution, many of the slipperiest sharks, led by George Roberts, left the sinking market and moved their operations to New York. There, with no impediments to their trade, they would resurrect San Francisco's mining stock mania. There too the expatriate Californians were joined by several other notorious western sharks, including Colorado senator Jerome Chaffee, his confederate David Moffat, and New Mexico politician Stephen Elkins. Roberts was their leading spokesman, and he soon dominated the new mining market. He piously complained to a *Times* reporter that California's "iniquitous law" had "practically put stock speculators on a level with criminals," and they were vilified by the "continued and baseless attacks" of newspapers and politicians who failed to see that they were just "capitalists . . . honestly striving to build up the country with their own fortunes." So, "when war was made on the great industry of the Golden State," he and his friends "gathered their household gods and came to New York," and he promised "all the good mines of the West will in a short time be organized here."[1]

New York investors and speculators were ripe for the picking. With the eastern states finally recovering from the Panic of 1873, the Californians found a burgeoning interest in western mines, which they quickly shaped into a new mining mania to work eastern pockets. They also found little effective competition. Brokers on the New York Stock Exchange were still obsessed with railroad stocks, and although George Hearst had placed two solid dividend payers, the Homestake and the Ontario, on it, its handful of other western mines were old flotsam like Mariposa and Quicksilver. The old New York Mining Stock Exchange had revived in 1876 at 62 Broadway to list many of the Pacific Coast stocks and satisfy local demands to get in on the Comstock boom. But its brokers were too far removed from the action to have any impact on the game, and they had to rely on San Francisco manipulators to make the market.[2]

The Californians would soon change all of that. They wanted an exchange of their own, and in December of 1879 George Roberts and his friends formed an aggressive rival, the American Mining Stock Exchange. New York brokers also clamored to get in, and the initial two hundred seats, selling at $1,000 each, were oversubscribed by a factor of two, so an additional three hundred were "rented" at $5,000. The new exchange was organized closely after the San Francisco Stock Exchange, whose former president and caller, George Smiley, came to New York to fill the same positions, and even its posh interior fittings down to the spittoons "imitated as closely as possible" those of its patron saint. It opened on Broadway at number 63, right across the street from its rival, with great fanfare on June 1, 1880, and after windy speeches and a brief call of stocks, the brokers closed with a lavish champagne lunch at Delmonico's. Since New York banks didn't consider mining shares good collateral, the Roberts syndicate also formed a new bank, the Mining Trust Company, later renamed the Mutual Trust, with a capital of $5,000,000 to carry 20 percent margin sales at 18 percent per annum to open the speculative market to a wider audience. One unappreciative cynic, however, later likened it to a "fly-trap, pure and simple." Roberts, nonetheless, also shrewdly courted members of the New York Stock Exchange, particularly "Deacon" Stephen V. White, a "bold and fearless operator," later hailed in the *Wall Street Journal* as "the most trusted man in Wall Street." In return for his influence, Roberts cut White in on profitable mining schemes and helped elect him president of the New York Mining Stock Exchange.[3]

Not all the California expatriates, however, had come to feed on gudgeons. Darius Mills, who by then in his mid-fifties had drawn over $10,000,000 out of his quicksilver and mill ring schemes, also arrived in 1880 to invest some of his millions in New York real estate. Planning to spend his summers in the East, he bought a mansion of "Oriental

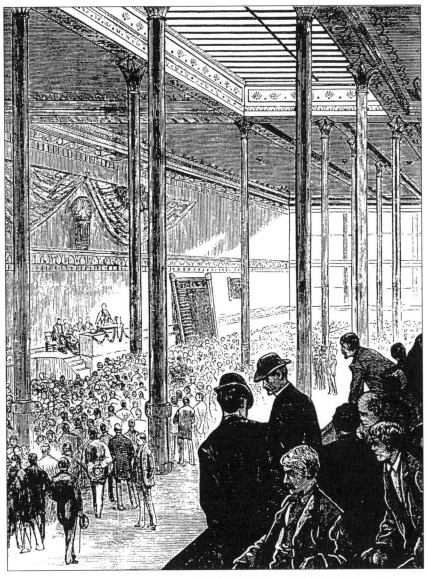

The American Mining Stock Exchange, No. 63 Broadway
Frank Leslie's Illustrated Newspaper, June 19, 1880

magnificence" on Fifth Avenue and kept his estate at Millbrae in California for the winters. The following year he bought another mansion on Madison Avenue as a wedding gift for his daughter when she married Whitelaw Reid, who ran the New York *Tribune* for Jay Gould, who had taken over the paper after Horace Greeley died. But Mills made his biggest splash by buying up most of the block across from the New York Stock Exchange for over $1,000,000 and tearing down all the old buildings. There in 1881, with another million of his Comstock loot, he put up an "immense new structure," the Mills Building, which the New York *Times* later call "our first skyscraper." Built by architect George B. Post, it was twelve stories high, resting on 17,000 pilings and "towering over the skyline of New York." It was heralded as "the highest business building in the world," and "many of the nation's powerful men," from J. P. Morgan on down, all rushed to be among its eight hundred tenants. Perhaps in a repentant mood, Mills also built a string of "Mills House" hotels on Bleecker Street and elsewhere to offer clean, cheap rooms to the less affluent, and he indulged in countless philanthropies. His fortune topped $36,000,000 before he died in 1910.[4]

BUT IT WAS the rush of California stock sharks who stirred up a feeding frenzy in the mining market and caused the greatest excitement. Eastern interest had been piqued by the phenomenal new silver-lead discoveries in Leadville, and they led the way. Up at the crest of the Rockies, two miles above the sea and eighty miles west of Denver, the new boss camp leaped past the fading Comstock in 1879 with an annual production of over $11,000,000, compared to the latter's $7,000,000. At the same time it eclipsed Virginia City as the largest mining metropolis in the West, boasting a population of 15,000, three daily papers and several weeklies, a railroad, an opera house, and of course a "bull pen"—the Leadville Mining & Stock Exchange opened in April. In the East many still looked upon mining stock as an investment rather than a speculation. So most of the action on the eastern exchanges was in making a market for new shares with hype and "wash sales," rather than working up stock deals in established shares, as was done on the San Francisco exchanges. The results, however, were the same for the public, for even Colorado's new instant millionaire, "honest miner" Horace A. W. Tabor, helped fleece them.[5]

Tabor was a none-too-prosperous storekeeper in his late forties who suddenly turned lucky when he grubstaked the discoverers of what proved to be one of Leadville's richest and most spectacular mines, the Little Pittsburg. Tabor's and Leadville's early fortunes were closely linked. He had come to the area with the first prospecting party in 1860, briefly working the gold placers in California Gulch, then he returned with his wife, Augusta, in 1868 to run a small store in the gulch for nearly a decade before the discovery of rich silver-lead deposits and the birth of the boom. Tabor's rapid rise to fame began in April of 1878, after a local restaurant keeper, George Fryer, first tapped into the richest deposits on what became known as Fryer Hill. The following month Tabor outfitted two hopefuls, August Rische and George Hook, with grub and tools at a cost of only $17 in exchange for a one-third share in whatever they might find nearby. Within weeks they found a bonanza that would pay millions, making Tabor's grubstake one of the greatest investments in the history of the West. Their first wagonload of ore paid them nearly $300, an immediate profit of 500 percent for Tabor. In September he and Rische bought out Hook for $98,000, which they recovered from the mine in just three weeks![6]

Returns like that soon attracted the attention of the ever-vigilant Jerome Chaffee, who had by then been dubbed by some the "Jesse James of Colorado mining." Always looking for a quick buck from a "fighting claim," Chaffee bought a half interest in Fryer's earlier claim for $50,000, and with the threat of litigation he quickly convinced Tabor and Rische to buy him out for $125,000. But when Chaffee boasted to a Leadville friend of how he'd just made a $75,000 profit from the sale, his friend shot back, "You have, have you! Well, you old fool, you have lost $3,000,000!" He was soon convinced of his error, and within weeks he and David Moffat had bought out Rische's half interest in the new bonanza for $262,500. But a Leadville doctor, Charles Bissell, and a couple of friends had bought the other half of Fryer's claim together with other rival claims, and he wasn't ready to sell out yet. So in November Chaffee, Moffat, and Tabor cut him in for a 27 percent share and combined all the claims as the Little Pittsburg Mining Company. By then they had taken out $900,000 worth of silver at a profit of over $500,000, and in the next five months they nearly doubled that return. At the same time the "magic city" mushroomed from 300 to 15,000 fortune seekers.[7]

The Little Pittsburg was Leadville's boss mine, and it was soon hailed as "the richest silver depository the world could show," as they took out over $10,000 a day and in January of 1879 officially began paying themselves regular monthly dividends of $100,000. In April they reincorporated in New York as the Little Pittsburg Consolidated Mining Company with an extravagant offering of $20,000,000 in 200,000 shares at $100 par. Tabor got 73,000 shares, Chaffee and Moffat split a like amount, and the remaining 54,000 went to Bissell. But Chaffee and Moffat now wanted all of it, so before the shares went on sale they bought out Bissell for $330,000 at a bargain price of only $6 a share, and Tabor for an even $1,000,000 at just under $14 a share, both on time from the output of the mine. Then Chaffee launched a dramatic promotional blitz, chartering a luxury railroad car with a well-stocked bar to bring a crowd of over thirty selected easterners out on a grand excursion to see the great wonder. These were mostly speculators, such as their Maxwell Grant partner, Steve Elkins, who could help make a market for the shares, and Brayton Ives, president of the New York Stock Exchange on which they would soon place the shares. But they also included Rossiter Raymond, the *Engineering & Mining Journal* editor, who was paid $5,000 for his expert appraisal and then hired as a consultant, and a few reporters, among them Charles H. Dow, later founder of the *Wall Street Journal*, to help spread the word. In Leadville David Moffat awaited them, offering to show that he could take out as much as $100,000 a day if they wanted![8]

The usually conservative Raymond relied heavily on the hands-on experience of his old Freiberg mining school classmate William Keyes, who had opened the similar silver-lead deposits of Eureka for Hearst and Roberts, and who had also been hired by Chaffee as an expert and consultant. Together they confidently estimated fully $2,000,000 worth of ore in sight with a comparable amount partially exposed. They also emphasized that barely 3 percent of the ground had been open, so the giddy young Dow quickly calculated its total worth at $54,000,000 and touted it as "the biggest pot of money in the world." The awed speculators snapped up the first 50,000 shares offered at $20 a share and subscribed for an additional 30,000 at $25 before they left Leadville. In the following weeks Chaffee signed what he touted as "the largest smelting contract ever made," to work the entire output of the mine at a rate of 150 tons a day until the end of the year, and he promised steady dividends of $100,000 a month or more "for many years to come." In the fall the shares were listed on the New York Stock Exchange at last, and an eager public was let in. Worked up to a fever pitch, they clamored for all they could get, giving the early speculators a tidy profit of 50 percent or more and taking 15,000 more from Chaffee and Moffat as the price climbed to $34.[9]

Amid this swelling excitement, in the summer of 1879 George Roberts arrived in Leadville on his way east, hunting for plums to offer in New York. He immediately began negotiations for the purchase of the camp's two other stellar producers, the Chrysolite and the Little Chief, which held the continuation of the Little Pittsburg bonanza. Boarding with the miners, he also got tips on other paying mines, and he picked up over half a dozen additional mines that had already shown a profit and were ripe for promotion, claiming they were the best out of eight hundred prospects in Leadville. Then he headed on to New York to begin a new blitz of mining promotion from his new offices on Broadway. He opened with the Chrysolite. Although it was located very close to the Little Pittsburg, it had shown no ore, aside from some "salt" that its discoverer, "Chicken Bill" Lovell, had planted in the bottom of a shallow shaft to attract buyers. But Horace Tabor had been attracted by its location, and in late 1878 he had happily bought it for $3,600 in partnership with a young promoter from Chicago, William Borden, who brought along wealthy merchant Marshall Field. They sank Lovell's shaft another couple of dozen feet and hit the real bonanza. They had taken out over $1,000,000 by the time Roberts arrived, so he had to offer them nearly $3,000,000, which gave them a total profit of close to a thousandfold before they finally agreed to sell.[10]

Despite the record-breaking price, Roberts worked out a "novel" contract whereby he actually put up no purchase money and got stock options that would net him well over $1,000,000. First he and Tabor formed the Chrysolite Silver Mining Company, in October, with 200,000 shares at a more modest par of $50, and Tabor took 65,000 shares for his interest. Then, since Borden and Field wanted their share, totaling $1,875,000, in cash, Roberts placed the remaining 135,000 in trust to offer to underwriters at $13.89 a share with only $4.26 down and the balance due on the Ides of March. That minimum price was exactly equal to the purchase price of the mine, needed to pay off the sellers in cash without any markup for Roberts. For his share, however, Roberts could call

back half of the shares in trust and their dividends for eight months by refunding the payment with slight interest and then sell them on the market for all the profit he could make above $13.89. Thus, Roberts and Tabor began eagerly to rip out the ore to pay themselves dividends and push the market price of shares as high as they could. This purchase contract worked so well for Roberts, letting the underwriters carry all the risk at no cost to himself until he was ready to take half the profit, that he used it again and again. With several schemes going all at once, however, Roberts also had to cut in his mine managers and a few selected brokers for added shares of the profits, to ensure their dedication.[11]

Roberts talked New York book publisher Daniel Appleton into serving as president of the Chrysolite, and he named Borden, Tabor, Nevada senator Jones, and a few other friends as trustees, together with some of the eastern bankers who had become underwriters. Although Roberts personally refrained from being part of the official management, all of them were quite happy to follow his lead. He also immediately hired his old Eureka superintendent as general manager for a share in the take. Keyes soon opened dazzling pockets of horn silver running up to $20,000 a ton and steadily boosted the prospective profits ever higher. Starting with a cautious estimate of only $1,500,000 worth of ore in sight, he soon boasted that he was opening $3 worth of new ore for $1 worth removed. Then, after taking out $530,000 in three months from just over 3,300 square feet of the bonanza, he speculated that the adjacent 465,000 square feet of new ground he had just blocked out could pay over $74,000,000! By the end of the year he swore that the mine "stands today without a rival amongst the silver-lead deposits of America." As if to prove it, Roberts had trumped the Little Pittsburg with Chrysolite dividends of $200,000 a month, mostly paid, of course, to Tabor and himself. With enticements like this, would-be investors waited anxiously for the shares to come on the market in the spring.[12]

At the same time, Roberts was starting to push an additional $40,000,000 in shares of other Colorado ventures. Foremost among these was $10,000,000 in the adjoining Little Chief. It was owned by another Chicago merchant, John V. Farwell, who had paid its discoverers $300,000 at the end of 1878 and aggressively opened it to take out nearly $2,000,000 the following year, clearing at least $1,000,000. But Tabor had brought suit, claiming the ground for his Little Pittsburg, so Roberts succeeded in buying out Farwell at the end of 1879 for a $1,750,000 stock deal like the Chrysolite. That gave Farwell a total profit of over 800 percent in a year, and he took his winnings to Texas, putting them into a land speculation that became the famous XIT Ranch, the largest in the country. Roberts took over the Little Chief in January of 1880 and talked harvester inventor Cyrus McCormick into being president, while he became vice president, and among the other trustees he brought in such kindred spirits as Abel Breed of Caribou notoriety. Wasting no time, Roberts ordered Keyes to start gutting the mine, and in just thirty days he declared its first monthly dividend of $100,000, grabbing more quick profits and dressing the stock for the market.[13]

In addition Roberts had obtained controlling interests in two other leading mines, the Iron Silver on nearby Iron Hill, owned by Field's partner, Levi Leiter, and Michigan copper mine owner William H. Stevens, and the Robinson Consolidated in the neighboring Ten Mile district, owned by Leadville merchant and future Republican lieutenant governor George B. Robinson. Since the exploitation of both these mines was being hindered by claim litigation, Roberts was able to step in much more cheaply as a white knight to try to quiet the disputes. He also stocked these at $10,000,000 each, openly assumed the presidency of both, and promptly put Keyes to work dressing them up for the market as well. Roberts had also taken control of a few older gold and silver mines around Central City, Idaho Springs, and elsewhere that he floated at more modest offerings of $1,000,000 to $5,000,000 apiece. By then Roberts was very much the man of the hour in Colorado, being welcomed as "the great California operator" and addressed as "Colonel." With a "flourish of trumpets" and high praise from Rossiter Raymond, Roberts's rather pompous mine managers also vowed to revolutionize Colorado's primitive "digging for potatoes" with modern mining methods and show the locals how to mine "without fear of getting out of the light of the sun!" One critic would later complain, however, that they also succeeded in getting the financial affairs of the mines "beyond the reach of sunlight."[14]

By early 1880 the Little Pittsburg and Chrysolite seemed capable of surpassing even the Con Virginia and California with a heady projected output of $130,000,000. Moreover, the Little Pittsburg was seen as a paragon of mining virtue with a "reputation of being a purely business enterprise and outside of the domains of speculative movements." By then too Chaffee and Moffat had collected roughly half of its $1,350,000 dividends and sold the other half of the stock for over $2,000,000 on the basis of those dividends. But they had nearly gutted the mine to maintain the dividends, and it was all about to collapse. With only enough ore for one more month's dividend, they suddenly announced the discovery of an "immense" new ore body and sent out a reassuring circular to stockholders, claiming $5,000,000 in ore still in sight to sustain dividends for at least five more years and predicting that the still "undeveloped portion of the property will yield a net profit of $25,000,000." At the same time Chaffee and Moffat began frantically unloading the last of their stock, driving the price down from its long-held $30 a share to $20 in just over a month before it crashed to $10 early in March, when it became obvious that the dividends had ceased.[15]

The crash of the Little Pittsburg, dubbed the "Little Big Fraud" by the New York weekly Puck, shattered the faith of investors who cried out angrily over the "villainous falsity" and breach of trust by the management. As company president Chaffee, piously claiming that he had been "perfectly straight and square with the public," tried to refute the charges by carefully pointing out that he had actually held more shares in early February than he had when the company was organized, and that he withheld word that the ore was exhausted only because the news would have unnecessarily depressed stock prices, while he "fully believed" that more ore would soon be found. The company stock ledger, however, clearly revealed that his earnest belief hadn't stopped him and his partner from dumping over 100,000 shares in February and early March to rid themselves of all but 12,000 of their remaining stock and grab a final $1,000,000 from their unsuspecting and mislead investors. In all, after paying off Bissell and Tabor, Chaffee and Moffat got away with a profit of nearly $3,000,000 from their actual cash outlay of about $300,000. But it was later rumored with much satisfaction that they had to quietly "disgorge" about $1,000,000 to their last large purchasers rather than face an ironclad fraud suit for "willful misrepresentation." The Little Pittsburg did at least cost Chaffee a coveted cabinet post in James Garfield's administration after his fellow Coloradans repeatedly opposed his appointment, insisting that they "don't want a mining swindler to be considered one of our representative citizens."[16]

Roberts, of course, was about to work the same scheme with the Chrysolite before the easterners got wise. By March it had become Leadville's biggest dividend payer at $1,000,000, paying back more than a third of its well-publicized purchase price in just five months. Roberts confidently began pushing his shares on the market at a hefty $40 a share and above, as he paid out the fifth dividend. Then suddenly the Little Pittsburg's collapse broke the market, forcing him to go slow until mid-April, when, after the worst of the shock had passed, he finally lured investors back again as shares climbed above $20. The slowdown, however, caused new problems, because Roberts too had nearly gutted the Chrysolite to pay its extravagant dividends, and there was not even enough ore for another full dividend without stripping it bare. So a couple of weeks later, blaming a reduced output due to bad roads and increased expenses of $88,000 for partial repayment of a $250,000 loan by Tabor, Roberts cut the dividend in half for April. Despite such excuses, the shares began to slide and hit bottom at a profitless $13, at which point he skipped the May dividend and called for a shift to quarterly payments.[17]

Undaunted, Roberts confidently called in Rossiter Raymond, who for a $5,000 fee had just appraised the mine in March for the original push, to go back for a reconfirmation. Although the Chrysolite was nearly gutted, Keyes had very skillfully concealed that fact. What he showed Raymond was a mining engineer's dream, a methodically opened mine with neatly blocked out sections holding an estimated $7,000,000 worth of ore, fully exposed and just waiting to be extracted for a profit of more than $5,000,000. The trusting Raymond enthusiastically reaffirmed that the Chrysolite "never looked so well; and beyond all question it has never been in so good condition for systematic, regular, economical, and large scale production"! He further lavished the highest praise on his old friend Keyes for his "extraordinary energy and executive skill" in its "rapid and successful development." As was later revealed, what Raymond did not realize was that the silver-lead carbonate ores had replaced the bedded limestone for only about five feet to either side of almost invisible vertical fractures that formed a fairly regular geometric network. Keyes, with years of experience in such deposits, had carefully run drifts and crosscuts along these fractures, taking out the rich ore from the center and leaving only a veneer of ore on the walls. The remaining rock thus gave the appearance of being solid blocks of ore, when in fact beneath the thin shell it was entirely barren.[18]

With Raymond's reassuring endorsement the shares immediately began to rise. But in his enthusiasm he also predicted that the mine was so rich that it would "easily" pay off the remaining $180,000 due Tabor without missing its June dividend. That created a new dilemma, for the whole sham would be exposed if they stripped off the last veneer to meet Raymond's predicted dividend. Either Roberts or the equally resourceful Keyes, however, quickly hit upon a new and even more devious ploy. Just two days after Raymond's report, Keyes and Tabor gave local mining men a tour of the mine to also reassure them of its richness, and Keyes blamed the recent slowdown on a sudden laziness of the miners. He then made a show of castigating the three shift bosses and issued new orders prohibiting smoking and even talking during the shift. The surprised bosses denied the charges and vowed to resign if the orders were not rescinded. Instead Keyes accepted their resignations, posted the orders, and briefly left for Denver, leaving in his place the hotheaded son of a Hawaiian missionary, George Daly, who had been driven out of Bodie by the miners' union for his violent provocations, to confront the inevitable protest. The Chrysolite miners promptly struck for reinstatement of the shift bosses and rescinding of the prohibition. Other miners joined in and soon broadened the demands to include a wage increase and a reduction of hours to Comstock rates, and a general strike began. Daly immediately hired gunmen and fortified not only the Chrysolite, but the Little Chief and Iron Silver as well, making them a "veritable Gibraltar," and a delighted Roberts telegraphed him to "close mines indefinitely."[19]

The strike served Roberts very well. Chrysolite stock climbed back up to $22, and Roberts and Tabor steadily unloaded their shares while they steadfastly refused any compromise to the miners' demands. But when the disenchanted editor of the Leadville *Democrat* finally suggested that the deadlocked strike was "merely a stock jobbing operation, with Messrs. Roberts, Keyes and Daly in the wood-pile," the stock began to slip again, and dramatic new action was called for. Keyes and Tabor, who was by then the Republican lieutenant governor, prodded Leadville merchants into organizing a vigilance committee and the governor into declaring martial law and sending in troops to help protect the mines, while Roberts declared that he would ask the president for help if necessary. With this impetus Chrysolite stock climbed back to better than $20 a share, and they unloaded the rest of their stock before the miners finally capitulated and the bubble burst.[20]

As soon as the strike ended, Keyes left for a month's vacation in California, and when the bonanza ore shipments and dividends failed to resume, its shares began to fall as the "coterie" of underwriters also dumped the last of their stock on the market and headed for the exit. Tabor and Keyes soon resigned, and most of the other trustees quickly sailed for Europe for the summer, leaving their resignations as well. Roberts, on the other hand, feigning great surprise, came out to Leadville to make a "personal examination" before declaring that the Chrysolite's profitable days were "finished" and the company was $400,000 in debt from the strike. Investors finally realized that the Chrysolite too had been gutted, and the stock collapsed, becoming the most active and demoralized stock on the New York exchanges. Trading reached 116,000 shares, or over half the total issue, in a week as everyone tried to get out from under before it bottomed out at $4 a share. In all, Roberts, Tabor, and their coterie had collected about $5,000,000 from their Chrysolite stock sales and dividends, out of which Borden and Field were each paid their $937,500 with interest, while Tabor most likely got about $1,500,000 from his share of the stock, Roberts took close to $1,000,000 out of his deals, Keyes also got away with enough to buy an "elegant" estate in San Francisco, and the rest of the coterie of early speculators shared about $500,000 for marketing it, while the public who purchased it lost about $4,000,000.[21]

A shocked and badly embarrassed Raymond suddenly turned the *Engineering & Mining Journal* against the "Roberts Combination" and their "disgraceful and criminal proceedings," warning investors too late that "it makes no difference how intrinsically valuable a mine may be, when it is managed by tricksters and stock-gamblers it is always a bad investment." In cynical retaliation, George Daly blasted Raymond's "rose-colored estimate" of $7,000,000 in sight, claiming there had really been only $700,000 in it! Although Raymond wasn't yet ready to listen, Daly soon revealed for the first time that the "solid block of ore," on which such extravagant estimates were based, was "simply a shell of ore enclosing a large core of barren porphyry." Nonetheless, the brother of Roberts's former company secretary, Drake De Kay, further pressed the attack with a suit against Raymond, claiming to have been deceived into buying stock by his false appraisal of the mine. Rival mining journal editors also delighted in pointing to the often imperious Raymond's faulty appraisal as the "source of any misconception," and he filed many columns of his *Engineering & Mining Journal* with indignant defenses. Raymond still didn't realize that he too had been deceived, and he eagerly took the presidency of the company under new management to try to prove its worth. But even with the discovery of new ore pockets, which paid an additional $500,000 in dividends, the subsequent production of the Chrysolite fell far short of his calculated $7,000,000, and he was finally forced to face just how badly he had been fooled by his former friend.[22]

On the heels of Chrysolite, the Little Chief also came tumbling down. Roberts had started working his Little Chief shares on the market at around $10 in late April, after its third dividend, as he also began to push Chrysolite again. He probably unloaded his shares by mid-July as sales passed $500,000, just before he let Chrysolite go down. Well before that, however, McCormick had smelled a bad odor and bailed as president, to be replaced by the former Republican carpetbagger, governor, and senator from Mississippi, Adelbert Ames, and Keyes had turned over the Little Chief gutting to Daly. There was just enough ore for one last

50 cent dividend in August, to help the speculators and underwriters as they scrambled to unload. Roberts meanwhile had gone short, and Daly dutifully declared the Little Chief "exhausted" early in September and hypocritically resigned to protest the final looting of the last ore reserves. Loyally defending Roberts, however, he claimed that if the "despised California operators" were still running the mine "it would not have been played out." Instead he blamed it all on the "New York sharks," who had "forced" him to "gut the mine most mercilessly"! The stock plummeted from $4 to 80 cents while Roberts picked up an added bundle on short sales. The final looting of the Little Chief soon followed when rumors of $25,000 still left in the treasury excited the directors of the impoverished Little Pittsburg to quickly file suit for all of it, as damages for ore allegedly stolen from their claims. With that, shares dropped nearly out of sight to barely 50 cents.[23]

Not all of Roberts's schemes paid off as he had hoped, however, for not only were he and his confederates overextended but some of his plans simply went awry. He found that the Iron Silver deposit lacked the obliging fracture structure of the Chrysolite, so he couldn't fake much "blocked out ore" there. Still he cleaned out $100,000 from the treasury, ran it into debt, and unloaded what he could before Leiter and Stevens rallied the stockholders to regain control in August. Roberts also probably cleared less than $100,000 from the older gold mines. On the other hand, he did do much better with the Robinson Con, but it took much longer than he had expected. As president and quarter owner, Roberts had put the aggressive Daly in as manager to start shipping out as much as $5,000 in high-grade silver ore a day, and in June of 1880 he paid the first dividend of $75,000, just three months after the formation of the company. With estimated ore reserves of $3,000,000 and already paying dividends, it was all ready for the market, but Roberts was thwarted by the stubborn young Robinson, who still held the remaining three-fourths of the shares and refused any compromise to buy off a conflicting claim that clouded the title. Instead Robinson fortified the mine and hired gunmen to fight to the last. This would drag on until late November, when shortly after he won election to be the next Republican lieutenant governor, Robinson was accidentally killed by one of his own guards. By then, however, Roberts was busy with new and even bigger schemes.[24]

BY THE SPRING OF 1880 the great Leadville boom, led by the Little Pittsburg and Chrysolite, had created such a rampant mining fever in the East that even the collapse of its first leaders had no immediate effect on it. For Leadville's bullion production rocketed to over $15,000,000 a year in 1882, and more new bonanzas elsewhere in the West were also offering promising investments. Indeed, by the end of 1881 there were three dozen western mines paying between $100,000 and $1,000,000 a year in dividends. In all, more than 10,000 mining companies were formed in the eastern states between 1879 and 1883 either to work mines in the West or, more commonly, just to work investors in the East, offering another outrageous $20,000,000,000 in paper hopes of a share in western riches. Action on the mining exchanges was furious, with sales running to 200,000 shares a day and totaling nearly $100,000,000 a year during the boom. But they were mostly fictitious "wash sales." Roberts's furious manipulations alone made up two-thirds of that amount, with the entire stock in one of his ventures changing hands sixteen times in a year![25]

The new companies formed in the East steadily escalated their stock offerings to well over $1,000,000 each, while they cut the par value of shares to around $10 as they tried to hook an ever-increasing number of smaller investors. They also tarted up the shares themselves to make them more seductive—bigger and brightly colored, with enticing scenes of prospectors striking ore, miners furiously digging it out, and smoke-belching mills and smelters turning out bullion. The shares of the new companies in the East were also proudly free of assessments, since after all the bad press about the "California system" of milking shareholders dry with assessments, eastern investors were afraid to touch them.[26]

Although the new mining mania had started in New York City, it soon spread to cities and town all along the eastern seaboard and out into the Midwest. The virulent craze for floating mining shares even infected such quiet little communities as Fond du Lac, Wisconsin; Norfolk, Virginia; Sioux City, Iowa; Xenia, Ohio; and Woonsocket, Rhode Island. In fact, only about a third of the new western mining companies formed in the East were incorporated in New York State. More were formed in the midwest, particularly Illinois, Ohio, and Missouri, and nearly as many in Massachusetts, Pennsylvania, and the Eastern Seaboard states. A score of new mining exchanges also sprang up like fever blisters during the boom. Chicago and Philadelphia each suffered four, while Boston,

The Wild-Cat Mining Swindle
Puck, March 31, 1880

Kansas City, St. Louis, and Topeka each had at least one. In the West, Denver began to have pretensions as a financial center with three successive mining exchanges, and token exchanges popped up in Albuquerque and Socorro, as well as the emerging camps of Butte, Rico, and Tombstone. Not to be left out, an enterprising con artist, Mrs. Marion E. Warren, also opened a Ladies' Mining and Stock Exchange on Union Square, which flourished briefly before she skipped out to Philadelphia with her customers' funds. There was, in fact, such a proliferation of exchanges that yet another exchange was opened in New York just to deal in seats on all the "other exchanges all over the country"![27]

The mania also brought forth a rash of new mining journals, seeking to satisfy the would-be investor's itch for the latest word from the mines and the promoter's compulsive needs for advertising space. At least three dozen were started during the boom, mostly in 1879 to 1881; nearly half were published in the East. Two dailies, the *Mining News* and the *American Exchange*, and half a dozen weeklies devoted exclusively to western mines were launched in New York, and one or more floated in Boston, Chicago, and Philadelphia. The remainder were regional promotion sheets, published principally in Denver, El Paso, Salt Lake City, San Francisco, Santa Fe, and Tucson. Some, like *Hadley's Pointers*, the *Mining Investigator*, and the "crisp, racy and gossipy" *Lode & Gulch* of New York, offered their readers investment "tips," but most provided little more than scissored items from mining camp papers and paid plugs from promoters.[28]

WHILE ROBERTS HAD moved on to new schemes elsewhere, Tabor, who had learned a trick or two, couldn't resist running a few more shameless scams with another dazzling new Leadville strike, to shed the last shreds of his cloak of the "honest miner." It had all started in August 1879, when a startlingly rich pocket of ore was found on the Matchless, a few claims east of the Little Pittsburg. Tabor quickly bought out the three discoverers for $117,000, but the pocket petered out after yielding only about $30,000, and Leadvillians dubbed it "Tabor's Mistake." But within weeks even richer ore was found in the Robert E. Lee, the next claim to the east, which proved to be a real bonanza, with the richest rock ever seen in Leadville. The mine had been purchased just a few months before the discovery by a group led by a bankrupted New York banker, Lorenzo Roudebush, who had come west to make a new start. They had paid $290,000 for it, and they had offered a controlling five-eighths to senators Chaffee and Jones and restaurateur Charles Delmonico for $300,000, but Chaffee suddenly soured on it and killed the deal just before the bonanza discovery, truly making it his biggest mistake. The Lee bonanza was so rich that they were able to take out as much as $118,500 in high grade in a single day, a record for any mine outside of the Comstock. Within two years they had taken out over $3,500,000, and even after steep smelting fees, the lucky investors had collected at least $1,900,000 in profits, or over 500 percent on their investment. When the high grade finally gave out, they gave Tabor a quarter share in exchange for a like share in a leaching works of his, so they could work the lower-grade ore themselves.[29]

Meanwhile, Tabor had eagerly bought interests in the neighboring Big Pittsburgh and Hibernia mines, which joined the Matchless and Lee on the south. Early in 1880, together with those tricky ex-California sharks James Keene and "Uncle Billy" Lent, he floated the Big Pittsburgh Consolidated Silver Mining Company in New York for an outrageous paper capital of $20,000,000 on claims that had cost them just $323,000. Offering some shares at very low prices, however, they were rumored to have taken in even that rapacious railroad shark William "The Public Be Damned" Vanderbilt. Then they hired George Daly part time from the then-booming Little Chief to see if he could find any extension of the Lee ore. All the while Tabor had continued to probe his Matchless, and in August of 1880 he finally struck a bonanza of his own in the southeast corner joining the Lee, Hibernia, and Big Pittsburgh. In the next year and a half the Matchless produced over $1,500,000, paid him a profit of at least $750,000, and gave him the last laugh on all who had laughed at it.[30]

But Tabor wanted even more, so he promptly incorporated the still-undeveloped Hibernia and a couple of other neighboring claims as the Hibernia Consolidated Mining Company with himself as president and $7,500,000 more in $25 shares to work in New York. To make a quick showing, he quietly tapped into some of his Matchless ore from the Hibernia side, taking out $2,000 a day, and on New Year's Eve he paid his first dividend of 10 cents a share. Then he went east to personally help work the price up from a bargain-basement 75 cents to at least $1.50

on the New York mining exchanges, as his wash sales dominated trading in the spring. With two more monthly dividends he held the price
there, while he quickly unloaded for close to $500,000. Then, skipping
two dividends, he cleaned up some more on short sales as the price fell to
barely 50 cents, bought the shares back, and paid two more dividends as
he unloaded them once more at $1.50. In March he and his Big Pittsburgh
partners also began the same game with it, taking out nearly $60,000
from the Matchless through its shaft to run it up to $4.75 a share, while
they unloaded the last of their shares for a few hundred thousand more.[31]

In a final grab the shameless Tabor, after having the underground
workings "resurveyed," suddenly slapped suits on both companies for
$200,000 for all the ore that they had taken from his Matchless under
his management! All work stopped, and the stock spiraled to a low of 16
cents as he probably picked up the shorts again. When the stockholders
finally took control, they discovered that the Hibernia was $15,000 in
debt on a loan Tabor had arranged from Chaffee for a final dividend,
which he had then piously rescinded to help "bear" the stock, before
he finally foreclosed on the loan to get the property back. Moreover,
even if they could hold onto the mine, they also found that Tabor's
Hibernia superintendent had signed over any rich ground that they
may have had to his Matchless to quiet his suit for the ore. In the midst
of all this, Tabor returned to Colorado, happily boasting that he had
just "had a little fun with the boys" on Wall Street. But the New York
Mining Record blasted his "inordinate greed" and "his vulgar success"
in deceiving all of the shareholders who trusted him as their president.
Tabor, however, a millionaire by then, scoffed at such criticism and was
so full of himself that he dropped $200,000 on a bid for a Colorado sen-
ate seat and got only the last thirty five days of an unexpired term. At
the same time, he also dumped Augusta, his faithful wife of twenty-five
years, for a trophy bride, Elizabeth McCourt "Baby" Doe, barely half
his age, and embarked on a lavish lifestyle that would nearly bankrupt
him before he suddenly died of appendicitis in 1899. His impoverished
widow, Baby Doe, was left with little more than the worked-out carcass
of the Matchless, which Horace had still imagined could be worked for
one last "deal." Unfortunately Baby Doe too shared that delusion, and
for the next thirty-five years she would carry on an obsessive and melo-
dramatic struggle to hang onto the Matchless. After she finally froze to

death, alone in a shack at the mine, her poignant fantasy was turned into a romantic opera, *The Ballad of Baby Doe*, starring Beverly Sills.[32]

Meanwhile, after the Hibernia steal, Tabor had been deemed "too tricky to be trusted" even on Wall Street, so he and his fellow pariahs, Chaffee and Moffat, who were still battling Little Pittsburg investors' suits, turned to London to push their next scheme in a mine called the Henriett. This time they ran it with a new front man, the Republican's "Plumed Knight" and soon-to-be presidential candidate, the senator from Maine, James G. Blaine, whom they had previously cut in for a piece of the Little Pittsburg deal. This mine actually turned out to be far richer than they thought, but that wasn't immediately obvious, and in their rush for quick profits they almost lost it. Only through a twist of fate did they reluctantly take it back and finally discover what they had. In 1879 Chaffee and Moffat had picked up the Henriett just south of their other bonanzas for $25,000 and sold Tabor three-eighths and Blaine one-eighth to recover their costs. Its ore was not spectacular, but some of it was up to half lead and contained enough silver to run $100 a ton, so in 1881 they made a show of taking out more than thirty tons a day. Then, with those helpful old British mining engineers John Taylor & Sons, reporting fully $1,700,000 worth of ore in sight and predicting annual profits of nearly $500,000, Chaffee and his pals finally sold the mine to London investors in the fall of 1882. They got $800,000 cash plus $450,000 in shares in the newly formed Henriett Mining & Smelting Company, capitalized at $1,500,000.[33]

The London directors immediately went to work, taking out seventy-five tons a day, and confidently started paying 2 percent quarterly dividends. But they only paid out two before the "ore in sight" all vanished. Once again investors began to howl, and Chaffee, Moffat, and Tabor all turned a deaf ear until nasty charges of fraud threatened Blaine's upcoming bid for the Republican presidential nomination. Then, in the summer of 1883, they suddenly did a turnabout and paid back the startled investors in full! Such unprecedented action won them high praise from the press. Blaine went on to win the Republican nomination in the summer of 1884, and Chaffee spread cash for him in the southern states. But his unpopular choice of his other close business partner, Steve Elkins, as his chief lieutenant, and their involvement in an earlier coal mining scheme and the old "Star Route" scandal, all helped bring about his defeat by Democrat Grover Cleveland that fall, the first Republican presidential loss in a quarter of a century. But ironically, by taking back the Henriett, they had at least won a chance to milk even more from the British. For after the election Chaffee pushed new exploration with leasers, who soon struck an enormous low-grade deposit that would prove to be a real bonanza.[34]

Steve Elkins and a couple of his other Star Route cronies, Richard Kerens, a Republican National committeeman and St. Louis mail contractor, and Preston Plumb, the Republican senator from Kansas, also worked a few less-spectacular Leadville mines in New York. Their biggest play was the Amie Consolidated Mining Company on the claim immediately east of the Little Pittsburg that shared a little of its rich neighbor's bonanza. They picked it up in June of 1879, and working it as a close corporation with Elkins's brother, John, as superintendent, they took out over $400,000 worth of silver by the following spring for $255,000 in dividends, which more than covered their purchase price. Then, as the ore was running out, they too went public and put it on the New York exchange in March of 1880 as a solid dividend payer with ore reserves of over $1,500,000. Starting at a bargain $2.50 for its half-million $10 par shares, they worked up a sales fury as they paid one more dividend of 10 cents and unloaded most of their shares at better than $1 apiece. Then they too jerked the shareholders around mercilessly. First they announced that the ore was exhausted, which knocked the price down to 37 cents while they cleaned up on short sales. Then they announced the discovery of new bonanza ore running into the thousands a ton, which kicked the price back up to 64 cents while they let all the shares out again. In all, they made it the most active mining stock on the exchange for several months running, surpassing even Roberts's machinations, as their total transactions topped 4,500,000 shares in a year and a half, turning over the entire stock every couple of months. By the time they were through they must have cleared close to $1,000,000 from it, and they earned Rossiter Raymond's condemnation. All the stockholders ever got was one last dividend of a nickel three years later, after their shares had dropped to barely a dime.[35]

IN SHARP CONTRAST to such investor disasters, however, a couple of Leadville's early bonanzas did pay some of their investors very well and ranked among the camp's biggest dividend payers. The most enduring of these was the Iron Silver, which William Stevens and Levi Leiter retrieved from Roberts's clutches in August of 1880 and made it Leadville's most steady producer. They had taken out $250,000 at an 80 percent profit margin before they sold Roberts an interest, and under Stevens's management they soon made it pay again. They also sued the neighboring mines, claiming ore beyond their boundaries as the dip of their "lode" under Stewart's mining law. But Colorado juries clearly saw the difference between limestone "blanket" deposits and quartz "fissures" and rejected their claims in May of 1881. Disgusted with both the lawsuits and Stevens's slow, conservative development of the mine, which so far had yielded only one dividend of just 20 cents a share, Leiter suddenly dumped all of his stock, nearly 125,000 shares in just a few days, breaking the price from $3 to $1.60, as he got out with barely $250,000. Stevens promptly paid a second dividend, which brought the price back above $2, and he continued regular dividends for the next eight years, paying out a total of $2,500,000, or $5 a share, and making an honest million from his share. Then, after a brief lapse, he and later his son kept right on for the next thirty years, making it Leadville's workhorse and more than doubling its dividends to $5,300,000 and their own fortunes by World War I. The other big winners were those investors who with Stevens had bought Leiter's dumped shares at only $1.60 and could have cleared over 30 percent a year, if they sold out by 1889 for $1 or more. Even if they'd held on to the very end in 1916, they still could have gotten a fat 15 percent a year.[36]

More profitable yet for its early investors was the Evening Star, but it also made a whole lot of mischief. It showered such lavish profits on its small group of investors that some became so confident of their Midas touch that they giddily rushed on into totally disastrous new schemes. The Evening Star lay halfway between the Iron Silver and Chrysolite–Little Pittsburg bonanzas, and it held a large mass of ore paying over $50 a ton. The mine was picked up late in 1878 for $60,000 by a seasoned New York broker, Watson B. Dickerman, his green young produce exchange broker, Ferdinand Ward; the latter's older brother from the U. S. Assay Office, William S. Ward; the president of the Marine National Bank, James Dean Fish; and Ulysses S. "Buck" Grant Jr., son of the former president. The mine had apparently been recommended to Buck as a good investment by Chaffee as a favor to his father, and the others

came in through Buck's acquaintance with the elder Ward brother. They
formed the Evening Star Mining Company early the following year, with
a relatively modest stock issue of $500,000 at $10 par, and let in a few
friends at a discount to recover their actual costs of only $1.20 a share.
Dickerman served as president, and William Ward, who had attended the
Columbia School of Mines, became vice president and general manager.
Ward immediately moved to Leadville, and by the fall he had made the
mine "one of the heaviest producers in the camp." He also let it pay its
own way from the start, and after thoroughly opening the ore body, he
began paying a steady stream of dividends in October of 1880. They
averaged over $1 a share, or nearly 100 percent return per month, for
two years. That even surpassed the Iron Silver for a time and made the
Evening Star the top dividend payer in Colorado and the fourth largest in the West. In all, it paid $1,300,000, or a profit of more than 2,000
percent for its ecstatic investors. While the older Ward hunted up new
mines, Buck married Chaffee's daughter and joined with Ferdinand Ward
and Fish to start a bank and brokerage of their own under the name of
Grant & Ward. Evening Star stock made up a third of the firm's assets,
and it seemed so promising that President Grant himself was lured into
becoming a partner, but it would all come to a disastrous end.[37]

STILL, THE STEADY YIELD of the Iron Silver and several dozen other,
smaller mines, plus a few later, even bigger discoveries, would keep up
investor hopes and give the camp a solid production of around $10,000,000
a year well into the twentieth century, to actually surpass the Comstock
at over $400,000,000 by the end of the First World War. This enormous
output of silver-lead ores also generated a steady income for several new
smelters at Leadville and made new fortunes for investors. Charging
roughly $40 a ton on the typical $100-a-ton ore taken from the mines,
they made a profit of $25 a ton on such ore, or around $2,500,000 a year,
and cleared a bigger return than most of the mines themselves. Gustav
Billing and Anton Eilers had soon moved in from Salt Lake City, and
other established smeltermen and their backers also came in for a share
of the profits. But the largest and most profitable smelter was that built
by an energetic young newcomer to the business, James B. Grant, and
his wealthy uncle of the same name, a prominent Iowa judge and first

president of the Rock Island railroad. Young Grant, the son of an Alabama doctor and plantation owner, had moved to Iowa after the Civil
War, where his generous uncle, whose own children had died in infancy,
put him through college, sent him on to Freiberg for two years, and then
on a tour around the world. On his return at the age of twenty-nine,
Grant came to Colorado as an assayer in 1877, just as the rush to Leadville began, and he quickly saw the profits to be made there in smelting.
The following spring, confident of his metallurgical skills, he persuaded
his uncle to invest in a smelter.[38]

The first furnace was fired up in October of 1878, and as the Leadville
boom took off, they rapidly expanded until the judge had put $160,000
into a giant works with nine furnaces. But they were amply rewarded.
In 1879, aided by Chaffee's record-setting contract for smelting all the
Little Pittsburg's output, Grant successfully worked 25,000 tons of ore
to produce $2,500,000 in bullion, for a heady profit of roughly $600,000.
By the end of the year the judge, having gotten his protégé off to a great
start and having made a profit that exceeded all expectations, sold out
his interest to his nephew and two new partners, Edward Eddy and William H. James, who ran a sampling mill and ore buying business. Young
Grant got half of both businesses, and they formally organized the Grant
Smelting Company early in 1880. In the ensuing year, as Roberts and
others also went into heavy production, they boosted bullion production to $4,000,000 for an even heftier profit of about $1,000,000 a year.
In just two years Grant was nearly a millionaire, and his smelter had
far surpassed even Nathaniel Hill's in Denver, becoming the biggest in
the Rockies. Even a disastrous fire that destroyed the works in May of
1882 was turned to advantage, when within months he and his partners
built an even bigger new works in Denver to handle ores from other new
camps as well. By then Grant was heralded as the new "smelter king,"
and in the fall of 1882 he was elected the first Democratic governor of
Colorado, as disgusted Republicans bolted the party after Hill's right-
hand man, Henry Wolcott, lost the Republican nomination in a nasty
fight with Chaffee and others.[39]

Grant further expanded in the summer of 1883, merging with Charles
Balbach's pioneering Omaha works, which also drew ores from Idaho,
Montana, and Utah, to form the Omaha & Grant Smelting & Refin-
ing Company, with a privately held capital of $2,500,000, as "the largest

silver-lead smelting works in the country." He and his partners also briefly joined with the other smelters in a pact to round down all ore assays, counting only full half ounces of silver and full percentages of lead. This device increased the smelters' profits by about $1 a ton on the average ore, but the outcry from the mine owners was so loud that Grant withdrew from the scheme after six months and as a result picked up more smelting contracts from angry owners. He was also aggressive in going after a larger share of Leadville's remaining lead carbonate ores, which were needed as "flux" for the recovery of gold and silver from Colorado's other, more rebellious ores. To get the lead ore he smelted it at $2 to $3 a ton less than cost and passed the loss on to charges for the more rebellious ores. He was also aided by railroad officials who invested in the smelter and reduced the shipping rates for lead ores to Denver, while increasing those for smelted bullion from Leadville. Lastly, Grant and his partners personally invested in the major lead mines to ensure their smelting contracts. They did so well that they had to buy one of the largest remaining smelters at Leadville to handle all the ore. Even with increasing competition, Grant and his partners still averaged roughly $600,000 a year in profits, for a grand total of about $10,000,000 by 1899, all the while helping to choke the air of Denver.[40]

DURING THE HEADY DAYS of the Leadville excitement, Chaffee and Moffat had also unloaded the old Caribou on the eager investors. Early in May of 1879, as they started pushing the dazzling new Little Pittsburg, they offered that old, but "proven," mine as Caribou Consolidated at a much more conservative $1,000,000 in shares of only $10 each. With Chaffee's original partner, Eben Smith, as superintendent, they began shipping over $30,000 a month in bullion to pay $10,000 a month in dividends at an attractive annual rate of 12 percent. Then with their new partner, New York Stock Exchange president Brayton Ives, they listed it on his exchange at a bargain $6 a share, raising the dividend return to 20 percent. New York investors gobbled up 40,000 shares within a month and another 20,000 soon after, as Chaffee and his friends paid out $40,000 in four straight dividends and took in over $360,000.[41]

But the old Caribou bonanza had previously been nearly worked out within the company's property, and most of this production had been secretly stolen underground from an extension of the ore deposit in the neighboring No Name claim, owned by Robert G. Dun of the prominent New York credit rating agency but not being worked at the time. Dun, nonetheless, suspecting the theft, brought suit for an injunction and damages, and to prove the charges he sank a shaft near the edge of his claim. His shaft did indeed drop into Caribou's secret workings, and he sent down hired gunmen to drive out the thieves, as other neighbors also filed suit and Chaffee fired back with countersuits. In the midst of these battles a fire, started by camping tourists, threatened the neighboring mines but was repulsed, and a sudden shift in the wind carried it over the Caribou, destroying its hoisting works, and on down the hill, leveling half the town below. The Caribou loss, however, was only $10,000, all covered by insurance, so the works were soon rebuilt.[42]

In January Chaffee and his partners quietly dug further into the No Name and resumed monthly dividends, as they unloaded most of their remaining shares on brokers for another $200,000 as prices began to fall below $5. But they still faced damage suits by Dun for a comparable amount, and because they were about to make their final play with the Little Pittsburg, they wanted a quick settlement. So late in February Moffat struck a relatively cheap but byzantine deal with Dun to kill the suits and merge the companies, giving Dun the controlling interest. Moffat knew the brokers, still holding about 40,000 shares, would go along with the merger to revitalize the stock, so he proposed to temporarily "smash" the stock to $1 a share and buy back enough shares for control to approve the merger. To this end the conspirators immediately stopped the dividends, and with an added punch from the Little Pittsburg crash, they easily knocked the Caribou shares down to $2 as they picked up the needed 10,000 shares to pass the merger in May. They also created 50,000 new shares for Dun in exchange for his mines, and they pushed shares on down to $1.25 in September, buying another 15,000 shares to give Dun the rest of the 75,000 shares needed for control. Dun happily took over as president, eagerly expecting to make millions from the mines, and it took him several years to realize that Chaffee and Moffat had practically gutted not only the Caribou but also his own No Name as well. They, on the other hand, had likely cleared over $500,000 from the mines, and they had bought off Dun's suits for a large piece of that at less than a tenth of their take. Dun eventually sank over $300,000 into trying to make the mines pay, and when he at last gave up, he found that he couldn't even sell them. Only after his death, twenty years later, did his heirs finally dump them, for a pittance.[43]

THE FEEDING FRENZY stirred up by the silver sharks around the Leadville bonanzas also attracted other opportunistic feeders into the fray. One of the more innovative and talented of these was "Brick" Pomeroy, a controversial and charismatic champion of the working man. He would become the Adolph Sutro of the Rockies, and he introduced widespread advertising to push mining stock on the mass market and pile its woes upon his faithful followers. The centerpiece of his efforts was the Atlantic-Pacific Tunnel, a grand scheme to drill nearly five miles through the crest of the Rockies, "Cutting the Backbone of the American Continent," he boasted, and "Opening the Vaults of Gold and Silver."[44]

Marcus M. Pomeroy was born in upstate New York on Christmas Day in 1833, raised from infancy by an aunt and uncle after his mother died, and apprenticed in a newspaper office at seventeen, starting at a scant $30 a year and board. After a few years he headed west to Wisconsin, started his own paper, and was soon hailed as a " perfect brick" for his outspoken, sarcastic, and often bombastic editorials. During the Civil War he gained a national following as a fiery Copperhead and claimed a circulation of 100,000 for his paper, the La Crosse *Democrat*. With growing celebrity as a lecturer and author, he moved his paper to New York after the war and carried on a crusade against the "great army of persons who live by plundering the people," particularly the capitalist who "fastens himself like a leech to the earnings of others." But when he also took on Democratic "Boss" Tweed, he was quickly crushed and forced to return to the West. Late in 1879, in his mid-forties and nearly broke, he moved his paper to Denver to cynically pursue all the rapacious practices that he still railed against.[45]

He began simply, with an old abandoned claim, appropriately called the Quandary, and resurrected it as the far more enticing Monte Cristo Mining & Milling Company to push $2,000,000 in $10 par shares on the 47,000 still-faithful subscribers of his wandering *Democrat*. He unblushingly touted the mine, just over the crest of the Rockies north of Leadville, as a bonanza with $50,000,000 worth of "ore in sight," while he generously

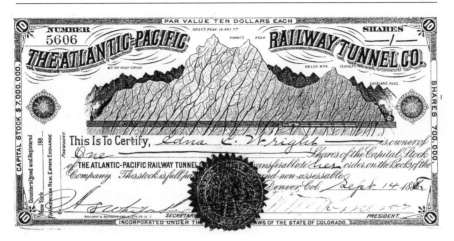

offered its shares to his flock at a bargain price of $1.50 and promised dividends soon. But several Colorado miners and editors, led by the scribe of the Fairplay *Flume*, quickly came down hard on the scheme, going so far as to "emphatically pronounce it a swindle," a "humbug" and a "fraud." So the fast-footed Pomeroy quickly changed course, to win local support first, by launching a lively, illustrated booster paper, *The Great West*, in July of 1880, advertising all the mining and farming opportunities of Colorado. He sent it to his old subscribers all over the country, and he soon boasted a paid readership in over 10,000 towns, which won him the immediate praise of local press. Eager miners also offered him large interests in their mines if he would promote them, and he floated more than a dozen new companies all around the state, this time with strong, loyal local supporters. One miner sought his help in promoting a 3,000-foot tunnel to tap a vein west of Georgetown, and although he turned down that offer, it gave him the much grander idea of running a 25,000-foot tunnel of his own all the way through the mountains to tap over two hundred veins, and his great tunnel scheme was born.[46]

That fall Pomeroy formed the Atlantic-Pacific Tunnel Company with $7,000,000 in $10 par certificates, enticingly illustrated with a cross-section of the Rockies with a treasure trove of rich veins, all cut by the tunnel, and great wagons hauling it out. He made himself its president and took about 300,000 shares in exchange for the tunnel sites and other property.

He also became the financial agent, handling all the stock sales, offering the company's shares, and quietly his own as well, at a bargain price of just $2.50 for a limited time only: "To All Who Dare to Be Rich." Pulling out all the stops in an advertising blitz, he filled *The Great West* with adoration of the fabulous tunnel, praising it as "the greatest enterprise of its kind in the world!" "It is an enterprise that men, women and children can invest in to a *certainty of steady profit.*" Indeed, he said, it would tap a new vein every hundred feet, drawing out "untold millions of wealth" and paying out dividends *"for all time to come!"* It was, he vowed, nothing less than *"God's Great Savings Bank!"*[47]

Brick's fabulous tunnel shares, and those of his other mines as well, all sold like hotcakes, and the money poured in at as much as $20,000 a day and averaged more than $600 a day for the first year, as dedicated admirers in every state of the union all eagerly bought in. Young and old were seduced, from his youngest investor, a nine-year-old boy who had saved up $25 in nickels and dimes, earned by picking fruit, to buy ten shares, to his oldest, an eighty-one-year-old man who bought shares to "benefit his grandchildren." In fact, the once racist Brick now generously reached out to the pockets of one and all, offering shares "to men and women, boys and girls, youth and old age, rich and poor, native and foreign born, white and black, residents of this or of any other country." The sudden flood of mail and money so startled the Denver postmaster and founder of the *Rocky Mountain News*, William N. Byers, that he took it upon himself to illegally open some of the letters and return the checks with a note of warning not to invest! But after Brick threatened his job, disdained him as a mere "louse on the tail of a political lion," exposed one of Byers's own questionable promotions, and then sweetly gave him a tour of the great tunnel, the postmaster retreated and meekly wrote a letter of endorsement.[48]

To keep up a sense of urgency and stimulate stock sales, Brick kept threatening to boost the price of shares and to stop all sales as soon as they struck their first bonanza and started paying dividends, which was always just a few feet farther ahead. To further help sales, he sent a few canvasers throughout the Midwest to give illustrated public lectures on the fabulous fortunes to be made in mining, and to take orders for stock, quietly giving a cut rate on large blocks to local leaders, whose purchases were promptly publicized to lure their followers. In addition he set up the independent-sounding Colorado Real-Estate & Mining Bureau to guide more investors into his own "safe, vigorously worked and honestly managed" companies, guaranteeing annual returns of "at least 10%." He also paid the Denver dailies to run his puff pieces as news, which he then quoted as endorsements in *The Great West* and such prospectuses as his airy *When the World Was Young* and his simple *Make Money, A Great and Safe Enterprise*.[49]

Still, Brick certainly must have hoped that the tunnel would indeed strike a bonanza that could pay a fortune to his followers, but he didn't count on that for his own fortune, which he made sure he took right off the top, sooner than later. By the end of the first year he had taken in about $250,000 from stock sales, the lion's share of this, or roughly $150,000, coming apparently from unloading his own shares. In addition, the former champion of the working man took a fat salary of $6,000 a year from the tunnel alone and only modestly smaller salaries from each of his dozen other companies for a total that was nearly a hundred times what he paid one of his hard-working miners. In addition, he charged the tunnel company about $13,000 a year in advertising for all his puffery in *The Great West*, and he collected several thousand apiece from each of his other promotions. Working other angles as well, he apparently cleared about $250,000 that year, and a similar amount the next. Out of this he put $65,000 into a "magnificent mansion," the "Pomeroy Villa," in Denver and even more into other lavish indulgences.[50]

So confident was Brick in his followers' loyalty that he even promised them their full money back if they were dissatisfied, and for the first year and a half, before he decided to postpone further repayments until the work was completed, only 129 tunnel certificates out of over 9,300 were actually returned, for a total of $6,370. But by October of 1882, after two years with barely 2,000 feet of the 25,000-foot tunnel completed, no bonanzas, and none of the promised dividends, three of his shareholders finally filed a class-action suit, charging him with fraud, and everything began to unravel. They claimed that he had "fraudulently represented the value of the property," luring "widow ladies and poor persons" to buy stock. They also accused him of "stealing over $100,000 of the company's funds," by inflating printing charges and taking kickbacks from contracts he awarded for the company, especially the tunnel contract, for which he charged the company $35 a foot for work that could have been done for $15! Brick's lawyers stalled as long they could, but the judge finally granted an injunction against further sales and forced Brick to surrender all the company books. The Denver *Republican*, which had originally carried some of his puffs, published a lengthy exposé of his schemes, and he sued, futilely, for libel. But he soon shut down all work, sold his paper and half-finished mansion, and fled to New York before the Denver grand jury finished going over the books and indicted him on charges of fraud and perjury.[51]

In the end even his old *Great West* turned against him, in a scathing piece with the banner headline:

> Brick Pomeroy—How a swindler, a libertine, and a villain has duped the public—A miscreant with no principle, honor or regard for life-long friends—How he made paupers of most of them and spent the money on fast horses, fast women and building palaces—How he has sacrificed men and women who have stood between him and ruin many a time—How he ran away from Denver to avoid the wrath of many of his victims, and how those he left behind have stood the brunt of all his rascalities—How he basely betrayed men who have served him for years—Documentary evidence from sources that cannot be disputed—Nearly half a million dollars received, and not a mine to show for it![52]

But Brick escaped extradition and trial, some said only because those in power also feared exposure, and he still managed to keep control of the tunnel, the stock, and most of his still-trusting stockholders. Following the traditional path, he simply let it go into default and let a few creditors set a low price on all the property at a sheriff's sale to clear all claims. Then, in the name of the shareholders, he redeemed the property cheaply and changed the name to the Atlantic-Pacific Railway Tunnel Company, with talk of added profits from tolls. He also replaced the ore wagons on the old shares with a transcontinental train and handed out the new shares in exchange for the old, putting a new face on the large batch of shares that he was still eager to sell. But since the country was sinking into a few years of depression in 1884, Brick, like the other seasonal sharks, had to bide his time, so he launched still another paper, the *United States Democrat*, in New York while he waited for the next raging market to strike again.[53]

MEANWHILE IN THE SOUTHWEST, the Southern Pacific Railroad began building east from Yuma late in 1878, opening the territory to cheaper transportation and triggering a new wave of prospecting and speculation, wildly promising even bigger fortunes than Leadville or the Comstock. With James Barney already turning his Silver King into a steady dividend payer that was attracting investors, Arizona's Republican governor, Anson P. K. Safford, and a former Comstock mining superintendent, William G. Boyle, got off to a rousing start, promoting some of Samuel Butterworth's old mines south of Tucson. They grandly touted these as the somehow "lost and famous Tumacacori mine of the Jesuits," which "many a gallant man has lost his life in vain attempts to find," even though the lode stretched boldly for several miles, sprouting bonanza rock of as much a $1,000 a ton! Following Butterworth's path with J. Ross Browne, they also hired an eastern photographer and journalist, Enoch Conklin, to further glamorize them in a *Picturesque Arizona* travel adventure, generously illustrated by his Continental Stereoscope Company. With the help of Thomas A. Hendricks, the former Democratic senator and governor of Indiana and soon-to-be vice president of the United States, and John C. New, the former Republican treasurer of the United States, they floated part of the mines in Philadelphia as the Aztec and Toltec syndicates for a highly discounted $1,000,000. At the same time, with the aid of Williston Blake, the former governor of the Bank of England, and other, lesser peers, they unloaded the remainder in London as the Tubac Mining & Milling Company for an added $250,000 fully paid. That was the last their investors saw of their money, of course, but before they had much time to worry, Safford would be back with a much better deal in the lively Tombstone excitement.[54]

The Tombstone mines, discovered just twenty miles south of the projected line of the Southern Pacific, were to become the Southwest's biggest silver bonanzas of the century, producing over $25,000,000 by its end. Despite its latter-day reputation as a lawless hell, it was initially an investor's heaven. For in spite of George Roberts's boast of grabbing all the good mines in the West for New Yorkers, Tombstone's biggest mines all escaped them and made big profits for early investors elsewhere, although later investors still did no better. The Tombstone mines were also one of those rare instances where the discoverers actually succeeded in making a fortune from their find rather than selling out too cheaply too soon like most, but it was a close call. In the summer of 1877 Ed Schieffelin, a thirty-year-old, perennially penniless but compulsive prospector from Pennsylvania, had ventured into Apache country just a couple of days east of Tubac with a couple of miners doing assessment work on the old Brunchow mine, and he stayed to prospect despite warnings that all he would find was his own tombstone. He had picked up some rich-looking pieces of loose rock, or "float," but he couldn't find anyone to assay them and he hadn't yet discovered the ledge from which they came. He did, however, jokingly locate one nearly worthless claim as the Tombstone, before he headed off with a few of the rocks to show his younger brother, Al, who was working at a mine north of Yuma. Al finally talked the local assayer and millman, Richard Gird, into testing the samples, and they proved to be far richer than Ed had even imagined, running up to $2,000 a ton. The three of them formed a partnership, and as soon as they had saved enough to buy grub, they quietly returned to look for the bonanza.[55]

They reached the Tombstone country in late February of 1878 and set up camp in Brunchow's abandoned adobe. Within a month, with Gird's on-the-spot assaying, Schieffelin had located two rich outcroppings half a dozen miles to the north—the Lucky Cuss, a pocket that ran to an astounding $15,000 a ton, and the much larger but not so spectacular Tough Nut, a mile away. Before they finished, another prospector, Henry Williams, and a partner came along and eagerly accepted Gird's offer to do all his assaying for a half interest in whatever he found. But when Williams did find a third bonanza, the Grand Central, another mile beyond, he located it just in his and his partner's names and he only relinquished a small fraction, about a hundred feet of the rich outcrop, at gunpoint. Gird relocated that ground as the Contention, but he and the Schieffelins were so angry with Williams that they didn't want to work alongside him, and they sold the claim to the next party who came along, six weeks later, for $10,000 without bothering to explore it. At the same time they bonded their own bonanzas for $90,000 to another speculator, who tried to get George Hearst's partner, James Haggin, to take the claims, but three months later he finally declined, which gave Gird and the Schieffelins a second chance to make real fortunes.[56]

In the meantime, a Tucson gun dealer, John Vosburg, who had loaned them $300 after they told him of their discoveries, had sent the exciting news to his old friend, ex-Governor Safford. The governor hurried back to Arizona in September with a couple of his Aztec scheme investors, an enthusiastic young Philadelphia insurance broker, Elbert Corbin, and his much wealthier older brother, Frank, who with other brothers ran the family hardware factory in New Britain, Connecticut. By then, however, the discoverers had opened a lot more rich ore in their claims, so they consented to sell only a quarter interest for $40,000 to put up a mill. With their new partners they then formed two companies, the Tombstone Gold & Silver Mill and Mining Company around the Tough Nut claims, stocked at $5,000,000 in $10 par shares, and the more modest Corbin Mill and Mining Company around the Lucky Cuss, stocked at only $500,000. The discoverers each kept a quarter of all the shares, Safford and Vosburg got one-sixteenth apiece, and the remaining one-eighth was taken by the Corbins, who actually put up the cash.[57]

With outside money coming in, the rush to Tombstone began. By the time Gird's mill started turning out bullion in June of 1879, over a thousand fortune seekers had crowded in; when the Southern Pacific Railroad finally passed just to the north a year later, Tombstone's population had more than doubled; and the following year it doubled again to 7,000 as the camp boasted a mining exchange and an opera house, plus an infamous reputation for gratuitous bloodshed. By early 1880, with the mines already producing about $200,000 a month, Safford and the Corbins were anxious to start pushing stock. At the same time Ed Schieffelin, who had quickly tired of urban life, had already taken off on another prospecting trip and wanted out, so he and Al happily agreed to sell their half interest for $600,000 cash in March. Gird was also anxious to sell, but the easterners didn't think they could handle that much stock at once—as well as the appearance of all the discoverers getting out. So they officially sold only the 250,000 shares in the Schieffelins' names, and Gird agreed to remain for another year as a director, superintendent, and major shareholder. But privately Ed and Al gave Gird an equal share of the money in exchange for their getting equal shares when he eventually sold out. The Corbins, Safford, and his principal mining-scheme partner, a sharp young Philadelphian and wealthy saw manufacturer's heir, Hamilton Disston, paid out the purchase money in six monthly installments from their returns as they started selling off the stock. They promptly combined the two companies into the more streamlined Tombstone Mill and Mining Company, with a puffed-up $12,500,000 in $25 par shares, and immediately started paying regular monthly dividends of 10 cents a share, as they pushed them on the Philadelphia mining exchange at $5. At that price the buyers could look forward to annual earnings of nearly 25 percent in dividends, while the sellers grabbed an instant profit of 100 percent from the sales![58]

Since the Corbins had put up the original $40,000 for the mill, they still wanted something extra, so their new partners and fellow directors quietly let them put up a second mill and take half of the production from the Lucky Cuss. Then, when that rich pocket was sudden exhausted, as directors they simply gave themselves a note on the company treasury for their additional expected profits of close to $250,000. When the shareholders later found out, the Corbins claimed the note was for "selling the mine" to the company, never mind that the company had always claimed they already owned it! Now, since the Tough Nut alone couldn't maintain the rate of dividends, Gird and the others also secretly began loaning the company money to help keep up the dividends and sustain the prices while they all unloaded. Gird finally, in the spring of 1881, sold out his shares at an average of around $5 apiece after collecting over $1 a share in dividends. That gave him and the Schieffelins an additional $800,000 for an overall profit of nearly $500,000 apiece, a near record for western prospectors.[59]

Ed Schieffelin, however, had little use for money, except to buy homes and farms for all of his relatives and to finance ambitious prospecting adventures. He soon made news again, in 1882, when he bought a steamboat, which he christened his *New Racket*, and headed up the Yukon River for a year and a half, looking for gold. Even though he found pay dirt all along the banks, he failed to find any quartz lodes and finally quit, disgusted not only with the mosquitoes and the cold, but also with the fact that it had taken sixteen months to get a letter from the states. He married a widow but couldn't settle down, and he was still prospecting the day he died, at the age of forty-nine, on a trip in Oregon in 1897, just as the Yukon was finally opening up. His brother, who had gone to Philadelphia to watch after their stock sales, quickly lost most of his fortune and his health in other speculations, and came back west to die in 1885, at barely thirty-six. Gird put half of his fortune into 50,000 acres of the old Chino Rancho and other speculations in the great Southern

California land boom, and he spent more than $100,000 of it paying off old creditors, who were happy to hear of his good fortune. He sold lots, started a town, and pioneered the sugar beet industry until he lost heavily in the Panic of 1893 and retired to Los Angeles, where he died in 1910 at seventy-four.[60]

Only when the dividends, totaling $1,250,000, suddenly stopped in April of 1882 did all the unsuspecting investors learn that the company had gone $350,000 into debt to the directors to help pay for such generous returns. Then the stock crashed from $4.75 to 80 cents a share in just a couple of months, and the stunned shareholders scrambled to take over the management. They found that there was still some ore in the mines, but to hang onto it they also had to pay off the debt to the old directors to keep them from foreclosing, and that consumed all the earnings from the remaining ore. So they never got another dividend, although they kept trying for well over a decade, and most ended up losing more than half of their money, not counting lost interest. For their part, of course, the promoters got away with about $2,000,000. The Corbins put their part of the take into new Connecticut enterprises, while Disston and Safford, who had also pushed stock in a couple of adjacent mines for extra cash, put $1,000,000 of their loot into 4,000,000 acres of Florida swampland. There they pushed $10,000,000 more in stocks and bonds in a scheme to drain the Everglades for real estate development and sugar cane plantations, working that until they died, Safford in 1891 at the age of sixty-one and Disston in 1896 at only fifty-one.[61]

Ironically, however, the greatest profits came from the Contention, so cheaply disposed of for $10,000, which became Tombstone's biggest producer and biggest dividend payer. It had been purchased by a group of San Francisco speculators, led by Albion K. P. Harmon, an old argonaut in his late sixties who had already made a modest fortune from the Comstock, and his far less successful friend, forty-year-old Walter E. Dean. In October of 1878 they had incorporated it as the Western Mining Company, as Tombstone's second stock venture, apparently intending to simply dump its inflated $10,000,000 in $100 par shares when the excitement mounted. But they soon struck a real bonanza and decided to hang onto all their shares for a while. In the spring of 1880 they even turned down an offer of $1,500,000 for 60 percent of the stock at $25 a share. Instead they started turning out $100,000 a month in bullion and taking out $75,000 a month in dividends, to be hailed as "the best mine yet opened in Arizona." They were also aggressive in acquiring profitable neighboring claims, insisting their ore dipped into them and filing such heavy damage suits that they intimidated their neighbors into cheaply merging. Then they reincorporated in the fall of 1881 as the Contention Consolidated Mining Company, boosting the stock to $12,500,000 at $50 par. By the end of 1882 they had taken out well over $4,000,000 in silver and paid themselves over $2,000,000 in dividends. But by then the rich ore was about worked out, so they too finally decided to let the public in. They placed the shares with Philadelphia speculators, who unloaded all they could at $5 before the price dropped to $2 after the dividends stopped in June of 1883. In all, they cleared more than $3,000,000, giving them a truly magnificent profit of 300-fold on their $10,000 investment in just five years—the biggest yield of the boom! Harmon used some on his profits to build a gym for the young University of California.[62]

Lastly, Tombstone's third bonanza, Henry Williams's Grand Central, also paid well for a string of early speculators and investors. In May of 1878, just over a month after staking the claim, Williams sold his two-thirds interest to a forty-year-old speculator from New Orleans, W. Frank Witherell, and his Yankee partner, Eliphalet B. Gage, for $6,500, all of which he blew on a drunken spree in Tucson. In the meantime, Witherell and Gage had to buy out Willliams's Tucson lawyers for another $15,000 or more. A few weeks later the post trader Frederick Austin and a couple of officers at Camp Huachuca picked up the other third interest for a mere $1,000 and a farm from Williams's partner, Oliver Boyer, as he was fleeing to Mexico after killing a man in a bar. Austin and his friends sold out to Witherell and Gage for a quick and hefty profit of close to $50,000, nearly fiftyfold in just six months! To raise the money for the deal, they sold off roughly a quarter interest to a Pennsylvania investor, Thomas Struthers, for $50,000, which gave them title to most of the mine for about $20,000. But they later had to buy off an ex-Indian agent, Henry Hart, and his lawyers for $25,000 to settle a grubstake claim for yet another "third." Finally, in July of 1879, they organized the Grand Central Mining Company in St. Louis with a conservative capital of $400,000 in just eight hundred shares at $500 par, of which they each held nearly three hundred shares, which they priced at about ten times their investment.[63]

Six months later, after demonstrating the potential of the mine, Witherell, backed to the wall by other debts, sold out his shares in a convuluted
deal for only about $50,000. That gave him a profit of about 150 percent
in a year and a half on his $20,000 outlay, but even that wasn't enough
to extricate him. The formal buyer was Charles D. Arms, a forty-year-old
Youngstown, Ohio, coal and iron speculator who had formed a $25,000
partnership with a younger fellow speculator, Peter L. Kimberly, to try
their luck in western mines. But all the purchase money was put up by
a fifty-year-old Chicago millionaire, the "Lord of Lard," Nathaniel K.
Fairbank, after Arms had also promised him half of the stock. In all,
they bought about 350 shares, and Arms signed 60 percent over to Fairbank for his half plus repayment, making him the largest stockholder.
Arms, however, refused to split the remainder with Kimberly, claiming
the Grand Central was a separate, personal deal, even though he had
used their partnership money on trips to investigate the mine. Kimberly
sued for his half clear to the Supreme Court and finally won in 1889.
In the meantime, Arms reincorporated the company in Ohio, splitting
the old shares 125 to 1 for a new capital of $10,000,000 at $100 par, and
became president. Although Gage had also sold about a third of his
shares, he kept the rest and remained as superintendent. Together the
shareholders paid in another $200,000 in assessments to open the mine
and put up a mill. But they finally took out $2,500,000 in bullion and paid
themselves back $1,000,000 before the dividends stopped in 1882. This
was an excellent return on all their investments, and then they made
even more by finally sharing their stock with the public. Fairbank was so
fired up by this successful mining venture that he soon joined with other
Chicagoans to try working a deal on the old Maxwell grant as well.[64]

Tombstone's mines were so profitable that even the "knightly" James
Keene had turned to them for quick cash after a spate of bad luck in
the winter of 1880–81, when his Derby-winning racehorse drowned and
his Newport mansion burned to the ground. He and his confederates
picked up some cheap claims on the outskirts near the old Brunchow
mine and worked a couple of scams, the Bradshaw Mining Company
with $2,250,000 in shares and the Washington with $1,000,000. They
put out a pack of lies, claiming that the full capital was already paid up,
there was over $800,000 worth of silver in sight, and there was "no rea-
son, either natural or theoretical," why the mines shouldn't soon rival

the Contention. They pushed the stock furiously and got away with a reported $600,000. But they were soon sued on the grounds that the mines were "fraudulently contrived" for the sole "purpose of carrying out a swindle," and they took the Fifth Amendment.[5]

IN THE MEANTIME, back up the tracks in California another bonanza was being opened near the Southern Pacific's new branch across the Mojave Desert that would give the "Golden State" its own "silver king" and, belatedly, its first actual mine-owner governor, Robert W. Waterman. Like the Tombstone bonanza, the Waterman bonanza was also worked by two brothers and an experienced millman, but though the cast was similar, the play was very different. A friendly, burly fellow in his early fifties, Bob Waterman had spent thirty years seeking elusive fortune in failing ventures from Illinois to California. He was bailed out, however, and loaned money for a new start every time by his older, Illinois banker brother, James, who had set him up most recently with a small resort hotel at Arrowhead Springs east of Los Angeles. During off seasons, Bob had also tried his hand at prospecting, without luck until the fall of 1880, when he decided to take up the so-called Quicksilver claim of an old eccentric, George Lee, who had disappeared in the desert the year before. Bob recruited a younger millman, John Porter, for a quarter share to help him locate and explore the claim, and they soon unearthed a "monster ledge" of ore running as much as $200 a ton and averaging $50. By April Bob was excitedly writing his brother that they could "make a million a year," if he would just put up the money for a mill—and, of course, forgive all of Bob's old debts of nearly $12,000. In return Bob would give him a 24 percent share out of his 75 percent interest and Porter would throw in another 3 percent—plus, Bob promised, "all the money" James put up would be paid back promptly "from the first earnings of the mill." James hurried west to see the bonanza and soon agreed to the deal. In the next few months he paid out over $28,000, as Porter hauled in an old mill from the Panamints and started turning out bullion at as much as $30,000 a month by the time the Southern Pacific tracks passed by the following year.[66]

The opening of the Waterman bonanza spurred a wild rush to the surrounding hills that spawned the new boom camp of Calico just to the east. But the opening of the bonanza also brought suits from old Lee's heirs and his partner, who claimed that Waterman had jumped the claim illegally, because Lee had done the required work to hold it in 1879, so that it couldn't be legally relocated until New Year's Day of 1881, if no work were done in 1880. Since the suits not only sought possession of the mine but also heavy damages for all the ore extracted, James wanted to wait to officially take title to his share until they were settled, to protect his own assets. By March of 1883, however, the suits were still dragging on, and James as yet hadn't received even a penny of the promised repayment of his outlays, even though the mine by then had produced over $400,000. So James, though critically ill, came out to California to demand his long-overdue payments and a further extension of his title rights. But Bob refused, offering only to sign over James's share right then, if he would waive all repayments. Feeling "tricked and humbugged," James angrily returned home and died of a stroke just a few months later.[67]

Meanwhile, Bob had new worries about the Lee suit. A drunken prospector, Lewis Hoffman, boasted of bashing in Lee's skull and then led a deputy right to the body. Apparently fearing a wave of sympathy for Lee's heirs, if it were proven that he had been murdered, Bob startled everyone by hiring lawyers to defend the confessed murderer. Even though Hoffman had claimed the body was Lee's and his dentist had identified a protruding chipped tooth as Lee's, Bob's lawyers brought in an Indian who claimed the bones were his mother's, and the jury split six to six. Hoffman went free and the Lees, failing to prove that any work had been done in 1879, finally lost their case a few months later. Right after James returned to Illinois, Bob had finally started paying back the $28,000 his brother had put up, and as soon as Bob heard of his death, he had hastily sent his widowed sister-in-law, Abbie, a final payment, hoping she wouldn't continue to press "a demand for the deed," although James had signed over the deed obligation to her before he died. Then, when Bob learned that James had forgiven all of his brothers' and sisters' debts in his will, he demanded that Abbie give his money back! At the same time, of course, he refused her request for her husband's share of the mine, so he shouldn't have been too surprised that she too soon sued him. Bob argued at first that James had told him he wanted Bob's sons to have his share, but then Bob claimed that James had never intended to have a share at all and that their agreement was only intended as collateral for the loans that he had now paid back. Despite, or perhaps because of such arguments, Abbie won her suit before a federal judge in San Francisco in the spring of 1886, but she too died soon after, while Bob appealed the decision on up to the Supreme Court.[68]

By the summer of 1886 Waterman's bonanza was finally worked out, but it had produced $1,400,000 in bullion and paid its two owners a modest, though still contested, fortune of $400,000. Bob put part of his money into a San Diego gold mine that he almost sold to a Chicago investor for $2,000,000, but the deal fell through when he learned that the promoters were going to clear an additional $1,000,000 for themselves, and he greedily demanded all of that too. He had also become such a heavy contributor to the Republican presidential campaign in 1884 that he was dubbed their "Golden Dustman," and to keep the money coming in for the governor's race in 1886, he was made the party's candidate for lieutenant governor. He ran as the "Honest Miner," which invited scathing exposes of all his legal battles, his "vile plot to rob his brother's widow," and even veiled suggestions that he may have hired Hoffman to murder Lee for the mine! But his easy, friendly manner won him a narrow victory, while his running mate for governor narrowly lost. Then when the Democratic victor, Washington Bartlett, died of Bright's disease a year later, Bob Waterman suddenly became governor of California. He was not politically adept, and his inclinations were too progressive for party stalwarts like Harrison Gray Otis of the Los Angeles *Times*, who derided him as "His Accidency," so he didn't seek a second term. In 1889, however, Waterman easily won Supreme Court justice Stephen Field's favor for the pending appeal of the Abbie Waterman case, by promptly dismissing all charges of complicity against Field after his bodyguard had shot and killed his long-time adversary, David Terry, in a California railroad station. When Waterman retired in January of 1891, he was eager to return to his mining ventures, but he too died suddenly of pneumonia just three months later. The following year the Supreme Court finally ruled, reversing Abbie's lower court victory, to give the dead man clear title to the abandoned mine.[69]

THE BIGGEST NEW silver-lead bonanza and biggest dividend payer of the entire boom, however, was not in Leadville or Tombstone. It was the great Horn Silver mine in the remote little camp of Frisco, in southern Utah.

It paid its investors over $1,000,000 a year for three years, eclipsing even Hearst's reliable old Ontario, and it made several fortunes, though most would still be lucky just to get their money back. The massive bonanza was discovered in the fall of 1875, and although its ore averaged about 40 percent lead and $50 a ton in silver, its discoverers quickly realized that even that wouldn't pay after hauling it by wagon over two hundred miles to Salt Lake City for smelting. So within just a few months they sold the claim for $25,000 to a successful Colorado and Montana miner turned merchant, Allen G. Campbell, and three partners, who tried putting up a local smelter at a cost of $25,000. But their smelter couldn't recover more than 80 percent of the value, which still wouldn't pay on any but the richest ore. What they really needed was a railroad to haul the ore cheaply to the big smelters, so in the fall of 1878 they succeeded in making a deal with former Philadelphia banker and railroad promoter Jay Cooke, who was down on his luck. The aging Cooke was once hailed as the "savior of the nation" for floating the $500,000,000 in Civil War bonds needed by Lincoln to pay the troops, and after the war he had turned his talents to floating another $100,000,000 in bonds for the Northern Pacific Railroad in a flamboyant campaign that made him the model for Mark Twain's "Col. Sellers" in his satire *The Gilded Age*. But Cooke lost the competition for government support to the Central and Union Pacific companies, and although he struggled on, he and his investors were taken down in the Panic of 1873. Still trying to make a new start, Cooke was steered by a friend into a deal with Campbell in which he promised to provide a railroad and market the stock in the mine, all in exchange for 10 percent of the stock and commissions on the sales.[70]

Cooke had no money to build the railroad, of course, but he formed the Utah Southern company, and with the lure of heavy freight he got an old friend, Sidney Dillon, president of the Union Pacific, to have his company take up a half interest by providing old rails and rolling stock; then he talked officials of the Mormon church into putting about $400,000 in church funds into a quarter interest; and finally he persuaded Campbell and his partners to subscribe for the remaining quarter. Cooke pushed the work quickly, and early in 1880 the first train rolled out of Frisco with ore. In the meantime, Cooke had hired Columbia School of Mines professor John Newbury, who effusively declared the Horn Silver to be "the most valuable body of silver ore known to exist in any mine in the world," and several other enthusiastic mining experts, who estimated as much as $45,000,000 worth of "ore in sight" and predicted profits of $18,000,000 or more. The owners thus confidently incorporated the Horn Silver Mining Company for $10,000,000 in 400,000 shares at $25 par, and they unleashed a whirlwind of elegant prospectuses, attractive supplements in the *Engineering & Mining Journal*, and enticing features in the New York dailies, plus, of course, a premature dividend. In the midst of all that, in the fall of 1879 Cooke succeeded in signing up a group of New York investors for 200,001 shares for a grand total of $3,900,022.50 cash, an average of $19.50 a share. Then, with the help of another dividend on the completion of the railroad, Cooke unloaded his own shares, collected his commissions, and retired with a clear profit of $1,000,000 to buy back his former mansion. Campbell soon moved to Southern California with his share of the fortune, but he and his partners held onto their remaining stock for dividends for a while.[71]

The New York investors were led by Charles G. Francklyn, the ambitious American agent of the Cunard steamship line and a grandson of its founder. He assumed the presidency of the Horn Silver company and pushed eagerly to increase profits by building their own giant smelter, the largest in the territory, at the new town of Francklyn, just outside of Salt Lake City, for a cost of several hundred thousand dollars. He also hired a new superintendent, who worked every possible, though not always wise, economy at the mine. By 1882 he had the mine efficiently producing over $3,000,000 a year and paying out over a third of that in steady annual dividends of $3 a share, for a solid 15 percent a year on the initial investment. That made it the new idol of eastern mining investors, whose faith was only briefly shaken in the spring of 1883 when the scantily timbered, but nearly worked out, top two levels of the mine slowly settled down to the third, but work vigorously continued on the four rich levels below, and the shareholders missed only one month's dividend. Campbell and his partners, however, may have decided it was time to get out, having taken another $1,000,000 by then from their share of the dividends, for increasing sales pushed the shares down briefly to $5 before the end of the year, even though most investors still held on confidently, steadily earning $3 a year. By the end of 1884 the Horn Silver had produced nearly $12,000,000 in bullion and delighted faithful investors with $4,000,000 in dividends, having already returned $10 a share, or half of their investment, and they were confidently looking forward to much more. But on February 12, 1885, their confidence was shattered when with only a few minutes warning, as the miners were hoisted out, all the remaining levels suddenly collapsed, with a violent shock that broke windows nearly twenty miles away and wiped out investments all the way to England, as the shares crashed from $7.50 to $1.70.[72]

Franckiyn had financed his investment in the Horn Silver by drawing heavily from the estate he was managing for his younger English cousin, Sir Bache Cunard. Heady with success as the dividends poured in, he began to live the life of a bonanza king, speculating in other mines and seemingly everything else from gas lights in New York to a cattle ranch in Texas. But as his spending soon outpaced the dividends, he also began, quietly, to borrow from the company treasury as well, and when the mine collapsed he was in debt $650,000 to his shareholders and much deeper to his cousin. He frantically sank a new shaft, trying to find a salvable portion of the bonanza, but by the fall of 1887, after the shareholders finally learned of his debts, they brought suit demanding immediate payment. At the same time his cousin, who had finally looked into his own affairs, sued him for $3,000,000 and had him arrested for embezzlement. By then the shares only fetched 75 cents, and all the shareholders ever got from him was some near-worthless land in Kentucky. His cousin, who became the head of the Cunard Line, still pressed his case, on and off for more than a decade, up to the New York Supreme Court, where he won a judgment for $1,657,700, but he too never recovered much. Although faithful investors who still held on eventually eked out another $4 a share over the next twenty years, they never got their money back. Even after the faithful finally reorganized as the Horn Silver Mines Company, slashing par from $25 to just $1, during the speculative frenzy of the First World War they got back only 20 cents a share. Once again the underlying problem for investors was not only crooked management but extravagant capitalization. So while the promoters made instant fortunes, investors had little chance of ever getting their money back, let alone turning a profit, even from the biggest dividend payer of the boom.[73]

THE BEST INVESTMENTS, of course, were still the handful of steady producers. George Hearst's Ontario consistently paid out $900,000 or more a year, for a 20 percent annual return on the market value of the mine, and its dividends reached nearly $12,000,000 by the end of the 1880s. Hearst's Homestake paid out $4,500,000 during the same period, on its way to building an even more impressive record, as it averaged about $350,000 a year, or 25 percent on its market value. At the same time the Eureka and Richmond Consolidated companies each paid their shareholders over $4,000,000 at similar rates. In California's deep gold mines, the Coleman brothers' Idaho paid its 240th monthly dividend by the end of the decade, for a total of over $5,000,000 to its private investors, while Alvinza Hayward's newly formed Plymouth Consolidated averaged over $400,000 a year in dividends, or again 25 percent on its market price, for a total of $2,300,000, before it was destroyed by a fire in the late 1880s. The Cook brothers' Standard Consolidated also paid out $3,500,000 at the same rate of return before its bonanza ores were exhausted and it began levying assessments in 1884.[74]

Even that great hydraulic bonanza, the North Bloomfield, was finally paying dividends after absorbing almost $5,000,000 in assessments and bonds, and it was at last being hailed as a triumph of the determination and perseverance of private enterprise. Visitors stood in awe before its "great amphitheatre," an open pit larger than the Roman Colisseum, and were deafened by the "incessant roar" of water from the nozzles of its monitors, imagining even that it rushed out "with a wicked, vicious, unutterable indignation." The great mine was consuming over 10,000,000,000 gallons of water a year, and tearing down 5,000,000 cubic yards of dirt and gravel, which yielded 15 cents a cubic yard to produce $750,000 a year in gold and $250,000 in dividends. But all that debris from the old channel was being flushed into the Yuba River, choking its channel and overflowing into the surrounding farms. Eventually, over 45,000 acres of farmland were destroyed by the debris, or "slickens" as they were commonly called, and the farming towns couldn't build levees fast enough to keep the Yuba from flooding them every year. The farmers began filing suits for damages, but the North Bloomfield and other hydraulic companies fought the cases successfully in the courts for years.[75]

In January of 1884, however, despite claims that the slickens didn't hurt the farms and were actually beneficial, a federal judge, Lorenzo Sawyer, finally granted an injunction against the North Bloomfield and others that ended all hydraulic mining on the Yuba. By then the North

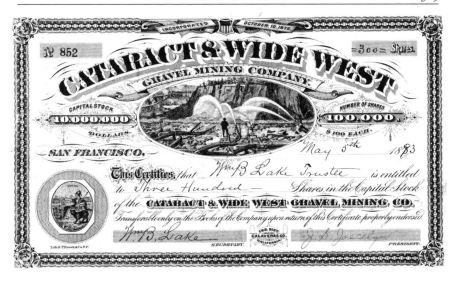

Bloomfield had produced over $4,500,000 and paid back $1,500,000 in dividends, which left a net loss of $3,500,000 to its investors. The mine owners complained that they had been cut off in their prime, having worked less than 10 percent of the channel, and that there was still $40,000,000 left that would have returned a great profit. But judging from the returns of even their most productive years, it would have taken them about seventy years to have worked all the channel. So, even though the expected dividends of $15,000,000 would indeed have given a profit of 200 percent, if spread over seventy years that would still have amounted to an average interest rate of less than 3 percent per year, which was less than they could have earned just leaving their money in the bank. Thus, even if they hadn't been cut short, it was still a losing investment![76]

Yet, while the North Bloomfield still roared, its seeming success enticed a myriad of investors into such splashy, but profitless, hydraulic ventures as the Cataract & Wide West Gravel Mining Company and countless others throughout the West. But even the biggest dividend payer of all these public hydraulics, the very conservative-looking Excelsior, proved to be a startling "disappointment." James Haggin, who had won the confidence of eastern investors after placing such well-paying ventures as the Ontario, Homestake, and Standard Consolidated on the New York

Stock Exchange, brought out the Excelsior Water & Mining Company as "the biggest thing going." It had been formed in the spring of 1877 by Haggin and others as a private venture for a nominal $5,000,000, to run three miles of drain tunnels to finally work the famous old Blue Gravel bonanza down to bedrock. But they had taken out only about $1,000,000 in the next two years and gotten back only $192,000 in dividends. So by the spring of 1879, as the rich gravel was giving out, the company was going into debt, and the Yuba farmers' suits were threatening, Haggin was able to get a "call" on the bulk of the stock for $14 a share, and he took control to work it on the market instead.[77]

Haggin promptly doubled the stock to 100,000 shares at $100 par, resumed regular dividends, and listed it too on the New York Stock Exchange in November, as he paid its twentieth dividend of 25 cents a share. He generously claimed that the Excelsior had already produced $6,000,000 in gold and would pay dividends of $400,000 a year for years. Through the brokerage firm of his son, Ben Ali Haggin, and his son-in-law, Richard P. Lounsbery, he offered to let eager investors in "on the bottom" at $20 a share. With steady monthly dividends, he soon worked the price up to $29, and over the next ten months he successfully unloaded over 84 percent of the stock at $20 or more, taking in nearly $2,000,000 while he paid out $250,000. But then the dividends stopped, the price dropped to $7, and he began levying assessments for a total of $3 a share that not only took back all of the previous dividends but knocked the stock on down to just 25 cents. Some claimed that he was just "bearing" the price to regain control of the mine, but the farmers' injunctions soon shut it down completely. Only then did the investors finally discover that Haggin too had betrayed them, loaning the company $286,000 to pay all those enticing dividends while he had unloaded. All of this was revealed in 1885, when a few irate shareholders took him and his relatives to court for conspiracy to defraud. The judge dismissed the conspiracy charges against the others, but a jury found Haggin guilty of defrauding the investors with false dividends and awarded them restitution, though he tied that up in appeals.[78]

EVEN THE "WIZARD OF MENLO PARK," Thomas Alva Edison, still in his early thirties, got caught up in the fever of hydraulic mining. In 1879, looking for a cheap source of platinum for the filaments in his new light bulb, he had tested metal-rich "black sands." He quickly found that he could easily extract such sands by using powerful electromagnets to pull the iron-enriched grains out of a flow of dry sand falling through a chute, and he patented the idea the following year. Then, hoping that there might be platinum, as well as gold, in the "black sands" that had slipped through the sluices with the tailings in all the California hydraulic operations, Edison sent a young engineer, Frank McLaughlin, out to the mines to get samples and contracts for their tailings. Although the sands that McLaughlin collected showed little platinum, Edison found that they were startlingly rich in gold, so he decided to also "revolutionize" the mining business by recovering all the "waste gold from the tailings of mines."[79]

With three of his partners in the Edison Electric Light Company, he formed the Edison Ore Milling Company, with a modest capital of $350,000, for the purpose of "extracting metals from ores by electricity." He gave the company the patent rights in exchange for about a third of the stock and 30 percent of any profits after 20 percent a year was paid on the stock, plus an annual salary of $10,000. His partners divided about half the stock, and 20 percent was put into the treasury to pay all expenses of his experiments. At the same time, after securing options on millions of tons of tailings, they also organized the companion Miocene Mining Company, with McLaughlin as manager, to erect an Edison Reduction Works on the Feather River opposite Oroville and begin mining as well. To push the stock, Edison talked grandly of recovering as much as "$1,400 per ton at an expense not exceeding $5 per ton," and of having "at least $50,000,000 in the tailings"! His publicist and former bartender, William McMahon, trumpeted it all across the country as the "latest Edisonian sensation," driving the shares up from their $100 par to $250. But the mining editors scoffed at such "ridiculously extravagant" claims that sound like "tales of the Arabian Nights" and only "make a fool of him." Others "with an eye of suspicion" more pointedly noted that the only profitable extraction by the process so far was coming from "the sale of its stocks, which fluctuate with the pulse-beats of the public, influenced by the reports of Edison's successes."[80]

Nonetheless, Edison and McLaughlin diligently tried to make it work, but after another year's experimentation and $500,000 in expenditures, they simply found that the marvelous "black sands" weren't nearly as rich as they had claimed; the magnets just couldn't recover enough gold to

make them pay. Then, in a final blow, a court injunction against flushing debris into the river shut down everything just two months after they finally started running. Disappointed investors wailed even louder after the boastful McMahon claimed that the whole "ore milling scheme was 'cooked'—intended purely for the Street and not for a mine." But Edison, who remained the major stockholder in the ore milling company, kept it alive, boosting its stock to $2,000,000, and spent another twenty years trying to find a profitable extraction scheme at least for iron. McLaughlin stayed with hydraulic mining and eventually made a small fortune from selling mines to the British, buying a mansion in Santa Cruz and becoming a power in the California Republican Party. But soon after the turn of the century, "crazed" by the sudden loss of both his fortune and his wife, he killed his step-daughter and himself.[81]

As THE EAGER FLOCK of new pigeons rushed into the frenzied mining stock market, an equally eager crowd of brokers had also rushed in to help pluck them, and many quickly set up traps of their own. One of the most notorious was a newcomer, James Madison Seymour, a slim Texan in his mid-thirties, who had speculated in cotton and grain in Chicago before moving east to buy a seat on the New York Stock Exchange. He got a crash course in the ways of mining promotion from that devious shark Bethuel Phelps, reviving the famous old Vulture mine. Phelps had let the mine go for taxes, but found that he could pick it up again for $50,000 to work in the new boom. So, claiming that he still owned it, he sold the newly arrived Seymour a quarter interest for the $50,000 needed to secretly buy it back. They formed the Central Arizona Mining Company with $10,000,000 in $100 par shares, which they launched on the New York Stock Exchange in the summer of 1879. With wash sales and claims of $2,000,000 in sight, they ran the price up to $23 a share by the end of the year, but it crashed to barely $4 after Phelps tried to dump most of his shares. When Seymour finally realized that he had been taken, he confronted Phelps at knifepoint to extract a bigger share of the take and force him out. Since there was still gold in the mine, Seymour pursued the scheme and spent a year and $258,000, he claimed, putting up a giant eighty-stamp mill at a new town he proudly named for himself. In the flurry of activity he worked the stock back up to $10 while he unloaded what he could, but even though the old Vulture did produce well more than $500,000 in the next three years, it never paid a dividend, and by then the shares had dropped to just 37 cents.[82]

By the beginning of 1881, however, Seymour had finally realized that actually working a mine cut too deeply into the sure profits of stock sales. So for his next venture he turned to a little coyote hole of a mine hidden away at Salt Spring in the south end of Death Valley. Since hardly anyone knew anything about it, and it was so remote that no one would know what he was doing, it was ideal for a quick and dirty play on the Broadway exchanges. Seymour picked it up for $22,500, gave it the exotic stage name of the South Pacific Mining Company, a conservative-looking paper capital of $500,000 in $1 par shares, and a respectable-looking cast of trustees, led by the Republican leader of the New Jersey Senate, Garret A. Hobart, who was hailed as a man who "never makes a mistake." Seymour opened the stock on the Broadway exchanges in October, and within a month, in a pool with friends feverishly trading among themselves, he teased the price up from its modest $1 par to a bawdy $14.63 a share! No other mining stock on the New York exchanges had ever been so far above par, and none had ever been more vacuous.[83]

He then brought out a rousing chorus of experts to praise the mine to the high heavens in the pages of the New York dailies. One claimed it had produced the richest gold quartz ever found, and he envisioned profits of precisely $145,800 a month, 350 percent a year on par. Another imagined millions more in gold not only in the little vein itself, but in the granite on either side, swearing that "the whole gulch, 40 feet wide, would pay." An aptly named "Dr. G. Wiss" even vowed that the great German geologist Baron von Richthofen had once predicted that the little range of hills at Salt Spring would someday "surpass California and Nevada combined in the production of the precious minerals!" With such publicity as that and steady washing of the shares, the stock only slipped back to around $4 by the end of the year as he flooded the market with its shares, and he was able to keep the price above par until late April, by which time he had dumped nearly all 500,000 shares for over $1,000,000. By then too the volume of both real and imaginary trading had passed 3,000,000 shares, shoving Seymour's own little South Sea Bubble well on its way to becoming the next most active mining stock of 1882.[84]

Not all of the press, of course, had been so accepting of Seymour's new hit as the New York dailies. The western papers and the mining journals had panned the play from the start. But the most cutting critic came from the New York *Daily Stockholder*, which summed him up neatly. "There is no investment nonsense about Seymour. He can't wait for dividends. He doesn't calculate interest on the money he puts into mining stocks; multiplying the principal by 500 and pocketing the product is good enough for him," its editor snapped. "He can buy a mine for less and sell it for more; he can get more of the public's money and hold it better; he can look you square in the face and talk less truth and more taffy; he has the most cheek and the least conscience; he can profess more sincerity and possess less; he has brought less to the mining interests and taken more from them—well, than any other schemer on a large scale who has made a big bank account and lost only a small reputation!" But Seymour still found an ample audience.[85]

He finally lowered the curtain on South Pacific in May, and its shares plummeted from 65 cents to 5 cents in two weeks. Only then did the shareholders discover that he had hired a skeleton crew of just five men at the mine and then stiffed them too for their wages. The miners sued, but all they got was the mine. The shareholders only got to keep some pretty paper, which was shuffled about, mine or no mine, for another year until it was finally pulled off the exchanges. Seymour meanwhile basked in his notoriety and was even called upon by the New York Stock Exchange's president, Brayton Ives, to do the dirty work in staging another con. He did so well that he eventually opened nineteen branch offices to push all his paper, before he finally retired to New Jersey. His prize director, Garret A. Hobart, went on to become a Republican vice president of the United States.[86]

IN THE HEAT of the new speculative fever, all the old warhorses were also trotted out once again. The old Maxwell grant was put through its paces after John Collinson and Stephen Elkins finally obtained a patent in 1879 on Elkins's brother's fraudulent survey of the grant. Then, just as it appeared they had successfully grabbed the property from the bondholders, a sharp new speculator, Frank Remington Sherwin, moved in on the scheme. Sherwin, a forty-year-old Bostonian and former New York broker with a decidedly "checkered" career, had bought up large blocks of the shares and bonds of the company at barely 4 to 5 cents on the dollar to become a major holder. Then he went to the Dutch bondholders, promising to resurrect their defunct investments, and they hailed him as their savior. "They gave me absolute power to do as I pleased," he claimed, and since the interest had never been paid on their bonds, he promptly foreclosed on the mortgage, wresting the grant back again. In the exuberance that followed, he ran the paper up from 5 to 55 cents on the dollar in three months of furious trading on the Amsterdam exchange, turning over the shares a couple times and cleaning up a tenfold profit. Then he reorganized as the Maxwell Land Grant Company in the spring of 1880, issuing $5,000,000 in new income bonds, exchanged dollar for dollar at par for the $3,425,000 in old first-mortgage bonds and 60 cents on the dollar for the $1,375,000 in second-mortgage bonds, and another $5,000,000 in new shares, which he exchanged for old shares at only 4 cents on the dollar despite the rise! In payment for all of his good deeds, Sherwin took all the rest, $4,800,000 in shares and $750,000 in income bonds.[87]

Sherwin naturally made himself president and successfully sold his shares and bonds, doubtless at a generous discount, to an array of prominent Chicago mining investors, including Marshall Field, Levi Leiter, Cyrus McCormick, Nathaniel Fairbank, and George Pullman.

He also quietly and quickly abandoned his Dutch bondholders. Instead he joined with Santa Fe railroad officials to give them a half interest in a subsidiary company to operate the grant's biggest asset, the Raton coal beds, probably in exchange for personal shares in the railroad. He then made a sweetheart contract with the railroad for most of the coal, charging them barely half of what it cost to extract it, leaving the grant's coal subsidiary to run at a loss. So while the new investors and old bondholders anxiously awaited their promised returns, Sherwin "lived in a princely style" in Maxwell's mansion, filled his stables with race horses from England, and eloped with former Colorado governor William Gilpin's eighteen-year-old stepdaughter. The old governor was furious and may have brought his political influence to bear, along with others who saw a chance to settle old debts. For, just six weeks after the elopement, Sherwin was arrested, promptly convicted, and sentenced to two years in prison for contempt of court in refusing to testify ten years earlier in the case of a New York state treasury embezzlement from which he allegedly received $120,000. But the conviction was reversed on a technicality two years later, and Stephen Elkins sued him for over $13,000 in unpaid loans. By then the Dutch bondholders had had enough. In August of 1883 they had agreed to buy out his interests for $750,000, but two months later, learning of the his arrest on earlier embezzlement charges, they backed out and charged him with embezzlement too. They subsequently dropped the charges after he agreed to resign, and they held out for a much cheaper buyout. Since they still had no money, only heavy debts, they put the company into bankruptcy and at last sent out a Dutch manager, Martinus Pels, to try to "unravel and settle the mismanagement." After four years of haggling and threats, the bondholders finally bought out Sherwin in 1888 for $75,000 cash and $90,000 in bonds of yet another reorganization. But Sherwin had already gotten away with enough to make him a millionaire.[88]

Meanwhile, the settlers had raised such a protest over what had come to be known as "Elkins' steal," that even his own party's administration under President Chester Arthur distanced itself from Star Route–tarnished "Smooth Steve" and at last brought suit in 1882 to void the three-year-old Maxwell grant patent because it had been "falsely, fraudulently and deceitfully" surveyed. And after the 1884 election, the new Democratic president, Grover Cleveland, appointed George W. Julian surveyor general of New Mexico to thoroughly investigate the case. Julian further concluded that the grant was an "inexcusable and shameful surrender to the rapacity of monopolists," and the suit was pushed harder. But Elkins and the old Republican Santa Fe Ring fought the suit all the way to the Supreme Court, where that Mariposa-blinded old justice Stephen Field and enough of his brethern could still see no evil. So in 1887 they finally confirmed the title, to widespread disbelief, a new call to arms by the settlers, and renewed bloodshed. Cleveland urged the settlers to comply with the court ruling, but the populist governor of Colorado, Davis Waite, still denounced the grant as "a fraud from the ground up."[89]

The old Mariposa and Quicksilver shares were also run up again on the old New York Stock Exchange after years of tangled litigation and to very different ends. The long battle for control between the Mariposa's two major creditors, bankers Mark Brumagim in the West and Eugene Kelly in the East, finally drew to a close in the early 1880s, but there were few investors left who cared. After being hit by Brumagim's repeated assessments, totaling nearly $10 a share, most of the Mariposa shareholders had forfeited their stock to the company treasury. So only 20,000 of its 150,000 shares were still in the hands of the faithful when Brumagim staged one last rally, raising the old stock up from 60 cents to a giddy $9 a share in the spring of 1881 on rumors of victory in the courts, as he too finally slipped out just before the judge actually ruled against him and the shares crashed. Kelly foreclosed for about $300,000 and officially took over the property in 1882, after the stockholders failed to come forward with enough money to redeem it, and even the most dedicated at last wrote it off as a total loss. But by then the boom had passed, so Kelly couldn't turn a profit either and had to wait for the next round to unload.[90]

The Quicksilver shareholders, on the other hand, finally got their long-awaited dividends after the courts in a historic ruling rejected Daniel Drew's outrageous attempt to grab most of the preferred profits at the very last minute. Freed at last, the company paid its accumulated dividends, amounting to a handsome $15 a share on its preferred stock, which was suddenly so sought after that it climbed to $62 in 1882. So even with the long delay, some of its most dedicated investors were finally able to realize the profits of their dreams. The biggest winners were those lucky enough to have bought the depressed stock in the spring of 1870 for $10 a share and promptly converted it to preferred for another $5. Twelve years later, if they still held it, they could have netted truly impressive profits of over 400 percent, or an average return of 35 percent a year, by taking the accumulated dividend and quickly selling out at $62 a share. Still, most of them doubtless hung in for more dividends, which came fairly regularly until 1892. These added another $26 a share to their returns, but by then their shares had fallen to around $40, so their net was just slightly more and their average return was only about half. That was, nonetheless, still a very nice return of nearly 20 percent a year for almost a quarter century. But these were the lucky few.[91]

Most of the stockholders bought their shares in the excitement of the 1860s at prices as high as $99 and averaging around $60. So even if they had converted their shares, only those who sold out in 1882 right after taking the dividend could have gotten their money back with a possible profit of a percent a year. But the majority hadn't even converted their shares, and all they got in 1882 was $2.25 a share in dividends on their common stock, for the preferred stock drained off all of the remaining profits. By the time the preferred dividends stopped in 1892, the common stock had dropped to just $7 a share, so for all their patience most of the stockholders lost nearly all of their investment. Still, the old company continued for another quarter century, but it paid only another $2.50 a share in occasional dividends, while its shares were shuffled back and forth on the New York Stock Exchange until they finally folded at the end of World War I. The New Almaden mine, nonetheless, was a huge bonanza, producing a total of more than $60,000,000 worth of quicksilver. That was equal to a gross of $600 a share, but nearly all the profits were taken one way or another by its original operators, Barron, Bell, and Mills, while only a small fraction of its public investors actually got out with a profit.[92]

MEANWHILE, DESPITE the end of Big Bonanza dividends, many San Francisco speculators, who just couldn't shake the habit, stayed in the game even without margins, steadily paying in over $5,000,000 a year in assessments in the hopes of finding just one more bonanza. Others, still hooked on the rush of market manipulations, were also given a fix in a long-running and elaborate shell game in Tuscarora stocks, which

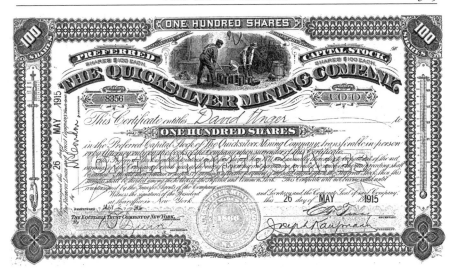

even attracted eastern addicts. Several promising new silver mines had been opened in late 1876–77 in the old camp of Tuscarora in northern Nevada. The richest were the Grand Prize, Independence, Navajo, and Belle Isle, which were all quickly bought up for under $100,000, together with a couple of other prospects, by a fifty-year-old Kentucky argonaut and seasoned speculator, George Washington Grayson, and half a dozen friends, soon dubbed the "Tuscarora Ring." They stocked each at 100,000 shares and stirred up an excitement, pulling out $1,500,000 in high grade from the Grand Prize, paying themselves $400,000 in dividends and unloading its shares for over $1,000,000 before the dividends stopped and the stock fell from $15 to $3 in 1878. Then they worked the whole batch in a skillful string of "bait and switch" with two mills and six mines, to run the various shares up and down for the next four years, raking in cash on every rise and fall. "The game is to stack up ores at different mines," one critic observed, "and sell or resell, lease or release a mill, or two mills, or a half a mill from one mine to another, or from another to one, and sometime lease a mine, or two to a mill for variety's sake, thus mixing up the babies, and keeping things at sixes and sevens so far as the money-supplying public is concerned." Together with these shuffles, they also shuffled assessments and dividends back and forth, repeatedly knocking each in turn down to pennies a share with assessments to buy it up cheaply and then kicking it up to a dollar or so as they paid themselves dividends and unloaded again! In all, the mines produced over $6,000,000 and the managers paid out $1,500,000 in dividends and took in $2,500,000 in assessments, and countless more in stock sales. By the time they finally ran out of players, Grayson had become a millionaire and his partners had done nearly as well.[93]

ALSO HOPING TO CASH IN on the eastern excitement, Bonanza King John Mackay had offered George Roberts a controlling interest in the old Savage at a low price, if he would work a new deal with it on the New York exchanges. But Roberts had much bigger plans. He had targeted the fat flock of new brokers eager to get on the inside, and he was ready to stage what would be his biggest and most successful play, the State Line mining companies, out on the very brink of Death Valley. The State Line lode was a "monster" gold-quartz cropping, a dozen feet wide and over a mile long, at the north end of the dreaded valley. Discovered in 1868 by a young Texas prospector, Tom Shaw, it had briefly created an excitement when he broke into a dazzling crystal grotto whose gold-studded walls mirrored candlelight in a myriad golden flashes and rivaled the fabled cave of Aladdin. But most of the lode was low-grade ore, and after cleaning out the high grade, Shaw finally sold it for only $6,500. It had passed without success through various hands by the fall of 1880, when Roberts bought it for a highly publicized $100,000 from Austin, Nevada, mine owner and banker Allen Curtis. Roberts hadn't even bothered to visit the mine, but he had sent his old confederate, Ashbury Harpending, and a terse Comstock superintendent, Isaac Taylor. Despite such obvious drawbacks as generally poor ore, utter lack of water, and an exorbitant asking price, they knew that Roberts would want it for its ominous location and its promotional potential, and they reported back favorably.[94]

Roberts paid out only $33,333 in cash for the mine, with notes for the remainder, and promptly incorporated it for $20,000,000 in four separate ventures, prosaically named the State Line Mining Company No. 1, No. 2, No. 3, and No. 4, although each had its own very distinctive and seductive certificates. To these he added another $10,000,000 in two more investor traps, the Oriental and Miller mining companies. They were formed on the nearby Oriental lode, which had once sported a magnificent boulder "alive with golden nuggets." The State Line, unlike Roberts's Colorado promotions, was not a prosperous paying mine, but he would sell it as something much better, a bonanza of unlimited promise. It was, in fact, a bonanza of unlimited promises, and anyone who dared challenge them would have to brave Death Valley, and none did. Also unlike the Colorado companies, Roberts held virtually all of the stock in these companies, taking 199,994 of the 200,000 shares in each, sparing just one share apiece to his token directors. He took the position of treasurer in all the companies and named Harpending's cousin, Lee R. Shryock, as president; New York Mining Stock Exchange president "Deacon" Stephen White as vice president of the State Lines; and a new cohort, the old coffin maker and Caribou schemer Abel Breed, as president of Oriental and Miller. With six closely tied companies, Roberts could throw his net wide with an enormous stock offering of 1,200,000 shares at $30,000,000 par, while still focusing their promotion into a single consolidated effort. That proved to be both very efficient and very profitable, and it eliminated the confusion and competition that had complicated his promotion of the array of companies in Colorado. Finally, even after he had unloaded the first time, Roberts would repeatedly manipulate the shares for a little added profit in the finest Comstock tradition.[95]

The campaign began with the usual parade of experts, ranging from practical miners to such easily bought pedagogues as Benjamin Silliman Jr. Straining investor credulity to the limit, they dutifully saw choice ore that ran to $15,000 a ton and imagined that the whole lode averaged $50 to $100 a ton, or better yet that it was Midas's own, with "not a pound of waste rock in the entire mine." They even accused one another of being too conservative, while all agreed that this tired old lode was nothing less than the "greatest gold mine on this continent"! All it lacked was water for milling. That was what would make it pay, and that would be the key to the whole deal. Roberts promptly ordered up twelve miles of six-inch spiral-riveted pipe from New York to bring water from the nearest big spring, and he started buying secondhand machinery for two forty-stamp mills. Taylor, who had become Roberts's general manager, expected to complete the pipeline in May or soon after, and he confidently predicted that they would start making profits of $1,000,000 a year when the first mill started up and double that after the second was running.[96]

Although Rossiter Raymond and a few other mining editors warned that a new swindle was brewing, the barrage of publicity attracted a wide audience of would-be investors. They were all the more enticed by Roberts's insistence that "the mine is not on the market," although they must have known that the genial little man would eventually relent. Late in March, as the work was well under way, Roberts began to move the stock, giving "Deacon" White and a few selected brokers 10 percent of the $25 par shares at "hard pan" prices of only $1 apiece and forming a pool to make a market. Roberts packaged the shares into three combinations of two companies each, suited to the tastes of the buyers. For the conservative investor, he offered the combination of State Line Nos. 2 and 3, which held the center of the lode, as well as the first mill and the pipeline, and as the leaders they commanded the highest prices. Speculators had their choice of two cheaper, long-shot combinations, Nos. 1 and 4, which had the tail ends of the lode and the promise of a mill, and Oriental and Miller, which had only the promise of a promise. By mid-May Roberts's pool had worked the lead stocks, Nos. 2 and 3, up to an impressive $14 a share and the others were up to $4. At the same time he was unloading directly on the public and selling large blocks to an expanded pool of knowing brokers at new bargain prices of about half of these peaks to hold until the finale when the mill started up. But at least one of the brokers, perhaps becoming suspicious, jumped the gun and the prices started dropping. With barely half of the stock out, the leaders were dipping below $7 a share and the outliers were down proportionately. With the expected completion of the pipeline and mill drawing near, the late-buying brokers were getting anxious that they'd make no profit, so it was time for Roberts's next trick.[97]

It was time for the stall, that annoying trifle that would delay the startup of the mill with its expected flood of riches, suspending expectations just long enough for Roberts to finish unloading. By then he had become a master of the stall, from the diamond swindle to the Chrysolite, and he was ready this time with the startling news early in June that the pipeline leaked and would have to be replaced, postponing the mill's startup by just a few more months. Although the need for new pipe was privately disputed by a disinterested party at the mine, it found a believing audience in the East, even among such critics as Rossiter Raymond, who had already questioned its adequacy. Meanwhile, in preparation for a new round of "experting," Taylor's assistant, Charles Kaufman, privately assured Roberts that the "mine is now in a condition that it will fool the smartest and best of experts, and if you would come out yourself we could, if you desired, fool you as you were never fooled in your life!" Thus Roberts confidently trotted out a new troop of experts, who were indeed fooled and vowed that the mine was still just as rich as ever, obviously undiminished by the "failure" of the pipeline. Roberts packed the popular papers with their puffery, particularly the New York *Tribune*, run by Horace Greeley's successor and Darius Mill's son-in-law, Whitelaw Reid. At the same time, blame for the "failure" was obligingly shouldered by Taylor, who stepped down to make way for a new manager and quietly retired to San Francisco with a fat bonus of $100,000, it was said.[98]

With these reassurances the lead stocks rallied slightly, as the knowing brokers eagerly "bulled up" the price again, once more confident of unloading in the flush of excitement when the mill finally started pouring out its flood of gold. What the knowing didn't know was that Roberts had no intention of starting up the mill, because he knew there would be no golden flood. He had the price he wanted, and while the pool helped keep it up, he was diligently divesting himself of his stock. Despite their best bulling, however, the stock soon began to fall again, for Raymond, still smarting from the Chrysolite, began a steady attack on the "Roberts plan of selling stocks." By the end of July Roberts appears to have dumped about three-fourths of his shares in Nos. 2 and 3 at an overall average of around $7 a share, and those of the other companies at about a fifth of that. But by then prices had fallen to barely half of that, and Raymond reported that the decline "nearly ruined a great many members of the Mining Exchange who loaded up with these stocks." Most, however, were still holding out for the grand finale, and Roberts still had a few more tricks.[99]

Early in August, hoping to end Raymond's attacks, it was suddenly announced that Roberts had "retired" from the State Lines and a new management had taken over. Stock prices quickly rebounded, nearly doubling in a week! The new company president, however, was John B. Alley, a former Republican congressman from Massachusetts and a director of the infamous Crédit Mobilier, which had swindled the American people out of tens of millions in fraudulently inflated self-dealing contracts to build the Union Pacific Railroad, a role that should hardly have inspired investor confidence. The new management was, of course, simply another pool, this one made up of Boston speculators to whom Harpending had quietly sold 100,000 shares, about a third of Roberts's remaining shares, at no more than half that paid by the previous New York pool. So with Roberts still pulling the strings, Alley sent out a new mine manager, a seasoned mining engineer, William Farish, to report on the state of the mine and the still uncompleted pipeline. Alley also promised him a large block of stock if he sent back word that the mine looked good. Farish got the idea, sent the message, and got the stock. He also soon sent word that he had the mill ready to go, had stockpiled a six-month supply of ore for it, and expected to complete the pipeline and start crushing ore on December 15. With this new round of assurances, Roberts worked the pool to try to run up the price once more in anticipation of the big day, while he finally unloaded the last of his shares.[100]

The bears, however, sensed that the price would soon crash, and they tried hard to precipitate it. They spread all sorts of rumors—that Farish had telegraphed his wife to sell his stock, and that Roberts had been scalped by the Sioux! The bulls, on the other hand, puffed so hard that the New York *Stock Report* jokingly imagined that next Roberts would be personally sending a dispatch straight from the mine proclaiming, "Three men are attempting to hold me down to prevent my telling all. . . . Biggest thing on Earth . . . Tell Harpending that it beats diamonds all hollow. Our great grandchildren will be able to play with gold ten pins if we hang onto this stock. . . . P.S.—Keep this mum. It might affect the market."[101]

By mid-November Roberts had at last shed all of his remaining shares, and he stepped out of the pool. But he promptly became a bear, shorting the stock for just a few more quick bucks. The bears then easily beat the startled bulls, the lead stocks crashed from $3.55 to $1.70 in a week, and the papers reported that "the slaughter pens are reeking with the blood of the innocents." After a "secret meeting" with Roberts, Farish further fed the panic with alarming new assays indicating average ore values of less than $10 a ton, not the wonderful $100 touted by all thirty previous "experts." As the shares fell further to as little as 90 cents, Roberts cleaned up on the shorts again. Then Alley fired Farish and sent out a new manager,

John Selover, the brother of one of Trenor Park's broker accomplices
in the Emma swindle, to start up the mill and clearly prove the value of
the mine. As the stocks climbed back briefly to $1.85 on this good news,
Roberts doubtless unloaded some more shares he'd just bought at the
low, doubling his money in yet another quick turn around. He certainly
couldn't have resisted selling short again in February, as the pipeline was
finally finished and the mill at last started crushing ore. The knowing
brokers, who still thought they were on the inside, eagerly snapped up
the shorts, expecting to make an even bigger killing. But Roberts then
broke the stock from $1.35 to 67 cents and cleaned up once more just by
spreading the rumor that he was going to take control again! Finally the
first mill run was made, and Selover had to telegraph Alley that Farish
was right! The long-anticipated finale was a flop, and the stock crashed
again. For the deceived brokers and the exhausted investors the seemingly endless roller-coaster ride was finally over. Before the year was out
the best that they could get for their once-promising State Lines was a
nickel a share, and by then the entire property had been put on the
sheriff's auction block for a debt of just $17,702.96.[102]

Roberts likely got away with around $3,000,000 in all, at a profit of
nearly hundredfold on his initial outlay! Although he undoubtedly
divided as much as a third of the loot among his loyal accomplices, the
State Line swindle was definitely his biggest haul, netting more than all
of his Colorado schemes combined. Raymond simply called it "down-
right robbery" and "the most barefaced operation of the year, if not the
late mining boom!" It was also by far the biggest operation, for Roberts
had thoroughly dominated trading on the New York mining exchanges
with State Line sales of over 6,000,000 shares in 1881, more than a sixth
of the total volume, and all of Roberts's stocks together made up more
than a quarter! After his repeated warnings, even Raymond had no
sympathy left for all the "fools" who would still "come up again and
again" to "offer themselves as pigeons ready for a new plucking." But
another mining editor laid part of the blame on the New York *Tribune*,
with its proud boast of giving only "the news, all the news, and nothing
but the news," for complicity in actually giving only "deceptive 'news,'
which beguiled its readers into disastrous investments" in the "delusive
schemes" of "unscrupulous speculators," and he wondered if Reid and
his editors really took their own advice.[103]

ALL THE WHILE that he was pushing State Lines, Roberts was also
reviving the stalled Robinson Con to grab all he could there too. In
December of 1880, immediately after George Robinson's death, Roberts had begun aggressive work in the mine again, with George Daly
as manager. But two months later he dispatched Daly on a new, even
grander scheme and brought in another confederate, a seasoned San
Francisco mining shark, Thomas Ewing, to take over as manager and,
together with Wilson Waddingham, buy Robinson's 150,000 shares from
his heirs for a reported $800,000, or just over $5 a share. They also
settled the title for an additional $160,000, to clear the last obstacle to
the promotion. As he did in the State Lines to try to dodge Raymond's
continuing assaults on the "Roberts combination," Roberts also officially
retired from the management. In his place he recruited New York Stock
Exchange president Brayton Ives as president and promptly placed the
stock on his exchange, while New York Mining Stock Exchange president
"Deacon" White became the treasurer. Once again touting ore reserves
of $3,000,000, Roberts resumed regular monthly dividends of $50,000 in
June and started running up the market price to unload, as Ewing tore
out ore as fast as he could to pay not only regular, but extra, dividends.
They also began such "incessant puffing" in the New York *Tribune* and
other papers that Raymond, still branding this as another "Roberts
mine," soon warned readers that "the insiders are working their way
out." The company's new secretary defensively countered that Roberts
"never owned but one share of the stock, and doesn't own any now!"
Although that might in fact have been technically true, either simply
because Roberts carried his stock under street names, or because he had
let out his shares on his favorite callback contracts, it still didn't quiet
Raymond, who also noted that the secretary denied that the ore was
refractory, which was indisputably false.[104]

Despite Raymond's warnings and the secretary's denials, Roberts and
Ewing kept up the dividends and successfully worked the shares up close
to $14 by August. Then, boosting the "ore in sight" to an extravagant
$5,600,000 and issuing extra double dividends at a rate of 45 percent
per annum on the market value, they kept the price there for over two
months, while they did indeed slowly work their way out. Roberts probably cleared about $800,000, counting dividends, for his quarter of the
stock, and Ewing likely made a profit of about half that. Waddingham,

who held roughly half the shares, could have unloaded most of his stock at the same time. But he held back because he actually believed that the mine was as rich as claimed, and he feared that Roberts and Ewing were just taking a quick profit and trying to lure him out before they broke the market to buy it all back cheaply and gain full control. So he secretly hired an expert, William Ashburner, to look at the mine. Ewing knew the man, however, and generously helped with the ore samples and assays. The results assured Ashburner that there was indeed at least $2,000,000 worth of ore still there, seemingly confirming Waddingham's fears.[105]

Late in November, once he and Roberts were out, Ewing resigned with an unexpected blast at the New York management. At the same time it was revealed that the smelter, owned by Republican senator Nathaniel P. Hill, had advanced $90,000 to pay the last dividend. The board had to suspend further dividends, and adding to the panic, the president Brayton Ives conceded that there had been a "good deal of pretty tall lying going on!" All that really broke the market, and the stock crashed to barely $3.75 in a matter of days, while Roberts and Ewing happily cleaned up on short sales. A month later, after Ashburner admitted that he had been conned and accused Ewing of "deceiving and robbing the public," Robinson Con shares crumbled to $1.90. Whether or not Waddingham finally realized what had happened, by then it was too late to unload profitably, since he still had about $2.50 a share in it even after dividends. A few months later, however, Robinson Con shares rallied temporarily to over $4 in a frenzy of trading that made it the most active mining stock on the New York Stock Exchange, and it was later said that Ives, who, on Ashburner's assurances, had also apparently held his shares, arranged a deal with the notorious new shark James Seymour to do some more "tall lying" to make a market again, so that he and Waddingham could finally work their way out. Ironically, they became the target of an irate stockholder's suit for "false statements" and "nefarious practices," while the press again cried out against the "vampire brood of stock-jobbing operators" that "sap the life-blood" of mining.[106]

MOST OF THE PRESS CRITICISM still targeted Roberts, of course, and he would even be damned as "his Satanic Majesty," but he was undaunted, for he still had one or two more devilish traps to spring on still-unsuspecting investors in Boston and Philadelphia. He baited the Bostonians with a wonderfully romantic tale of the "Nine Mines of St. Mary," Las Nueve Minas de Santa Maria, a silver treasure of the Sierra Madre in old Mexico, just seventy miles below the border south of Tombstone. They were truly a "second Comstock," he crowed, already having produced over $80,000,000 just from the surface, with much more at depth. All that they needed was "Yankee push and energy"—and, of course, a little money! That hapless Comstock superintendent Philipp Deidesheimer had picked up an option on all of them for a mere $17,000 in the spring of 1881, and he offered them to Roberts for $62,500. But since neither wanted to be directly connected with the scam, Roberts offered an additional $50,000 to a hungry young mining engineer, Pierre Humbert, Jr., son of a New York auctioneer, to serve as the official vendor, write up a glowing report on the mines, and prepare a deed showing the purchase price as $940,000. Roberts then had Harpending front for him, and to help bring in the Bostonians, they offered the helpful John Alley a big piece of the scam, telling him it cost $250,000! Finally, they incorporated Las Nueve Minas de Santa Maria Gold & Silver Mining Company in the fall to float a spectacular $25,000,000 in a million $25 shares.[107]

The returns from this scam would be pure gravy for Roberts, since he arranged it so that he didn't even have to put any money into it. The Mexican owners, Deidesheimer, and Humbert were all so happy to get some money out of the scheme that they agreed to wait for payment until the stock was sold. Only Alley, whom Roberts made the nominal president, actually paid in anything for his interest, a bargain 25 cents a share for probably about a third of the stock, which gave Roberts more than enough promotional money. Alley and Harpending then put together a pool of speculators, led by that venerable Wall Street veteran Henry Clews, plus President Grant's son Fred and Fred's father-in-law, Henry Honoré, to take up most of the shares at $1.10 a share, with the goal of selling them at $5. To hook Honoré on the deal, they even got the still-green Humbert to make a show of signing a $25,000 note to Alley for an option on 5,000 shares at their target price.[108]

Early in 1882 they started the public shares at $2 and sent down a crew of fifty seasoned Comstock miners to begin work under another pliable young mining engineer, who boasted that he could produce over $1,000,000 a year with only a seven-stamp mill and talked of immediately putting up a sixty-stamp mill to rival that of the great Con Virginia. No one bothered to mention, of course, that the property hadn't even changed hands yet! At the same time Roberts also enlisted "Podgers," the New York *Times*'s popular and witty San Francisco correspondent and speculator, Richard L. Ogden, to tour the mines, doubtless for a block of shares, and do a series of very lively and colorful puff pieces that ran for five weeks in the Sunday edition of the paper. He dazzled his readers with seductive ores, running from an average of $1,000 a ton to a "fabulous" $14,000, and far more fabulous yields of $52,000,000 and $28,000,000, taken out by Spanish, French, and Mexican miners from just the two biggest of the fabled nine mines. With returns like that from only the most sporadic and primitive work between "siestas and fiestas," he left it to his readers to just try to imagine how much Yankees could reap![109]

By then Roberts, Harpending, and Alley had placed nearly all their shares with speculators at their "insider" price of $1.10 for a total of roughly $935,000. With that the deal was actually done, and Roberts quietly cashed out with his share of the take. But his two partners still kept up appearances to keep the price above $2 for a while and allow the speculators to think a big play was still to come, as they tried to unload on the public. But all that the "Yankee push" actually managed to produce was an embarrassing $19,000, so the following year the Yankees came home, and prices tumbled until the shares could be picked up at just three for a penny! In his greed to get every last dollar, Alley even double-crossed Humbert and sold his $25,000 note, which was supposed to be sham, cheating him out of half his ill-gotten loot. The angry young novice was afraid to protest the fraudulent note publicly because he was an accomplice, so he fumed in silence for a time. But he finally got revenge by coming clean a few years later, when some of the stockholders brought suit against the most conspicuous promoters. Humbert was their star witness, exposing most of the inner workings of the scam, which convinced the judge to declare the company a "gross fraud" and order Alley and Harpending to pay back all the investors. They eventually settled out of court with just a few, however, and only gave up $30,000. By then Alley had retired from active business, and he died quietly in 1896 at the age of seventy-nine, while some of the stockholders were still trying to get their money.[110]

ROBERTS, MEANWHILE, HAD saved his last, flashiest, and most seductive bait for the Quakers, a fabulous silver grotto he called the Bridal Chamber. It was tucked away in a low limestone hill just north of the border in New Mexico, not far from where his great diamond swindle had begun a decade before. He began to bait that trap early in 1882, as he slipped out of the Nine Mines, but he had been preparing it for a year. He was first attracted in February of 1881 by reports of a rich silver-lead deposit on a ranch in Lake Valley, about thirty miles east of the great Santa Rita and just north of the Southern Pacific tracks. He promptly dispatched George Daly from the Robinson Con to investigate. The mine was already said to have paid out $100,000 in exceedingly rich high grade, and Daly recognized that it was perfect for promotion. In April he bought up all of the promising claims for Roberts for a well-publicized total of $450,000, although the actual value was much less than that because almost all of it was to be paid in stock.[111]

By then Roberts had trapped nearly all the pigeons he could on the New York and Boston markets, both brokers and public, so he went looking for a fresh flock in Philadelphia, where the Tombstone bonanza was still exciting interest. There he recruited a thirty-six-year-old one-time Methodist minister from England, John Whitaker Wright, who was working a couple of Leadville mines, without much success, for Quaker investors. Roberts, still playing the numbers game, organized the claims into a quartet of companies along the line of the State Lines, as the Sierra Apache, Sierra Bella, Sierra Madre, later renamed the Sierra Grande, and the Sierra Plata. Each was also capitalized at $5,000,000 in 200,000 shares at $25 par, and Roberts gave Daly a quarter of the stock and apparently gave Wright the same. The companies were all based in Philadelphia, where Wright handled the marketing as vice president of each, and some of his early investors served as the nominal presidents and directors. They formed what came to be known as the "Philadelphia Syndicate." Its most prominent member was the well-known geologist and prodigious dinosaur hunter, Edward D. Cope. He had inherited $500,000 from his father and was eager to increase it. As one of the first to buy in, he was made president of the Sierra Apache. The Syndicate was let in, probably again on about a tenth of the stock, at bargain pool prices of $2 a share for the richer-appearing middle companies, Sierra Bella and Sierra Grande, and half that for those at either end, grouped like the State Lines. Wright also launched the new Philadelphia weekly *Mining Journal*, dubbed Roberts's "court-journal" by critics, in which they persistently and unblushingly pushed the Sierras on the Quakers as nothing less than "the Richest Mine on Earth."[112]

Daly, meanwhile, was opening the mines, and in early August he officially reported that the entire hillside was "a mass of extremely rich ore," so rich that "even the weeds growing over the mine give large assays in silver!" He also claimed he had a crew sacking ore averaging over $300 a ton just from its outcroppings, while others were sinking nearly a hundred shafts to probe its depth. But what he did not report was that the rich ore was only in small pockets, so he was having second thoughts and had already begun unloading most of his stock on Wright's friends at the pool prices. Just two weeks later, however, he was gunned down in an ambush while chasing Apaches. Ironically, shortly after his death his right-hand man, Bernard MacDonald, discovered the pocket of all pockets, the magnificent Bridal Chamber. But on orders from Roberts he kept the discovery "in obscurity," carefully hidden behind padlocked doors until it was "needed in a bull movement." Roberts, nonetheless, doubled the share prices to $4 or $5 for the leaders and $2 for the trailers, as he began slowly to put out his stock in private sales to a wide circle of Wright's friends. Some were investors who bought it to hold for dividends, but most were speculators who bought it to unload, expecting to at least double their money in all the excitement once the dividends began and the stock was listed on the exchanges.[113]

Early in 1882 one wag jokingly announced that the "new litter of untamed kittens, which Mr. George D. Roberts has been suckling down in New Mexico for the Philadelphia market, have about got their eyes open, and are beginning to play with their tails." By then Roberts's new manager, an old hand in Comstock scams, David H. Jackson, had carefully dressed the mines for their debut. He had sunk all the shafts and prospect holes further into the ore body, but he had taken great care not to sink them too far "for fear of running through the vein and exposing its weakness." In the main shaft, he apparently probed ahead with a diamond drill, so he was able to stop just before he got into barren ground, leaving just a few inches of ore in the bottom. That way he could claim they were down nearly twenty-five feet in solid ore and still hadn't reached the bottom. Thus could Roberts confidently send out

Benjamin Silliman Jr. and other easy experts, who sent back "flaming reports" of $29,000-a-ton ore and an absolutely outrageous "$144,000,000 in sight"! The more conservative Cope, who had also eagerly visited the mines, estimated only $70,000,000, but unfortunately for him, he actually seemed to believe it.[114]

The pièce de résistance, however, was the fantastic Bridal Chamber, which was unveiled at last, as the trap was sprung in the spring. When Jackson unlocked the door and with an "open sesame" ushered visitors in, they were all bedazzled. "We gazed in awe and wonder," one wrote. "We thought of Aladdin, Croesus, and Solomon's Temple." One jabbed his pocketknife into the virgin silver ore and "cut it in a dozen places." Others held their candles under the sparkling crystals "until the silver would melt and fall in drops," and at least one burned his hand trying to catch them. It was variously reported that the governor of Arizona offered $50,000 to work in the chamber for just twelve hours and keep all he took out, that a prominent mining engineer offered $500,000 to work it for thirty days, and that a California capitalist offered $500,000 for only ten square feet of it. Eventually it was claimed that "over 200 of the leading capitalists, mine owners and experts of America and Europe have visited these mines, and all say that it is the richest ore and the largest bodies ever discovered in the world!" How could anything top that! It was a "Mineral Mecca," "the Eighth Wonder of the World!"[115]

Such rave reviews sustained demand for the stock at $5 or more for the leaders throughout the spring, allowing Roberts to privately unload the bulk of his shares. Wright doubtless shed much of his as well, as the speculative buyers feverishly loaded up in anticipation. Roberts even chartered two private railroad cars early in June to personally chaperone a grand tour for Wright's Syndicate, plus new prospective investors and speculators, and of course the press, so they could all see it for themselves. They were also entertained by the official tour expert, the ever-ready junior Silliman, who had been lured back with a block of stock in the Sierra Plata. Once there Roberts, as father of the bride, "blushingly" introduced his dazzling Bridal Chamber to "the demure Philadelphia Quakers," and they too all swooned in awe. A new wave of wonderful stories of the "Aladdin's Cave" flooded the mining sections of the faithful New York *Tribune* and other papers east and west, swelling the excitement.[116]

By then even the critics were awed at how Roberts "without the assistance of Aladdin's wonderful lamp or any other known agency, save his seemingly magnetic attraction to mineral, is able to find more riches from seeking than any other man or combination of men in the country!" Yet even though the critics all agreed that the ore was "certainly phenomenal and the mines are undoubtedly valuable," they again warned that so were other Roberts bonanzas that he "eventually wrecked to his own advantage and the heavy loss of many who blindly followed him." Another critic noted that those who had gotten into Roberts's previous schemes "at the commencement and sold out in time, are understood to have made money," but he emphasized that "there will be a time to get out as well as get in," and he warned that "Roberts is not accustomed to inform his friends when he is going to leave them high and dry." And even though Raymond was also warning investors to stay away from the Sierras, as a deadly combination of Silliman's endorsement and Roberts's management, the advertising department of his *Engineering & Mining Journal* was happily selling Roberts half-page advertising space for them for months![117]

Roberts also continued to tease speculators by still not listing the shares of his "four beautiful sisters" on the Philadelphia exchanges, while he used the rush of excitement to push prices to new highs. Shares of Sierra Grande, which held the prized grotto, reached $8, and the adjacent Sierra Plata, which became the new runner-up, climbed to $6.25, while Sierra Bella and Sierra Apache went up to $3.25 and $2.25, boosting the total market price to $4,000,000. So while the buyers still loaded up in anticipation of a big rise when the shares went public, Roberts already had his price, and unloaded on the speculators who only thought they were on the inside, just as he had done with the State Lines. Indeed, he may have unloaded the last of his shares privately right there in the Bridal Chamber, for immediately afterward he casually went on to California for a six-week vacation, while the shares fell off in Philadelphia by nearly $1,000,000.[118]

With Roberts out, Wright was next to leave. He immediately set Jackson to work gutting the Bridal Chamber, taking out over $800,000 by the end of the year to lure more investors with the mine's richness and keep up the demand, while he now fed them his shares. Jackson even claimed a new record for the largest single day's production from any mine, with $130,000 taken out by eight men in just ten hours. As a further publicity grabber, he carved out an immense block of lustrous horn silver, dubbed the "Jackson Baby," six feet long, four feet wide, and two feet thick, weighing five tons and worth close to $80,000. Wright sent it to an exhibition in Denver to outshine even Leadville. Wright also tried to unload a majority of the stock in London for $10 a share. Although the British eventually failed to bite, just the talk of the deal set a new target price for speculators to dream of and helped Wright get rid of the rest of his stock at $5 or more to eager American brokers, whom he doubtless assured that the deal was going through. Early in November he also started paying monthly dividends of 25 cents a share on the Sierra Grande and the Sierra Plata, which had merged into it, doubling its shares, to keep up the price as he dumped the last of his. At the same time he finally listed Sierra Grande on both the Philadelphia and New York exchanges when it briefly reached $4.90.[119]

This was the moment the speculators had all been waiting for, to run up the price and unload. But the bears were hungry, and contrary to all expectations the stock was ravaged as the bears slaughtered the bulls in the Bridal Chamber. Shares fell to $2.20 in just a few weeks, and the bears, including both Roberts and Wright, who had joined in the attack, feasted on the shorts. "Not since the days of the drop in State Line has the street seen such a clean scoop," one reporter noted. "Even those who were supposed to be on the inside seem to have been scorched, and the most 'intimate friends' of the manipulators have been dosed just as badly as if they had been mere 'suckers'"! Despite six regular monthly dividends totaling $1.50 a share, the beleaguered Syndicate speculators were barely able keep prices above their original purchase of $2 a share until the end of March, while they scrambled furiously to dump their stock. Then the dividends stopped. Wright "beared" the shares again, as his own *Mining Journal* now joked that the "bride had eloped and will never return" and cynically charged that insiders were bulling the stock to sell out. He picked up more shorts as the Sierra Grande fell even further, ending the year at 60 cents. With the help of other large shareholders, the bewildered and betrayed Cope, still hoping to recoup both his fortune and his reputation, won control of the board in July on promises of a seventh dividend and loyally repaid his supporters with another 25 cents a share. But after thoroughly exploring the mines, Cope found just a few small pockets of rich ore that gave them only 40 cents more in dividends before he gave up four years later, finally realizing just how badly he had overestimated the worth of the mine.[120]

In all, Roberts probably cleared around $1,500,000 for his half of the Sierra shares, which brought his total take from his eastern scams to at least $5,000,000, some claimed as much as $8,000,000, and made him by far the biggest shark of the boom. Roberts by then was so full of himself that he had brashly challenged Western Union's seemingly monolithic monopoly, worked by some of the country's richest and most notorious sharks, William Vanderbilt, Darius Mills, James Keene, Russell Sage, and Jay Gould. In the summer of 1882 Roberts, together with Oberlin College professor and founder of the Western Electric company Elisha Gray, formed an ambitious new rival, the Postal Telegraph Company, with an expansive paper capital of $21,000,000 in $1,000 shares, to string new lines all across the country and a new cable across the Atlantic. Roberts promoted it widely, stressing the need for competition and even testifying before Congress. He finally sold Bonanza King John Mackay on buying over half the shares and spinning off a separate company, the Commercial Cable Company, to lay the Atlantic cable, while he, true to form, unloaded his shares and slipped out in all the excitement, when the cable was completed in 1884. Western Union fought back in a furious price war, and for a time Mackay wailed, "It makes me mad every time I think how I was roped in by thieves and confidence men like Roberts." But Mackay was determined, and cutting consumer rates to less than half, he finally won the war, broke the monopoly, began paying dividends, and then laid the first cable across the Pacific, all of which grandly enhanced his own reputation.[121]

Wright must also have gotten away with about $700,000 from the Lake Valley scam. That started him on the road to fortune. Even more important, he had picked up many tricks from Roberts that would eventually make him millions. For he returned to England, and in the booming market of the late 1890s he launched a flurry of bloated West Australian and Canadian mining ventures on the London Stock Exchange. He united them all, in the fashion of the times, under a gigantic umbrella, the London & Globe Finance Corporation, and rigged their assets so they "could each show the same bank balance, . . . used like the stage army many times over." But what Wright had failed to pick up from Roberts was his keen sense of when to slip out of a deal. Moreover, unlike his mentor, Wright vainly made himself the flamboyant head of all his schemes. So when he stayed too long and the whole house of cards came crashing down around him, the furious shareholders all wanted his head and had him arrested when he belatedly fled to New York, claiming American citizenship. There, perhaps fondly recalling Lake Valley, he complained, "when Americans speculate and lose, . . . they never 'squeal'," but "those in London . . . cannot take their medicine." In the ensuing uproar even the royal family was "smirched" by their involvement, and two of his dummy-director lords died of disgrace. But Wright alone was finally convicted of larceny early in 1904 and sentenced to seven years in prison. Just minutes after his sentencing, he took his medicine, quickly ending it all with a cyanide tablet. "Whitaker Wright died game," one American editor wrote, "and the tragic act won him tons of vulgar admiration, if not sympathy!"[122]

The Lake Valley deal had also marked the end of George Roberts's thorough fleecing of all the overly eager eastern speculators, who were his primary marks. Most never really had a chance. Even the original Philadelphia Syndicate members, who had actually gotten in at such low prices on their first shares that they could have taken a handsome profit, were also hoodwinked by Roberts to buy more. So even after he doubled the prices, Cope and the rest loaded up more, instead of selling. The biggest loser, in fact, was Cope, who may have lost as much as $300,000, wiping out most of his inheritance, which had comfortably financed his geological research. The loss forced him to sell both his home and his enormous fossil collection and left him "soured" and quarrelsome. He died ten years later at the age of fifty-six.[123]

Roberts's richest pigeons were, in fact, his easiest marks, because they were much more confident of their shrewdness, which they thought went with their wealth. So even that seasoned schemer Wilson Waddingham had been hooked on Robinson Con, and Abel Breed, that cagey old coffin maker and one-time Caribou shark, had been taken in on Oriental and Miller, both sure that they were insiders. Waddingham survived his losses, but Breed was forced into bankruptcy in 1884 for liabilities of over $800,000 by creditors who included Asbury Harpending. His assets amounted to only $30,000, although he still held piles of worthless mining stock whose par "values" topped $6,500,000. At seventy-three he was too old to make a comeback, and he died on Christmas Eve four years later.[124]

Most of those who had dived into the whirlpool of mining stock speculation were far less prosperous, and their losses often hit them even harder. Although most suffered in silence, one modestly prosperous unfortunate recounted all the "painful" details in a popular magazine article, "My Mining Investments," as a cautionary tale for the benefit of others to "read and heed." In just a few years he had sunk at least $40,000 of his savings into a couple dozen ventures, and even though he had made a steady profit in the Ontario and Bodie Standard, he lost it all and much more in the likes of Chrysolite, Little Pittsburg, and others. He carefully itemized how each had failed through all the time-honored tricks of the trade, from greedy promoters gutting the mine, taking flashy dividends and then dumping the shares, through crooked managers siphoning off the profits to their own mills and insatiable stock manipulators playing dividends against assessments, to the simple greenhorn enthusiasts erecting extravagant mills without actually having any ore, and all the shades and mixes in between. He and a few friends had even tried going partners with a couple of "honest miner types," contributing $15,000 for surface works and a road, expecting to haul out ore at a profit of $12,000 a month, before their rustic partners disappeared![125]

He finally thought to ask a successful speculator for some advice on mining investments and was promptly told, scoldingly, "No man should consider the purchase of mining stocks as an investment. They are gunpowder securities! . . . My rules, if you choose to call them such, have been mostly negative," he added, warning, "Especially watch the *action* of 'insiders,' and disregard their *statements*. Shrewd rodents, you know, leave the sinking ship, and mining-directors are all rats by nature. . . . Finally," he concluded, "do the opposite of what your brokers advise. Their opinion is based on popular views, which are formed by reports from 'insiders.' Do what the insiders don't want you to do, and you will make money, because they want you to lose."[126]

Others offered similar advice, urging prospective investors not only to beware of "mining liars" but to take a little time to learn something about the business before risking their money, instead of foolishly betting on mining shares like lottery tickets and being constantly conned. They were soon joined, of course, by a rising young opportunist, Henry B.

Clifford, still in his early twenties and anxious to make a reputation as the champion of the small investor, who published a pious pamphlet, dubbed *Years of Dishonor; or, The Cause of the Depression in Mining Stocks, Also the Remedy*, in 1883. He bravely blamed Roberts and kindred schemers long after Raymond and many others had thoroughly denounced them. The dapper and articulate young Clifford then toured small towns throughout the East, lecturing over 250,000 potential investors, he claimed, on how to find safe mining investments. Naturally, he also offered to provide just such shares through his philanthropic United Mining and Investment Company of North and South America.[127]

Words of warning came too late, however, for most enthusiasts as they dived in over their heads. Some, like the wealthy widow Cordelia Atwell and her spinster sister and niece, Catherine and Caroline Tallman, didn't even realize that they were taking the plunge. They had long entrusted their estate to their well-respected financial trustee and former banker, Gilbert L. Crowell, only to suddenly discover in February of 1883 that he had exuberantly sunk $595,000 of their fortune into stock and assessments in an illusory "hole in the ground" called the Empire in Park City, Utah. Crowell had confidently started putting their money, plus $100,000 of his own, into the mine to become its president just three years before, after no less eminent an authority than William P. Blake had expressed "no doubt" that the Empire was the western extension of the great Ontario vein, and the *Engineering & Mining Journal* seemed to endorse that opinion. Thus Crowell started, certain of rich rock and dividends, with shares briefly reaching $14.50, but he met with fire and water instead, and was soon forced to levy assessments totaling $3.50. In all he took out only a few hundred tons of paying rock before he finally struck "a deluge," which flooded the mine, swept away all hope, and smashed the shares to 10 cents early in 1883. But the ladies still had more than half their trust funds, and after the initial shock they took pity on him and didn't press charges, although they did find a new trustee, while he retired a broken man. For others less fortunate, however, their mining stock losses were far more than they could bear. In New York City, one "enthusiastic dabbler in mining stocks" gave his last wildcat shares to his landlady as "security" for overdue rent, then walked to Central Park and drowned himself in the lake. Another former alderman in Troy, New York, bankrupted by "constant assessments," threw himself from a train six miles out of Utica.[128]

But by far the biggest catastrophe was the collapse of the whirlwind banking and brokerage firm of Grant & Ward at 2 Wall Street, founded on Leadville dividends and Ulysses's name. Its collapse not only marked the end of the silver boom but took the whole New York stock market down with it. Once the Evening Star dividends had started pouring in at an intoxicating 100 percent return per month, its two youngest shareholders, Buck Grant and Ferdinand Ward, both still in their late twenties, were convinced that the sky was the limit and they couldn't fail. The former president was also "a perfect child in financial matters," but their sixty-year-old partner, banker James Dean Fish, should have known better. They immediately started putting money into various other hopeful bonanzas, but none paid off, and they couldn't make a market for the shares. Ward and Fish realized, however, that they could still get their money out of them as collateral for loans from Fish's Marine National Bank. By the time the Evening Star bonanza was exhausted and their wonderful dividends ceased in the summer of 1882, they had become so jaded by its rich profits that the simple bank interest they were earning on deposits and loans, which Grant's name had attracted, seemed pitifully inadequate. So, frantically searching for another rich source of revenue, Ward, the sharp young minister's son, finally concocted such a clever new scheme that he would soon be heralded as "the Young Napoleon of Finance!"[129]

Ward, charmingly persuasive and "deceitfully unassuming," began to quietly confide to his major depositors that through the president's friends he had an opportunity to make loans to certain government contractors at very profitable rates, which he would split, if the depositors put up the money. Since these contracts were to be paid in installments at the completion of each phase, the contractors needed short-term loans of $50,000 or more to carry them through the first phase, until the contracts began to pay. Because of the "delicate character" of these arrangements, Ward said, the contractors also wanted strict secrecy and were willing to give a share of their overall profits that was far above normal interest rates. Although these wonderful contracts were all imaginary, the deals sounded credible enough to his avaricious investors, and as soon as they got 30 percent or more in just a couple of months on their first contract deal, they abandoned all doubts and voraciously turned all their original money and more back into new deals. That was just what Ward wanted, of course, in order to keep the money coming in and the pyramid a-building. Their banking returns soon leaped tenfold, while Ward and Fisk, "paying out deposits as profits," were also taking out over $30,000 apiece in cash monthly, or about 400 percent a year on their nominal capital investments of $100,000. At the same time Ward succeeded in keeping the clueless Grants happy with monthly payments of only $3,000 in cash and the rest in "credits." To further fatten the take, Ward also borrowed heavily for the firm from Fisk's bank with little or no collateral, and from other banks as well, illegally reusing collateral that his own borrowers had put up. All went surprisingly smoothly, and the wonderful profits poured in steadily for two years, until the whole airy fabrication collapsed at last in the spring of 1884.[130]

On May 6 shocked depositors and investors learned that the fabulous firm of Grant & Ward was $5,000,000 in debt and had reached the limit of its credit. The Marine National Bank failed the same day from all its bad loans to the firm. In the "distrust and fear, aroused in financial circles by the ugly circumstances" of the failure, other dominoes began to fall until the failures became widespread. Before it was over the "Grant & Ward panic" had triggered the collapse of several more banks and over a dozen brokers, a loss of 20 percent in stocks, a tumble in oil prices, and a depression that would last a couple of years. Ulysses S. Grant and his sons escaped public condemnation for the swindle, being blamed only for being "incredibly careless and credulous." But they lost more than $750,000, deposited and invested in the bank. "We are all ruined," Grant said, and some later claimed that "the infamous knavery of Ferdinand Ward" was the real cause of Grant's death the following year at the age of sixty-one. Chaffee also lost $500,000 in bonds he had entrusted to Buck. In all, Fisk had taken out $1,000,000 from the contract swindle and Ward tried to grab over twice that, transferring an extra $1,300,000 in cash and nearly $600,000 in property to a friend, William S. Warner, in the final months before the failure, to get it out of the hands of creditors. The scheme had paid even better than the Evening Star dividends, but it had also triggered such a financial disaster that Ward and Fisk were both promptly indicted and convicted on charges of fraud and grand larceny. Ward was sentenced to ten years "at hard labor" in Sing Sing and Fish to four. Warner was also indicted, but the charges were dismissed on a technicality and he fled to England with the money to escape a court

THE PANIC IN WALL STREET ON THE MORNING OF MAY 14, 1884
Harper's Weekly, May 24, 1884

order to return it to the investors. There he died, two years before Ward got out of prison, and the "young Napoleon" never did get the money. Many years later Ward tried to exonerate himself and make a little more money off Grant by selling to the *Sunday Herald* his insider's "Recollections of General Grant" as "an Easy Prey for the Wolves of Finance."[131]

THE COLLAPSE OF THE BOOM raised once again the perennial question, "Does mining pay?" The answer, as always, was that the profits of mining were very highly concentrated. A handful of highly publicized mines paid fabulously well, but most mines did not pay a profit to investors, and the bulk of all the mining stocks were simple scams. Indeed, one broker sarcastically remarked, "The public must get over the old fogyish idea that mining stocks should represent mines," since even "the most careless observer of the business at the Boards should be able to see that the question of mine, or no mine, has little bearing on the price of a stock!" Others, nonetheless, insisted that mining was, in fact, "one of the most profitable—if not *the* most profitable—industries of the day, when conducted upon strict business principals." Although they conceded that many had been "deliberately robbed of their money" by the "unscrupulous methods" of mining sharks, they argued that mining stock losses were still "insignificant" compared with the "colossal" loss of nearly $1,000,000,000 by railroad investors through ceaseless manipulations and crooked foreclosures.[132]

There was no question that the western mines were enormously productive. By 1884 they had produced a spectacular $2,300,000,000 in gold, silver, and copper since 1848—roughly $230,000,000,000 today—and they had paid dazzling dividends of nearly $300,000,000, or about $30,000,000,000 today. During just the last boom years of 1879 to 1884 they produced over $400,000,000, and as the placers declined, three-quarters of that had come from hardrock mines. But nearly all of that wealth came from less than a tenth of those mines—less than a thousand producers with a nominal capital of around $4,000,000,000. They employed more than 40,000 underground miners and surface hands at $3 to $4 a day, who got about half the total yield of the mines, leaving the remaining $200,000,000 for supplies, machinery, and profits. The reported dividends during those years totaled $50,000,000, but three-fifths of that came just from the dozen major mines—including the Ontario, the Standard, the Horn Silver, and the Homestake—and the Leadville and Tombstone bonanzas and went into the pockets of a relatively small fraction of all the investors and speculators.[133]

Excited by this new outpouring of treasure, however, investors had again clamored for a share, and the sharks had again flooded the market with a scandalous $20,000,000,000 in enticing shares of over 10,000 ventures promising a piece of that pie. That once more rivaled the total wealth of the nation, and it was fifty times the whole gold and silver output from all the western mines in the boom! But even though only a small fraction of all that enticing bait could possibly be taken up, it still added up. By 1882 an estimated $50,000,000 of "live Philadelphia capital" was said to have been invested in western mining stock in less than four years. As the nation's second largest city, with a population of around 850,000, that amounted to an average of about $15 a year for every man, woman, and child, and if an equal enthusiasm was exhibited among the other 4,000,000 souls in New York, Boston, and two other big eastern cities, the total drawn into western dreams would have topped $300,000,000—still close to the entire output of the mines—and six times their total dividends! Although most steered clear of the frenzy, at least several hundred thousand hopefuls must have nibbled at the bait, dropping several hundred dollars apiece. The bulk of their "investment," as usual, never got anywhere near the mines, but it generously fed the large pool of sharks at the top of the mining stock feeding chain. Much of the catch was made on the exchanges, not only with the few flashy though fleeting bonanzas but with a hundred or so other seductive kittens as well, tarted up with a few tempting little dividends, that worked the street to hook hungry speculators. Although the sharks failed to match the Big Bonanza's frenzy, they did run up sales of over $300,000,000 on the New York exchanges, and they put a price of $100,000,000 just on the traded shares before they crashed. Then thousands of shares once bought for over $10,000 could all be bought for only $1![134]

Once again the shocking collapse of even the big dividend payers, plus the failure of countless other promised profits, had begun to chill the fervor of mining investors in 1883, and the Grant & Ward market crash the following year finally broke the great eastern silver fever. Some investors, however, still hoping to share in other profits being won in the West, turned to the booming cattle and wheat businesses on the high plains, sinking their savings into cattle companies in Wyoming and bonanza farms in the Dakotas. But most of these ventures proved to be as profitless as most mining schemes, and interest chilled in the big winter of 1886–87. But by then copper was booming from the demands of the electrification of the nation and would become the primary product of the western mines. Revolutionary new technologies would also slash mining and processing costs to open enormous low-grade bonanzas that would quadruple the total western production by the end of World War I, while profits increased tenfold and thunderous new waves of speculation would crash again.

APPENDIX

WESTERN MINING DIVIDENDS, 1803–1884
(Companies paying $1,000,000 or more by 1884)
*(In dollars at the time, $1 of which is
roughly equal to $100 in present-day dollars)*

	Par Issued	Years Paid	Dividends
Consolidated Virginia Mining Co.	54,000,000	1874–80	42,400,000
California Mining Co.	54,000,000	1875–79	31,200,000
Belcher Silver Mining Co.	10,400,000	1872–76	15,400,000
Crown Point Gold & Silver Mining Co.	10,000,000	1868–75	11,900,000
Union Mill & Mining Co.	1,500,000	1867–75	10,000,000
Pacific Mill & Mining Co.	8,000,000	1874–81	9,000,000
Ontario Silver Mining Co.	10,000,000	1877–84	6,100,000
Bolton, Barron & Co.	—	1852–58	5,500,000
Hayward's Badger–Eureka Mine	—	1857–68	~5,000,000
Boston & Colorado Smelting Co.	500,000	1867–84	~5,000,000
Selby Smelting & Lead Co.	200,000	1875–84	~5,000,000
Eureka Consolidated Mining Co.	5,000,000	1872–84	4,800,000
Savage Mining Co.	1,600,000	1863–69	4,500,000
Standard Gold Mining Co.	5,000,000	1878–84	4,500,000
William E. Barron & Co.	—	1860–71	~4,000,000
Omaha Smelting & Refining Co.	60,000	1871–83	~4,000,000
Gould & Curry Silver Mining Co.	2,400,000	1862–70	3,800,000
Horn Silver Mining Co.	10,000,000	1882–84	3,700,000
Idaho Quartz Mining Co.	310,000	1860–84	3,600,000
Chollar-Potosí Mining Co.	2,800,000	1861–72	3,100,000
Raymond & Ely Mining Co.	3,000,000	1871–73	3,100,000
New Idria Mining Co.	33,000	1858–84	~3,000,000
Richmond Consolidated Mining Co. Ltd.	1,350,000	1872–84	3,000,000
Keystone Consolidated Mining Co.	600,000	1873–84	3,000,000

	Par Issued	Years Paid	Dividends
New Almaden Mine	~300,000	1850–58	2,800,000
Contention Consolidated Mining Co.	12,500,000	1881–84	2,600,000
Selby Lead & Silver Smelting Works	100,000	1867–75	~2,500,000
Northern Belle Mill & Mining Co.	5,000,000	1877–84	2,500,000
Homestake Mining Co.	25,100,000	1879–84	2,500,000
Lewisohn Bros.	—	1879–84	~2,500,000
Grant Smelting Co.	—	1880–83	~2,500,000
Yellow Jacket Silver Mining Co.	1,200,000	1863–74	2,200,000
Eureka Quartz Mining Co.	500,000	1865–73	2,100,000
Plumas-Eureka Mine Ltd.	1,406,250	1873–84	2,100,000
Tuolumne County Water Co.	550,000	1852–60	~2,000,000
Nevada Mill & Mining Co.	—	1871–75	2,000,000
Germania Separating & Refining Works	58,000	1872–84	~2,000,000
Robert E. Lee Mining Co.	5,000,000	1879–81	1,900,000
Chrysolite Silver Mining Co.	10,000,000	1879–84	1,700,000
Ophir Silver Mining Co.	5,040,000	1860–84	1,600,000
Hale & Norcross Silver Mining Co.	400,000	1865–84	1,600,000
Bodie Mining Co.	5,000,000	1878–84	1,600,000
Santa Rita del Cobre Mine	—	1803–37	~1,500,000
Whitlatch Union Mine	—	1865–71	1,500,000
North Bloomfield Gravel Mining Co.	5,000,000	1877–84	1,500,000
Comstock Mill & Mining Co.	—	1884	1,500,000
Quicksilver Mining Co.	10,000,000	1864–84	1,400,000
Sierra Buttes Gold Mining Co. Ltd.	1,225,000	1871–84	1,400,000
Hite's Cove Gold Quartz Mining Co.	2,500,000	1874–78	1,400,000
Silver King Mining Co.	10,000,000	1877–84	1,400,000
Little Pittsburg Consolidated Mining Co.	20,000,000	1879–80	1,400,000
Evening Star Mining Co.	500,000	1880–84	1,400,000
Kentuck Mining Co.	3,000,000	1866–84	1,300,000
Meadow Valley Mining Co.	6,000,000	1870–73	1,300,000
Tombstone Mill & Mining Co.	12,500,000	1880–82	1,300,000
Iron Silver Mining Co.	10,000,000	1881–84	1,300,000
Copper Queen Mining Co.	2,500,000	1881–84	1,300,000
Allison Ranch Mine	—	1855–67	1,200,000
Reises' Sierra Buttes Mine	—	1856–69	1,100,000
Imperial Mining Co.	1,000,000	1863–76	1,100,000
Redington Quicksilver Mining Co.	31,500	1864–78	1,100,000
Carsons Creek Consolidated Mining Co.	1,000,000	1850–51	~1,000,000
Bear River & Auburn Water & Mining Co.	550,000	1853–60	~1,000,000
Union Copper Mining Co.	13,000	1861–67	1,000,000
Vulture Mining Co.	250,000	1865–71	~1,000,000

	Par Issued	Years Paid	Dividends
Emma Silver Mining Co. Ltd.	5,000,000	1871–72	1,000,000
Black Bear Quartz Gold Mining Co.	3,000,000	1873–82	1,000,000
Rocky Bar Mining Co.	22,000	1877–80	˜1,000,000
Walker Bros.' Alice Mine	170,000	1877–80	˜1,000,000
Grand Central Mining Co.	400,000	1879–81	1,000,000
Plymouth Consolidated Gold Mining Co.	5,000,000	1883–84	1,000,000
Omaha & Grant Smelting & Refining Co.	2,500,000	1883–84	˜1,000,000
5 companies paying $10M or more	—	1867–84	110,000,000
24 companies paying $3M or more	—	1852–84	195,000,000
72 companies paying $1M or more	—	1803–84	270,000,000

Sources: The dividends and profits for all of the companies and partnerships discussed in the text are taken from the notes for each company. The principal sources for many of those, as well as small companies not discussed, are *Mining & Scientific Press*, Jan. 23, 1875; *Engineering & Mining Journal*, Jan. 1, 1881, Dec. 30, 1882, Jan. 9, 1886; *Chicago Mining Review*, June 24, 1882; *Report of the Director of the Mint*, (1881) 517–527, (1882) 680–687, (1884) 510–517; Richard Rothwell, *Mineral Industry*, v. 1 (1892), 475–478.

Note: Par issues are listed merely for general comparison, although they usually do not represent what either inside or outside investors paid.

ABBREVIATIONS
USED IN THE NOTES

AJM	*American Journal of Mining*
BDAC	*Biographical Directory of the American Congress, 1774–1961*
BE	Brooklyn, N.Y., *Eagle*
CMR	Chicago *Mining Review*
EMJ	*Engineering & Mining Journal*
LD	Leadville, Colo., *Democrat*
LH	Leadville, Colo., *Herald*
LAT	Los Angeles *Times*
LMJ	London *Mining Journal*
LT	London *Times*
MSP	*Mining & Scientific Press*
NCAB	*National Cyclopaedia of American Biography*
NSJ	*Nevada State Journal* (Reno)
NYB	New York *Bullion*
NYG	New York *Graphic*
NYH	New York *Herald*
NYMR	*Mining Record* (New York)
NYT	New York *Times*
NYTr	New York *Tribune*
PMJ	Philadelphia *Mining Journal*
PO	Portland *Oregonian*
RMN	*Rocky Mountain News* (Denver)
SacU	Sacramento *Union*
SFAC	*Alta California* (San Francisco)
SFBul	San Francisco *Bulletin*

SFC	San Francisco *Call*
SFCh	San Francisco *Chronicle*
SFEx	San Francisco *Exchange*
SFSR	San Francisco *Stock Report*
SLGD	St. Louis *Globe-Democrat*
SLT	Salt Lake City *Tribune*
UMG	*Utah Mining Gazette* (Salt Lake City)
VCTE	*Territorial Enterprise* (Virginia City, Nev.)
WNI	*National Intelligencer* (Washington, D.C.)

NOTES

PREFACE

1. Charles P. Kindleberger, *Manias, Panics, and Crashes; A History of Financial Crises* (New York: John Wiley, 2000), 76.

2. Peter H. Burnett, *Recollections and Opinions of an Old Pioneer* (New York: Appleton, 1880), 414.

3. U.S. Bureau of the Census, *Historical Statistics of the United States: Colonial Times to 1970* (Washington, D.C.: G.P.O., 1975), 163–168, 200–202, 212; U.S. Bureau of Mines, *Metal Prices in the United States through 1991* (Washington, D.C.: G.P.O., 1993), 57–63; Robert Twigger, *Inflation: The Value of the Pound 1750–1998* (London: House of Commons Library, 1999), 18–22; U.S. Bureau of the Census, *Comparing Employment, Income, and Poverty: Census 2000 and the Current Population Survey* (Washington, D.C.: G.P.O., 2003), 47; *Historical Statistics of the United States: Millennial Edition* (Cambridge: Cambridge University Press, 2006), 3-158–3-159.

4. Clark Spence, *British Investments and the American Mining Frontier, 1860–1901* (Ithaca, N.Y.: Cornell University Press, 1958); Rodman W. Paul, *Mining Frontiers of the Far West, 1848–1880* (New York: Holt, Rinehart and Winston, 1963); William S. Greever, *The Bonanza West; The Story of the Western Mining Rushes, 1848–1900* (Norman: University of Oklahoma Press, 1963); Marian V. Sears, *Mining Stock Exchanges, 1860–1930: An Historical Survey* (Missoula: University of Montana Press, 1973); Joseph E. King, *A Mine to Make a Mine: Financing the Colorado Mining Industry, 1859–1902* (College Station: Texas A&M University Press, 1977); Richard H. Peterson, *The Bonanza Kings: The Social Origins and Business Behavior of Western Mining Entrepreneurs, 1870–1900* (Lincoln: University of Nebraska Press, 1977); Charles K. Hyde, *Copper for America: The United States Copper Industry from Colonial Times to the 1990s* (Tucson: University of Arizona Press, 1998); Dan Plazak, *A Hole in the Ground with a Liar at the Top* (Salt Lake City: University of Utah Press, 2006); Richard E. Lingenfelter, *The Mining West: A Bibliography & Guide to the History & Literature of Mining in the American & Canadian West*, 2 vols. (Lanham, Md.: Scarecrow Press, 2003).

CHAPTER 1. MINES AND MONEY

1. John M Guilbert and Charles F. Park, *The Geology of Ore Deposits* (New York: Freeman, 1985), 13–22, 55–71, 109–110, 120–133.

2. Andrew V. Corry and Oscar E. Kiessling, *Grade of Ore* (Philadelphia: W.P.A., 1938), 1–39.

3. Corry and Kiessling, op. cit., n. 2, 1–13.

4. Keith R. Long, John H. DeYoung, and Steve Ludington, "Significant Deposits of Gold, Silver, Copper, Lead, and Zinc in the United States," *Economic Geology* 95 (May 2000), 632–633.

5. "Letters Patent from Queen Elizabeth, 11th June, 1578," quoted in William Stith, *The History of the First Discovery and Settlement of Virginia* (Williamsburg: William Parks, 1747), 4; "First Charter of Virginia, April 10/20, 1606," in *Documentary Source Book of American History, 1606–1926*, ed. William MacDonald (New York: Macmillan, 1926), 2.

6. "Capitulations of Santa Fe, April 17, 1492," in *Journals and Other Documents on the Life and Voyages of Christopher Columbus*, trans. and ed. Samuel Eliot Morison (New York: Heritage Press, 1963), 27–28; Samuel Eliot Morison, *Admiral of the Ocean Sea, a Life of Christopher Columbus* (Boston: Little, Brown, 1942), xiii, 71–73, 92–95, 102–110, 118–119, 143. Morison (pp. 103–104) reviews the costs and their gold equivalents (p. xiii), while the nineteenth-century equivalent dollars are determine by William Prescott (see note 9 below).

7. Letter of Christoforo Colombo to Luis de Santangel, Feb. 15, 1493, in *Journals*, op. cit., n. 6, 182–186; Morison, op. cit., n. 6, 103–104, 431–433, 491–494, 568–571, 661–662; Carlos Prieto, *Mining in the New World* (New York: McGraw-Hill, 1973), 20; Helmut Waszkis, *Mining in the Americas* (Cambridge: Woodhead, 1993), 8–13; Noble David Cook, "Sickness, Starvation, and Death in Early Hispaniola," *Journal of Interdisciplinary History* 32 (Winter 2002), 349–386.

8. William H. Prescott, *History of the Conquest of Mexico* (New York: Harper, 1853), v. 1, 238; James Lockhart and Stuart B. Schwartz, *Early Latin America: A History of Colonial Spanish America and Brazil* (Cambridge: Cambridge University Press, 1983), 79; Rafael Varon Gabai, *Francisco Pizarro and His Brothers* (Norman: University of Oklahoma Press, 1997), 12–13; Shirley C. Flint, "The Financing and Provisioning of the Coronado Expedition," in *The Coronado Expedition from the Distance of 460 Years*, eds. Richard and Shirley Flint (Albuquerque: University of New Mexico Press, 2003), 42–44.

9. Hernán Cortés, *Cartas de Relación de la Conquista de la Nueva España Escritas por Hernán Cortés al Emperador Carlos V, y Otros Documentos Relativos a la Conquisita, Años de 1519–1527*, English translation by Anthony Pagden, as *Hernán Cortés: Letters From Mexico* (New Haven, Conn.: Yale University Press, 1986), xliv–lii, 3–11; Bernal Díaz del Castillo, *Historia Verdadera de la Conquista de la Nueva España*, 1534, English translation by A. P. Maudslay, as *The Discovery and Conquest of Mexico, 1517–1521* (Cambridge, Mass.: Da Capo Press, 2003), 31–36; Prescott, op. cit., n. 8, v. 1, 230–265;

Salvador de Madariaga, *Hernán Cortés, Conqueror of Mexico* (Chicago: Regnery, 1955), 87–101. Prescott (pp. 320–321) calculated the commercial value of the early sixteenth century peso de oro to be worth $11.67 in mid-nineteenth-century dollars, based on wheat prices, which was about four times the simple metal value. This valuation will be used in all sixteenth-century monetary conversions instead of present-day values, because it provides an easy comparison with all the other nineteenth- and early twentieth-century dollar values that will follow. Similarly, other foreign currencies at early periods are also converted to nineteenth-century dollars, as discussed later, and for nineteenth- and early twentieth-century currencies they are converted at what were then their nominal values, such as five dollars to the British pound and twenty cents to the French franc.

10. Cortés, op. cit., n. 9, 99–101; Prescott, op. cit., n. 8, v. 1, 299–303, 319–321; v. 2, 199–205, 357–366; v. 3, 233–254.

11. Cortés, op. cit., n. 9, 100; Díaz, op. cit., n. 9, 248–251; Prescott, op. cit., n. 8, v. 2, 202–205, 357–366; v. 3, 233–235; Prieto, op. cit., n. 7, 20–25. In Cortés's letter to the king (p. 100), the royal fifth was stated to be 32,400 pesos de oro in melted-down gold ingots and 100,000 ducats, equal to 75,000 pesos de oro, in jewelry, which Prescott (p. 203) multiplied by five to get a total value of 537,000 pesos de oro for the entire treasure. Taking the sixteenth-century peso de oro to be worth $11.67 (n. 9), he thus determined the nineteenth-century value of the treasure to be $6,300,000. However, in Cortés's accounting he speaks of the jewelry as pieces he "set aside" for the crown beyond the royal fifth, rather than being a fifth of all the unmelted jewelry. Thus the minimum value of the treasure is only five times the ingot gold plus the crown's jewelry, which comes to 237,000 pesos de oro, or about $2,800,000. However, Cortés and others also helped themselves to some of the better jewelry before it was melted down, so the actual amount of the treasure that they got away with was probably around $4,000,000. In addition, they also recovered another 200,000 pesos de oro, or $2,400,000, when they recaptured Tenochtitlán (Cortés, p. 159). This coincidentally brings the total back very close to Prescott's value, but this added treasure was not included in the first division of the loot.

12. William H. Prescott, *History of the Conquest of Peru* (New York: Harper, 1847), v. 1, 202–210, 230–240; v. 2, 189–200; James G. Wilson and John Fiske, *Appleton's Cyclopaedia of American Biography* (New York: Appleton, 1900), v. 2, 374; v. 5, 35–36; Varon Gabai, op. cit., n. 8, 10–25.

13. Prescott, op. cit., n. 12, v. 1, 303–318, 395–399, 407–434, 458–459, 466–468; v. 2, 490–497; Wilson and Fiske, op. cit., n. 12, 36–37; Varon Gabai, op. cit., n. 8, 36–40. Prescott (pp. 466–468) again calculates the value of the treasure in nineteenth-century dollars, using the commercial rate of $11.67 for a peso de oro, based on wheat prices.

14. Prescott, op. cit., n. 12, v. 1, 463–473, 517, 524–527; James M. Lockhart, *The Men of Cajamarca* (Austin: University of Texas Press, 1972), 90–102.

15. Prescott, op. cit., n. 12, v. 2, 27–35, 90–107, 119–143, 171–200; Prieto, op. cit., n. 7, 28–34, 58; Varon Gabai, op. cit., n. 8, 111–112.

16. Prescott, op. cit., n. 8, v. 3, 333–338, again using his rate of one sixteenth-century peso de oro to $11.67 of the nineteenth century; Hubert H. Bancroft, *History of Arizona and New Mexico, 1530–1888* (San Francisco: History Co., 1889), 7–9, 35; Stanley H. Ross, "The Prehistoric Use of Copper in Arizona" (Ph.D. diss., University of California, Los Angeles, 1963), 88–117, 151–152; Edwin Gudde, *California Place Names* (Berkeley: University of California Press, 1969), 48–49; Jeremiah F. Epstein, "Cabeza de Vaca and the Sixteenth Century Copper Trade in Northern Mexico," *American Antiquity* 56 (July 1991), 474–482; William K. Hartmann and Richard Flint, "Before the Coronado Expedition," in *The Coronado Expedition from the Distance of 460 Years* (Albuquerque: University of New Mexico Press, 2003), 20–41.

17. Pedro de Castañeda de Najera, *Relación de la Jornada de Cíbola* (1596), English translation by George P. Hammond and Agapito Rey, in *Narratives of the Coronado Expedition, 1540–1542* (Albuquerque: University of New Mexico Press, 1940), 1–8, 16–18, 26–27, 54–57, 83–108, 196–208, 220–221, 233–244; Prescott, op. cit., n. 8, v. 3, 336–337; Bancroft, op. cit., n. 16, 35–70; Flint, op. cit., n. 8, 42–56.

18. Bancroft, op. cit., n. 16, 110–136, 147–148; George P. Hammond and Agapito Rey, *Don Juan de Oñate, Colonizer of New Mexico, 1595–1628* (Albuquerque: University of New Mexico Press, 1953), 1–10, 42–57, 320–321, 408–415, 420–424, 581, 584–589, 597–607, 622, 630–632, 641–642, 653–654, 821–822, 829, 1001–1002; George P. Hammond and Agapito Rey, *The Rediscovery of New Mexico, 1580–1594* (Albuquerque: University of New Mexico Press, 1966), 96, 138; John M. Townley, "Mining in the Ortiz Mine Grant Area" (M.A. thesis, University of Nevada, Reno, 1968), 21–28; Homer E. Milford, "History of the Los Cerrillos Mining Area," in *Cultural Resource Survey for Real de Los Cerrillos Project* (Santa Fe: Mining and Minerals Division, 1995), v. 1, 9–37; Charles David Vaughan, "Taking the Measure of New Mexico's Colonial Miners, Mining, and Metallurgy" (Ph.D. diss., University of New Mexico, 2006), 1–13, 151–156.

19. Jacques Cartier, *The Voyages of Jacques Cartier; Published from the Originals with Translations, Notes and Appendices* (Ottawa: F. A. Acland, 1924), and Henry P. Biggars, ed., *A Collection of Documents Relating to Jacques Cartier and the Sieur de Roberval* (Ottawa: Public Archives of Canada, 1930), reprinted as *The Voyages of Jacques Cartier* (Toronto: University of Toronto Press, 1993), 96–105, 126–129, 135–139, 144–155, 171–176; *Dictionary of Canadian Biography Online* (Toronto: University of Toronto, 2004), Jacques Cartier and Jean-François de la Rocque de Roberval. As previously, the equivalent dollars are all given in nineteenth-century values at about $2 to the sixteenth-century livre, including the factor of four decline in the commercial value of gold from the sixteenth to the nineteenth-century, from Prescott n. 9.

20. George Best, *A Trve Discovsre of the Late Voyages of Discouerie, for the Finding of a Passage to Cathaya, by the Northvveast, vnder the Conduct of Martin Frobisher Generall* (London: Henry Bynnyman, 1578), 51–52; William R. Scott, *The Constitution and Finance of English, Scottish and Irish Joint-Stock Companies to 1720* (Cambridge: Cambridge University Press, 1912), v. 1, 20, 64, 75–77, 86–87; Thomas Arthur Rickard, *The Romance of Mining* (Toronto: Macmillan, 1944), 99–121; Vilhjalmur Stefansson, *The*

Three Voyages of Martin Frobisher (London: Argonaut, 1938), v. 1, lxxxviii–cxxiii, v. 2, 99–118, 137, 251–252; Donald D. Hogarth, Peter W. Boreham, and John G. Mitchell, *Martin Frobisher's Northwest Venture, 1576–1581: Mines, Minerals, Metallurgy* (Hull, Que.: Canadian Museum of Civilization, 1994), 24–52, 74–75, 103–107, 149–151; James McDermott, "Michael Lok, Mercer and Merchant Adventurer" and "The Company of Cathay: The Financing and Organization of the Frobisher Voyages," in Thomas H. B. Symons, Stephen Alsford, and Chris Kitzan, *Meta Incognita: A Discourse of Discovery; Martin Frobisher's Arctic Expeditions, 1576–1578* (Hull, Que.: Canadian Museum of Civilization, 1994), 119–178; Robert Baldwin, "Speculative Ambitions and the Reputations of Frobisher's Metallurgists," in Symons et al., ibid., 401–476. Again, the equivalent dollars are all given in nineteenth-century values at $20 to the sixteenth-century pound sterling, including the factor of four decline in gold discussed in n. 9.

21. John Smith, *A Map of Virginia with a Description of the Country, the Commodities, People, Government and Religion* (Oxford: J. Barnes, 1612), 21–22; Stith, op. cit., n. 5, 60–62; B. Henry Latrobe, *American Copper-Mines* (Philadelphia, 1800), 3–8; Herbert L. Osgood, *The American Colonies in the Seventeenth Century* (New York: Macmillan, 1904), 59; J. H. Granbery, "The Schuyler Mine," *Journal of the Franklin Institute* 164 (July 1907), 13–128; Scott, op. cit., n. 20, v. 1, 86; v. 2, 246–259; James A. Mulholland, *A History of Metals in Colonial America* (Tuscaloosa: University of Alabama Press, 1981), 17–54; Mira Wilkins, *The History of Foreign Investment in the United States to 1914* (Cambridge, Mass.: Harvard University Press, 1989), 3–4; Charles K. Hyde, *Copper for America* (Tucson: University of Arizona Press, 1998), 3–10.

22. Daniel Defoe in *Mist's Journal* (London), Nov. 28, 1719, reprinted in *Great Bubbles*, v. 2: *The Mississippi Scheme and Bubble*, ed. Ross B. Emmett (London: Pickering and Chatto, 2000), 171; Charles Mackay, *Memoirs of Extraordinary Popular Delusions and the Madness of Crowds* (London: National Illus. Library, 1852) v. 1, 1–45; Josiah D. Whitney, *The Metallic Wealth of the United States* (Philadelphia: Lippincott, Grambo, 1854), 27; H. Montgomery Hyde, *John Law* (London: Allen, 1969), 113–163; Antoin E. Murphy, *John Law* (Oxford: Clarendon Press, 1997), 164–264. Isaac Newton in 1712 valued the livre at 1.04 shillings, or roughly 4 livres to the nineteenth-century dollar; see William A. Shaw, *Selected Tracts and Documents Illustrative of English Monetary History, 1626–1730* (London: Wilsons, 1896), 160–163.

23. Bancroft, op. cit., n. 16, 86–89, 207, 258, 399–401; Townley, op. cit., n. 18, 43–47, 51–62; Paige W. Christiansen, *The Story of Mining in New Mexico* (Socorro: New Mexico Bureau of Mines, 1974), 17–27; Patricia Herring, "The Silver of El Real de Arizonac," *Arizona & the West* 20 (Autumn 1978), 245–258; Austin N. Leiby, "Borderland Pathfinders: The 1765 Diaries of Juan María de Antonio de Rivera" (Ph.D. diss., Northern Arizona University, 1985).; James E. Officer, "Mining in Hispanic Arizona: Myth and Reality," in *History of Mining in Arizona*, v. 2 (Tucson: Mining Club of the Southwest, 1991), eds. J. Michael Canty and Michael N. Greeley, 1–7; Milford, op. cit., n. 18, 37–53; Donald T. Garate, "Who Named Arizona? The Basque Connection," *Journal of Arizona History* 40 (Spring 1999), 53–82; Vaughan, op. cit., n. 18, 151–180.

24. William Taylor, "The Pueblos and Ancient Mines Near Allison, New Mexico," *American Antiquarian* 20 (Sept. and Oct. 1898), 258–261; Alfred B. Thomas, *Forgotten Frontiers* (Norman: University of Oklahoma Press, 1932), 273–290; Thomas Arthur Rickard, *A History of American Mining* (New York: McGraw-Hill, 1932), 252–254; Francisco R. Almada, *Resumen de Historia del Estado de Chihuahua* (Mexico City: Libros Mexicanos, 1955), 137; James H. Gunnerson, "Southern Athapaskan Archeology," in *Handbook of North American Indian*, v. 9, *Southwest* (Washington, D.C.: Smithsonian Institution, 1979), 162–169; Billy D. Walker, "Copper Genesis: The Early Years of Santa Rita del Cobre," *New Mexico Historical Review* 54 (Jan. 1979), 5–8,

25. Zebulon M. Pike, *An Account of Expeditions to the Sources of the Mississippi* (Philadelphia: Conrad, 1810), in *The Journals of Zebulon Montgomery Pike* (Norman: University of Oklahoma Press, 1966), v. 2, 48; William H. Emory, *Notes of a Military Reconnaissance* (Washington, D.C.: Wendell, 1848), 58–59; Rickard, op. cit., n. 24, 254–255; Almada, op. cit., n. 24, 138; Walker, op. cit., n. 24, 8–20; Elinore M. Barrett, *The Mexican Colonial Copper Industry* (Albuquerque: University of New Mexico Press, 1987), 8, 47–49, 61; Rick Hendricks, "Spanish Colonial Mining in Southern New Mexico," *Mining History Journal* 6 (1999), 152–162; Robert L. Spude, "The Santa Rita del Cobre, New Mexico: Early American Period, 1846–1886," *Mining History Journal* 6 (1999), 11.

26. James O. Pattie, *The Personal Narrative of James O. Pattie* (Cincinnati: Wood, 1831), x, 52–53, 71–81; Bancroft, op. cit., n. 16, 338; Walker, op. cit., n. 24, 15–17; Richard Batman, *James Pattie's West* (Norman: University of Oklahoma Press, 1986), 39–45, 52, 144–158; Sweeney, op. cit., n. 25, 38–42.

27. Dr. Willard, "Inland Trade with New Mexico," *Western Monthly Review* 2 (May 1829), 651–652; Robert W. H. Hardy, *Travels in the Interior of Mexico, in 1825, 1826, 1827, and 1828* (London: Colburn and Bentley, 1829), 462–463; Pattie, op. cit., n. 26, 81, 102–103, 112–115, 123–124, 129–132, 183, 276; Stephen C. Foster, "A Sketch of the Earliest Kentucky Pioneers of Los Angeles," *Annual Pub. of the Hist. Soc. of Southern California* 1 (1887), 30; Hubert H. Bancroft, *History of California, 1825–1840* (San Francisco: History Co., 1889), 162–172; J. J. Warner, "Reminiscences of Early California from 1831 to 1846," *Annual Pub. of the Hist. Soc. of Southern California* 7 (1907–08), 183; Raymond W. Settle, "Nathaniel Miguel Pryor," in *The Mountain Men and the Fur Trade of the Far West* (Glendale, Calif.: Arthur H. Clark, 1971), v. 2, 285–288; William C. McGaw, *Savage Scene; The Life and Times of James Kirker* (New York: Hastings House, 1972), 78–79; Batman, op. cit., n. 26, 81, 159, 192–199, 237, 337.

28. Emory, op. cit., n. 25, 58–59; Louis Houck, *A History of Missouri* (Chicago: Donnelley, 1908), v. 3, 88–91; Rickard, op. cit., n. 23, 255–256; *Index to Records of Aliens' Declarations of Intention and/or Oaths of Allegiance, 1789–1880* (Harrisburg: Pennsylvania Historical Commission, 1940), v. 2, 287–288 (Curciers); Rex W. Strickland, "Robert McKnight," in *The Mountain Men and the Fur Trade of the Far West* (Glendale, Calif.: Arthur H. Clark, 1971), v. 9, 259–268; McGaw, op. cit., n. 27, 36, 78–79; Batman, op. cit., n. 26, 192–199; *Généalogie et Histoire de la Caraïbe Bulletin* 24 (Feb. 1991), 279 (Curciers);

Edgar O. Gutiérrez, "Esteban Courcier: Un Empresario Franco Estado Unidense en Chihuahua, 1826–1846," in *Los Inmigrantes en el Mundo de los Negocios Siglos XIX y XX* (Mexico, D.F.: Instituto Nacional de Antropología e Historia, 2003), 17–23.

29. Willard, op. cit., n. 27, 651–652; Hardy, op. cit., n. 27, 462–463, 477; Santa Fe *Republican*, Nov. 20, 1847; Emory, op. cit., n. 25, 58; Frederick Adolph Wislizenus, *Memoir of a Tour to Northern Mexico* (Washington, D.C.: Tippin and Streeper, 1848), 57–58; John R. Bartlett, *Personal Narrative of Explorations and Incidents in Texas, New Mexico, California, Sonora, and Chihuahua* (New York: Appleton, 1854), 227–235; *NYT*, Aug. 4, 1875; *Diccionario Porrúa de Historia, Biografía y Geografía de Mexico* (Mexico, D.F.: Editorial Porrua, 1965), 417; Strickland, op. cit., n. 28, 267–268; McGaw, op. cit., n. 27, 87, 108–121; Rex W. Strickland, "The Birth and Death of a Legend: The Johnson 'Massacre' of 1837," *Arizona & the West* 18 (Autumn 1976), 257–286; Eduardo Flores Clair, Cuauhtemoc Velasco Avila, and Elia Ramírez Bautista, *Estadisticas Mineras de Mexico en el Siglo XIX* (Mexico, D.F.: Instituto Nacional de Antropología e Historia, 1985), 25, 29; Sweeney, op. cit., n. 25, 42–52, 70–80; Long, et al., op. cit., n. 4, 629, 636–637; Gutiérrez, op. cit., n. 28, 22–33.

30. Frederica De Laguna and Catharine McClellan, "Ahtna," *Handbook of North American Indians*, v. 6, *Subarctic*, ed. June Helm (Washington, D.C.: Smithsonian Institution, 1981), 641–652; Andrei V. Grinev, "On the Banks of the Copper River: The Ahtna Indians and the Russians, 1783–1867," *Arctic Anthropology* 30 (1993), 54–66; Kenneth L. Pratt, "Copper, Trade, and Tradition Among the Lower Ahtna of the Chitina River Basin: The Nicolai Era, 1884–1900," *Arctic Anthropology* 35 (1998), 77–98.

31. William R. Abercrombie, *Alaska. 1899. Copper River Exploring Expedition* (Washington, D.C.: G.P.O., 1900), 110, 118; Frank C. Schrader and Arthur C. Spencer, *The Geology and Mineral Resources of a Portion of the Copper River District, Alaska* (Washington, D.C.: G.P.O., 1900), 22, 86–87; James W. VanStone, "Exploring the Copper River Country," *Pacific Northwest Quarterly* 46 (1955), 115–123; William C. Douglass, *A History of the Kennecott Mines* (Fairbanks: Alaska Division of Mines and Minerals, 1964), 11; Grinev, op. cit., n. 30, 54–66; Jim and Nancy Lethcoe, *Valdez Gold Rush Trails of 1898–99* (Valdez, Alas.: Prince William Sound Books, 1996), 113; Pratt, op. cit., n. 30, 77–98; see Ronald N. Simpson, *Legacy of the Chief* (Anchorage: Publication Consultants, 2001).

32. Nicholas I. Roosevelt, et al., *Papers, Relative to an Application to Congress for an Exclusive Right of Searching for and Working Mines, in the North-West and South-West Territory* (Philadelphia: Samuel Harrison Smith, 1796), 1–28; *Journal of the House of Representatives* (Washington, D.C.: Gales and Seaton, 1826), v. 2, 430, 442, 628, 640; *American State Papers* (Washington, D.C.: Gales and Seaton, 1834), 140–141; Granbery, op. cit., n. 21, 23–24; Devereaux Dunlap Cannon, "Nicholas I. Roosevelt," 2000, http://64.235.34.221/rosehill/genroosevelt.htm.

33. Adam Smith, *An Inquiry into the Nature and Causes of the Wealth of Nations* (London: Strahan and Cadell, 1776), v. 2, 154–155; Roosevelt, op. cit., n. 32, 12–13; *Journal of the House*, op. cit., n. 32, 640.

34. Raleigh, N.C., *Register*, Dec. 5, 1803, Dec. 30, 1805, May 12, 1806; New York *Advertiser*, Feb. 14, 1804; Albany, N.Y. *The Balance and Columbian Repository*, Feb. 21, 1804, May 21, 1805; *WNI*, May 12, June 20, Oct. 27, 1806; William Thornton, *North Carolina Gold-Mine Company* (Washington, D.C., 1806), 1–20; Stephen Ayres, "A Description of the Region in North-Carolina Where Gold Has Been Found," *Medical Repository* 10 (Aug.-Oct. 1806), 148–151; Fletcher Green, "Gold Mining: A Forgotten Industry of Ante-Bellum North Carolina," *North Carolina Historical Review* 14 (Jan. 1937), 7–8; Richard F. Knapp, "Golden Promise in the Piedmont: The Story of John Reed's Mine," *North Carolina Historical Review* 52 (Jan. 1975), 1–19; Richard F. Knapp and Brent D. Glass, *Gold Mining in North Carolina, a Bicentennial History* (Raleigh: North Carolina Division of Archives and History, 1999), 10, 43, 50–51.

35. Thornton, op. cit., n. 34, 1–20; William Thornton, "Letter from W. Thornton, Esq. to the Members of the North Carolina Gold Mine Company," *Philosophical Magazine* 27 (Apr. 1807), 261–264; *Maryland Gazette*, Apr. 22, 1824; John H. Wheeler, *Historical Sketches of North Carolina, from 1584 to 1851* (Philadelphia: Lippincott, Grambo, 1851), v. 2, 63–64; Willard L. Thorp, *Business Annals: United States, England, &c* (New York: National Bureau of Economic Research, 1926), 94, 115–116; Knapp and Glass, op. cit., n. 34, 43.

36. *Niles' Weekly Register*, Feb. 1, 1834; Josiah D. Whitney, *The Metallic Wealth of the United States* (Philadelphia: Lippincott, Grambo, 1854), 145; Thorp, op. cit., n. 35, 94, 120–121; Fletcher Green, "Georgia's Forgotten Industry: Gold Mining," *Georgia Historical Quarterly* 19 (June 1935), 99–101, 106–107, 214–215; Green, op. cit., n. 34, 13; Fletcher Green, "Gold Mining in Ante-Bellum Virginia," *Virginia Magazine of History and Biography* 45 (July 1937), 232–234; David Williams, *The Georgia Gold Rush* (Columbia: University of South Carolina Press, 1993), 11–12, 48–56, 74–79; Knapp and Glass, op. cit., n. 34, 13–19.

37. Legislature of North Carolina, *Report on Incorporating the Mecklenburg Gold Mining Company* (Raleigh: Lawrence and Lemay, 1831), 1–4; *Act of Incorporation of the Mecklenburg Gold Mining Company* (New York: William Tolefree, 1833), 3–16; Daniel A. Tompkins, *History of Mecklenburg County and the City of Charlotte from 1740 to 1903* (Charlotte, N.C.: Observer, 1903), v. 2, 123; Green, op. cit., n. 34, 15–16, 135; Knapp and Glass, op. cit., n. 34, 24–31.

38. *Merchants' Magazine* 11 (Nov. 1844), 482; *History of Allegheny County Pennsylvania* (Chicago: Warner, 1889), v. 2, 254–258; Thorp, op. cit., n. 35, 94, 123–124; Robert James Hybels, "The Lake Superior Copper Fever, 1841–1847," *Michigan History* 34 (Sept. 1950), 226, 233, 241–243, 324–326; Charles K. Hyde, "From 'Subterranean Lotteries' to Orderly Investment: Michigan Copper and Eastern Dollars, 1841–1865," *Mid-America* 66 (Jan. 1984), 6–11; Hyde, op. cit., n. 21, 32–34.

39. *Report of the Committee of Stockholders of the Pittsburgh and Boston Copper Harbor Mining Company* (Boston: S. N. Dickson, 1847), 3–7, 28–30; *Charter and By-Laws of the Pittsburgh and Boston Mining Company of Pittsburgh* (Pittsburgh: Geo. Parkin, 1848), 3–6; Whitney, op. cit., n. 36, 275–279; Joseph G. Martin, *Seventy-three Years' History of the*

Boston Stock Market (Boston: printed by author, 1871), 91–92, 98; "Curtis G. Hussey," *Magazine of Western History* 3 (Feb. 1886), 329–348; William B. Gates, *Michigan Copper and Boston Dollars* (Cambridge, Mass.: Harvard University Press, 1951), 3–6, 10–13, 33–35, 216–217; Donald Chaput, *The Cliff, America's First Copper Mine* (Kalamazoo, Mich.: Sequoia Press, 1971), 18–37; Hyde, op. cit., n. 21, 32–40.

40. Martin, op. cit., n. 39, 38–40, 91–92, 96–98.

41. Hybels, op. cit., n. 38, 233–243; Gates, op. cit., n. 39, 5–6; C. Harry Benedict, *The Red Metal* (Ann Arbor: University of Michigan Press, 1952), 40–102; Hyde, op. cit., n. 21, 34–35; Long, et al., op. cit., n. 4, 636 (total production of Keweenaw 4.9 and Santa Rita 5.0 million tons of copper).

CHAPTER 2. THE GOLD LUST

1. San Francisco *Californian*, Mar. 15, 1848; *NYH*, Sept. 17–18, 21, 27, 1848; Daniel Webster, *The Writings and Speeches of Daniel Webster* (Boston: Little, Brown, 1903), v. 10, 26; Rodman Paul, *The California Gold Discovery* (Georgetown, Calif.: Talisman, 1966), 32–40.

2. *NYH*, Dec. 2–3, 6–8, 11, 21, 1848; *WNI*, Dec. 4, 6, 1848; *Saturday Evening Post*, Dec. 9, 1848; Philadelphia *Public Ledger*, Dec. 15, 1848; "The Golden Land," *Scientific American* 4 (Dec. 16, 1848), 98; Boston *Cultivator*, Dec. 16, 23, 1848; James Polk, *Message of the President of the United States . . . December 5, 1848* (Washington, D.C.: Wendell and Van Benthuysen, 1848) 3–44; Peter Browning, ed., *To the Golden Shore* (Lafayette, Calif.: Great West, 1995), 34–42, 48–49, 53–58, 69–70.

3. *NYTr*, Jan. 12, 23, Feb. 3, 10, 17, Mar. 1, 3, 5, 23, Apr. 18, May 3, 18, June 1, 1849; Marshall Hughes, "The Argonaut Mining Companies of 1848–1850," (M.A. thesis, University of California, Berkeley, 1939), 1–14; R. W. G. Vail, "Bibliographical Notes on Certain Eastern Mining Companies of the California Gold Rush 1849–1850," *Papers of the Bibliographical Society of America* 43 (1949), 247–278; Maureen A. Jung, "Capitalism Comes to the Diggings," *California History* 77 (Winter 1998/99), 52–77.

4. *NYTr*, Jan. 12, 23, Feb. 3, 10, 17, Mar. 1, 3, 5, 23, Apr. 18, May 3, 18, June 1, 1849; Octavius T. Howe, *Argonauts of '49; History and Adventures of the Emigrant Companies from Massachusetts 1849–1850* (Cambridge, Mass.: Harvard University Press, 1923), 187–217; Vail, op. cit., n. 3, 253, 264–287; Hughes, op. cit., n. 3, 3–10.

5. A sampling of cooperative company charters can be found in *Constitution of the Bunker Hill Trading and Mining Association* (Boston, 1849) broadside; *Rev. Dr. Worcester's Address before the Naumkeag Mutual Trading & Mining Company* (Salem, 1849) 1–11; "Articles of Association and Agreement" of the Wolverine Rangers, in Marshall, Mich., *Statesman*, Jan. 17, 1849; "Washington City Mining Company," Jan. 30, 1849, in *Gold Rush: The Journals of J. Goldsborough Bruff* (New York: Columbia University Press, 1949), xxxii–xxxiii; *Constitution of the the New-York Commercial & Mining Company, in California* (New York: John Belcher, 1849) 1–12; *By-Laws of the Montague Mining and Trading Assoc'n, of New Haven, Conn.* (New Haven, Conn.: 1849) broadside; *Rules*

and *Regulations of the Mutual Protection Trading and Mining Company* (Boston, 1849) broadside; *Articles of Agreement and By-Laws of the New-York Excelsior Trading & Mining Association* (New York: H. Cogswell, 1849) 1–20; Cayuga Joint Stock Company, *Articles of Association* (Auburn, N.Y., 1849), 1–8; *The Brothers' Mining and Trading Co. of New Haven* (New Haven, Conn.: 1849), 1–4; "Constitution and By-Laws of the Knox County Company, Illinois," in the Kanesville, Iowa, *Frontier Guardian,* May 30, 1849, Howe, op. cit., n. 4, 4–7; Hughes, op. cit., n. 3, 7–14; see also John Phillip Reid, *Policing the Elephant* (San Marino, Calif.: Huntington Library, 1997).

6. Robert S. Morrison, *Digest of the Law of Mines and Minerals* (San Francisco: A. L. Bancroft, 1878), 298–300; Howe, op. cit., n. 4, 187–217; Hughes, op. cit., n. 3, 78–85.

7. *LT,* June 8, 1846; Gordon's California Association, *To California on Shares* (Philadelphia, 1848) broadside; *NYH,* Dec. 28, 1848, Jan. 10, 16, 30, Feb. 1, 10, 21, 1849; *SFAC,* May 23, 1869; Albert Shumate, *The California of George Gordon* (Glendale, Calif.: Arthur H. Clark, 1976), 3, 50–69, 88.

8. *SFBul,* May 22, 1869 (obit); *SFAC,* May 23, 1869; Shumate, op. cit., n. 7, 76–86.

9. *LT,* Feb. 6, 1849; Morrison, op. cit., n. 7, 298–300; Howe, op. cit., n. 4, 4–5; Willard L. Thorp, *Business Annals: United States, England, &c* (New York: National Bureau of Economic Research, 1926), 94, 124–125; "A Gold Rush Document," *California Historical Society Quarterly* 23 (Sept. 1944), 226; Rodman W. Paul, *California Gold* (Cambridge, Mass.: Harvard University Press, 1947), 119–120.

10. *LT,* Jan. 2–12, 1849, Nov. 10, 19, 21, 1851; Francis Palmer, *Company Law* (London: Stevens, 1909), 7–9; Gilbert Chinard, "When the French Came to California," *California Historical Society Quarterly* 22 (1943), 289–314; Henry W. Ballantine, *Ballantine on Corporations* (Chicago: Callaghan, 1946), 34–38; Henry Blumenthal, "The California Societies in France, 1849–1855," *Pacific Historical Review* 25 (1956), 251–260; Richard E. Towey, "British Investment in California 1849–1860," (M.A. thesis, University of California, Berkeley, 1957), 26–29, 75, 109–110.

11. *LT,* Jan. 2, 4, 8, 1849, Nov. 10, 1851; Paris *Journal des Débats,* July 9, Sept. 15, 1850; "The California Mania in Paris," *Merchants' Magazine* 23 (Dec. 1850), 698–699; *La Toison d'Or Cie. des Placers de la Californie, 10 Mai 1850, Capital Social 1,500,000 Francs Divise en 300,000 Titres d'Action de Cinq Francs,* certificate no. 4332; Chinard, op. cit., n. 10, 296–309; Abraham P. Nasatir, *French Activities in California* (Stanford, Calif.: Stanford University Press, 1945), 402–421, 444–445; Towey, op. cit., n. 10, 26–30.

12. *LT,* Jan. 2–4, 8, 12, 1849.

13. *LT,* Jan. 3, 8, 29, 1849; Paris *Journal des Débats,* June 11, July 9, Aug. 2, 15, Sep, 15, 1850; Blumenthal, op. cit., n. 10, 252–253; Chinard, op. cit., n. 10, 299–301.

14. *LT,* Jan. 8, 10, 12, 16, 20, 1849, Jan. 21, 1875; Blumenthal, op. cit., n. 10, 252; Chinard, op. cit., n. 10, 296. *The History of the Times: The Tradition Established, 1841–1884* (London: *Times,* 1939) 543, 595–596.

15. Blumenthal, op. cit., n. 10, 256–258.

16. Blumenthal, op. cit., n. 10, 259–260; Towey, op. cit., n. 10, 109–110; Paul, op. cit., n. 9, 118.

17. *SacU*, Apr. 28, Sept. 11, 1852; *York Mining Company, Capital $20,000, 200 Shares, $100 Each*, stock certificate no. 92, May 17, 1852, Little York, Nevada Co., Calif.

18. Daniel B. Woods, *Sixteen Months at the Gold Diggings* (New York: Harper, 1851), 169–175; *SacU*, Mar. 19, 1852.

19. Hubert H. Bancroft, *Annals of the California Gold Era* (San Francisco: Bancroft, 1886), 414–416; Paul, op. cit., n. 9, 131–146.

20. *A Pile; or, A Glance at the Wealth of the Moneyed Men of San Francisco and Sacramento City* (San Francisco: Cooke and Lecount, 1851), 1–15; San Francisco *Herald*, Oct. 2, 1851; *Annual Report of the Trustees of the Merced Mining Company. Organized March 4, 1851* (San Francisco: Alta California, 1852) 1–16; *MSP*, Nov. 12, 1864; Morrison, op. cit., n. 6, 277; R. S. Lawrence, ed., *Pacific Coast Annual Mining Review and Stock Ledger* (San Francisco: Francis and Valentine, 1878), 122–123; William B. Clark, *Gold Districts of California* (San Francisco: California Division of Mines and Geology, 1970), 29, 94–95; Jung, op. cit., n. 3, 66.

21. *NYTr*, Feb. 8, 18, Mar. 9, 20, Apr. 24, 1850; *Notes on California and the Placers* (New York: Long and Bro., 1850), 62–66, 86–87; *Rocky Bar Mining Company, Circular, Articles of Association, Resolutions, etc* (New York, 1850), 10; Carl Wheat, "The Rocky Bar Mining Company," *California Historical Society Quarterly* 12 (Mar. 1933), 65–68, 71; James Delavan, *The Gold Rush: Letters of Dr. James Delavan from California to the Adrian, Michigan, Expositor, 1850–1856* (Mount Pleasant, Mich.: Cumming, 1976), ii–xi,19–21, 89–95.

22. *NYTr*, Nov. 11, Dec. 11, 17–18, 1850, Jan. 3, Feb. 10, 26, Mar. 31, 1851; Rocky Bar, op. cit., n. 21, 1–12; *LT*, Feb. 16, 1852; Wheat, op. cit., n. 21, 67–69; Towey, op. cit., n. 10, 43; Delavan, op. cit., n. 21, xi–xvi, 89–92.

23. Rocky Bar, op. cit., n. 21, 4–5; *Union Quartz Mountain Mining and Crushing Co.* (New York, 1851), 18, 22; *Charter of the Grass Valley Gold Mining Company* (New York: Winchester, 1851), 12; *SacU*, Oct. 22, 1857.

24. *NYTr*, Feb. 28, Mar. 8–Apr. 5, 8–19, 1852; Concord *New Hampshire Statesman*, Mar. 13, 1852; *Facts Concerning Quartz and Quartz Mining; Together with the Charter of the Manhattan Quartz Mining Company* (New York: Burroughs, 1852), 1–30; Jung, op. cit., n. 3, 67; Fred R. Shapiro, *The Yale Book of Quotations* (New Haven, Conn.: Yale University Press, 2006), 322–323.

25. *SFAC*, Dec. 15, 1849; C. Gregory Crampton, "The Opening of the Mariposa Mining Region, 1849–1859, with Particular Reference to the Mexican Land Grant of John Charles Fremont" (Ph.D. diss., University of California, Berkeley, 1941), 25–30; *BDAC*, 546; Mary Lee Spence and Donald Jackson, *The Expeditions of John Charles Fremont* (Urbana: University of Illinois Press, 1973), v. 2, 297–300; Pamela Herr and Mary Lee Spence, *The Letters of Jessie Benton Fremont* (Urbana: University of Illinois Press, 1993), 3–10.

26. *NYH*, Apr. 29, 1850; *Congressional Globe*, 31st Cong., 1st Sess., Sept. 27, 1850, 2045–2047, Appendix, 1360–1373; *WNI*, Feb. 7, 1853; *NYT*, Feb. 8, 1853; *SFAC*, Nov. 8, 1867 (Jones obit); Crampton, op. cit., n. 25, 203–222; Spence and Jackson, op. cit.,

n. 25, v. 3, xxxvi–xxxvii, lix–lxi, 118–119, 200–205; Paul W. Gates, "Adjudication of Spanish-Mexican Land Claims in California," *Huntington Library Quarterly* 21 (May 1958), 218–225; Paul W. Gates, "The Fremont-Jones Scramble for California Land Claims," *Southern California Quarterly* 56 (Spring 1974), 14–27.

27. *LT*, Nov. 8, 1851, Mar. 12, 1852; David Hoffman, *The Fremont Estate* (London: C. Richards, 1851), 1–63; *SFAC*, Jan. 15–16, Feb. 9, 1852; *Articles of Association of the Philadelphia and California Mining Company* (Philadelphia: Clark, 1852), 18; *WNI*, Feb. 21, 24, 1852; David Hoffman, *California, Fremont Estates, and Gold Mines Non-sale to Mr. T.D. Sargent* (London: Richards, 1852), 1–16; *NYT*, Apr. 29, May 11, 1852; San Francisco *Herald*, May 30, 1852; Thomas Allsop, *California and Its Gold Mines* (London: Groombridge, 1853), 101–103; *NYH*, May 19, 1856; Crampton, op. cit., n. 25, 174–187; Towey, op. cit., n. 10, 30–39; Spence and Jackson, op. cit., n. 25, v. 2, 299, v. 3, xlv–lii, lxix–lxx, 146–148, 154–159, 180–198, 220–236, 242–375; Mary Lee Spence, "David Hoffman: Fremont's Mariposa Agent in London," *Southern California Quarterly* 22 (Fall 1978), 379–403; Virginia Bell, "Trenor Park, New Englander in California," *California History* 60 (1981), 158–171; Jung, op. cit., n. 3, 63.

28. *Le Nouveau Monde Compagnie en Commandité, Pour l'Exploitation des Mines d'Or de la Californie. Capital Social, Fr. 5,000,000. Cinq Actions de Vingt-Cinq Francs*, shares nos. 127401 à 127405, Paris, 21 Aout 1851; *LT*, Nov. 10, 1851; *Black Republican Imposture Exposed! Fraud Upon the People. Fremont and His Speculations* (Washington, D.C.: H. Polkinhorn, 1856), 10; Edwin F. Bean, *Bean's History and Directory of Nevada County* (Nevada City: Gazette, 1867), 186; Crampton, op. cit., n. 25, 215–217; Towey, op. cit., n. 10, 36, 61–63, 69–72; *BDAC*, op. cit., n. 25, 916; Spence and Jackson, op. cit., n. 25, v. 3, lxi–lxx; Gates, op. cit., n. 26, 36–38.

29. *LT*, Nov. 19, 1851; "Schedule to the Deed of Settlement of the Anglo-Californian Gold Mining Company, Part 3. The Names, &c., of the Subscribers, and Shares Held by Each in the Company, 5 Nov. 1851," in Great Britain, Board of Trade, "Archive of the Companies Registration Office," microfilm, Bancroft Library, University of California, Berkeley; *Quartz Mining. The Burns Ranche Gold Mining Company* (New York: John F. Trow, 1851), 1–20; *Articles*, op. cit., n. 27; *York Mining Company*, op. cit., n. 17.

30. *NYTr*, Feb. 4, 1851; *LT*, June 18, 1852; New York *Mining Magazine* 1 (July 1853), 56–58, and (June 1854), 642.

31. Josiah D. Whitney, *The Metallic Wealth of the United States* (Philadelphia: Lippincott, Grambo, 1854), 142; Leland Jenks, *The Migration of British Capital to 1875* (New York: Alfred Knopf, 1927), 161.

32. Allsop, op. cit. n. 27; U.S. Bureau of the Census, *Seventh Census of the United States: 1850* (Washington, D.C.: Armstrong, 1853), 985; *SacU*, May 23, 1857; Bancroft, op. cit., n. 19, 417.

33. *NYTr*, Aug. 20, Oct. 13, Nov. 2, 1852, Jan. 3, 17, Feb. 22, 1853, Apr. 18, 1855, Oct. 13, 1857, Mar. 27, 1858; New York *Mining Magazine* 2 (Apr.–May 1854), 427, 542–543, and (Oct. 1854), 423–425; *SFBul*, July 27, 1857, Feb. 24–26, June 19, 1858; *SFAC*, Feb. 24, 26, Mar. 2, 1858; *NYH*, Mar. 28, 1858; Bean, op. cit., n. 28, 49, 207–208; *NYT*, Dec.

15, 1868; Harry L. Wells, *History of Nevada County* (Oakland, Calif.: Thompson and West, 1880), 111, 192–193; Wheat, op. cit, n. 21, 70–72; Delavan, op. cit., n. 21, xxii–xxiv, 36–37, 92–97.

34. Whitney, op. cit., n. 31, xxvii–xxix.

35. *SFBul*, Mar. 21, 1863; Bancroft Scraps, v. 52 (California Mining Stocks), 14–15, Bancroft Library, Berkeley; Hayward Chaplin, "The History of Gold Mining in California" (M.A. thesis, Stanford University, 1946), 106; Paul, op. cit., n. 9, 131–146; Towey, op. cit., n. 10, 77–84.

36. Bean, op. cit., n. 28, 115, 223; Oscar T. Shuck, ed., *Sketches of Leading and Representative Men of San Francisco* (San Francisco, 1875), 1001–1002; Lawrence, op. cit., n. 20, 48–49; John Daggett, "Scrapbook, no. 1," 27, California State Library, Sacramento; Alonzo Phelps, *Contemporary Biography of California's Representative Men* (San Francisco: A. L. Bancroft, 1882), v. 2, 9–13; *NYTr*, Nov. 28,1887; George W. Starr, "The Empire Mine, Past and Present," *MSP* 81 (Aug. 4, 1900), 120; George Hearst, *The Way It Was* (New York: Hearst Corp., 1972), 5–11; Fremont and Cora Older, *The Life of George Hearst* (San Francisco: John Henry Nash, 1933), 23, 36–38, 94–95; Louis J. Rasmussen, *San Francisco Ship Passenger Lists* (Colma, Calif.: Historic Record, 1965), v. I, 89; Clark, op. cit., n. 20, 13; Jack R. Wagner, *Gold Mines of California* (Berkeley, Calif.: Howell-North, 1970), 213–224; Frank W. McQuiston, *Gold: The Saga of the Empire Mine, 1850–1956* (Grass Valley, Calif.: Empire Mine Park Assoc., 1986), 14–15, 61–65, 76, 83; Judith Robinson, *The Hearsts* (Newark: University of Delaware Press, 1991), 36–39; Keith R. Long, John H. DeYoung and Steve Ludington, "Significant Deposits of Gold, Silver, Copper, Lead, and Zinc in the United States," *Economic Geology* 95 (May 2000), 635–637.

37. SFAC, Dec. 7, 1850, Dec. 24, 1851, Feb. 2, Apr. 16, 1852, Nov. 30, Dec. 1, 1854, Apr. 27–28, 1860 (Morgan obit), June 17, 1867; *SacU*, Dec. 20, 24, 1851; *NYT*, Jan. 19, 1855 (nugget); *MSP*, Jan. 7, 1888; Wagner, op. cit., n. 36, 67–72; Julia G. Costello and Judith Cunningham, *History of Mining at Carson Hill* (Carson Hill Mining Co., 198–?), 1–15; Ronald H. Limbaugh and Willard P. Fuller, *Calaveras Gold* (Reno: University of Nevada Press, 2004), 50–51.

38. SFAC, June 13, 16, Dec. 22–24, 29, 1851, Jan. 1, 4, 7, Apr. 16, June 3, 1852, Nov. 30, Dec. 1, 1854, June 17, 1867; London *Globe & Traveller*, Feb. 3, 1852; *LT*, Feb. 5, Oct. 5, 1852, June 1, 1855, Feb. 18, 20, 26, Mar. 10, 12, Apr. 2–3, June 17–19, July 7, Sept. 9, 1856, Apr. 23, 1877; *Carsons Creek Consolidated Mining Company*. £2.10. shares nos. 83206 to 83210, London, Oct. 29, 1852; Charles Dickens, "Little Dorrit," *Harpers New Monthly Magazine* 13 (June 1856), 106ff; *Dictionary of National Biography* (London: Oxford University Press, 1922), v. 17, 591–592; Towey, op. cit., n. 10, 40–41, 59–60; Wagner, op. cit., n. 36, 67–72; Costello and Cunningham, op. cit., n. 37, 1–15; Limbaugh and Fuller, op. cit., n. 37, 50–51.

39. *SFBul*, Oct. 27, 1857; *MSP*, June 11, 1864, July 8, 1865, Mar. 3, May 19, 1866, Jan. 6, 1867, Mar. 14, 1868; J. Ross Browne and James Taylor, *Report Upon the Mineral Resources of the United States* (Washington, D.C.: G.P.O., 1867), 66; J. Ross Browne, *Report on the*

Mineral Resources of the States and Territories West of the Rocky Mountains (Washington, D.C.: G.P.O., 1868), 75–76; Rossiter W. Raymond, *Mineral Resources of the States and Territories West of the Rocky Mountains* (Washington, D.C.: G.P.O., 1869), 16–19; Austin, Nev., *Reese River Reveille*, July 15, 1871; *SFCh*, July 18, 1871, Mar. 3, 1872, Mar. 31, 1895, Feb. 15, 1904 (Hayward obit); Jesse Mason, *History of Amador County* (Oakland, Calif.: Thompson and West, 1881), 150–152; *Mining & Engineering Review*, Feb. 20, 1904; Thomas Arthur Rickard, "The Re-Opening of Old Mines Along the Mother Lode," *MSP* 112 (June 24, 1916), 935–938; Richard E. Lingenfelter, *The Hardrock Miners* (Berkeley: University of California Press, 1974), 23; Wagner, op. cit., n. 36, 129.

40. *LT*, Nov. 10, 1851, Aug. 30, 1852, Aug. 23, 1853, Feb. 13, June 10, 1854, Jan. 13, Sept. 21, 1855; *Placer Times* (San Francisco), Aug. 18, 21, 1852; Merced Mining Company, op. cit., n. 20, 1–16; *San Joaquin Republican* (Stockton, Calif.), Aug. 29, 1856; Crampton, op. cit., n. 25, 197–200; Towey, op. cit., n. 10, 66–69.

41. *NYT*, Mar. 20, 1854; *LeCount & Strong's San Francisco City Directory for the Year 1854* (San Francisco: Herald, 1854), 221; Fremont vs. United States, 58 U.S. 542 (1855); *San Joaquin Republican* (Stockton, Calif.), Apr. 1, 1855, Feb. 12, July 14, 1858; *WNI*, Mar. 17, 1856; *SFBul*, Aug. 25, 1856, Jan. 30, 1857, May 7, 25, June 8, 1858; Mariposa *Gazette*, July 14, 1858; William Godfrey, *The Fremont Grant* (San Francisco: O'Meara and Painter, 1858), 1–31; Browne, op. cit., n. 39, 21–30; *The Mariposa Estate . . . With a Review of the Causes Which Led to the Late Failure* (New York: Russells, 1868), 43–44; Carl B. Swisher, *Roger B. Taney* (New York: Macmillan, 1935), e.g. 273, 284, 289; Crampton, op. cit., n. 25, 188–235; *BDAC*, op. cit., n. 25, 1791; Paul W. Gates, "California's Embattled Settlers," *California Historical Society Quarterly* 41 (June 1962), 113; Spence and Jackson, op. cit., n. 25, v. 3, lix–lxx; Carl B. Swisher, *History of the Supreme Court of the United States*, v. 5, *The Taney Period 1836–64* (New York: Macmillan, 1974), 778–781; Robin W. Winks, *Frederick Billings: A Life* (New York: Oxford University Press, 1991), 120.

42. *Georgia Telegraph* (Macon), Mar. 4, 1856; *Baltimore Sun*, June 20, 1856; *Ohio Statesman* (Columbus), July 12, 1856; *SFBul*, Aug. 25, 1856, Jan. 28, June 29, Oct. 3, 1857, May 25, June 2, 4, July 15–16, 24, Aug. 21, Dec. 3, 1858, June 16, 1859, Oct. 7, 1862; *San Joaquin Republican* (Stockton, Calif.), Nov. 15, 1856, June 24, July 4, 1857, Jan. 31, July 4, 14, 17, 31, 1858; John Bigelow, *Memoir of the Life and Public Services of John Charles Fremont* (New York: Derby and Jackson, 1856), 379–387; Charles W. Upham, *Life, Explorations and Public Services of John Charles Fremont* (Boston: Ticknor and Fields, 1856), 299, 324–325, 334–335; *NYH*, Feb. 28, 1857, Aug. 12–13, 1858; *NYT*, Feb. 28, 1857; Mariposa *Gazette*, June 19–26, 1857, July 14, Aug. 24–31, 1858; *WNI*, Nov. 3, 1857, Aug. 14, 20, 1858; San Francisco *Globe*, June 3, 1858; *SFAC*, Aug, 2, 1858; Godfrey, op. cit., n. 41, 1–31; Crampton, op. cit., n. 25, 235–265; Herr and Spence, op. cit., n. 25, 40, 208–209.

43. *SFBul*, Jan. 31, Oct. 3, 1857, Mar. 17, 26, 1858; Mariposa *Gazette*, Mar. 25, 1858; *SFAC*, Nov. 18, 26, Dec. 5, 9, 24, 1859; Biddle Boggs v. Merced Mining Company, 14 Cal. 279 (1859); Ex-Supreme Court Broker, *The Gold Key Court; or, The Corruptions of a Majority of It* (San Francisco: 1860), 1–4; Stephen J. Field, *Personal Reminiscences of Early Days in California* (Washington, D.C.: printed by author, 1893), 132–133; Carl B.

Swisher, *Stephen J. Field* (Washington, D.C.: Brookings, 1930), 84–87; Crampton, op. cit., n. 25, 258–264; Winks, op. cit., n. 41, 120–121; Paul Kens, "John C. Fremont and The Biddle–Boggs Case," *Mining History Journal* 5 (1998), 8–21; Jack Beatty, *Age of Betrayal: The Triumph of Money in America, 1865–1900* (New York: Alfred Knopf, 2007), 148–191.

44. SFAC, Nov. 18, 26, Dec. 5, 9, 24, 1859; SFBul, Nov. 19, 1859; *Biddle Boggs*, op. cit., n. 43.

45. SFAC, Nov. 26, 1859.

46. *The Merced Mining Company, Appellant, vs. Biddle Boggs* Washington, D.C.: 1865), 1–42; David Dudley Field, *The Merced Mining Company, Appellant, agst. Biddle Boggs, Respondent* (New York: M. B. Brown, 1865), 1–13; Merced Mining Company vs. Boggs, 70 U.S. 304; *The Great Libel Case. Geo. Opdyke agt. Thurlow Weed* (New York: American News, 1865), 15, 59, 129; NYT, Sept. 4, 1866; Glyndon G. Van Deusen, *Thurlow Weed* (Boston: Little, Brown, 1947), 315, 383; Swisher, *History of the Supreme Court*, op. cit., n. 41, 830.

47. Alexander Forbes, *California: A History* (London: Smith, Elder, 1839), 352 pp.; LT, Sept. 6, 1839; *Andres Castillero, Claimant, vs. The United States, Defendant, for the Place Named New Almaden* (San Francisco: Whitton, Towne, 1858), 55–77, 90–91, 493–494, 510–511, 870–871, 1102–1109; *The United States vs. Andres Castillero, No. 420. "New Almaden." Transcript of the Record* (San Francisco, 1860–1861), 1823–1826, 1830; Bancroft, op. cit., n. 19, 551, 555–561; Leonard W. Ascher, "The Economic History of the New Almaden Quicksilver Mine, 1845–1863," (Ph.D. diss., University of California, Berkeley, 1934), 7–18; Mary L. Coomes, "From Pooyi to the New Almaden Mercury Mine: Cinnabar, Economics, and Culture in California to 1920" (Ph.D. diss., University of Michigan, 1999), 29–30; John Mayo, *Commerce and Contraband on Mexico's West Coast in the Era of Barron, Forbes & Co., 1821–1859* (New York: Peter Lang, 2006), 388–393.

48. C. S. Lyman, "Mines of Cinnabar in Upper California," *American Journal of Science and Arts.* 6, no. 17 (Sept. 1848), 270–271; SFAC, Mar. 29, 1851, Sept. 15, 1853; *Andres Castillero*, op. cit., n. 47, 172–175, 381, 493–511; Browne and Taylor, op. cit., n. 39, 174; Thomas Egleston, *The Metallurgy of Silver, Gold, and Mercury in the United States* (New York: John Wiley, 1890), v. 2, 901; Ascher, op. cit., n. 47, 17–27, 32–40.

49. SFAC, Sept. 15, 1853, Aug. 25, 1854; NYT, Sept. 10, 1860; Browne and Taylor, op. cit., n. 39, 174; MSP, June 5, 1875; Egleston, op. cit., n. 48, 901; Oscar T. Shuck, *Historical Abstract of San Francisco* (San Francisco, 1897), v. 1, 20, 28; Ascher, op. cit., n. 47, 21–22, 41.

50. *Andres Castillero*, op. cit., n. 47, 172–175, 189–191, 514–515; NYT, Mar. 12, 1863; Browne and Taylor, op. cit., n. 39, 174; MSP, June 5, 1875; Lawrence, op. cit., n. 20, 32; Phelps, op. cit., n. 36, v. 1 (1881), 222–223; Egleston, op. cit., n. 48, 901; Ascher, op. cit., n. 47, 41–44; Cecil G. Tilton, *William Chapman Ralston* (Boston: Christopher, 1935), 296–297.

51. San Francisco *Californian*, Feb. 16, 1848; Thomas O. Larkin, *The Larkin Papers* (Berkeley: University of California Press, 1951–1968), v. 7, 143, 232, 261–264; *Andres*

Castillero, op. cit., n. *47, 376–377*; *SFBul*, July 23, 1863; *NYT*, Nov. 12, 1869; Ascher, op. cit, n. *47*, 62, 102–110; Leonard Ascher, "Lincoln's Administration and the New Almaden Scandal," *Pacific Historical Review* 5 (Mar. 1936), 39–40; Milton H. Shutes, "Abraham Lincoln and the New Almaden Mine," *California Historical Society Quarterly* 15 (Mar. 1936), 4; Frank Tick, "The Political and Economic Policies of Robert J. Walker" (Ph.D. diss., University of California at Los Angeles, 1947), 1–5, 277–281; James P. Shenton, *Robert John Walker* (New York: Columbia University Press, 1961), 11–21, 127–128; *BDAC*, op. cit., n. 25, 1766; Harlan Hague and David J. Langum, *Thomas O. Larkin* (Norman: University of Oklahoma Press, 1990), 191–192; Coomes, op. cit., n. *47*, 125–130, 148–149.

52. *SFBul*, July 23, 1863; J. Ross Browne, "Down in the Cinnabar Mines, A Visit to New Almaden in 1865," *Harper's New Monthly Magazine* 31 (Oct. 1865), 550; *SFAC*, Oct. 26, Nov. 14, 1871; Ascher, op. cit, n. *47*, 102–113, 138; Shutes, op. cit., n. 51, 4; Shenton, op. cit., n. 51, 128, 178.

53. *SFAC*, Jan. 9, 1856, May 3, 1857, Oct. 31, Nov. 7, 1858, Jan. 27, Feb. 3, 5, 1859, June 3, 1860; *SFBul*, June 16, Nov. 16, 1855, May 21, Dec. 19, 1857, May 1, 1858, July 23, 1863; *Andres Castillero*, op. cit., n. *47*, 376–379, 864–865; Browne and Taylor, op. cit., n. 39, 176; *AJM*, Feb. 2, 1867; Bancroft, op. cit., n. 19, 554–555, 558; Ascher, op. cit, n. *47*, 75–87, 94–95; Ascher, op. cit, n. 51, 39–40; Shutes, op. cit., n. 51, 4–6; Samuel C. Wiel, *Lincoln's Crisis in the Far West* (San Francisco, 1949), 24.

54. *SFBul*, Nov. 26, Dec. 1, 1858, Jan. 13, 1859; *SFAC*, Feb. 3, 1859; *WNI*, Sept. 5–6, 1859; George Gordon, *Mining Titles: Are There Any, What Are They* (San Francisco: Towne and Bacon, 1859), 1–24; Quartz Miner, *Smart and Cornered, How to Get a Mine Without Finding, Opening or Working One* (San Francisco: Commercial, 1860), 1–31; United States vs. Andres Castillero. Andres Castillero vs. United States, 67 U.S. 17; Egleston, op. cit., n. 48, 901; Swisher, op. cit., n. 41, 786–791; Coomes, op. cit., n. *47*, 132–147.

55. *Bear River & Auburn Water & Mining Company. Incorporated May 1st. 1851. Capital Stock $650,000. Twenty Six Hundred Shares, Par Value $25*, stock certificate no. 2051, no date; *Tuolumne County Water Company. Incorporated September 1, 1852. Capital Stock 550,000 Dollars. 2200 Shares $250 Each*, stock certificate no. 560, Columbia, Apr. 21, 1854; *SacU*, Dec. 15, 1851; *SFAC*, Dec. 1, 1852, Jan. 21, June 14, 1854; *Placer Times* (San Francisco), Sept. 13, 1853; New York *Mining Magazine* 1 (Dec. 1853), 261, and (Aug. 1854), 187; Bancroft Scraps, op. cit., n. 35, 14–15; *SFBul*, Mar. 1, 1859; Lawrence, op. cit., n. 20, 32; Phelps, op. cit., n. 36, v. 1 (1881), 222–223; *NYT*, Nov. 27, 1898, Sept. 24, 1905, Apr. 18, 1914; Washington *Post*, Feb. 14, 1905; William Robert Kenny, "History of the Sonora Mining Region of California, 1848–1860" (Ph.D. diss., University of California, Berkeley, 1955), 298–303; Paul, op. cit., n. 9, 161–168; Towey, op. cit., n. 10, 85–86.

56. *SFAC*, Jan. 21, June 14, 1854, Mar. 17, 27–28, 1855, May 21, 1857, Jan. 23, 1860; *Placer Times* (San Francisco), Mar. 19, 1855; *SacU*, Apr. 9, 1856; *SFBul*, Apr. 15, Nov. 25, Dec. 1, 6, 1856, Feb. 9, 20, 1858, Nov. 23, 1859, Apr. 26, Dec. 10, 1860; San Francisco *Globe*, Apr. 12, June 2, 1857; San Francisco *Herald*, Nov. 17, 1858; *MSP*, Nov. 30, 1860; Herbert Lang, *A History of Tuolumne County* (San Francisco: Alley, 1882), 162–171; Paolo Sioli, *Historical Souvenir of El Dorado County* (San Francisco: printed by author, 1883),

101–103; Kenny, op. cit., n. 54, 443–452; Paul, op. cit., n. 9, 325; Susan Lee Johnson, *Roaring Camp* (New York: Norton, 2000), 250–258.

57. United States Census, 1850, California, El Dorado County, Dutch Creek, 4th December 1850, p. 426, line 25, Edward E. Mattison; "Mining for Gold in California," *Hutchings' California Magazine* 2 (July 1857), 12–13; "The Gold Fields of California," *United States Democratic Review* 40 (July 1857), 34–37; Bean, op. cit., n. 28, 62; Browne and Taylor, op. cit., n. 39, 22–23; Wells, op. cit., n. 33, 179; *Second Report of the State Mineralogist of California* (Sacramento: State Office, 1882), 115; Robert Kelley, *Gold vs. Grain* (Glendale, Calif.: Arthur H. Clark, 1959), 27–29; Philip May, *Origins of Hydraulic Mining in California* (Oakland, Calif.: Holmes, 1970), 40–41; Wagner, op. cit., n. 36, 23, 27; Powell Greenland, *Hydraulic Mining in California* (Spokane, Wash.: Arthur H. Clark, 2001), 33–36, 49–52, 157.

58. *State Register & Year Book of Facts for the Year 1857* (San Francisco: Langley, 1857), 230–236, and (1859) 275–284; *United States Democratic Review*, op. cit., n. 57, 33; Taliesin Evans, "Hydraulic Mining in California," *Century Illustrated Magazine* 25 (Jan. 1883), 331–334.

59. *The New Mexico Mining Company. Capital $500,000. No. of Shares 5,000. Par Value $100 per Share*, stock certificate no. 6, Washington City, June 6, 1855; *NYT*, Nov. 29, Dec. 24, 1856; *Private Land Claims in New Mexico* (Washington, D.C.: G.P.O., 1861), 51–65 (35th Cong., 2nd Sess., House Ex. Doc. 28); *The New Mexico Mining Company: Its Condition and Prospects* (Washington, D.C.: W. Koch, 1862), 2–7; *The New Mexico Mining Co. Preliminary Report* (New York: Baker and Godwin, 1862), 1–21; *BDAC*, op. cit., n. 25, 1413, 1510; John M. Townley, "Mining in the Ortiz Mine Grant Area" (M.A. thesis, University of Nevada, Reno, 1968), 67–73; John M. Townley, "The New Mexico Mining Company," *New Mexico Historical Review* 46 (Jan. 1971), 57–73; William Baxter, *The Gold of the Ortiz Mountains* (Santa Fe: Lone Butte, 2004), 64–77.

60. John R. Bartlett, *Personal Narrative of Explorations and Incidents in Texas, New Mexico, California, Sonora, and Chihuahua* (New York: Appleton, 1854), 227–235; *Journal of the Executive Proceedings of the Senate*, v. 10 (Jan. 24, 1856), 31 (Mar. 13, 1857), 254; *NYT*, Aug. 5, 1857; J. J. Bowden, *Spanish and Mexican Land Grants in the Chihuahuan Acquisition* (El Paso: Texas Western Press, 1971), 59–60; Edwin R. Sweeney, *Mangas Coloradas* (Norman: University of Oklahoma Press, 1998), 229–240, 282, 305; Robert L. Spude, "The Santa Rita del Cobre," *Mining History Journal* 6 (1999), 9–12.

61. *NYT*, May 11, 1853, Sept. 16, 1858; *Report of the Mowry City Association, Territory of Arizona, for 1859* (Palmyra, Mo.: J. Sosey, 1859), 6–7; United States Census, 1860, New Mexico, Dona Ana County, Hanover Copper Mines, 24th August 1860, p. 151, line 34, Sofio Henkel; "Mines and Mining Companies of Arizona," *The Merchant's Magazine and Commercial Review* 44 (Feb. 1861), 243; *Preliminary Report on the Eighth Census, 1860* (Washington, D.C.: G.P.O., 1862), 173; Sylvester Mowry, *Arizona and Sonora* (New York: Harper, 1864), 24; Browne and Taylor, op. cit., n. 39, 325; Francisco R. Almada, *Diccionario de Historia, Geografía y Biografía Chihuahuenses* (Chihuahua: Ediciones Universidad de Chihuahua, 1968), 503; Bowden, op. cit., n. 60, 59–60; Spude, op. cit.,

n. 60, 13–17; *The Handbook of Texas Online* (Texas State Historical Association, 2002), Simeon Hart, John Batiste LaCoste, and James R. Sweet.

62. *NYT*, Feb. 15, 28, Apr. 26, June 23, 1854; San Diego *Herald*, Sept. 29, 1855, May 3, Aug. 23, Sept. 27, Dec. 6, 1856; *NYT*, Apr. 17, Dec. 1, 1856, Apr. 15, 1857; "The Arizona Copper Mines," *The Merchant's Magazine and Commercial Review* 34 (June 1856), 759–760; Sylvester Mowry, *Memoir of the Proposed Territory of Arizona* (Washington, D.C.: H. Polkinhorn, 1857), 11, 19; Edward E. Dunbar, "Arizona and Sonora," letters in *NYT*, Jan. 5–6, 11, 1859; "Mines and Mining," op. cit., n. 61, 242; Thomas Edwin Farish, *History of Arizona* (Phoenix: 1915), v. 1, 278–279; Dan Rose, *The Ancient Mines of Ajo* (Tucson: Mission, 1936), 21–24; Morris Elsing and Robert Heineman, *Arizona Metal Production* (Tucson: University of Arizona, 1936), 97; Robert L. Spude, "Mineral Frontier in Transition: Copper Mining in Arizona, 1880–1885" (M.A. thesis, Arizona State University, 1976), 7–8; Stanley B. Keith et al., *Metallic Mineral Districts and Production in Arizona* (Tucson: Arizona Bureau of Geology, 1983), 16; Forrest R. Rickard, *Exploring, Mining, Leaching, and Concentrating of Copper Ores as Related to the Development of Ajo, Arizona* (Ajo, Ariz.: Rickard, 1996), 5–8.

63. *Railroad Record* (Cincinnati), Apr. 19, 1857; *Sonora and the Value of Its Silver Mines. Report of the Sonora Exploring and Mining Co.* (Cincinnati: Railroad Record, 1856), 1–7; *First Annual Report of the Sonora Exploring and Mining Company* (Cincinnati: Railroad Record, 1857), 1–2; *Second Annual Report of the Sonora Exploring & Mining Co.* (Cincinnati: Railroad Record, 1858), 3–12; *NYT*, Jan. 11, 1859, Aug. 3, 1860; "Mines and Mining," op. cit., n. 61, 242; Charles D. Poston, "Building a State in Apache Land, II. Early Mining and Filibustering," *Overland Monthly* 24 (Aug. 1894), 203–204, 210; Joseph F. Park, "The History of Mexican Labor in Arizona during the Territorial Period" (M.A. thesis, University of Arizona, 1961), 47–50; Mario De Blasio, "Sonora Exploring and Mining Co., 1856–1861" (M.A. thesis, University of San Diego, 1971), 7–12, 23–26, 33–49; Diane North, *Samuel Peter Heintzelman and the Sonora Exploring and Mining Company* (Tucson: University of Arizona Press, 1980), 10–11, 19–45.

64. *Sonora Exploring and Mining Co., Report of Frederick Brunckow* (Cincinnati: Railroad Record, 1859), i–iv, 16–18, 44; *NYT*, Nov. 10, 1859, Aug. 3, 1860, June 18, 1861; "Mines and Mining," op. cit., n. 61, 242; Mowry, op. cit., n. 61, 36, 58–59, 79–80; Phelps, op. cit., n. 36, v. 1 (1881), 272–280; North, op. cit., n. 63, 171–174; De Blasio, op. cit., n. 63, 63–66, 71–75, 83–95, 196; Diane North, " 'A Real Class of People' in Arizona: A Biographical Analysis of the Sonora Exploring and Mining Company, 1856–1863," *Arizona & the West* 26 (1984), 261.

65. *NYT*, Apr. 15, Dec. 2, 1857, June 21, Sept. 17, Oct. 7, 1859; *SFBul*, Mar. 25, 1860; "Mines and Mining," op. cit., n. 61, 242; Mowry, op. cit., n. 62, 1–30; Sylvester Mowry, "Arizona and Sonora," *Journal of the American Geographical and Statistical Society* 1 (Mar. 1859), 66–75; *Charter and By-Laws of the Arizona Land and Mining Company* (Providence, R.I.: State Printers, 1859), 1–18; Mowry, op. cit., n. 61, 37, 62, 73–84; Hiram C. Hodge, *Arizona As It Is* (New York: Hurd and Houghton, 1877), 126–128; Hubert H. Bancroft, *History of Arizona and New Mexico, 1530–1888* (San Francisco: History Co.,

1889), 505–506; Park, op. cit., n. 63, 17–18, 50; Benjamin Sacks, "Sylvester Mowry, Artilleryman, Libertine, Entrepreneur," *American West* 1 (1964), 14–24, 79; Constance W. Altshuler, "The Case of Sylvester Mowry: The Mowry Mine," *Arizona & the West* 15 (1973), 161–163; North, op. cit., n. 63, 32, 198–199.

66. "Mines and Mining," op. cit., n. 61, 242–243; *NYT*, Aug. 10, Oct. 5, 25, Dec. 20, 1861, Jan. 11, 13, July 15, 1862; *Information in Relation to the Seizure of the Silver Mine of Sylvester Mowry, in Arizona* (Washington, D.C.: G.P.O., 1864), 1–2; Mowry, op. cit., n. 61, 61–64; Bancroft, op. cit., n. 65, 511–517, 649, 688–700; Park, op. cit., n. 63, 61, 86–95; De Blasio, op. cit., n. 63, 96–104; Constance W. Altshuler, "The Case of Sylvester Mowry: The Charge of Treason," *Arizona & the West* 15 (1973), 63–82; Altshuler, op. cit., n. 65, 149–156; Spude, op. cit., n. 60, 17–20.

67. Howe, op. cit., n. 4, 187–217; Hughes, op. cit., n. 3, 78–85; Paul, op. cit., n. 9, 23–25, 43, 120–122,345–352; Blumnethal, op. cit., n. 10, 255–260; Towey, op. cit., n. 10, 109–115; U.S. Bureau of the Census, *Historical Statistics of the United States: Colonial Times to 1970* (Washington, D.C.: G.P.O., 1975), 993.

Chapter 3. The Silver Rage

1. William Wright, *History of the Big Bonanza* (Hartford, Conn.: American, 1876; reprinted New York: Alfred Knopf, 1947), 19–33; Eliot Lord, *Comstock Mining and Miners* (Washington, D.C.: G.P.O., 1883), 33–55; Ovando Hollister, *The Mines of Colorado* (Springfield, Mass.: Samuel Bowles, 1867), 59–88; Frank Fossett, *Colorado: A Historical, Descriptive and Statistical Work* (Denver: Tribune, 1878), 208–213; William Vickers, et al., *History of Clear Creek and Boulder Valleys, Colorado* (Chicago: Baskin, 1880), 206–211; Rodman W. Paul, *Mining Frontiers of the Far West, 1848–1880* (New York: Holt, Rinehart and Winston, 1963), 56–63, 109–114.

2. *SacU*, Aug. 2, Oct. 5, 28, Nov. 9, 1859; *SFBul*, Oct. 26, Nov. 11, 1859, Jan. 12, Feb. 25, Apr. 21, May 2, 1860, Feb. 1, 1861; *SFAC*, Apr. 13, 1860; *San Joaquin Republican* (Stockton, Calif.), May 5, 1860; *Gould & Curry Silver Mining Company. Capital Stock $2,400,000. 4,800 Shares, $500 Each*, stock certificate no. 1580. San Francisco, March 31st, 1864; J. Wells Kelly, *First Directory of Nevada Territory* (San Francisco: Valentine, 1862), 14–16; Myron Angel, *History of Nevada* (Oakland, Calif.: Thompson and West, 1881), 60–61; Alonzo Phelps, *Contemporary Biography of California's Representative Men* (San Francisco: A.L. Bancroft, 1882), v. 2, 9–13; Lord, op. cit., n. 1, 59–62; George Hearst, *The Way It Was* (New York: Hearst Corp., 1972), 16–18; Fremont and Cora Older, *The Life of George Hearst* (San Francisco: John Henry Nash, 1933), 96, 103–108; Grant H. Smith, *The History of the Comstock Lode, 1850–1920* (Reno: Nevada State Bureau of Mines, 1943), 80–86; Judith Robinson, *The Hearsts: An American Dynasty* (Newark: University of Delaware Press, 1991), 46–47; Sally Zanjani, *Devils Will Reign: How Nevada Began* (Reno: University of Nevada Press, 2006), 111–115, 125.

3. *SFBul*, Feb. 27, Apr. 21, May 11, 1860, Feb. 1, 1861; *SFAC*, Apr. 11, 15, May 13, 1860; *Saturday Evening Post*, Nov. 14, 1863; Thomas Fitch, "Nevada," *Harper's New Monthly*

Magazine 31 (Aug. 1865), 321; J. Ross Browne and James Taylor, *Report Upon the Mineral Resources of the United States* (Washington, D.C.: G.P.O., 1867), 30–31; Lord, op. cit., n. 1, 415.

4. Browne and Taylor, op. cit., n. 3, 30, 111–117; Bancroft Scraps, v. 52 (California Mining Stocks), 16, 35, 48–49; Lord, op. cit., n. 1, 416–418; Bertrand Couch and Jay Carpenter, *Nevada's Metal and Mineral Production (1859–1940)* (Reno: University of Nevada, 1943), 13, 133.

5. Kelly, op. cit., n. 2, 14–16: *SFBul*, Mar. 21, 1863; *SFAC*, Oct. 29, 1863; Bancroft Scraps, op. cit., n. 4, 11, 14, 23; Franklin Tuthill, *The History of California* (San Francisco: H.H. Bancroft, 1866), 607–608; Charles H. Shinn, *The Story of the Mine* (New York: Appleton, 1896), 144–145; Maureen A. Jung, "The Comstocks and the California Mining Economy, 1848–1900" (Ph.D. diss., University of California, Santa Barbara, 1988), 72–73.

6. *MSP*, Mar. 16, July 13, 1863, July 2, 1864; Browne and Taylor, op. cit., n. 3, 294; Bancroft Scraps, op. cit., n. 4, 19; Alfred Doten, *The Journals of Alfred Doten, 1849–1903* (Reno: University of Nevada Press, 1973), 696.

7. *SFAC*, Aug. 3, 1863; *MSP*, Jan. 2, 30, 1864; Browne and Taylor, op. cit., n. 3, 30; Bancroft Scraps, op. cit., n. 4, 16, 29; William R. Balch, *The Mines, Miners and Mining Interests of the United States in 1882* (Philadelphia: Mining Industrial Publishing Bureau, 1882), 85; Lord, op. cit. n. 1, 416; Shinn, op. cit., n. 5, 145; Couch and Carpenter, op. cit., n. 4, 133; U.S. Bureau of the Census, *Historical Statistics of the United States: Colonial Times to 1970* (Washington, D.C.: G.P.O., 1975), 25, 31.

8. *SFBul*, May 5, 1860; California Secretary of State, Division of Archives, "General Index of Corporations Filed and Recorded 1851–1889, Book 1, A–Z," Sacramento; *Webster Gold and Silver Mining Co. Incorporated May 1863. Capital Stock $280,000, 1400 Shares $200 Each*, stock certificate no. 64, San Francisco, June 12, 1863; "List of Mining Companies Having their Offices in San Francisco," published in *MSP*, 1863, reprinted by Douglas and Gina McDonald, *Mines of the West, 1863* (Helena, Mont.: Gypsyfoot, 1996), 14–64. A large sampling of certificates is reproduced in Frank Hammelbacher, *A Treasury of Mining Stocks from the Nevada Territory, 1861–1864* (Flushing, N.Y.: Norrico, 1996) 184 pp, including the *Fly By Night Gold & Silver Mining Company. Incorporated March 28, 1863. Capital Stock, $180,000, One Foot to the Share, 1,800 Shares, $100 each*, stock certificate no. 56, May 12, 1863.

9. *MSP*, Feb. 23, Nov. 9, 1863, Jan. 28, Feb. 4, 1865; San Francisco *Golden Era*, Feb. 5, 1865; Frederick Henry Smith, *Rocks, Minerals and Stocks* (Chicago: Railway Review, 1882), 185–188; also on Silliman, see Clark Spence, *British Investments and the American Mining Frontier, 1860–1901* (Ithaca, N.Y.: Cornell University Press, 1958), 142–143; Clark Spence, *Mining Engineers and the American West* (New Haven, Conn.: Yale University Press, 1970), 101, 133.

10. *SFAC*, Aug. 20, 1863; Tuthill, op. cit., n. 5, 607–608; Smith, op. cit., n. 9, 185–188.

11. Doten, op. cit., n. 6, 740, 889, 899, 952, 973, 987, 998, 1008, 1015, 1034, 1036.

12. *SFBul*, May 11, 1860, Oct. 23, 1862, Jan. 15, May 21, Nov. 4, 6, 1863, May 23, June

13, 1864; Browne and Taylor, op. cit., n. 3, 32; Lord, op. cit., n. 1, 146–178; Hubert H. Bancroft, *History of Nevada, Colorado, and Wyoming, 1540–1888* (San Francisco: History Co., 1890), 122–129, 172–175; Shinn, op. cit., n. 5, 123–135; Smith, op. cit., n. 9, 64–74; Bruce Alverson, "The Limits of Power: Comstock Litigation, 1859–1864," *Nevada Historical Society Quarterly* 43 (Spring 2000), 74–99.

13. R. S. Lawrence, ed., *Pacific Coast Annual Mining Review and Stock Ledger* (San Francisco: Francis and Valentine, 1878), 30–31; Phelps, op. cit., n. 2, v. 1 (1881), 165–169; Wiliiam M. Stewart, *Reminiscences of Senator Wiliiam M. Stewart of Nevada* (New York: Neale, 1908), 138, 152–163; Clement T. Rice's appraisal of Stewart, quoted in Samuel L. Clemens, *Mark Twain of the Enterprise: Newspaper Articles & Other Documents, 1862–1864*, ed. Henry Nash Smith (Berkeley: University of California Press, 1957), 62–63; BDAC, 1658; Russell Elliott, *Servant of Power: A Political Biography of Senator William M. Stewart* (Reno: University of Nevada Press, 1983), 18–41.

14. *SFBul*, July 27, 1864; *MSP*, July 30, 1864; *SFC*, Aug. 19, 1864; Browne and Taylor, op. cit., n. 3, 32; Lord, op. cit. n. 1, 172–173; Samuel L. Clemens, *Roughing It* (Hartford, Conn.: American, 1872), 194; *Clemens of the Call, Mark Twain in San Francisco* (Berkeley: University of California Press, 1969), 239–240; Jung, op. cit., n. 5, 87–88.

15. *SFAC*, July 25, 1863; *SFBul*, Aug. 3–4, 1863; *MSP*, Sept. 28, 1863; Lawrence, op. cit., n. 13, 181, 186, 200; Phelps, op. cit., n. 2, v. 1 (1881), 169–172; Lord, op. cit., n. 1, 132–133, 418–420; *LAT*, Aug. 10, 1903 (Nickerson obit); J. Edward Johnson, *History of the Supreme Court Justices of California*, v. 1, 1850–1900 (San Francisco: Bender-Moss, 1963), 13–30.

16. Allen and Hosea Grosh, letters to their father, Aaron Grosh, July 31 [Dec. 3], 1853, Nov. 8, 1854, Mar. 31, Nov. 3, 22, 1856, and "Articles of Association of the Utah Enterprize Mining Company," (Jan. 8, 1857), transcribed from the originals in the collection of Charles T. Wegman, courtesy of Fred Holabird; Francis J. Hoover, letter Aug. 7, 1860, in *SFAC*, Mar. 16, 1884, and "True History of the Discovery of Silver in Washoe," Sept. 9, 1863, in *SFC*, May 6, 1894; Aaron Grosh, letters to Maurice Bucke, June 22, 1864, transcript in "Letters of A. B. Grosh to R. M. Bucke," pp. 146–147, MSS P-G 267, Bancroft Library, University of California, Berkeley; Wright, op. cit., n. 1, 14–16; Angel, op. cit., n. 2, 51–54, 498; Lord, op. cit., n. 1, 26–27; Shinn, op. cit., n. 5, 26–30; Smith, op. cit., n. 2, 4; Zanjani, op. cit., n. 2, 44–48.

17. Allen and Hosea Grosh, letters to their father, May 21, June 8, Aug. 16, 23, Sept. 7, 11, 1857, and R. Maurice Bucke, letter to Aaron B. Grosh, Feb. 10, 1858, in Wegman collection, op. cit., n. 16; E. Allen Grosh, letter to "Dr. Gov. [Francis J. Hoover] & Friend," Dec. 12, 1857, in "Letters," op. cit., n. 16, 1–4; *SFBul*, Dec. 29, 1857; Hoover, op. cit., n. 16; *MSP*, July 22, 1876; Angel, op. cit., n. 2, 51–54; Lord, op. cit., n. 1, 27–32; Richard M. Bucke, "Twenty Five Years Ago," *Overland Monthly* 1 (June 1883), 553–560; Bancroft, op. cit., n. 12, 96–99; Shinn, op. cit., n. 5, 28–34; Zanjani, op. cit., n. 2, 83–91.

18. *SFBul*, Oct. 26, Nov. 11, 1859, Sept. 30, 1870 (Comstock obit); Wright, op. cit., n. 1, 19–33, 42–52; Angel, op. cit., n. 2, 55–60; Lord, op. cit., n. 1, 33–40, 57–61; Bancroft, op. cit., n. 12, 98–107; Shinn, op. cit., n. 5, 33–47; Smith, op. cit., n. 2, 5–11; Zanjani, op. cit., n. 2, 113–115.

19. Aaron Grosh, letters to Maurice Bucke, May 7, Dec. 14, 1860, Feb. 12, 1861, Nov. 18, 22, 1863, Mar, 31, 1864, in "Letters," op. cit., n. 16, 35–37, 48–53, 127–134, 140–145; Hoover, op. cit., n. 16; Grosh Consolidated Gold & Silver Mining Company, "Certificate of Incorporation," July 24, 1863, and "Certificate of Increase of Capital," Mar. 26, 1864, California Secretary of State, Sacramento, courtesy of Fred Holabird; SFAC, July 25, Aug, 5, Oct. 20, 1863; SFBul, Aug. 3–4, Oct. 19, 1863; SacU, Aug, 4, 17, Sept. 2, 1863; Benjamin R. Nickerson, Statement of the Grounds of the Claim of the Grosh Consolidated Gold & Silver Mining Company, to the Comstock Mine in Nevada Territory, Together with Their Reply to the Attacks of the Press (San Francisco: Towne and Bacon, 1863), 1–11; MSP, Sept. 28, 1863; Bancroft Scraps, v. 94 (Nevada Mining), 109–110, 126–127, 154–155; Lord, op. cit., n. 1, 132–133.

20. SFBul, Aug. 3, Oct. 19, Dec. 5, 1863; SacU, Aug. 4, 17, Sept. 2, 1863; SFAC, Aug. 5, Oct. 20, 1863; Nickerson, op. cit., n. 19, 10–11, 21–22; Bancroft Scraps, op. cit., n. 19, 109–110, 126–127, 154–155; Lord, op. cit., n. 1, 132–133.

21. Aaron Grosh, letters to Maurice Bucke, May 7, Dec. 14, 1860, Feb. 12, 1861, Nov. 18, 22, 1863, in "Letters," op. cit., n. 16, 30–37, 48–53, 127–128, 133–134; SFBul, Aug. 3, 1863; SacU, Aug. 4, 17, Sept. 2, Oct. 26, 1863; SFAC, Aug. 5, Oct. 20, 27, 1863; Nickerson, op. cit., n. 19, 11–22; MSP, Sept. 28, 1863; Bancroft Scraps, op. cit., n. 19, 109–110, 126–127, 154–155; Lord, op. cit., n. 1, 132–133.

22. E. Aaron Grosh, letter to "Dr. Gov.," op. cit., n. 16, 1–4; SacU, Aug. 17, Sept. 2, 1863; Nickerson, op. cit., n. 19, 1–22; Bancroft Scraps, op. cit., n. 19, 109–110, 126–127; Gold Hill, Nev., News, Jan. 6, June 28, 1865; SFBul, Mar. 9, 1865; Mountain Democrat (Placerville, Calif.), Aug. 3, 24, 1867; Angel, op. cit., n. 2, 54; Lord, op. cit., n. 1, 30–31, 132–133.

23. Henry G. Langley, comp., The San Francisco Directory for . . . 1861 (San Francisco: Valentine, 1861), 263, 353, (1863) 317, 376, (1864) 350, 415; SFBul, Aug. 27, 1862, Nov. 3, 6, 1863, Feb. 12, 1864, Feb. 14–18, 20–23, 25, 1865; SFAC, Nov. 3, 5, Dec. 6, 1863; Sacramento Bee, Nov. 4, 1863; Bancroft Scraps, op. cit., n. 19, 24–25, 162–163.

24. SFBul, Nov. 3, 1863, Feb. 14–18, 20–23, 25, Apr. 25, 27–28, 1865; SFAC, Nov. 3, 5, Dec. 6, 1863, Feb. 14–16, 18–19, 21–22, 24, 1865; Sacramento Bee, Nov. 4, 1863; Bancroft Scraps, op. cit., n. 19, 24–25, 162–163; NYT, Aug. 28, 1865 (Tuthill obit); MSP, Nov. 4, 1876.

25. Visalia Delta, Jan. 19, 1861; SFAC, Mar. 14, Apr. 11–13, 1861; Richard E. Lingenfelter, Death Valley & the Amargosa (Berkeley: University of California Press, 1986), 65–66.

26. SFAC, Apr. 11–13, July 18, 1861, Oct. 2, 1864; MSP, July 6–Aug. 20, 1861; SFBul, July 26–27, Aug. 6, 1861, Jan. 26, 1864; Los Angeles, Star, July 27, Aug. 10, Sept. 21–28, 1861; Proceedings of the California Academy of Natural Sciences (1862), 142; SacU, Mar. 6, 9, 11, 1863; SFCh, Nov. 25, 1865; J. Ross Browne, Adventures in Apache Country (New York: Harper, 1868), 517–518; Clemens, Roughing It, op. cit., n. 14, chap. 44; An Illustrated History of Southern California (Chicago: Lewis, 1890), 199–201; San Francisco Mining & Engineering Review, Sept. 5, 1903; Clemens, op. cit., n. 13, 311–312; Lingenfelter, op. cit.,

n. 25, 66–68; Samuel L. Clemens, *Mark Twain's Letters*, v. 1, 1853–1866, ed. Edgar M. Branch (Berkeley: University of California Press, 1988), 260–261.

27. *SFBul*, Sept. 8, 1866; Bancroft Scraps, op. cit., n. 4, 42; Doten, op. cit., n. 6, 721, 730; *Honest Miner Gold and Silver Mining Co.*, *1400 Shares, $1,000 Each*, stock certificate no. 108, San Francisco, Sept. 22, 1863, ; Hammelbacher, op. cit., n. 8, 1ff.

28. Bancroft Scraps, op. cit., n. 4, 13–14.

29. Clemens, op. cit., n. 14, 306–309; Clemens, op. cit., n. 26, 244–253, 260–261.

30. Langley, op. cit., n. 23, (1862) 63, 160, (1863) 62, 151, (1864) 63, 164; *MSP*, Oct. 22, 1864; *SFBul*, Oct. 29, Nov. 1, 1864, Nov. 9, 1865; Bancroft Scraps, op. cit., n. 19, 276–277.

31. *MSP*, Oct. 22, 1864; *SFBul*, Oct. 29, Nov. 1, 1864, Nov. 9, 1865; Bancroft Scraps, op. cit., n. 19, 276–277.

32. *SFAC*, Oct. 28, Nov. 2, 1864; *SFBul*, Oct. 29, Nov. 1, 1864; *MSP*, Nov. 5, 1864, Nov. 9, 1865; Bancroft Scraps, op. cit., n. 19, 276–277.

33. *SFAC*, Oct. 28, Nov. 2, 1864, Jan. 13–14, Mar. 25–26, 1865; *SFBul*, Oct. 29, Nov. 1, 1864, Nov. 9, 16, 1865; *MSP*, Nov. 5, 1864; Bancroft Scraps, op. cit., n. 19, 276–278.

34. *MSP*, Mar. 23, Aug. 27, 1863; *SFAC*, Oct. 28, 1863; Langley, op. cit., n. 21, (1863), 404–405; Lisle Lester, "The Brokers of San Francisco," *Pacific Monthly* 10 (Jan. 1864), 337–340; *SFBul*, Sept. 8, 1866; Bancroft Scraps, op. cit., n. 4, 42; Tuthill, op. cit., n. 5, 607.

35. *MSP*, Oct. 20, Dec. 21, 1860, Jan. 18, 1861; *SFBul*, Sept. 12, 1862; Bancroft Scraps, op. cit., n. 4, 10–11, 20–21; *Constitution and By-laws of the San Francisco Stock and Exchange Board. Organized September 11, 1862* (San Francisco: Agnew and Deffebach, 1865), 1–20; Lawrence, op. cit., n. 13, 5–8; Hubert H. Bancroft, *History of California*, v. 7, 1860–1890 (San Francisco: History Co., 1891), 666–669; Joseph L. King, *History of the San Francisco Stock and Exchange Board* (San Francisco: printed by author, 1910), 373 pp.; John P. Young, *Journalism in California* (San Francisco: Chronicle, 1915), 77; W. John Carlson, "A History of the San Francisco Mining Exchange" (M.A. thesis, University of California, Berkeley, 1941), 7ff; Marian V. Sears, *Mining Stock Exchanges, 1860–1930: An Historical Survey* (Missoula: University of Montana Press, 1973), 19; Jung, op. cit., n. 5, 76, 144.

36. *SFBul*, Sept. 12, Oct. 29, 1863; Bancroft Scraps, op. cit., n. 4, 20, 23; *VCTE*, Oct. 1863, reprinted in Jackson *Amador Ledger*, Oct. 31, 1863, and in Paul Fatout, *Mark Twain in Virginia City* (Bloomington: Indiana University Press, 1964), 93; Theodore H. Hittell, *History of California* (San Francisco: Stone, 1897), v. 4, 542–543.

37. *MSP*, Jan. 19, Mar. 9–23, Apr. 20, Aug. 24, Nov. 2, 1863, Jan. 2, 16–30, Feb. 13, 27, Apr. 16, 1864; Gold Hill, Nev., *News*, Feb. 6, Mar. 1–3, 1864; Virginia City, Nev., *Union*, Mar. 22, 1864; Grass Valley, Calif., *National*, Mar. 19, 1864, Jan. 17, 1865; *SFBul*, Apr. 8, 1864, Sept. 8, 1866; Bancroft Scraps, op. cit., n. 4, 42; Doten, op. cit., n. 6, 770–772; Fitch, op. cit., n. 3, 318; Sears, op. cit., n. 35, 9–10, 17–23.

38. *SFBul*, Apr. 25, 1860, Sept. 24, 1862; Lester, op. cit., n. 34, 338; Lawrence, op. cit, n. 13, 11; James K. Medbery, *Men and Mysteries of Wall Street* (New York: R. Worthington, 1878), 48.

39. Bancroft Scraps, op. cit., n. 4, v. 53, 520; Benjamin E. Lloyd, *Lights and Shades in San Francisco* (San Francisco: A. L. Bancroft, 1876), 44–45; Medbery, op. cit., n. 38, 54–63; Smith, op. cit., n. 9, 194–196; Hittell, op. cit., n. 36, v. 4, 552–553; Jung, op. cit., n. 5, 73.

40. Bancroft Scraps, op. cit., n. 4, 3–4.

41. *SFAC*, Apr. 14, May 13, 1860; San Francisco *Californian*, Aug. 13, 1864; Doten, op. cit., n. 6, 733.

42. Virginia City, Nev., *Union*, Mar. 22, 1864; Doten, op. cit., n. 6, 733.

43. *MSP*, May 4, 1863; Clemens, op. cit., n. 13, 320; New York *Mining Herald*, June 1904.

44. *American Mining Gazette* (New York) 1 (June 1864), 195–197; *MSP*, Dec. 3, 17, 1864; Hollister, op. cit., n. 1, 131–132; Medbery, op. cit., n. 38, 278–279; Fossett, op. cit., n. 1, 133; Willard L. Thorp, *Business Annals: United States, England, &c* (New York: National Bureau of Economic Research, 1926), 128–129; Sears, op. cit., n. 35, 10, 29–31; Joseph E. King, *A Mine to Make a Mine: Financing the Colorado Mining Industry, 1859–1902* (College Station: Texas A&M University Press, 1977), 12–13.

45. *NYT*, Mar. 25, Apr. 20, 22, Aug. 14, 1865, Jan. 22, 1866; Medbery, op. cit., n. 38, 283–285.

46. *Manhattan Silver Mining Company of Nevada. Capital $1,000,000. Shares $100 Each*, stock certificate no. 410, New York, Jan. 10, 1868; Angel, op. cit., n. 2, 465, 468–469, 472; Donald R. Abbe, *Austin and the Reese River Mining District* (Reno: University of Nevada Press, 1985), 25–28.

47. *NYT*, Feb. 17, 1866; J. Ross Browne, *Report on the Mineral Resources of the States and Territories West of the Rocky Mountains* (Washington, D.C.: G.P.O., 1868), 398; Rossiter W. Raymond, *Statistics of Mines and Mining in the States and Territories West of the Rocky Mountains* (Washington, D.C.: G.P.O., 1872), 3rd rpt., 112–114, 4th rpt. (1873), 170–171, 5th rpt. (1873), 136–142; *MSP*, Feb. 13, 1875; Lawrence, op. cit., n. 13, 214–215; *EMJ*, Feb. 22, 1879; Angel, op. cit., n. 2, 283–284, 468–469; Couch and Carpenter, op. cit., n. 4, 77; Abbe, op. cit., n. 46, 28–31.

48. *NYT*, Aug. 30, 1856, Oct. 7, 10, 1861; *SFBul*, Aug. 7, 1858, July 13–14, 1859; Baltimore *Sun*, Jan. 1, 1859; *San Joaquin Republican* (Stockton, Calif.), May 28, 1859; Browne, op. cit., n. 47, 21–30; *The Mariposa Estate: Its Past, Present and Future. . . . With a Review of the Causes Which Led to the Late Failure* (New York: Russells, 1868), 43–44; *Biographical Encyclopaedia of Vermont of the Nineteenth Century* (Boston: Metropolitan, 1885), 197–201; Lewis C. Aldrich, *History of Bennington County, Vt.* (Syracuse, N.Y.: Mason, 1889), 518–522; Mary Lee Spence and Donald Jackson, *The Expeditions of John Charles Fremont* (Urbana: University of Illinois Press, 1973), v. 3, lix–lxx; Pamela Herr and Mary Lee Spence, *The Letters of Jessie Benton Fremont* (Urbana: University of Illinois Press, 1993), 284, 295–298; see also Allan Nevins, *Fremont, Pathmarker of the West* (New York: Appleton-Century, 1939), 493–540.

49. *San Joaquin Republican* (Stockton, Calif.), Sept. 22, 1860; *SFBul*, June 25, 1862; *The Mariposa Company, 34 Wall Street, New York. Organized 25th June, 1863* (New York:

Bryant, 1863), 17–26; Benjamin Silliman, *A Report of an Examination of the Mariposa Estate* (New York: Bryant, 1864), 9–10; *The Mariposa Estate*, op. cit., n. 48, 3–10, 45–46; Browne, op. cit., n. 47, 21–30; Edwin L. Godkin, "How the Thing Is Done," *The Nation* 8 (June 24, 1869), 488–490; Nevins, op. cit., n. 48, 583–587; C. Gregory Crampton, "The Opening of the Mariposa Mining Region, 1849–1859, with Particular Reference to the Mexican Land Grant of John Charles Fremont" (Ph.D. diss., University of California, Berkeley, 1941), 268–271; Robin W. Winks, *Frederick Billings: A Life* (New York: Oxford University Press, 1991), 152–157; Pamela Herr and Spence, op. cit., n. 48, 281, 346–347.

50. *SFBul*, Feb. 12, 1863, Feb. 2, 1865; *SFAC*, Oct. 30, 1863, Feb. 27, 1864; *NYT*, Feb. 4, Apr. 14, May 3, Dec. 22, 1864, Aug. 21, Oct. 29, Dec. 1, 1865; *The Great Libel Case. Geo. Opdyke agt. Thurlow Weed* (New York: American News, 1865), 15, 57–64; *Proceedings of the Meeting of the Bondholders of the Mariposa Company, May 20th, 1868, and Report of the Committee of Investigation* (New York: New York Printing, 1868), 1–15; Browne, op. cit., n. 47, 21–30; *The Mariposa Estate*, op. cit., n. 48, 45–50; Godkin, op. cit., n. 49, 488–489; Charles F. and Henry Adams, *Chapters of Erie, and Other Essays* (Boston: James R. Osgood and Co., 1871), 100–134; *NYTr*, Feb. 5, 1876; Fort Wayne *Sentinel*, Oct. 21, 1890; Kenneth D. Ackerman, *The Gold Ring: Jim Fisk, Gould, and Black Friday, 1869* (New York: Dodd, Mead and Co., 1988); Frederick Law Olmsted, *The Papers of Frederick Law Olmsted*, v. 5, *The California Frontier* (Baltimore: Johns Hopkins University Press, 1990), 196–199, 219–221; Herr and Spence, op. cit., n. 48, 346–347. Although Fremont technically started with a five-eighths share of the company and Park and two other creditors, Frederick Billings and Abia Selover, each held one-eighth, Fremont had to give 29 percent of the total to Opdyke, Ketchum, Hoy, and others for the promotion, leaving him with roughly one-third. Of that Ketchum withheld a full one-fourth share, leaving Fremont with just 8.5 percent until he finally sued Ketchum. The company never got title to the Mariposa because Fremont's debts totaled nearly $2,500,000, payable in gold, and the bond sales yielded only $1,500,000 in greenbacks, then worth only about half their face value in gold.

51. Godkin, op. cit., n. 49, 488–489.

52. *SFAC*, Jan. 9, 1856, Jan. 17, 20, 1861; *SFBul*, Dec. 19, 1857, July 16, 1858; *Andres Castillero, Claimant, vs. The United States, Defendant, for the Place Named New Almaden* (San Francisco: Whitton, Towne, 1858), 864–869; *WNI*, Sept. 5–6, 1859; Henry Halleck, *A Collection of Mining Laws of Spain and Mexico* (San Francisco: O'Meara and Painter, 1859); *The United States vs. Andres Castillero, No. 420. "New Almaden." Transcript of the Record* (San Francisco, 1860–1861), 3580–3581; *NYT*, Sept. 6, 10, 1860, Feb. 13–14, 1861, Jan. 10, 1872; Leonard W. Ascher, "The Economic History of the New Almaden Quicksilver Mine, 1845–1863" (Ph.D. diss., University of California, Berkeley, 1934), 85–90.

53. *NYT*, Mar. 3, 11, Aug. 18, 1863; Aug. 20, 1883; *SFBul*, Mar. 11, July 22–23, 1863; *SFAC*, July 13, 23, 1863; Ascher, op. cit., n. 34, 131–132; Leonard Ascher, "Lincoln's Administration and the New Almaden Scandal," *Pacific Historical Review* 5 (Mar. 1936), 39–40;

Milton H. Shutes, "Abraham Lincoln and the New Almaden Mine," *California Historical Society Quarterly* 15 (Mar. 1936), 6, 14–18; Samuel C. Wiel, *Lincoln's Crisis in the Far West* (San Francisco, 1949), 93–98; Frank Tick, "The Political and Economic Policies of Robert J. Walker" (Ph.D. diss., University of California at Los Angeles, 1947), 277–281.

54. *NYT*, Jan. 29, Aug. 18, 1863, Aug. 20, 1883; *SFAC*, July 14, 23, 1863; *SFBul*, July 23, 1863; Edward Bates, *The Diaries of Edward Bates, 1859–1866* (Washington, D.C.: G.P.O., 1933), 338; Ascher, op. cit., n. 52, 131–132; Shutes, op. cit., n. 53, 6; William H. Brewer, *Up and Down California in 1860–1864* (Berkeley: University of California Press, 1960), 160.

55. *NYT*, Jan. 29, Mar. 3, 10–12, Aug. 16, 18, 1863; *SFAC*, Apr. 13, 21, July 11, 14, 23, 1863; *SFBul*, July 22–23, 1863; Hubert H. Bancroft, *Annals of the California Gold Era* (San Francisco: Bancroft, 1886), 551, 555–561; *Diary of Gideon Welles* (Boston: Houghton Mifflin, 1911), v. 2, 390, and v. 3, 307 (quoting Seward); Bates, op. cit., n. 54, 338–342, 354, 356; Ascher, op. cit., n. 52, 100–101; Shutes, op. cit., n. 53, 14–18, 20 (n.66); Wiel, op. cit., n. 53, 35–39, 94–95; James P. Shenton, *Robert John Walker: A Politician from Jackson to Lincoln* (New York: Columbia University Press, 1961), 192; Carl B. Swisher, *History of the Supreme Court of the United States*, v. 5, *The Taney Period 1836–64* (New York: Macmillan, 1974), 792–795.

56. *NYT*, Mar. 11–12, Apr. 9, 20, Aug. 16, 18, 1863, May 8, 1875; *SFAC*, July 11, 13–14, 23, 1863, May 6, 1875; *SFBul*, July 11, 14, 18, Aug. 5, 1863; United States vs. Andres Castillero. Andres Castillero vs. The United States, 67 U.S. 17; Abraham Lincoln, *The Complete Works of Abraham Lincoln*, eds. John G. Nicolay and John Hay (New York: Tandy-Thomas, 1905), v. 9, 85–86, 124; War Department, *The War of the Rebellion: A Compilation of the Official Records of the Union and Confederate Armies* (Washington, D.C.: G.P.O., 1897), ser. I, v. L, pt. 2, 514–518, 523–524, 574; Bates, op. cit., n. 52, 303–304; Ascher, op. cit., n. 52, 115–124; Ascher, op. cit., n. 53, 43–46; Shutes, op. cit., n. 53, 7–13; Carl Sandburg, *Abraham Lincoln, The War Years* (New York: Harcourt, 1939), v. 2, 613–617; Tick, op. cit., n. 53, 285–292; Wiel, op. cit., n. 53, 24–27, 36–41, 44–46, 51–70, 75–78; Swisher, op. cit., n. 55, 795–798; Mary L. Coomes, "From Pooyi to the New Almaden Mercury Mine: Cinnabar, Economics, and Culture in California to 1920" (Ph.D. diss., University of Michigan, 1999), 117–119.

57. *SFAC*, Apr. 21, July 13–14, 23, Aug. 29, 1863; *SFBul*, July 11, 14, 18, 1863; *NYT*, Aug. 29, Oct. 29, 1863; *United States v. Andres Castillero*, op. cit., n. 56; *The Quicksilver Mining Company* (New York: Amerman, 1864), 1–9, 20–29; War Department, op. cit., n. 56, 574; Ascher, op. cit., n. 52, 113–114; Shutes, op. cit., n. 53, 11–13; Wiel, op. cit., n. 54, 82–87; Coomes, op. cit., n. 56, 117–119.

58. *Quicksilver Mining Company*, op. cit., n. 57, 1–15; *SFAC*, Aug. 29, 1863; *NYT*, Aug. 29, Oct. 29, 1863, Jan. 16, 1865, May 3, 12, 18, 1866; Langley, op. cit., n. 23, (1864), xciii, 61, 91; J. Ross Browne, "Down in the Cinnabar Mines, A Visit to New Almaden in 1865," *Harper's New Monthly Magazine* 31 (Oct. 1865), 549–550; MSP, May 18, 1867; Titus Cronise, *The Natural Wealth of California* (San Francisco: A. L. Bancroft, 1868), 590–591; Raymond, op. cit., n. 47, 3rd rpt. (1872), 15–16; Lawrence, op. cit., n. 13, 40; Thomas Egleston, *The Metallurgy of Silver, Gold, and Mercury in the United States* (New

York: John Wiley, 1890), v. 2, 901; *LAT*, Oct. 17, 1892 (Bell obit); Jimmie Schneider, *Quicksilver: The Complete History of Santa Clara County's New Almaden Mine* (San Jose, Calif.: Zelda Schneider, 1992), 39–42, 115–118.

59. *MSP*, May 18, 1867, Jan. 16, 1869, May 4, 1872, Jan. 4, Sept. 6, 1873; *NYT*, May 27, Nov. 13, 1869, Mar. 25, 1870, May 8, 1875, Nov. 17, 1877, Sept. 17, 1879; *SFAC*, Oct. 26, Nov. 14, 1871, May 6, 12, 1875; Raymond, op. cit., n. 47, 3rd rpt. (1872), 15–16; Egleston, op. cit., n. 56, 901; Oscar T. Shuck, *Historical Abstract of San Francisco* (San Francisco: 1897), v. 1, 20; *SFCh*, Dec. 24, 1903, Apr. 8, 1904 (Randol obit); Rodman W. Paul, *California Gold* (Cambridge, Mass.: Harvard University Press, 1947), 274–277; Shenton, op. cit., n. 55, 207–216; Schneider, op. cit., n. 58, 42.

60. *MSP*, May 4, 1872; Raymond, op. cit., n. 47, 3rd rpt. (1872), 15–16, 8th rpt. (1877), 20–21; *CMR*, Apr. 2, 1881; Phelps, op. cit. n. 2, v. 1 (1881), 217–219.

61. *SFBul*, Dec. 31, 1858, Jan. 31, Apr. 25, 1860; *NYT*, Jan. 28, 1859, Feb. 25, 1860; *Merchants' Magazine and Commercial Review* 44 (June 1861), 702–703; Charlene Gilbert, "Prospectors, Capitalists, and Bandits: The History of the New Idria Quicksilver Mine, 1854–1972" (M.A. thesis, San Jose State University, 1984), 5–22.

62. *NYT*, July 7, 1854, Feb. 20, 1881, May 13, 1900; *The United States, Appellant, vs. Vicente P. Gomez* (Washington, D.C.: G.P.O., 1866), 1–13; *SFBul*, Jan. 21, May 26, July 15, 1868; *The History of the McGarrahan Claim, As Written by Himself* (Washington, D.C., 1878), i–vi, 395–411; *Irish World* (New York), Aug. 13, 1892; Robert J. Parker, "William McGarrahan's 'Panoche Grande Claim,' " *Pacific Historical Review* 5 (Sept. 1936), 214–217; Gilbert, op. cit., n. 61, 26–30.

63. *NYT*, Feb. 6, 1864, Feb. 20, 1881; *SFBul*, May 26, July 15, 1868; *Congressional Globe*, Feb. 18, 1871, 1404–1408; *Memorial of William McGarrahan* (Washington, D.C.: G.P.O., 1878) (45th Cong., 2nd Sess., Senate Misc. Doc. 85), 950–951; *History*, op. cit., n. 62, 73, 81, 128; Parker, op. cit., n. 62, 218–219; Gilbert, op. cit., n. 61, 31–32.

64. *American Mining Gazette* (New York) 1 (Aug. 1864), 333; *SFBul*, May 26, July 15, 1868; *Congressional Globe*, Feb. 18, 1871, 1409; *NYT*, Feb. 20, 1881; *Memorial*, op. cit., n. 63, 736–737; *History*, op. cit., n. 62, 83–109, 146–179, 202–215, 255; Gilbert, op. cit., n. 61, 34. Although a clerk did sign Lincoln's name, it was voided and the patent was never issued,

65. Los Angeles *Star*, Feb. 5, July 9, 1859, Sept. 29, 1860; *Supreme Court of the United States, December Term, 1866. No. 66. The United States vs. Josefa Montalva de Serrano et al.* (Washington, D.C.: 1866), 1–13, 132–136; Browne and Taylor, op. cit., n. 3, 187–188; *NSJ*, Jan. 15, 1876; Wallace W. Elliott, *History of San Bernardino County* (San Francisco: W. W. Elliott and Co., 1883), 109; *Transcript of Record, The United States vs. the San Jacinto Tin Company, et al.* (Washington, D.C.: 1887), 1369–1377, 1421–1427, [125 U.S. 273]; *An Illustrated History of Los Angeles County* (Chicago: Lewis, 1889), 369; Donald Chaput, "The Temescal Tin Fiasco," *Southern California Quarterly* 67 (Spring 1985), 1–3.

66. *MSP*, Nov. 18, 1865, Jan. 11, 1868, Jan. 16, Feb. 6, Mar. 13, 1869; *LT*, Jan. 18, 1869; San Jacinto Tin Company, "List of Shareholders August 20, 1869," in *Transcript of Record*, op. cit., n. 65, 1084–1085; *Serrano v. United States*, 72 U.S. 451; Elliott, op. cit.,

n. 65, 109; *Transcript of Record*, op. cit., n. 65, 1–16, 24–48, 160–162, 1137–1141, 1159–1165, 1268–1282, 1421–1427; Bancroft, op. cit., n. 35, 552, 660–661; Enoch Knight, "An American Tin Mine," *Overland Monthly* 19 (Feb. 1892), 169–172; Chaput, op. cit., n. 65, 4–8.

67. *SFBul*, Feb. 5, June 12, 1861, May 24, 1862; *SFAC*, June 12, 1861; *MSP*, Aug. 17, 1863; Browne and Taylor, op. cit., n. 3, 139–144; United States Census, 1870, California, Stanislaus County, Emory Township, 3rd August 1870, p. 22, lines 26 and 30, Hiram and Napoleon Hughes (including personal worth of $35,000 and $16,600, respectively); *A Memorial and Biographical History of the Counties of Fresno, Tulare and Kern* (Chicago: Lewis, 1892), 581; Melinda L. Cave, "A History of the Copper Mining Community of Copperopolis, 1861–1867" (M.A. thesis, California State University, Sacramento, 1970), 7–12; James J. Kroll, "The Copperopolis El Dorado" (M.A. thesis, California State University, San Francisco, 1970), 5–6; Charles K. Hyde, *Copper for America* (Tucson: University of Arizona Press, 1998), 57; Ronald H. Limbaugh and Willard P. Fuller, *Calaveras Gold* (Reno: University of Nevada Press, 2004), 76.

68. *MSP*, Dec. 21, 1860; Feb. 22, 1861, July 31, 1862, Aug. 17, 24, 1863; *SFBul*, June 12, Nov. 15, 1861, Jan. 9, May 24, July 14, 1862, Jan. 14, 1863; *Scientific American*, July 19, 1862, Oct. 3, 1862; *SFAC*, Dec. 17, 1862, Sept. 27, 1863; Browne and Taylor, op. cit., n. 3, 143, 148–149, 157–159, 166–168; Balch, op. cit., n. 7, 535; Cave, op. cit., n. 56, 16–22, 90; Kroll, op. cit., n. 67, 6–7, 20–21, 26; Limbaugh and Fuller, op. cit., n. 67, 76–77. Copper tonnages are given in English tons of 2,376 pounds.

69. *MSP*, Nov. 16, 1861, May 25, Aug. 17, 1863, Mar. 11, Apr. 1, 1865, Feb. 17, Apr. 7, Sept. 1, 1866; *SFAC*, Dec. 17, 1862, Sept. 27, 1863, Nov. 18, 1865; Browne and Taylor, op. cit., n. 3, 143, 148–149, 159, 166–168; Balch, op. cit., n. 7, 535; "Thomas Hardy," *A Memorial and Biographical History of the Counties of Merced, Stanislaus, Calaveras, Tuolumne and Mariposa* (Chicago: Lewis, 1892), 319; Cave, op. cit., n. 67, 23, 90; Kroll, op. cit., n. 67, 21–33, 67–70; Limbaugh and Fuller, op. cit., n. 67, 77–83.

70. *MSP*, Aug. 10, 1863; Browne and Taylor, op. cit., n. 3, 139, 147–149, 159, 167–168; *Napoleon Copper Mining & Smelting Co. Capital $1,200,000. 12,000 Shares $100 Each*, stock certificate no. 4, New York, Feb. 14, 1866; *A Memorial*, op. cit., n. 69, 319–320; Douglas and Gina McDonald, *Mines of the West, 1863* (Helena, Mont.: Gypsyfoot, 1996), 14–64.

71. *MSP*, Feb. 18, 1865, July 14, Aug. 25, 1866, Oct. 12, 1867, Oct. 5, 1872; *NYT*, Mar. 17, 1865; Edwin F. Bean, *Bean's History and Directory of Nevada County* (Nevada City, Calif.: Gazette, 1867), 212–214, 244–245; Browne, op. cit., n. 47, 131; *AJM*, July 11, 1868; Raymond, op. cit., n. 47, 1st rpt. (1869), 25, 4th rpt. (1873), 123; Harry L. Wells, *History of Nevada County* (Oakland, Calif.: Thompson and West, 1880), 192, 215; Jack R. Wagner, *Gold Mines of California* (Berkeley, Calif.: Howell-North, 1970), 177–179, 187; Frank W. McQuiston, *Gold: The Saga of the Empire Mine, 1850–1956* (Grass Valley, Calif.: Empire Mine Park Assoc., 1986), 62–65.

72. *SFBul*, July 24, 1857, Dec. 23, 1865; *MSP*, Feb. 13, July 2, Oct. 8, 1864, Nov. 11, 1865, Sept. 22, 1866, May 22, 1875; *SacU*, Apr. 29, 1864, Jan. 29, 1872; Bean, op. cit., n. 69, 208–209, 245–246; Browne, op. cit., n. 47, 131–132; *NYT*, Dec. 15, 1868; Wells, op. cit., n. 71, 192.

73. *SFBul*, July 2, Oct. 8, 1862; *MSP*, June 16, Nov. 10, 1866, Jan. 19, 1867, Nov. 14, 1868; Browne and Taylor, op. cit., n. 3, 66–67; Browne, op. cit., n. 47, 145–146; Cronise, op. cit., n. 58, 582–583; Henry Janin, *Report Upon the Sierra Buttes Quartz Mine* (San Francisco: E. Bosqui, 1869), 1–13; Rossiter W. Raymond, *Mineral Resources of the States and Territories West of the Rocky Mountains* (Washington, D.C.: G.P.O., 1869), 60–63; Frank T. Gilbert, et al., *Illustrated History of Plumas, Lassen and Sierra Counties* (San Francisco: Fariss and Smith, 1882), 481–482; *SFAC*, Aug. 7, 1888; Albin J. Dahl, "British Investment in California Mining, 1870–1890" (Ph.D. diss., University of California, 1961), 133, 183–186.

74. Jesse Mason, *History of Amador County* (Oakland, Calif.: Thompson and West, 1881), 153, 303; Bancroft, op. cit., n. 35, 545; George T. Clark, *Leland Stanford* (Stanford, Calif.: Stanford University Press, 1931), 69–70; Wagner, op. cit., n. 71, 127–128; Norman E. Tutorow, *The Governor: The Life and Legacy of Leland Stanford* (Spokane, Wash.: Arthur H. Clark, 2004), 74–76.

75. Mason, op. cit., n. 74, 153–154, 303; Phelps, op. cit. n. 2, v. 1 (1881), 184–190; Bancroft, op. cit., n. 35, 544–545; Clark, op. cit., n. 74, 69–70; Oscar Lewis, *The Big Four: The Story of Huntington, Stanford, Hopkins, and Crocker, and of the Building of the Central Pacific* (New York: Alfred Knopf, 1938), 156–167; Tutorow, op. cit., n. 74, 76–79.

76. *MSP*, Jan. 19, Feb. 16, Mar. 9, 23, Oct. 26, 1863, Apr. 23, May 14, Aug. 27, 1864; *Bodie Bluff Consolidation Mining Co. Incorporated January 26, 1863. Capital Stock, $1,110,000. 11,100 Shares, $100 Each*, stock certificate, signed by Leland Stanford but not issued; *SFAC*, Apr. 14, 1864 (adv.); *Prospectus of the Empire Gold & Silver Mining Co., of New York* (New York: Wm. H. Arthur, 1864), 1–43; J. Ross Browne, *The Bodie Bluff Mines . . . Belonging to the Empire Gold & Silver Mining Co.* (New York: Clark and Maynard, 1865), 1–15; J. Ross Browne, "A Trip to Bodie Bluff and the Dead Sea of the West," *Harper's New Monthly Magazine* 31 (Aug. 1865), 274–284; Joseph Wasson, *Bodie and Esmeralda* (San Francisco: Spaulding, 1878), 5–8; *NYT*, Sept. 8, 1878; H. S. W., "The Bodie Mining District," *Scientific American* 46 (Sept. 6, 1879), 148; William B. Clark, *Gold Districts of California* (San Francisco: California Division of Mines and Geology, 1970), 13; Warren Loose, *Bodie Bonanza* (New York: Exposition, 1971), 28–34; Michael H. Piatt, *Bodie* (El Sobrante, Calif.: North Bay, 2003), 22–23, 27–30.

77. *Miners' Register* (Central City, Colo.), Mar. 23, 1865; *NYT*, Feb. 17, 1866; Fossett, op. cit., n. 1, 209, 252–253.

78. *NYT*, Mar. 31, Apr. 1, 4, 1864; *MSP*, May 14, 1864; *American Mining Gazette* (New York) 1 (Aug. 1864), 327; Hollister, op. cit., n. 1, 195–196; Fossett, op. cit., n. 1, 133, 212–213, 216, 255, 277–280, 336–337; Medbery, op. cit., n. 38, 185–188; William W. Fowler, *Twenty Years of Inside Life in Wall Street* (New York: Orange Judd, 1880), 298–299, 309–311; Anita Leslie, *The Remarkable Mr. Jerome* (New York: Henry Holt, 1954), 65; King, op. cit., n. 44, 13–16.

79. Fossett, op. cit., n. 1, 222–229; Vickers, op. cit., n. 1, 223–228, 680–681; William B. Vickers, et al., *History of the City of Denver* (Chicago: O. L. Baskin, 1880), 361–363; Colorado Springs *Gazette*, Oct. 22, 1884; William M. Thayer, *Marvels of the New West* (Norwich, Conn.: Henry Bill, 1887), 531–532; Henry Hall, *America's Successful Men of*

Affairs (New York: Tribune, 1896), v. 2, 153–154; *BDAC*, op. cit., n. 13, 676–677; King, op. cit., n. 44, 151–153, 166.

80. Paul, op. cit., n. 1, 138–143, 150–151, 156.

81. *SFAC*, June 25, 1863; *PO*, Dec. 6, 14, 1864; *Idaho Statesman* (Boise), Oct. 12, Nov. 23, Dec. 12, 1865, July 28, 1866; *Idaho, the "Gem of the Mountains!" The Celebrated Ada Elmore Gold and Silver Mine, of South Boise, Idaho Territory* (New York: Robertson, 1866), 1–31; *The Waddingham Gold and Silver Mining Company. Organized May 27th 1865. Capital $600,000. Shares $25 Each. Capital Stock Increased April 16, 1866 to $1,200,000*, stock certificate no. 488, New York, Nov. 7, 1866; Stevens Point, Wis., *Gazette*, June 25, 1887; Atchison, Kans., *Globe*, Sept. 21, 1888; *Kansas City Star*, Dec. 13, 1889, May 17, 1899 (Waddingham obit); *Santa Fe New Mexican*, May 16–17, 1899; *NYT*, May 17, 1899; Robert L. Romig, "The South Boise Mines, 1863–1892" (M.A. thesis, University of California, Berkeley, 1950), 9, 14, 16, 28, 37–38, 45, 49–53; Merle W. Wells, "Wilson Waddingham in Idaho," *Denver Westerners Brand Book* 24 (1968), 256–278. reprinted in *Idaho Yesterdays* 44 (Winter 2001), 36–46; Merle W. Wells, *Gold Camps & Silver Cities: Nineteenth Century Mining in Central and Southern Idaho* (Moscow, Idaho: Bureau of Mines and Geology, 1983), 18–23; David A. Remley, *Bell Ranch* (Albuquerque: University of New Mexico Press, 1993), 70–73.

82. *Owyhee Avalanche* (Ruby City, Idaho), Sept. 16–30, Oct. 7, Nov. 4, 25, Dec. 23, 1865, Jan. 6, Feb. 10, June 16, 1866; *NYT*, Jan. 15, 1866; *PO*, Dec. 27, 1866; Albert D. Richardson, *Beyond the Mississippi* (Hartford, Conn.: American, 1867), 509–510; *The Poorman Gold and Silver Mining Company of Idaho* (New York: American Mining Index, 1867), 1–22; Browne and Taylor, op. cit. n. 3, 84; Browne, op. cit., n. 47, 523–528, 534–535; *MSP*, Dec. 5, 1868, Apr. 9, May 21, 1870, Apr. 1, 1871, Dec. 8, 1877; Hiram French, *History of Idaho* (Chicago: Lewis, 1914), 142–144; Paul B. Christian, "Idaho's Silver City, The Early Period, 1863 to 1878" (M.A. thesis, University of California, Berkeley, 1952), 22–26; Wells, op. cit., n. 81, 26–45, 160.

83. *Idaho Statesman* (Boise), Nov. 21, 1867, Apr. 21, 1868, Feb. 25, 1869; *MSP*, Nov. 14, 1868, May 22, 1869, May 16, 1870; Raymond, op. cit., n. 47, (1869) 163, (1870) 235, 4th rpt. (1873), 518; Hearst, op. cit., n. 2, 19; Older, op. cit., n. 2, 139–140; Wells, op. cit., n. 81, 38–39, 45.

84. Alexander K. McClure, *Three Thousand Miles through the Rocky Mountains* (Philadelphia: Lippincott, 1869), 286–287; Helena *Independent*, Nov. 25, 1883; Michael Leeson, *History of Montana, 1739–1885* (Chicago: Warner, Beers, 1885), 1359–1360; *PO*, Nov. 24, 1887; Hubert H. Bancroft, *History of Washington, Idaho, and Montana: 1845–1889* (San Francisco: History Co., 1890), 723–727; *SFBul*, Aug. 2, 1890 (obit); *SFCh*, Aug. 2, 1890; *BE*, Aug. 11, 1890; *LAT*, July 6, 1892; *NYT*, Jan. 22, 24, 1893; Patrick Henry McLatchy, "A Collection of Data on Foreign Corporate Activity in the Territory of Montana, 1864–1889" (M.A. thesis, Montana State University, 1961), 59–61; Kimberly Morrison, "James Whitlatch & the Last Chance Motherlode," *More from the Quarries of Last Chance Gulch* (Helena, Mont.: Independent Record, 1996), v. 2, 174–177.

85. *The Missouri and Montana Mining Company* (St. Louis: Studley, 1865), 1–26; *The*

St. Louis and Montana Mining and Discovery Company (St. Louis: George Knapp, 1865), 1–36; Leeson, op. cit., n. 82, 1217; Joaquin Miller, *An Illustrated History of the State of Montana* (Chicago: Lewis, 1894), 126–127; Donald L. Sorte, "The Hope Mining Company of Philipsburg" (M.A. thesis, Montana State University, 1960), 1–8, 74, 85–87; McLatchy, op. cit., n. 84, 59, 62, 72–75; John W. Hakola, "Samuel T. Hauser and the Economic Development of Montana" (Ph.D. diss., Indiana University, 1961), 38–48, 74–78.

86. Leeson, op. cit., n. 84, 1217; Miller, op. cit., n. 85, 126–127; Sorte, op. cit., n. 85, 7–19, 74, 76; McLatchy, op. cit., n. 84, 85, 88; Hakola, op. cit., n. 85, 48–58.

87. "West Mountain Mining District," Book A, 1–5, Sept. 17–Nov. 17, 1863. Salt Lake County Recorder's Office, Salt Lake City; *Union Vedette* (Camp Douglas, Utah), Nov. 20, 27, 1863, Jan. 7, 21, 29, Feb. 4, 18, Mar. 15, 22, Sept. 24, Dec. 12, 1864; *Jordan Silver Mining Company. Capital Stock, $104,000. 520 Shares, $200 Each*, stock certificate no. 266, Salt Lake City, Mar. 5, 1864, courtesy of Geff Pollock; John R. Murphy, *The Mineral Resources of the Territory of Utah* (San Francisco: A. L. Bancroft, 1872), 1–2, 14; Catherine V. Waite, *Adventures in the Far West and Life Among the Mormons* (Chicago: Waite, 1882), 132–136; Thomas Arthur Rickard, *A History of American Mining* (New York: McGraw-Hill, 1932), 184–188; Leonard J. Arrington and Gary B. Hansen, *"The Richest Hole on Earth"* (Logan: Utah State University Press, 1963), 11–12; William Fox, "Patrick Edward Connor, 'Father' of Utah Mining" (M.A. thesis, Brigham Young University, 1966), 45–59, 67–70, 109–114; Lynn R. Bailey, *Old Reliable: A History of Bingham Canyon* (Tucson: Westernlore, 1988), 13–17; Brigham D. Madsen, *Glory Hunter: A Biography of Patrick Edward Connor* (Salt Lake City: University of Utah Press, 1990), 105–107; Sally Zanjani, *A Mine of Her Own: Women Prospectors in the American West, 1850–1950* (Lincoln: University of Nebraska Press, 1997), 117–118; Keith R. Long, John H. DeYoung and Steve Ludington, "Significant Deposits of Gold, Silver, Copper, Lead, and Zinc in the United States," *Economic Geology*, 95 (May 2000), 629, 635–636.

88. .*Union Vedette* (Camp Douglas, Utah), Sept. 24, 29, Dec. 12, 1864, Feb. 17, Mar. 18, 1865, Jan. 6, 1866, May 9, June 5, Nov. 12, 1867; Browne and Taylor, op. cit., n. 3, 130; *The New-York and Utah Prospecting and Mining Company* (New York: Amerman, 1867), 1–38; Murphy, op. cit., n. 87, 2–4, 7–8; Hubert H. Bancroft, *History of Utah, 1540–1886* (San Francisco: History Co., 1889), 741–742; Rickard, op. cit., n. 87, 188–189; Arrington and Hansen, op. cit., n. 87, 12–13; Fox, op. cit., n. 87, 67–82; Bailey, op. cit., n. 87, 17; Madsen, op. cit., n. 87, 107–119.

89. *Private Land Claims in New Mexico* (Washington, D.C.: G.P.O., 1861), 51–73 (35th Cong., 2nd Sess., House Ex. Doc. 28); *The New Mexico Mining Company* (Washington, D.C.: W. Koch, 1862), 2–7; *The New Mexico Mining Co. Preliminary Report* (New York: Baker and Godwin, 1862), 1–21; *The New Mexico Mining Company. Capital $2,500,000. 25,000 Shares. Shares $100 Each*, stock certificate no. 452, Washington City, Oct. 5, 1864; *EMJ*, Aug. 21, 1880; Santa Fe *New Mexican*, Oct. 17, 1891; *Gildersleeve v. New Mexico Mining Co.* (1896) 161 U.S. 573; Fayette Jones, *New Mexico Mines and Minerals* (Santa Fe: New Mexican, 1904), 23–24; John M. Townley, "Mining in the Ortiz Mine Grant

Area" (M.A. thesis, University of Nevada, Reno, 1968), 71–85; John M. Townley, "The New Mexico Mining Company," *New Mexico Historical Review* 46 (Jan. 1971), 57–73; William Baxter, *The Gold of the Ortiz Mountains* (Santa Fe: Lone Butte, 2004), 76–100.

90. *SFBul*, July 6, 1865; *MSP*, Oct. 28, Dec. 9, 1865, Feb. 10, Aug. 11, 1866, Jan. 26, Feb. 16, Apr. 13, May 11, 25, June 8, Aug. 3, 24, Dec. 14, 21, 1867, Feb. 8, Mar. 21, 1868; Browne, op. cit., n. 47, 477–478; Raymond, op. cit., n. 73, (1869) 257–260; *The Vulture Mining Company of Arizona. $1000 First Mortgage Bond. $250,000*, bond no. 84, June 1, 1870; *SFEx*, May 11, 1881; Thomas E. Farish, *History of Arizona* (San Francisco: Filmer, 1915), v. 2, 211–214; Rickard, op. cit., n. 87, 270; Duane A. Smith, "The Vulture Mine," *Arizona & the West* 14 (Autumn 1972), 231–239, 250; Paul W. Gates, "The Land Business of Thomas O. Larkin," *CHQ* 54 (Winter 1975), 330, 335; Fred Holabird, *The James Garbani Arizona Mining Stock & Document Collection* (Reno: Holabird Americana, 2002), 105.

91. William M. B. Hartley, *The Arizona Mining Company* (Jersey City, N.J.: Davidson and Ward, 1863), 3–19; *NYT*, Mar. 12, Apr, 8, 1864; *SFBul*, Nov. 3, 1864; *AJM*, Sept. 29, 1866; Morris Elsing and Robert Heineman, *Arizona Metal Production* (Tucson: University of Arizona, 1936), 97; Constance W. Altshuler, "The Case of Sylvester Mowry: The Mowry Mine," *Arizona & the West* 15 (Summer 1973), 169–171; Diane North, *Samuel Peter Heintzelman and the Sonora Exploring and Mining Company* (Tucson: University of Arizona Press, 1980), 175.

92. *SFAC*, Mar. 14, May 25, 1863, Nov, 3, 1864; *NYT*, Mar. 12, Apr. 6, 1864; Samuel F. Butterworth, *Arizona Mining Company* (New York, 1864), 1–23; Sylvester Mowry, *Arizona and Sonora* (New York: Harper, 1864), 232–248; J. Ross Browne, "A Tour Through Arizona," *Harper's New Monthly Magazine* 30 (Feb.–Mar. 1865), 283–293, 409–417; William Keleher, *Turmoil in New Mexico* (Santa Fe: Rydal, 1952), 244–251, 269–271; Altshuler, op. cit., n. 91, 149–156, 161.

93. Browne, op. cit., n. 47, 444–448; Browne, op. cit., n. 26, 195–224, 264–275.

94. *BDAC*, op. cit., n. 13, 708; J. J. Bowden, *Spanish and Mexican Land Grants in the Chihuahuan Acquisition* (El Paso: Texas Western Press, 1971), 61; Robert L. Spude, "The Santa Rita del Cobre," *Mining History Journal* 6 (1999), 18–21.

95. *MSP*, May 14, June 11, 1864, Apr. 1, May 27, 1865; New York *American Mining Gazette* 1 (Oct. 1864), 440, (Feb. 1865), 95; *NYT*, May 4, 5, 8, 23, June 14, 1867; "Our Late Mining Enterprises," *Merchant's Magazine* 57 (Dec. 1867), 471; Hollister, op. cit., n. 1, 140, 195–196; Fossett, op. cit., n. 1, 134–136; Medbery, op. cit., n. 38, 285; Fowler, op. cit., n. 78, 311; King, op. cit., n. 44, 17–18.

96. Bayard Taylor, *Colorado: A Summer Trip* (New York: Putnam and Son, 1867), 61; *MSP*, Dec. 5, 1868; Fossett, op. cit., n. 1, 134–136, 212, 220, 255–257; Samuel S. Wallihan and T.O. Bagley, *The Rocky Mountain Directory and Colorado Gazetteer, for 1871* (Denver: Wallihan, 1870), 244–247; Medbery, op. cit., n. 38, 285; Vickers, op. cit., n. 1, 215; Charles W. Henderson, *Mining in Colorado; A History* (Washington, D.C.: G.P.O., 1926), 88, 122; King, op. cit., n. 44, 19–26.

97. Washoe United Consolidated Gold & Silver Mining Company, *Prospectus* (London, 1864) quoted in Spence, op. cit., n. 9, 62, plus other companies outside of California, 241–261.

98. *MSP*, May 24, Nov. 30, 1860, May 7, 1864; *AJM*, Mar. 3, 1866; *EMJ*, July 6, 1869; Wells, op. cit., n. 71, 149; Arthur J. Wilson, *The Pick and the Pen* (London: Mining Journal, 1979), 44–63.

99. John P. Young, *Journalism in California* (San Francisco: Chronicle, 1915), 81.

100. *SFC*, quoted in *MSP*, May 28, 1864; Boston *Commercial Bulletin*, quoted in *SFBul*, June 13, 1864.

101. *MSP*, Apr. 16, July 2, 30, Oct. 8, 1864; *SFBul*, July 7, 1864; Tuthill, op. cit., n. 5, 610–611; Bancroft Scraps, op. cit., n. 4, v. 55, 1551; Bancroft, op. cit., n. 35, v. 7, 671; Smith, op. cit., n. 2, 59–60; Jung, op. cit., n. 5, 83.

102. Clemens, *Roughing It*, op. cit., n. 14, 420.

103. *MSP*, Jan. 30, June 11, 1864; *SFBul*, Sept. 7, 1866; *AJM*, Sept. 29, 1866; "Our Late Mining Enterprises," op. cit., n. 95, 471; Browne and Taylor, op. cit., n. 3, 30; and see notes 7 and 95.

104. *SFBul*, Aug. 3, 1863, Sept. 1, 1863; Bancroft Scraps, op. cit., n. 4, 18, 48–49; Browne and Taylor, op. cit., n. 3, 78–79; Lawrence, op. cit., n. 13, 228; Smith, op. cit., n. 2, 80–86.

105. *MSP*, Aug. 6, 1864.

106. *SFCh*, Feb. 16, 18, 1872; Adolph Sutro, "Mr. Sutro's Argument," *Report of the Commissioners . . . in Regard to the Sutro Tunnel* (Washington, D.C.: G.P.O., 1872), 412; Wright, op. cit., n. 1, 401–403; Lawrence, op. cit., n. 13, 24–26; Lord, op. cit. n. 1, 244–246; Phelps, op. cit. n. 2, v. 2 (1882), 148–150; Hubert H. Bancroft, "Life of William Sharon," in *Chronicles of the Builders of the Commonwealth* (San Francisco: History Co., 1891), v. 4, 49–52; *EMJ*, July 2, 1892; Hittell, op. cit., n. 36, v. 4, 552–553; King, op. cit., n. 35, 209–210; Samuel P. Davis, *The History of Nevada* (Reno: Elms, 1913), 630; Thorp, op. cit., n. 44, 94, 129–130; Cecil G. Tilton, *William Chapman Ralston* (Boston: Christopher, 1935), 137–143; George D. Lyman, *Ralston's Ring* (New York: Scribner's Sons, 1937), 35–42, 54; Paul, op. cit., n. 1, 76; Jung, op. cit., n. 5, 107–116; Michael J. Makley, *The Infamous King of the Comstock: William Sharon* (Reno: University of Nevada Press, 2006), 21–32.

107. *SFCh*, Feb. 16, 18, 1872; Sutro, op. cit., n. 106, 412–414; John S. Hittell, *The Commerce and Industries of the Pacific Coast of North America* (San Francisco: A. L. Bancroft, 1882), 298–299; Lord, op. cit., n. 1, 246–249; Bancroft, op. cit., n. 106, 54–59, 195–196; Davis, op. cit., n. 106, 412–414; Tilton, op. cit., n. 106, 144–148; Lyman, op. cit., n. 106, 60, 85–87; Smith, op. cit., n. 2, 131; Paul, op. cit., n. 1, 76–77, 84–85; Jung, op. cit., n. 5, 119–128, 199–200; Makley, op. cit., n. 106, 32–38.

108. Sutro, op. cit., n. 106, 401–402; Wright, op. cit., n. 1, 165, 402–403; Angel, op. cit., n. 2, 280–283; Lord, op. cit. n. 1, 251–256; Bancroft, op. cit., n. 106, 236–238; Oscar T. Shuck, *History of the Bench and Bar of California* (Los Angeles: Commercial, 1901), 486; Tilton, op. cit., n. 106, 148–154; Smith, op. cit., n. 2, 122–125; Jung, op. cit., n. 5, 128–130; Makley, op. cit., n. 106, 45–54.

109. *SFCh*, Nov. 27, Dec. 4, 1870, Feb. 16, 18, 1872; San Francisco *Stock Report & California Street Journal*, May 25, 1872; Bancroft Scraps, op. cit., n. 4, 4; *MSP*, Jan. 18, 1873; *NSJ*, Sept. 1, 1883 (Robbins identification), Feb. 16, 1884.

110. *SFBul*, Feb. 15, July 28, 1865, Mar. 31, 1866, Feb. 20, Mar. 7, Apr. 5, 1867; *SFAC*, July 16, 1866; Gold Hill, Nev., *News*, Sept. 27, 1869; *Sutro Tunnel Company. $12,000,000 Capital Stock. 1,200,000 Shares, 10 Dollars Each Share*, stock certificate no. 1294, San Francisco, May 24, 1872; Adolph Sutro, "Speech of Adolph Sutro, on the Sutro Tunnel and the Bank of California," in *Report of the Commissioners*, op. cit., n. 106, 48–66; Sutro, op. cit, n. 106, 395–434; *EMJ*, Nov. 30, 1878; Lawrence, op. cit., n. 13, 67, 107–118; Angel, op. cit, n. 2, 504–509; Lord, op. cit. n. 1, 233–235; Bancroft, op. cit., n. 12, 141–143; Tilton, op. cit., n. 106, 211–214; Lyman, op. cit., n. 106, 47–53, 67–70; Smith, op. cit., n. 2, 107–109; Robert and Mary Stewart, *Adolph Sutro* (Berkeley, Calif.: Howell-North, 1962), 41–54; Paul, op. cit., n. 1, 82–83; Makley, op. cit., n. 106, 42–44.

111. Sutro, op. cit., n. 110, 50; Sutro, op. cit., n. 106, 396–400; *EMJ*, Nov. 30, 1878; Angel, op. cit, n. 2, 506–509; Lord, op. cit. n. 1, 237–243; Bancroft, op. cit., n. 12, 143–144; Tilton, op. cit., n. 106, 214–216; Lyman, op. cit., n. 106, 67–74; Smith, op. cit., n. 2, 109–110; Stewart, op. cit. n. 110, 54–58; Makley, op. cit., n. 106, 44–45.

112. *SFBul*, June 3, 1868, Oct. 20, Nov. 30, 1869, July 1, 10, 1878; *NYT*, July 15, 1868; *Congressional Globe*, Apr. 27–28, 1870, 3027, 3053–3054; Sutro, op. cit., n. 98, 64; Sutro, op. cit., n. 94, 400–406; Adolph Sutro *The Bank of California Versus the Sutro Tunnel* (Washington, D.C.: 1870), 1–14; *The California Monopolists Against the Sutro Tunnel* (Washington, D.C.: 1874), 1–6; *The Style of Warfare as Carried on by the California Bank Ring* (n.p., 1874), 1–5; Raymond, op. cit., n. 47, 8th rpt. (1877), 163; Angel, op. cit, n. 2, 506–511; Lord, op. cit., n. 1, 235–243, 342; Tilton, op. cit., n. 106, 214–216; Lyman, op. cit., n. 106, 70–74; Smith, op. cit., n. 2, 109–115; Stewart, op. cit., n. 110, 59–115; Elliott, op. cit., n. 13, 67.

113. *NYT*, Dec. 15, 1868; Bean, op. cit., n. 71, 205; *MSP*, Dec. 27, 1873, Feb. 7, 1874, Jan. 3, 1880, June 26, 1886; *SFAC*, Jan. 10, 1876, Dec. 27, 1878; Raymond, op. cit., n. 47, 8th rpt. (1877), 80–83; Wells, op. cit., n. 71, 189–191, 215; Balch, op. cit., n. 7, 1062–1063, 1142–1143; Wagner, op. cit., n. 71, 169–176, 187.

114. *NYT*, Mar. 17, 1865, Dec. 15, 1868; *MSP*, July 15, Sept. 30, 1865, Jan. 13, Apr. 14, Aug. 4, 1866, Nov. 1, 1873, July 18, 1874, Jan. 23, 1875; Bean, op. cit., n. 71, 204–205, 239–241; Browne and Taylor, op. cit., n. 3, 67; *AJM*, July 11, 1868; Browne, op. cit., n. 47, 130; Raymond, op. cit., n. 73, (1869) 23–29; *EMJ*, July 28, 1877; Raymond, op. cit., n. 47, (1877) 78–79; *SFAC*, July 8, 1878, May 20, 1879; Wells, op. cit., n. 71, 190–191; Wagner, op. cit., n. 71, 167.

115. *MSP*, Sept. 30, Oct. 7, 1865; Feb. 3, Mar. 3, 1866, Feb. 24, May 5, 1877, June 22, Aug. 10, 1878, May 1, 1880; *SFBul*, Aug. 17, 1871, Apr. 12, 1873; *NYT*, Apr. 14, 1873; Lawrence, op. cit., n. 13, 210; Mason, op. cit., n. 74, 155–158; *SFC*, July 16, 1885, June 15 (McDonald death), Dec. 20, 1907; Wagner, op. cit., n. 71, 141–149.

116. *SFBul*, Nov. 1, 1865, Sept. 19, 1866; *MSP*, Jan. 13, 1866, Oct. 26, 1867; Browne and Taylor, op. cit. n. 3, 68–69; Cronise, op. cit., n. 58, 585; Raymond, op. cit., n. 47, (1873)

70; Powell Greenland, *Hydraulic Mining in California* (Spokane, Wash.: Clark, 2001), 101–102.

117. MSP, Sept. 1, 1866, Feb. 22–29, 1868; Balch, op. cit., n. 7, 1160; Wells, op. cit., n. 71, 183–186; Egleston, op. cit., n. 58, 89–91; Tilton, op. cit., n. 106, 417; Robert Kelley, *Gold vs. Grain* (Glendale, Calif.: Clark, 1959), 37–52; Wagner, op. cit., n. 71, 33–35; Greenland, op. cit., n. 116, 198–226.

118. Bean, op. cit., n. 71, 201, 223–224; Browne, op. cit., n. 47, 131; MSP, Feb. 20, 1875; Wells, op. cit., n. 71, 191; George W. Starr, "The Empire Mine, Past and Present," *MSP* Aug. 4, 1900; Wagner, op. cit., n. 71, 213–214; McQuiston, op. cit., n. 71, 28–29.

119. *Nevada Gold, Silver and Copper Mining Co. Location of Mine, Nevada County, Cal. Incorporated January 19, 1863. Capital Stock $230,000. 460 Shares, $500 Each,* stock certificate no. 52, Nevada, Feb. 14, 1863; AJM, July 11, 1868; MSP, Sept. 14, Oct. 26, 1872, Mar. 9, 1873, Jan. 9, 1875, and various small dividend notices from 1875 through 1878; SFAC, July 25–26, Aug. 16, 1874; *SFBul*, July 25–27, Aug. 5, 12, 15, 1874; *NYT*, Aug. 3, 19, 1874; Ross Conway Stone, *Gold and Silver Mines of America* (New York: Scientific, 1878), 33; CMR, June 24, 1882; Wells, op. cit., n. 71, 191; Starr, op. cit., n. 118, 120; Wagner, op. cit., n. 71, 214–215; McQuiston, op. cit., n. 71, 31–34, 64; Roger P. Lescohier, *Gold Giants of Grass Valley* (Grass Valley, Calif.: Empire Mine Park Assoc., 1995), 57–61; Ferol Egan, *Last Bonanza Kings, The Bourns of San Francisco* (Reno: University of Nevada Press, 1998), 69–75.

120. *Population of the United States in 1860* (Washington, D.C.: G.P.O., 1864), 35, 405, 549, 565, 573; Walter R. Crane, *Gold and Silver, Comprising an Economic History of Mining in the United States* (New York: John Wiley, 1908), 563; Smith, op. cit., n. 2, 61; Paul, op. cit., n. 1, 62–68; U.S. Bureau of the Census, op. cit., n. 7, 606; Richard E. Lingenfelter, *The Hardrock Miners* (Berkeley: University of California Press, 1974), 4. The total western production of gold and silver from 1859 to 1869 at about $608,000,000, based on both Crane (p. 563) and the U.S. Bureau of the Census (p. 606). Lode mines produced roughly $195,000,000 of that, with Nevada contributing $136,000,000, California about $42,000,000, and Colorado $16,000,000, based on Henderson, op. cit., n. 96, 69; James M. Hill, *Historical Summary of Gold, Silver, Copper, Lead, and Zinc Produced in California, 1848 to 1926* (Washington, D.C.: G.P.O., 1929), 4 (lode production 14 percent of total for 1859–69); Couch and Carpenter, op. cit., n. 4, 13; Paul, op. cit., n. 59, 345.

121. Hittell, op. cit., n. 36, v. 4, 542.

CHAPTER 4. THE DIAMOND DREAM

1. *SFBul*, July 31, Dec. 21, 1868; MSP, Dec. 19, 1868; SFAC, Jan. 21, 1869; Myron Angel, *History of Nevada* (Oakland, Calif.: Thompson and West, 1881), 649, 660–661; Hubert H. Bancroft, *History of Nevada, Colorado, and Wyoming, 1540–1888* (San Francisco: History Co., 1890), 277–279; W. Turrentine Jackson, *Treasure Hill* (Tucson: University of Arizona, 1963), 5–10; Rodman W. Paul, *Mining Frontiers of the Far West, 1848–1880* (New York: Holt, Rinehart and Winston, 1963), 106–107.

2. *SacU*, Dec. 25, 1868, Jan. 22, Feb. 4, 1869; *MSP*, Jan. 30, Mar. 13, Apr. 3, 17, 24, May 8, June 5, 19, 1869; Hamilton, Nev. *Inland Empire*, Mar. 27, Apr. 4, 15, May 29, 1869; *NYT*, Feb. 22, Apr. 14, 1869; Bancroft Scraps, v. 52 (California Mining Stocks), 54–55; Angel, op. cit., n. 1, 661; Jackson, op. cit., n. 1, 10–17, 153–154; Norman E. Tutorow, *The Governor: The Life and Legacy of Leland Stanford* (Spokane, Wash.: Arthur H. Clark, 2004), 81–83.

3. *MSP*, May 8, 1869; Apr. 16–30, May 21, 1870; *SFAC*, July 25–26, 1874; *SFBul*, July 25–27, 1874; Rossiter W. Raymond, *Statistics of Mines and and Mining in the States and Territories West of the Rocky Mountains* (Washington, D.C.: G.P.O., 1877), v. 8, 467; Walter Ingalls, *Lead and Zinc in the United States, Comprising an Economic History of the Mining and Smelting* (New York: Hill, 1908), 178–180; Bertrand Couch and Jay Carpenter, *Nevada's Metal and Mineral Production (1859–1940)* (Reno: University of Nevada, 1943), 149, 152–153.

4. *LMJ*, Dec. 14, 1869, Jan. 20, Feb. 3, Mar. 26, 1870; *MSP*, Apr. 30, 1870; *LT*, Sept. 10, 1871; *EMJ*, Jan. 31, 1874; Angel, op. cit., n. 1, 661; Bancroft, op. cit., n. 1, 279; Ingalls, op. cit., n. 3, 178–180; Couch and Carpenter, op. cit., n. 3, 152; Clark Spence, *British Investments and the American Mining Frontier, 1860–1901* (Ithaca, N.Y.: Cornell University Press, 1958), 95, 246–247, 265; Jackson, op. cit., n. 1, 166–197.

5. *MSP*, May 21, 1870; Spence, op. cit., n. 4, 96, 257, 266; Jackson, op. cit., n. 1, 173–177, 184, 188–189; Tutorow, op. cit., n. 2, 83–86.

6. Henry DeGroot, "Mining on the Pacific Coast," *Overland Monthly*, 7 (Aug. 1871), 158.

7. *MSP*, July 16, 1870, Nov. 18, 1871, Oct. 18, 1873; Angel, op. cit., n. 1, 434; William R. Balch, *The Mines, Miners and Mining Interests of the United States in 1882* (Philadelphia: Mining Industrial Publishing Bureau, 1882), 1156; George Hearst, *The Way It Was* (New York: Hearst Corp., 1972), 19–20; Fremont and Cora Older, *The Life of George Hearst* (San Francisco: John Henry Nash, 1933), 140–142; Couch and Carpenter, op. cit., n. 3, 59–63; Paul, op. cit., n. 1, 104–105; Lynn R. Bailey, *Shaft Furnaces and Beehive Kilns: A History of Smelting in the Far West, 1863–1900* (Tucson: Westernlore, 2002), 60–61, 67, 74–79.

8. *LMJ*, Aug. 5, 1871, July 13, Nov. 16, 1878; *SFBul*, Aug. 22, 1877, July 22, 1878, Oct. 31, 1891; *EMJ*, July 13–27, 1878, Jan. 18–25, 1879; *LT*, Nov. 14, 1878, Nov. 18, 1879; Richmond Mining Co. v. Eureka Mining Co., 103 U.S. 839–847; Angel, op. cit., n. 1, 432–433; Balch, op. cit., n. 7, 1175; *SFCh*, Feb. 23, 1900 (Probert obit); Curtis H. Lindley, *A Treatise on the American Law Relating to Mines and Mineral Lands* (San Francisco: Bancroft-Whitney Co., 1914), 1276–1285; Spence, op. cit., n. 4, 98, 125–127, 241–266; Albin J. Dahl, "British Investment in California Mining, 1870–1890" (Ph.D. diss., University of California, 1961), 86–122, 130, 133, 135–138, 183–191, 221–229; Bailey, op. cit., n. 7, 68, 80, 87–88.

9. *MSP*, May 22, 1869, Jan. 14, Aug. 12, Oct. 7, 28, 1871, Jan. 27, 1872, Feb. 8, July 5, 1873; *SFCh*, Oct. 8, 1871; *SFBul*, May 2, 4, 1872; *SFAC*, May 3, 1872; Raymond, op. cit., n. 3, 4th rpt. (1873), 203–286, 520–521, 5th rpt. (1873), 183, 502–503, 6th rpt. (1874),

526–539; R. S. Lawrence, ed., *Pacific Coast Annual Mining Review and Stock Ledger* (San Francisco: Francis and Valentine, 1878), 233–234; *Lincoln County Record* (Pioche, Nev.), Feb. 1, 1901 (Ely obit); David G. Dalin and Charles A. Fracchia, "Forgotten Financier: Francois L. A. Pioche," *CHQ* 53 (Spring 1974), 17–24; Mel Gorman, "Chronicle of a Silver Mine: The Meadow Valley Mining Company of Pioche," *Nevada Historical Society Quarterly* 29 (Summer 1986), 69–88.

10. *MSP*, Aug. 31, Sept. 7, 1872; *SFAC*, Apr. 7–16, May 2–3, 1873; *SFBul*, Apr. 16, May 3, Aug. 25, 1873; Samuel P. Davis, *The History of Nevada* (Reno: Elms, 1913), 397–398; Older, op. cit., n. 7, 142–143; Jackson, op. cit., n. 1, 172; Hearst, op. cit., n. 7, 19–20; Lloyd K. Long, "Pioche, Nevada, and Early Mining Developments in Eastern Nevada" (M.A. thesis, University of Nevada, Reno, 1975), 87–88, 92–96.

11. *MSP*, Aug. 6, Oct. 15, 1870, May 2, 1874; *SFBul*, Nov. 1, 1870, Aug. 2, Dec. 23, 1872, Apr. 24, 1874; *SFCh*, Nov. 1–2, 1870; George Roberts, letters to Asbury Harpending, Mar. 13, 20, 1871, Asbury Harpending Collection, California Historical Society, San Francisco; *LMJ*, June 24, 1871; *LT*, Oct. 30, 1871, Jan. 13, 1875; Nevada State Mineralogist, *Biennial Report for the Years 1871 and 1872* (Carson City, Nev.: Putnam, 1873), 22; Lawrence, op. cit., n. 9, 31–32; Davis, op. cit., n. 10, 839; Older, op. cit., n. 7, 140; *Dictionary of National Biography*, v. 22 supp., 763–764; David J. Jeremy, ed., *Dictionary of Business Biography* (London: Butterworths, 1984), v. 2, 623–626.

12. *MSP*, July 20, Nov. 2, Dec. 7, 1872; *LMJ*, Sept. 21, 1872; *LT*, Jan. 3, 19, Mar. 8, 1872, Jan. 18, 1873, Dec. 22, 1874; *SFBul*, Aug. 2, Dec. 23, 1872; Nevada State Mineralogist, op. cit., n. 11, 22; Spence, op. cit., n. 4, 54, 265.

13. *SLT*, Nov. 16, 1871; *UMG*, Aug. 30, 1873; *Emma Mine Investigation* (Washington, D.C.: G.P.O., 1876), ii, 45–46, 443; James E. Lyon, *Dedicated to William M. Stewart* (n.p., n.d.), 1–10; Orson F. Whitney, *History of Utah* (Salt Lake City: Cannon, 1904) v. 4, 1263–1265; Asbury Harpending, *The Great Diamond Hoax and Other Stirring Incidents in the Life of Asbury Harpending* (San Francisco: James H. Barry, 1913), 170–172; W. Turrentine Jackson, "The Infamous Emma Mine," *Utah Historical Quarterly* 23 (Oct. 1955), 338–362; Leonard J. Arrington, "Banking Enterprises in Utah, 1847–1880," *Business History Review* 29 (Dec. 1955), 321–328; Spence, op. cit., n. 4, 139–182; Jonathan Bliss, *Merchants and Miners in Utah: The Walker Brothers and Their Bank* (Salt Lake City: Western Epics, 1983), 163–175; Charles L. Keller, *The Lady in the Ore Bucket* (Salt Lake City: University of Utah Press, 2001), 126–132; Dan Plazak, *A Hole in the Ground with a Liar at the Top* (Salt Lake City: University of Utah Press, 2006), 59–61.

14. *Emma Mine*, op. cit., n. 13, ii, 270–271, 439–445, 455, 587–588, 784–785, 805; George H. Fitch, "Early California Millionaires," *Californian Illustrated Magazine* 3 (Dec. 1892), 83; Spence, op. cit., n. 4, 140–141; Keller, op. cit., n. 13, 130–131.

15. *LMJ*, Nov. 11, 1871; *Wall-Street Journal*, Nov. 25, 1871, quoted in *LMJ*, Dec. 9, 1871; Samuel T. Paffard, *The History of the Emma Mine* (London: Collingridge, 1873), 5–14; *EMJ*, Sept. 4, Nov, 27, 1875, Mar. 18, 25, 1876; *Emma Mine*, op. cit., n. 13, 62–73, 125–126, 158, 229–230, 252–253, 279–280, 708; Spence, op. cit., n. 4, 142–153; *BDAC*, 1567–1568; Plazak, op. cit., n. 13, 61–68.

16. *LMJ*, Nov. 11, Dec. 2, 1871; *LT*, Nov. 10, Dec. 2, 7, 1871, Jan. 5, 12, 1872; *The Bull and Bear Theatre* (London, 1873), broadside; *EMJ*, Dec. 12, 19, 1871, Nov. 27, 1875; *MSP*, Mar. 29, Apr. 5, 1873; Paffard, op. cit., n. 15, 14–16, 22–29; *NYT*, Mar. 17, 1876; *Emma Mine*, op. cit., n. 13, 385, 503–505, 513–514, 536–537, 549–553; Edwin L. Godkin, "Mr. T. W. Park's Case," *The Nation* 22 (June 1, 1876), 345–347; Spence, op. cit., n. 4, 151–160; Plazak, op. cit., n. 13, 71–72.

17. *LMJ*, Apr. 13, 20, 1872; Paffard, op. cit., n. 15, 17–21; *EMJ*, Nov. 27, 1875; *NYTr*, Jan. 17, Feb. 1, 1876; *NYT*, Mar. 17, 1876; *Emma Mine*, op. cit., n. 13, 385, 500–503, 534–537, 549–553, 852–854, 874–875; Godkin, op. cit., n. 16, 345–347; Spence, op. cit., n. 4, 155–156; Plazak, op. cit., n. 13, 66–67.

18. *LT*, Jan. 16, 1873, Jan. 30, 1875, Oct. 17, 1881; *LMJ*, Jan. 25, 1873, Jan. 30, 1875; *MSP*, Apr. 5, 1873, Jan. 10, Aug. 15, 1874, Dec. 4, 1875; *SLT*, June 10, July 4, 1873, Nov. 30, 1875; Paffard, op. cit., n. 15, 32–38; *NYT*, Mar. 18, 1875, Mar. 4, Dec. 14, 1876, Apr. 29, 1877; *Emma Mine*, op. cit., n. 13, 520–522, 549–553, 632–634; Spence, op. cit., n. 4, 162–167, 171.

19. *Emma Mine*, op. cit., n. 13, 549–553, Godkin, op. cit., n. 16, 345–347; *LT*, Oct. 17, 1881, Dec. 22, 1882 (Park obit), Aug. 31, 1899 (Grant obit); *NYT*, Feb. 20, 1879, Dec. 22, 1882 (Park obit); *MSP*, Apr. 21, 1883; *Biographical Encyclopaedia of Vermont of the Nineteenth Century* (Boston: Metropolitan, 1885), 197–201; Spence, op. cit., n. 4, 178–181; Jeremy, op. cit., n. 11, v. 2, 626–628; Plazak, op. cit., n. 13, 73–76.

20. *LT*, Nov. 30, 1871, Apr. 27, June 15, 17, Nov. 12, Dec. 25, 1872, May 10, 1875, Nov. 20, 1876, July 28, 1877, Feb. 22, Oct. 3, Nov. 9, 1878; *UMG*, May 30, 1874; *SLGD*, May 22, 1876; Spence, op. cit., n. 4, 26, 109–112, 247–248, 265; W. Turrentine Jackson, "British Impact on the Utah Mining Industry," *Utah Historical Quarterly* 31 (Fall 1963), 352–371; Keller, op. cit., n. 13, 133–135.

21. *SLT*, Apr. 30, Aug. 21, 1871, May 25, Oct. 26, 1872, Dec. 30, 1888 (Lawrence obit); *UMG*, Jan. 3, 1874; *NYT*, Jan. 5 (Ward obit), Sept. 25, 1875; *Grand Traverse Herald* (Traverse City, Mich.), Jan. 7, 1875; *The Bullion, Beck and Champion Mining Company, Plaintiff, vs. The Eureka Hill Mining Company, John Q. Packard and J. H. McCrystal, Defendants. Statement on Motion for New Trial* (Salt Lake City: J. C. Parker, 1886), 330–331; Harry W. B. Kantner, *A Hand Book on the Mines, Miners, and Minerals of Utah* (Salt Lake City: R. W. Sloan, 1896), 318; *Transcript of Record. John N. Whitney vs. Moylan C. Fox, Executor of Joab Lawrence, Deceased* (Washington, D.C.: Judd and Detweiler, 1896), 33–36, 272–276; David Ward, *The Autobiography of David Ward* (New York, 1912), 3–4; Charles C. Goodwin, *As I Remember Them* (Salt Lake City: Commercial Club, 1913), 329–331; *NCAB*, v. 16, 411; Jesse W. Wooldridge, *History of the Sacramento Valley* (Chicago, Pioneer Historical, 1931), v. 2, 459–461; Beth Kay Harris, *The Towns of Tintic* (Denver: Sage, 1961), 48; Carlton Stowe, *Utah Mineral Industry Statistics through 1973* (Salt Lake City: Utah Geological and Mineral Survey, 1975), 44; Philip F. Notarianni, *Faith, Hope, and Prosperity: The Tintic Mining District* (Eureka, Utah: Tintic Historical Society, 1982), 17.

22. *SLT*, Nov. 4, Dec. 21, 27, 1873, Feb. 25, Sept. 21, 1874, Mar. 31, 1875; *Denver Times*, Dec. 16, 1873; *UMG*, Jan. 3, 1874; *NYT*, Jan. 6, 11, 20, 31, Mar. 26, 28, 1874, Jan. 5 (Ward obit), Sept. 25, 1875; *Grand Traverse Herald* (Traverse City, Mich.), Jan. 7, 1875 (Ward

obit); *Deseret News* (Salt Lake City), Jan. 7, 21, 1874, Aug. 2, 1876; *Bullion, Beck,* op. cit., n. 21, 31–36, 299–318, 392–395; Frederick Carlisle, *Chronography of Notable Events in the History of the Northwest Territory and Wayne County* (Detroit: Gulley, Bornman, 1890), 240–243; *Transcript of Record,* op. cit., n. 21, 3–4, 17–20, 28–31, 36–41, 104–107, 237–244, 264–295; *Whitney v. Fox,* 166 U.S. 637.

23. *Bullion, Beck,* op. cit., n. 21, 31–36; *PO,* July 5, 1888; *SLT,* Dec. 30, 1888 (Lawrence obit), Feb. 7, 1892, Sept. 12, 1894, Mar. 2, Apr. 20, 1897; Patrick Donan, *Utah, a Peep into a Mountain-Walled Treasury of the Gods* (Buffalo: Matthews-Northrup, 1891), 40–41; *Utahnian* (Salt Lake City), Nov. 7, 1896; Kantner, op. cit., n. 21, 318; *Transcript of Record,* op. cit., n. 21, 4–5, 17–23, 196, 236–244; *Whitney vs. Fox,* 166 U.S. 637; Salt Lake City Mining Review, Oct. 30, 1899, Oct. 15, 1908 (Packard obit); *LAT,* Oct. 7, 1908; *NCAB,* v. 16, 411.

24. *SLT,* Dec. 13, 1874, Aug. 23, 1876, June 3, Sept. 14, 1877, May 1, 1878, 9 May 1890; Lawrence, op. cit., n. 9, 34–35, 228; Alonzo Phelps, *Contemporary Biography of California's Representative Men* (San Francisco: A. L. Bancroft, 1881), 27–31, 325–328; Balch, op. cit., n. 7, 1174; Thomas J. Almy, "History of the Ontario Mine," *Transactions of the American Institute of Mining Engineers* 16 (July 1887), 35–37; Edward W. Tullidge, *Tullidge's Histories,* v. 2, *Containing the History of All the Northern, Eastern and Western Counties of Utah* (Salt Lake City: Juvenile Instructor, 1889), 505–510; Hearst, op. cit., n. 7, 20–21; Richard P. Rothwell, *The Mineral Industry, 1892* (New York: Scientific, 1893), v. 1, 477, v. 4, 69–70, v. 8, 798; Kantner, op. cit., n. 21, 271–276; Walter H. Weed, *The Mines Handbook* (New York: Stevens, 1916), v. 12, 888–889, v. 13 (1920), 1440–1441; Kenneth Ross Toole, "Marcus Daly, A Study of Business in Politics" (M.A. thesis, University of Montana, 1948), 5–8; George A. Thompson and Fraser Buck, *Treasure Mountain Home; A Centennial History of Park City* (Salt Lake City: Deseret, 1968), 9–17; Keith R. Long, John H. DeYoung, and Steve Ludington, "Significant Deposits of Gold, Silver, Copper, Lead, and Zinc in the United States," *Economic Geology* 95 (May 2000), 629, 635–636.

25. *Idaho Statesman* (Boise), Sept. 19, 1867, Apr. 2, 1868; *Owyhee Avalanche* (Silver City, Idaho), Feb. 1, 15–Apr. 11, July 18, 1868; Paul B. Christian, "Idaho's Silver City, the Early Period, 1863 to 1878" (M.A. thesis, University of California, Berkeley, 1952), 100–101; Merle W. Wells, *Gold Camps & Silver Cities: Nineteenth Century Mining in Central and Southern Idaho* (Moscow, Idaho: Bureau of Mines and Geology, 1983), 41–45; Dale M. Gray, "War on the Mountain," *Idaho Yesterdays* 29 (Winter 1986), 24–32.

26. *Idaho Statesman* (Boise), Nov. 21, 1867, Dec. 15, 1868, Oct. 18, Nov. 15, 1870, Apr. 25, Dec. 7, 1871; *MSP,* Apr. 4, Nov. 14, Dec. 5, 1868, Jan. 23, Mar. 6, 27, Apr. 10, 17, June 19, Oct. 9, 29, Dec. 3, 1869, Mar. 19, Dec. 10, 1870, Oct. 28, Nov. 18, Dec. 2, 1871, June 21, 1873, Mar. 7, 1874, Jan. 23, 1875; Raymond, op. cit., n. 3, (1869) 162–163, (1870) 235–236, 239–244, 4th rpt. (1873), 518; Lawrence, op. cit., n. 9, 198, 262; Ross Conway Stone, *Gold and Silver Mines of America* (New York: Scientific, 1878), 33; Waldemar Lindgren, "The Gold and Silver Veins of Silver City, De Lamar, and Other Mining Districts in Idaho," *20th Annual Report of the U.S. Geological Survey* (Washington, D.C.: G.P.O., 1900), pt. 3, 149–150.

27. *LMJ*, June 17, Nov. 18, 1871; Frank Fossett, *Colorado* (Denver: Tribune, 1876), 334–338; William Vickers, et al., *History of Clear Creek and Boulder Valleys* (Chicago: O. L. Baskin, 1880), 524–526; Charles W. Henderson, *Mining in Colorado* (Washington, D.C.: G.P.O., 1926), 109; Clark C. Spence, "The British and Colorado Mining Bureau," *Colorado Magazine* 33 (Apr. 1956), 81–92; Clark C. Spence, "Colorado's Terrible Mine: A Study in British Investment," *Colorado Magazine* 34 (Jan. 1957), 48–53; Spence, op. cit., n. 4, 44–45.

28. *Colorado Miner* (Georgetown, Colo.), Apr. 7, 1877, Jan. 12, 1878; Raymond, op. cit., n. 3, v. 8 (1877), 298; *EMJ*, Nov. 9, Dec. 28, 1878; Frank Fossett, *Colorado, Its Gold and Silver Mines* (New York: Crawford, 1880), 397; Spence, "Colorado's Terrible Mine," op. cit., n. 27, 53–61; Christine Bradley, *William A. Hamill* (Fort Collins: Colorado State University, 1978), 4–11.

29. Fossett, op. cit., n. 28, (1878) 341–344, (1880) 393; Liston E. Leyendecker, "The Pelican-Dives Feud," *Essays and Monographs in Colorado History*, 1985, 9–17, 72–75.

30. *NYT*, Oct. 21, 1874; Decatur, Ill., *Republican*, July 19, 1875; *PO*, July 31, 1875; *SFBul*, Dec. 24, 1875; Fossett, op. cit., n. 28, (1878) 345–347; Adam Badeau, *Grant in Peace* (Hartford, Conn.: Scranton, 1887), 365–368; George Dawson, *Life and Services of Gen. John A. Logan as Soldier and Statesman* (Chicago: Belford, Clarke, 1887), 217–218; *BDAC*, 1230; Elliot West, "Jerome B. Chaffee and the McCook-Elbert Fight," *Colorado Magazine* 46 (Spring 1969), 145–165; Leyendecker, op. cit., n. 29, 13–28, 75–78.

31. *RMN*, May 21–23, 27, 29, June 1, 6, 19, 1875, Nov. 15, 1879; Fossett, op. cit., n. 28, (1878) 346–347, (1880) 389–392; Frank Hall, *History of the State of Colorado* (Chicago: Blakely, 1891), v. 3, 220–221; John Horner, *Silver Town* (Caldwell, Idaho: Caxton, 1950), 206–212; Leyendecker, op. cit., n. 29, 34–57.

32. Central City, Colo., *Register*, June 8, 1875; *RMN*, Apr. 9, 1880; *NYT*, Apr. 12, 1880 (adv.); Fossett, op. cit., n. 28, (1880) 392–393; Leyendecker, op. cit., n. 29, 62–64, 87.

33. *RMN*, Apr. 30, 1873; *EMJ*, Nov. 21, 1871; Samuel S. Wallihan, *The Rocky Mountain Directory and Colorado Gazetteer, for 1871* (Denver: printed by author, 1870), 208–209, 414–415; *Out West* (Colorado Springs), May 30, 1872; Fossett, op. cit., n. 27, (1876) 378–382; *NYMR*, Dec. 13, 1877; *RMN*, Dec. 28, 1877; Bancroft Scraps, op. cit., n. 2, v. 53, 743; Vickers, op. cit., n. 27, 426–427; *NYT*, Dec. 25, 1888 (Breed); Duane A. Smith, *Silver Saga; The Story of Caribou* (Boulder: Pruett, 1974), 1–8, 24–28, 65, 116.

34. *RMN*, June 2, 1876; Fossett, op. cit., n. 27, (1876) 381–382; *NYT*, June 14, 1877; *NYMR*, Dec. 13, 1877; Bancroft Scraps, op. cit., n. 2, v. 53, 743; Vickers, et al., op. cit., n. 27, 427; Smith, op. cit., n. 33, 25–35, 64–68; Steven F. Mehls, *David H. Moffat, Jr.* (New York: Garland, 1989), 50–53.

35. Jerome B. Chaffee, *The Beaubien and Miranda Grant in New Mexico and Colorado* (New York: printed by author, 1869), 1–8; *NYT*, Jan. 19, 1869, June 28–29, 1870; Stanley G. Fowler, "The Maxwell Land Grant. Inside History," *Denver Mirror*, Mar. 29, 1874; *Transcript of Record. The United States vs. The Maxwell Land-Grant Company, et al.* (Washington, D.C., 1887), 1–11; "Maxwell Land-Grant Case," 121 U.S. 325; Leon Noel, "The Largest Estate in the World," *Overland Monthly* 12 (Nov. 1888), 480–494;

Gustavus Myers, *History of the Great American Fortunes* (Chicago: Charles H. Kerr, 1911), v. 3, 311–337; William Keleher, *Maxwell Land Grant* (Santa Fe: Rydal, 1942), 115, 147–151; Jim Berry Pearson, *The Maxwell Land Grant* (Norman: University of Oklahoma Press, 1961), 9–26, 48–49, 62, 76; Simeon H. Newman, "The Santa Fe Ring: A Letter to the "New York Sun," *Arizona & the West* 12 (Fall 1970), 269–280; Victor Westphall, *Thomas Benton Catron and His Era* (Tucson: University of Arizona Press, 1973), 98–102; Lawrence R. Murphy, *Lucien Bonaparte Maxwell* (Norman: University of Oklahoma Press, 1983), 152–154, 168–182; David A. Remley, *Bell Ranch* (Albuquerque: University of New Mexico Press, 1993), 74–76; Elizabeth Rogers, "S. B. Elkins: Business in New Mexico's Early Banking Era, 1873–1875," *New Mexico Historical Review* 70 (Jan. 1995), 67–76; Richard D. Loosbrock, "Managing a Gold Rush: Mining of the Maxwell Grant, New Mexico, 1867–1920," *Mining History Journal* 6 (1999), 1–7; Howard R. Lamar, *The Far Southwest, 1846–1912* (Albuquerque: University of New Mexico Press, 2000), 121–146; Maria E. Montoya, *Translating Property: The Maxwell Land Grant* (Berkeley: University of California Press, 2002), 90–95, 115–116.

36. Chaffee, op. cit., n. 35, 1–8; *MSP*, Aug. 13, Sept. 17, 1870; *NYT*, Sept. 10, 1870; Fowler, op. cit., n. 35; *RMN*, Sept. 30, 1882; Boston *Advertiser*, Sept. 2, 1884; *SLGD*, Nov. 15, 1886, July 19, 1887; Noel, op. cit., n. 35, 481; *Transcript of Record*, op. cit., n. 35, 427–429; Keleher, op. cit., n. 35, 116–118; Pearson, op. cit., n. 35, 49, 90–92; Murphy, op. cit., n. 35, 183–186; Montoya, op. cit., n. 35, 90–95.

37. John Collinson and William A. Bell, *The Maxwell Land Grant* (London: Taylor, 1870), 1–32; John Collinson and William A. Bell, *De Maxwell Land Grant and Railway Company* (Amsterdam: Blikman and Sartorius, 1870), 1–36; John Collinson, *Proposal to the Stock and Bondholders of the Maxwell Land Grant and Railway for Reorganisation* (Brussels, 1875), 7–11; *Transcript of Record*, op. cit., n. 35, 430–442, 446–456; Pearson, op. cit., n. 35, 49–52, 58–59; Westphall, op. cit., n. 35, 98–99; Montoya, op. cit., n. 35, 240n30.

38. *RMN*, May 7, 1873; Fowler, op. cit., n. 35; Collinson, op. cit., n. 37, 11–14; *NYT*, Mar. 30, 1884, June 13, 30, July 1, 5, 7, 9, 1885, Oct. 16, 1886, Apr. 19, 30, 1887; George W. Julian, "Land-Stealing in New Mexico," *North American Review* 145 (July 1887), 17–31; Stephen W. Dorsey, "Land-Stealing in New Mexico. A Rejoinder," *North American Review* 145 (Oct. 1887), 396–409.

39. Fowler, op. cit., n. 35; Collinson, op. cit., n. 37, 25–37; Myers, op. cit., n. 35, 326–327, 331; Keleher, op. cit., n. 35, 96, 128–129; Pearson, op. cit., n. 35, 72–78; Westphall, op. cit., n. 35, 109–112; Lamar, op. cit., n. 35, 126–128.

40. *NYTr*, Jul. 20–Aug. 5, 1880; *NYT*, Mar. 11, 1881, Feb. 12, Oct. 28, 1882, June 29, 1884, July 7, 1885, Oct. 6, 1892, May 17, 1899, Feb. 13, 1910, Jan. 5, 1911 (Elkins obit), Mar. 29, 1913; Henry Hall, *America's Successful Men of Affairs* (New York: Tribune, 1896), v. 1, 215–218; Santa Fe *New Mexican*, May 16–17, 1899 (Waddingham obit); Kansas City *Star*, May 17, 1899; Myers, op. cit., n. 35, 334–337; *BDAC*, 674–675, 783, 853; Westphall, op. cit., n. 35, 39–64; Lamar, op. cit., n. 35, 128–135; Remley, op. cit., n. 35, 122–124. For Elkins's own very selective recollections, see Oscar D. Lambert's *Stephen Benton Elkins* (Pittsburgh: University of Pittsburgh, 1955).

41. *NYT*, Apr. 10, May 3, 1872, May 16, 1874; *RMN*, June 17, 1874; John M. Sully, "The Story of the Santa Rita Copper Mine," *Old Santa Fe Magazine* 3 (1916), 139–140; Thomas Arthur Rickard, "The Chino Enterprise I. History of the Region and the Beginning of Mining at Santa Rita," *EMJ* 116 (Nov. 3, 1923), 757; Robert L. Spude, "The Santa Rita del Cobre, New Mexico, The Early American Period, 1846–1886," *Mining History Journal* 6 (1999), 22–25.

42. *NYT*, Jan. 5, 1875 (Ward obit), Sept. 25, 1875, Feb. 8, 1878; *Grand Traverse Herald* (Traverse City, Mich.), Jan. 7, 1875; Chicago *Inter Ocean*, Sept. 24, 1875; Carlisle, op. cit., n. 22, 240–243; Roberta Watt, *History of Morenci, Arizona* (M.A. thesis, University of Arizona, 1956), 4–10; W. Turrentine Jackson, *The Enterprising Scot* (Edinburgh: Edinburgh University Press, 1968), 164; Robert L. Spude, "Mineral Frontier in Transition: Copper Mining in Arizona, 1880–1885" (M.A. thesis, Arizona State University, 1976), 77–78; David F. Myrick, *Railroads of Arizona* (Glendale, Calif.: Trans-Anglo, 1984), v. 3, 12–14, 17–18; Charles K. Hyde, *Copper for America* (Tucson: University of Arizona Press, 1998), 115; Long et al., op. cit., n. 24, 636.

43. *SFBul*, Dec. 18, 1873; *Arizona Miner* (Prescott), Nov. 27, 1874; *RMN*, Dec. 5, 1874; *EMJ*, Dec. 12, 1874, Jan. 16, 1875; *NYT*, June 17, 1879, Apr. 25, 1924 (Lesinsky obit); Henry Lesinsky, *Letters Written by Henry Lesinsky to His Son* (New York: Albert R. Lesinsky, 1924), 46–51, 69–72; James Colquhoun, *The Early History of the Clifton-Morenci District* (London: William Clowes, 1935), 20–43; Watt, op. cit., n. 42, 10–13; Floyd S. Fierman, "Jewish Pioneering in the Southwest: A Record of the Freudenthal-Lesinsky-Solomon Families," *Arizona & the West* 2 (Spring 1960), 54–72; Joseph F. Park, "The History of Mexican Labor in Arizona during the Territorial Period" (M.A. thesis, University of Arizona, 1961), 189–191; Jackson, op. cit., n. 42, 163–165; Spude, op. cit., n. 42, 79–83; Myrick, op. cit., n. 42, 14–23; Hyde, op. cit., n. 42, 113–114.

44. *EMJ*, Dec. 1, 1877; *MSP*, June 8, 1878; Lesinsky, op. cit., n. 43, 51–53; Colquhoun, op. cit., n. 43, 40–43; Park, op. cit., n. 43, 202–203; Jackson, op. cit., n. 42, 164–165; Spude, op. cit., n. 42, 83–84.

45. *EMJ*, Dec. 1, 1877, Feb. 14, 21, 1880; *NYT*, July 27, 1878, June 17, 1879; *MSP*, Feb. 21, Mar. 27, 1880; Lesinsky, op. cit., n. 43, 53–54; Colquhoun, op. cit., n. 43, 44–47, 61, 79–80; Park, op. cit., n. 43, 195–196; Jackson, op. cit., n. 42, 164–166; Spude, op. cit., n. 42, 85–87, 90–97; Hyde, op. cit., n. 42, 116–117.

46. *MSP*, May 4, 1867, Nov. 21, 1868, Jan. 7, 1871; Titus Cronise, *The Natural Wealth of California* (San Francisco: H. H. Bancroft, 1868), 628; *SFAC*, Sept. 8, 1869, June 10, 1875; Oscar T. Shuck, *Representative and Leading Men of the Pacific* (San Francisco: Bacon, 1870), 411–420; Phelps, op. cit., n. 24, 420–423; Ingalls, op. cit., n. 3, 77–79, 203; Paul, op. cit., n. 1, 101–102, 125; Bailey, op. cit., n. 7, 43–45.

47. Raymond, op. cit., n. 3, (1869) 131, 4th rpt. (1873), 22–23, 180; *MSP*, Jan. 7, 1871, Nov. 27, 1875; *SFAC*, June 10, 19, Nov. 9, 18, Dec. 16, 22, 1875; *SFBul*, June 9, 1875; *EMJ*, Jan. 23, 1875; *LAT*, Feb. 3, 1897; Ingalls, op. cit., n. 3, 77–79; Mae Purcell, *History of Contra Costa County* (Berkeley: Gillick, 1940), 637–639; Bailey, op. cit., n. 7, 45–46.

48. Wallihan, op. cit., n. 33, 240, 384 adv.; *MSP*, Jan. 21, 1871; *Out West* (Denver), Apr.

20, 1872; *Early History of Omaha* (Omaha: Bee, 1876), 221; *NYT*, Oct. 15, 1890 (Balbach obit); *Biographical and Historical Memoirs, Adams, Clay, Hall and Hamilton Counties, Nebraska* (Chicago: Goodspeed, 1890), 100; Thomas W. Herringshaw, *Herringshaw's Encyclopedia of American Biography of the Nineteenth Century* (Chicago: American, 1902), 71, 415; Ingalls, op. cit., n. 3, 77–79, 188; James E. Fell, *Ores to Metals: The Rocky Mountain Smelting Industry* (Lincoln: University of Nebraska Press, 1979), 71–75; Bailey, op. cit., n. 7, 58–59.

49. *SLT*, Nov. 22, 1872, Jan. 1, 1875; *UMG*, Feb. 21, 1874; S. F. Emmons and G. F. Becker, *Statistics and Technology of the Precious Metals*, v. 13, *1880* (Washington, D.C.: G.P.O., 1885), 111, 431; Kantner, op. cit., n. 21, 128; Ingalls, op. cit., n. 3, 80, 188–189; NCAB, v. 14, 222, v. 23, 253, v. 30, 40; Fell, op. cit., n. 48, 95; Bailey, op. cit., n. 7, 110–118, 139–140, 143.

50. Ovando Hollister, *The Mines of Colorado* (Springfield, Mass.: Samuel Bowles, 1867), 159; James D. Hague, *Mining Industry* (Washington, D.C.: G.P.O., 1870) 584; Vickers, op. cit., n. 27, 452–453; William B. Vickers, et al., *History of the City of Denver* (Chicago: O. L. Baskin, 1880), 286–287, 453–454; Hubert H. Bancroft, "Life of Nathaniel P. Hill," in *Chronicles of the Builders of the Commonwealth* (San Francisco: History Co., 1891), v. 4, 377–398; William N. Byers, *Encyclopedia of Biography of Colorado; History of Colorado* (Chicago: Century, 1901), 473–474; Paul, op. cit., n. 1, 13–124; James E. Fell, "Nathaniel P. Hill," *Arizona & the West* 15 (Winter 1973), 315–329; Fell, op. cit., n. 48, 10–26.

51. Central City, Colo., *Register*, June 11, 1867, Dec. 19, 1871, June 11, 1873; Hague, op. cit., n. 50, 577–586; Wallihan, op. cit., n. 33, 227–228; Raymond, op. cit., n. 3, 4th rpt. (1873), 345–347; Fossett, op. cit., n. 27, 203–207; Fossett, op. cit., n. 28, 239, 244; Vickers, op. cit., n. 27, 452–453; Vickers, op. cit., n. 50, 287–288, 454; Byers, op. cit., n. 50, 473–474; Paul, op. cit., n. 1, 123–124; James E. Fell, "The Boston and Colorado Smelting Company" (M.A. thesis, University of Colorado, 1972), 44–77; Fell, op. cit., n. 50, 329–332; Fell, op. cit., n. 48, 25–41. The percentage return in 1869 is calculated from the smelter's average ores (e.g. Hague, p. 585, and Fell, pp. 62–63) which yielded 4 ounces of gold per ton, worth $20.67 an ounce, 10 ounces of silver, worth $1.33 an ounce, and 5 percent copper, worth $400 a ton, for a total value of $120 a ton. Hill returned to the mine owners 40 percent of the gold value, or $33, plus 75 cents an ounce for silver after reducing the number of ounces of silver by the number of percentage points of copper, which would leave just 5 ounces for an additional $3.75, and nothing for the copper, because he only paid $2 for each percentage point above 12 percent. Thus Hill returned to the mine owner a total of only $36.75 a ton on $120 ore, or 31 percent of the value, compared to the 69 percent that he took for the smelter. His smelter costs were not published, but since he worked ore yielding as little as 2 ounces of gold, or $41 a ton, without any silver and returned 20 percent, or $8 a ton, to the mine owners and took the remaining $33 a ton for smelting, it seems clear that the actual costs must have been less than $30 a ton to still pay at least a small profit. Thus Hill's net smelter profits must have been at least 40 percent or $48 a ton. No wonder the mine owners complained.

52. Raymond, op. cit., n. 3, 8th rpt. (1877), 379–382; Fossett, op. cit., n. 27, 203–207; Fossett, op. cit., n. 28, 239–244; Vickers, op. cit., n. 27, 453–454; Vickers, op. cit., n. 50, 287–289, 454–455; Thomas Egleston, *The Metallurgy of Silver, Gold, and Mercury in the United States* (New York: John Wiley, 1887), v. 1, 117–122; Byers, op. cit., n. 50, 233–235, 474–477; Paul, op. cit., n. 1, 124; Fell, op. cit., n. 50, 315–332; Fell, op. cit., n. 48, 41–54, 74–78, 84–89, 113–117.

53. *EMJ*, Jan. 23, 1875, Oct. 13, 1885; *MSP*, July 20, 1872, Nov. 25, 1876, July 7, 1883, Feb. 21, Mar. 14, 1885, June 24, 1916; *SFBul*, Apr. 29, May 11, July 28, 1881; Jesse Mason, *History of Amador County* (Oakland, Calif.: Thompson and West, 1881), 160; *SFCh*, Feb. 11, Apr. 27, Dec. 5, 1888; Rothwell, op. cit., n. 24, v. 1 (1892), 477; Thomas Arthur Rickard, "The Re-Opening of Old Mines Along the Mother Lode," *MSP*, June 24, 1916; Jack R. Wagner, *Gold Mines of California* (Berkeley, Calif.: Howell-North, 1970), 153–154.

54. *MSP*, Mar. 14, July 11, Sept. 26, Oct. 10–17, 1868, Jan. 6, 1869, Jan. 27, 1872, Jan. 23, 1875; *SFAC*, Jan. 9, 1870; *NYT*, Jan. 19, 1870, Mar. 23, 1873, Aug. 18, 1878, Mar. 20, 1902; *CMR*, June 24, 1882; *SFCh*, Mar. 31, 1895; Rickard, op. cit., n. 53; Wagner, op. cit., n. 53, 129–131.

55. Bancroft Scraps, op. cit., n. 2, v. 53, 779; Frank T. Gilbert, et al., *Illustrated History of Plumas, Lassen & Sierra Counties* (San Francisco: Fariss and Smith, 1882), 241–242, 481; *MSP*, Apr. 14, 1883; *LT*, Oct. 19, 1900; Thurman Wilkins, *Clarence King, A Biography* (Albuquerque: University of New Mexico Press, 1988), 172; W. Turrentine Jackson, "Lewis Richard Price," *Pacific Historical Review* 29 (Nov. 1960), 337–343, 348; Dahl, op. cit., n. 8, 183–191, 229; Clark C. Spence, "The Janin Brothers, Mining Engineers," *Mining History Journal* 3 (1996), 76–82.

56. *MSP*, July 30, 1870, July 23, 1881; *LT*, Jan. 29, 30, Dec. 2, 4, 1872; Mason, op. cit., n. 49, 158; *Black Range* (Chloride, N.M.), Dec. 7, 1883; Jackson, op. cit., n. 55, 344–347; Dahl, op. cit., n. 8, 86–93, 181–184, 129, 225, 229.

57. *NYT*, Feb. 1, July 3, 1872, Jan. 11, 1873, May 18, 1875, Dec. 20, 1894 (Kelly obit); *SFAC*, Feb. 5, 1872; *SFBul*, June 24, 1872; Mark Brumagim, *The Mariposas Estate* (New York: Russells, 1873), 191 ff.; Benjamin Silliman, *Review of . . . the Mariposa Estate* (New York, 1873), i–iv, 5–17, 54–57; Cecil G. Tilton, *William Chapman Ralston* (Boston: Christopher, 1935), 83–85.

58. *SFBul*, Jan. 26, Nov. 4, Dec. 16, 1874, May 8, 1875, Aug. 15, Oct. 17–18, Dec. 31, 1877, Jan. 7, July 26, 1878, Dec. 17, 1880, Jan. 16, 1883; *NYT*, Dec. 12, 1874, Jan. 6, Mar. 20, 27, May 18, 1875, Sept. 25, 1876, Apr. 20, 1877, Jan. 4, 11, Feb. 1, Mar. 18, Aug. 10, 1878; *The Mariposa Land and Mining Company of California* (New York: Evening Post, 1875), 1–15; *Semi-Annual Statement of the Mariposa Land and Mining Company of California* (New York: Evening Post, 1876), 148–149; *MSP*, Jan. 5, 1878, Dec. 25, 1880; *SFAC*, Oct. 1, 1881; *LAT*, July 24, 1887.

59. *NYT*, Feb. 25, 1871, Feb. 26, Nov. 25, 1874, Jan. 27, Apr. 14, 1875, Oct. 11, Nov. 17, 1877, Mar. 31, Apr. 14, Aug. 29, 1878, Sept. 17, 1879, Mar. 21, Apr. 27, July 25, 1880, Apr. 2, 1881; *MSP*, July 6, 1872, Jan. 24, 1874; Raymond, op. cit., n. 3, 4–7; Egleston, op. cit., n. 52, (1890), v. 2, 901.

60. *MSP*, Feb. 7, Aug. 22, 1874, Oct. 30, 1875, plus Mar. 14, 1874–Mar. 13, 1875 (New Incorp); Raymond, op. cit., n. 3, 6, 10, 17; C. Augustus Menefee, *History of Napa and Lake Counties* (San Francisco: Slocum, Bowen, 1881), 132; Egleston, op. cit., n. 52, 901; Jimmie Schneider, *Quicksilver: The Complete History of Santa Clara County's New Almaden Mine* (San Jose, Calif.: Zelda Schneider, 1992), 117–118.

61. *SFBul*, Feb. 20, Mar. 10, Nov. 24, Dec. 10, 1869, Mar. 9, Jul. 9, Aug. 15, Oct. 31, Nov. 10, 17, 21–22, Dec. 9, 12, 16, 1870, Jan. 17, Feb 2, 28, Mar, 9, 1871; *WNI*, Apr. 2, 10, 1869; *NYT*, Oct. 29, 1869, Oct. 26, Nov. 21, 1870, Jan. 11, 1871; New York *World*, Nov. 15, 1870; *Congressional Globe*, Feb. 18, 1871, 1402–1407; *The History of the McGarrahan Claim, As Written by Himself* (Washington, D.C., 1878), 109–213, 387; *Memorial of William McGarrahan* (Washington, D.C.: G.P.O., 1878) 765–766; *Congressional Record*, Jan. 16, 1893, 589–592; Egleston, op. cit., n. 52, 901; Theodore C. Smith, *The Life and Letters of James Abram Garfield* (New Haven, Conn.: Yale University Press, 1925), v. 1, 462–465; Claude G. Bowers, *The Tragic Era* (Boston: Houghton Mifflin, 1929), 327–328; Robert J. Parker, "William McGarrahan's 'Panoche Grande Claim,'" *Pacific Historical Review* 5 (Sept. 1936), 219–220; Charlene Gilbert, "Prospectors, Capitalists, and Bandits: The History of the New Idria Quicksilver Mine, 1854–1972" (M.A. thesis, San Jose State University, 1984), 37–38.

62. *Congressional Globe*, Feb. 18, 1871 (1404; App. 136); *SFBul*, Mar. 11–Apr. 27, 1878; *NYT*, Mar. 18, 1878; *Memorial*, op. cit., n. 61, iii–xiii, 949–953; *The History*, op. cit., n. 61, 109–213; Francis Bret Harte, *The Story of a Mine* (London: George Routledge, 1877), 188 pp.

63. Raymond, op. cit., n. 3, (1871) 109, (1872) 520–521; Lawrence, op. cit., n. 9, 174, 189–190; Eliot Lord, *Comstock Mining and Miners* (Washington, D.C.: G.P.O., 1883), 281–286; Davis, op. cit., n. 10, 414–415; Tilton, op. cit., n. 57, 221–223; Grant H. Smith, *The History of the Comstock Lode, 1850–1920* (Reno: Nevada State Bureau of Mines, 1943), 137–140.

64. *MSP*, Nov. 12–26, 1870, Apr. 15, 1871; *SFCh*, Mar. 19, 1871; William Wright, *History of the Big Bonanza* (Hartford, Conn.: American, 1876; reprinted New York: Alfred Knopf, 1947), 115–116; Lawrence, op. cit., n. 9, 26–28, 189–190; Phelps, op. cit., n. 24, v. 1 (1881), 223–230; Angel, op. cit, n. 1, 591–592; Lord, op. cit., n. 63, 281–283; George T. Marye, *From '49 to '83 in California and Nevada* (San Francisco: A. M. Robertson, 1932), 124–125; Tilton, op. cit., n. 57, 219–223; George D. Lyman, *Ralston's Ring* (New York: Scribner's Sons, 1937), 172–179; Smith, op. cit., n. 63, 128–131; Michael J. Makley, *The Infamous King of the Comstock: William Sharon* (Reno: University of Nevada Press, 2006), 81–82. With his carefully crafted statements to the directors and the brokers, Jones later claimed that he acted in good faith since he had told them that he believed the mine was promising.

65. *SFCh*, Mar. 19, 1871; *MSP*, Apr. 15, 1871, May 19, 1877; Alfred Doten, *The Journals of Alfred Doten, 1849–1903* (Reno: University of Nevada Press, 1973), 1118; Lord, op. cit. n. 63, 283–284; Hubert H. Bancroft, "Life of William Sharon," in *Chronicles of the Builders of the Commonwealth* (San Francisco: History Co., 1891), v. 4, 61; Davis, op.

cit., n. 10, 414–415; Marye, op. cit., n. 64, 126; Tilton, op. cit., n. 57, 223; Smith, op. cit., n. 63, 87, 129–130; Maureen A. Jung, "The Comstocks and the California Mining Economy, 1848–1900" (Ph.D. diss., University of California, Santa Barbara, 1988), 199; Makley, op. cit., n. 64, 82–83.

66. SFCh, Feb. 4, 7, Apr. 28, May 8, 16–19, June 5, 7, 1872; SFAC, Feb. 6, 13–16, May 17, 1872; SFBul, Feb. 7, 9–10, 1872, Apr. 8, 18, 23, 1879; SFC, May 16, 1872; MSP, Feb. 10, 17, May 4–25, 1872; Bancroft Scraps, op. cit., n. 2, 75–77; Raymond, op. cit., n. 3, (1873) 500–505; Doten, op. cit., n. 65, 1164–1165; SFSR, Dec. 22, 1879; Angel, op. cit, n. 1, 91; Lord, op. cit. n. 63, 290–293; Tilton, op. cit., n. 57, 274–277; Lyman, op. cit., n. 64, 211–218; Smith, op. cit., n. 63, 132–133; Jung, op. cit., n. 65, 146–149; Makley, op. cit., n. 64, 86–87. Note that the post-crash share price of Crown Point is adjusted for an 8.33 to 1 stock split.

67. SFCh, May 19, 1872; Tilton, op. cit., n. 57, 277–284.

68. SFAC, Feb. 16, May 19, 1872; SFCh, Dec. 29, 1872, Jan. 11, 1873; SFBul, Jan. 10, 1873; Lord, op. cit., n. 63, 286; Davis, op. cit., n. 10, 420–421; Smith, op. cit., n. 63, 134–135; BDAC, 1138, 1587; Jung, op. cit., n. 65, 147–150.

69. SFBul, Jan. 5, Mar. 21, June 9, Nov. 3, 1870, Dec. 7, 1872; Tucson Arizonian, Apr. 9, 1870; Arizona Citizen (Tucson), Nov. 19, 1870; MSP, May 7, July 16, Dec. 3, 1870; LMJ, Jan. 28, 1871; LT, Jan. 28–Feb. 1, 6, Mar. 24, 1871; NYT, Dec. 8, 1872; Henry G. Langley, The San Francisco Directory for 1873 (San Francisco: printed by author, 1873), 167, 194; Harpending, op. cit., n. 13, 175–179; Bruce A. Woodard, Diamonds in the Salt (Boulder, Colo.: Pruett, 1967), 11–16; Homer Milford, "Early History of the Lordsburg Mining District, The Burro Mines, 'Pater Argenti of the World,'" Gore Canyon Mine Reclamation Project Environmental Assessment, Appendix F (Santa Fe: Abandoned Mine Land Bureau, New Mexico Mining and Minerals Division, 2001), 11–14; Plazak, op. cit., n. 13, 78–96. There is an extensive popular literature on the diamond swindle that is nearly all based on Harpending's fallacious account, which he wrote to cover up his and Roberts's leading roles in the swindle and place the guilt solely on Arnold and Slack. Even Woodard's extensively researched study and Plazak's recent account suffer from their acceptance of Harpending's innocence and their failure to fully examine Roberts's career.

70. SFCh, Sept. 16, 1869, Oct. 22, 1871; George Roberts, letters to Asbury Harpending, Dec. 1, 1870, Jan. 4, 8, 20, Feb. 5, Mar. 20, 1871, in the Harpending Collection, op. cit., n. 11. These letters clearly indicate that Roberts was originally taken in by Arnold and Cooper. Harpending in England showed the letters to a friend, laughed, and said it was all "nonsense" and a "trick" (see Albert Rubery, LT, Dec. 19, 1874), but he didn't tell Roberts until he returned. Roberts then took over the scheme to blow it into a grand swindle and later gave the SFCh (Oct. 7, 1872) an appropriately deceptive account of how Slack brought the first diamonds to San Francisco.

71. "Agreement Made This Thirty-first Day of October 1871 between Philip Arnold and Asbury Harpending," and "Agreement Made between Asbury Harpending, William M. Lent, George S. Dodge, Philip Arnold, William Ralston, George D. Roberts

and Samuel L. M. Barlow," New York Oct. 31, 1871, plus a supplement agreement May 11, 1872, Philip Arnold's Power of Attorney to Asbury Harpending, July 20, 1872, its revocation in Oct. 1872, and George Roberts's letters to Harpending, Dec. 4, 13, 1871, Feb. 26, 1872, John Slacks's letter to Harpending, Apr. 26, 1872, William Ralston's telegrams to Harpending, Apr. 29, June 27, 1872, Harpending Collection, op. cit., n. 11; United States General Land Office, *United States Mining Laws and Regulations Thereunder* (Washington, D.C.: G.P.O., 1872), 1–19; *SFCh*, Nov. 27, 1872; *SFBul*, Dec. 6, 1872 (Lent), Jan. 20, 1875; *MSP*, Aug. 3, Dec. 14 (Lent), 1872; *LT*, Dec. 21 (Rubery), 23 (Rubery), 24 (Barlow, Dodge), 1874; Lawrence, op. cit., n. 9, 31–33; *BDAC*, 1562.

72. "Agreement," op. cit., n. 71; *SFCh*, Aug. 1–3, Nov. 25–27, 1872; *SFAC*, Aug. 1, Nov. 26, 1872; *SFBul*, July 31, Nov. 26, Dec. 6, 7, 1872, Jan. 20, 1875; *MSP*, Dec. 14, 1872; *LT*, Dec. 19 (Rubery), 22 (Janin), 1874; Lawrence, op. cit., n. 9, 50; Harpending, op. cit., n. 13, 215–216; Emmons, op. cit., n. 52; Clark Spence, *Mining Engineers and the American West* (New Haven, Conn.: Yale University Press, 1970), 113–114.

73. *SFBul*, July 31, 1872; *SFCh*, Aug. 1–3, 1872; *SFAC*, Aug. 1–2, 4, Nov. 26, 1872; *MSP*, Aug. 3, 1872; John Hay, letter to Samuel Barlow, July 16, 1872, Barlow Papers, Huntington Library, San Marino, quoted in Woodard, op. cit., n. 69, 46; *BDAC*, 1196.

74. *SFBul*, Aug. 2, Sept. 20, Oct. 7, 10, 1872; *MSP*, Aug. 3, 10, 24, Sept. 7 (adv.), 1872; *NYT*, Aug. 11, 1872; *SFCh*, Aug. 1–3, 17, Oct. 7, 8, 11, 15, Nov. 7, 10, 26–28, 1872; *LT*, Aug. 27, 29, 1872, Dec. 21, 24, 1874; Woodard, op. cit., n. 69, 85, 93.

75. *SFBul*, Nov 25–26, Dec. 5–7, 12, 16, 23, 1872, Jan. 20, 1875; *SFCh*, Nov. 25–28, Dec. 6–8, 12–13, 1872, Apr. 3, 1873; *SFAC*, Nov. 26–29, 1872; *MSP*, Nov. 30, Dec. 14, 1872; Clarence King, letter to the Board of Directors of the San Francisco and New York Mining and Commercial Company, San Francisco, Nov. 11, 1872, Appendix DD2 in *Report of the Secretary of War, Report of the Chief of Engineers* (Washington, D.C.: G.P.O., 1873), 1208–1210; Allen D. Wilson, "The Great California Diamond Mines: A True Story," *Overland Monthly* 43 (Apr. 1904), 291–294; Wilkins, op. cit., n. 54, 172–185; Robert Wilson, *The Explorer King: Adventure, Science, and the Great Diamond Hoax—Clarence King in the Old West* (New York: Scribner, 2006), 239–242, 247–250. Philip Arnold's net take from the diamond swindle can be constructed from the disposition of his quarter interest or 25,000 virtual shares in the proposed company, as detailed in the "Agreements" and other sources as noted below: Oct. 31, 1871, he received $100,000 from Lent and Ralston as their intial payment on 6,250 shares, and July 6, 1872, he received $150,000 as their final installment on those shares after Janin's report; July 17, 1872, he received $75,000 from Lent, plus a draft for $75,000 on Ralston and a draft for $150,000 from Roberts and Harpending on 15,000 shares, and he gave Harpending power of attorney to collect the latter two drafts and hold the money; Oct. 15, 1872, he received $197,183.10 from Harpending in payment of the two drafts (Woodard, op. cit., n. 69, 125) and revoked his power of attorney; this could have given Arnold a gross of as much as $522,000 plus a residual 3,750 shares to sell, if the scam had succeeded. But on Nov. 13, 1872, immediately after King first reported the fraud, Rubery retrieved "nearly $200,000" for Harpending and Roberts

(Rubery *LT*, Dec. 21, 1874) and on Mar. 24, 1873, Lent retrieved $150,000 in settlement of his suit for $350,000 (Woodard, 130), leaving Arnold at most about $172,000, but this probably also included Slack's share and any payoff to Cooper as well. Lent and Dodge's $100,000 for "Slack's interest" was paid directly to Roberts, not Slack (Dodge in *LT*, Dec. 21, 1874).

76. *SFCh*, Nov. 28, 1872; Edwin L. Godkin, "The Diamond Bubble and Its Bursting," *The Nation* 15 (Dec. 12, 1872), 379–380.

77. *SFCh*, Nov. 26, 28, 1872; *SFAC*, Nov. 27, 1872; Hubert H. Bancroft, *History of Arizona and New Mexico, 1530–1888* (San Francisco: History Co., 1889), 592; Harpending, op. cit., n. 13.

78. *LT*, Aug. 27, 29, Oct. 30, Nov. 18, Dec. 20, 1872, Dec. 18, 1874–Jan. 21, 1875, July 13, Nov. 10, 1876; *SFBul*, Oct. 20, 1876; *The History of The Times: The Tradition Established, 1841–1884* (London: Times, 1939), 543, 595–596.

79. *SFBul*, May 7, 1872; San Francisco *Stock Report & California Street Journal*, May 13, 1872; *MSP*, Jan. 25, 1873; Lawrence, op. cit., n. 9, 20–21; Balch, op. cit., n. 7, 520–521; California Secretary of State, Division of Archives, "General Index of Corporations Filed and Recorded 1851–1889, Book 1, A–Z," Sacramento; Smith, op. cit., n. 63, 132–133; Jung, op. cit., n. 65, 144.

80. Balch, op. cit., n. 7, 520; Emmons and Becker, op. cit., n. 49, 156; U.S. Bureau of the Census, *Historical Statistics of the United States: Colonial Times to 1970* (Washington, D.C.: G.P.O., 1975), 25, 31, and see notes 8 and 79 above.

81. *MSP*, Oct. 25, 1873.

CHAPTER 5. THE BIG BONANZA BOOM

1. Gold Hill, Nev., *News*, Mar. 1–2, 1873; *VCTE*, Mar. 2, 1873; *MSP*, Mar. 22, 1873; *NYT*, Jan. 23, 1874; *SFCh*, Dec. 9, 20, 1874, Jan. 5, 10, 1875; William Wright, *History of the Big Bonanza* (Hartford, Conn.: American, 1876; reprinted New York: Alfred A. Knopf, 1947), 363–367; Bancroft Scraps, v. 53 (California Mining Stocks), 648; Myron Angel, *History of Nevada* (Oakland, Calif.: Thompson and West, 1881), 617–619; Eliot Lord, *Comstock Mining and Miners* (Washington, D.C.: G.P.O., 1883), 310; Cecil G. Tilton, *William Chapman Ralston* (Boston: Christopher, 1935), 287–288; Grant H. Smith, *The History of the Comstock Lode* (Reno: Nevada State Bureau of Mines, 1943), 149–152; Michael J. Makley, *John Mackay* (Reno: University of Nevada Press, 2009), 69–72.

2. Wright, op. cit., n. 1, 399–405; *SFAC*, May 3, 1878; *NYT*, May 3, 12, 1878, Feb. 22, 1889, Dec. 30, 1894, July 21–24, 27, 1902; R. S. Lawrence, ed., *Pacific Coast Annual Mining Review and Stock Ledger* (San Francisco: Francis and Valentine, 1878), 21–24, 181–182, 186–188; Squire P. Dewey, *The Bonanza Mines and the Bonanza Kings of California. Their 5 Years Reign: 1875–1879* (San Francisco, 1879), 1–87; Angel, op. cit., n. 1, facing 48, 56, 591; Lord, op. cit., n. 1, 301–304, 309–311, 319–320, 419, 427; Alonzo Phelps, *Contemporary Biography of California's Representative Men.* (San Francisco: A. L. Bancroft, 1882), v. 2, 39–41, 72–73, 168–170, 195–196; Hubert H. Bancroft, "Life of James G. Fair," in

Chronicles of the Builders of the Commonwealth (San Francisco: History Co., 1891), v. 4, 209–236; Charles C. Goodwin, *As I Remember Them* (Salt Lake City: Commercial Club, 1913), 160–170, 178–184; George L. Upshur, *As I Recall Them* (New York: Wilson-Erickson, 1936), 82–124; Oscar Lewis, *Silver Kings* (New York: Alfred Knopf, 1947), 47–280; Smith, op. cit., n. 1, 116–119, 147–152, 254–255, 262–263; Ethel Manter, *Rocket of the Comstock* (Caldwell, Idaho: Caxton, 1950) 256 pp.; Allen Weinstein, "The Bonanza King Myth," *Business History Review* 42 (Summer 1968). 185–218; Maureen A. Jung, "The Comstocks and the California Mining Economy, 1848–1900" (Ph.D. diss., University of California, Santa Barbara, 1988), 199–202; Makley, op. cit., n. 1, 16–72.

3. Rossiter W. Raymond, *Statistics of Mines and Mining in the States and Territories West of the Rocky Mountains* (Washington, D.C.: G.P.O., 1872), 520–521, (1874) 534; *Annual Report of the Consolidated Virginia Mining Company* (San Francisco: Stock Report, 1877), 5–6; SFAC, Dec. 15, 1880; Lord, op. cit., n. 1, 309–315; Joseph L. King, *History of the San Francisco Stock and Exchange Board* (San Francisco: printed by author, 1910), 245; Smith, op. cit., n. 1, 159–160, 169–170; Manter, op. cit., n. 2, 166; Jung, op. cit., n. 2, 154–156, 222–223; Makley, op. cit., n. 1, 73–82.

4. SFAC, Jan. 7–8, 1875; VCTE, May 27, 1877; *Annual Report*, op. cit., n. 3, 5–6; Lawrence, op. cit., n. 2, 182, 188; SFSR, Dec. 22, 1879; Lord, op. cit., n. 1, 314–315, 433; Hubert H. Bancroft, *History of Nevada, Colorado, and Wyoming, 1540–1888* (San Francisco: History Co., 1890), 137–138; Upshur, op. cit., n. 2, 111–112; Smith, op. cit., n. 1, 157–158, 174; Jung, op. cit., n. 2, 155–157, 222–223; Makley, op. cit., n. 1, 75, 84–86.

5. NYT, July 31, 1877, July 17, 1881 (San Francisco assessment roll excerpts); SFBul, Jan. 10, 1878, May 16, 1879; Dewey, op. cit., n. 2, 57–58; John Trehane, *John H. Burke, et al., Plaintiffs, vs. James C. Flood, et al., Defendants. Opening Argument of John Trehane, Esq., on Behalf of the Plaintiffs* (San Francisco: A. J. Leary, 1880), 46–55; Lewis, op. cit., n. 2, 113; Smith, op. cit., n. 1, 259–263; Makley, op. cit., n. 1, 78, 205–206.

6. SFCh, Jan. 4, 1877; VCTE, May 27, 1877; SFSR, Dec. 22, 1879; Lord, op. cit., n. 1, 314–315.

7. SFCh, Dec. 20, 1874; Lawrence, op. cit., n. 2, 64–65; NYT, Mar. 14, 1880; G. Thomas Ingham, *Digging Gold Among the Rockies* (Philadelphia: Cottage Library, 1881), 406–412; James O'Leary, *Pacific Coast Mining Review 1888, Mines, Men, Money* (San Francisco: Journal of Commerce, 1889), 28; SFAC, Feb. 1, 11, 1890.

8. SFCh, Jan. 9, 1875; Benjamin E. Lloyd, *Lights and Shades in San Francisco* (San Francisco: A.L. Bancroft, 1876), 39; Dewey, op. cit., n. 2, 9–10; Angel, op. cit., n. 1, 618; Lord, op. cit., n. 1, 314; Bancroft, op. cit., n. 4, 137; Robert Louis Stevenson, *Works* (New York, 1906), v. 2, 194; Willard L. Thorp, *Business Annals: United States, England, &c* (New York: National Bureau of Economic Research, 1926), 94, 132–133; Smith, op. cit., n. 1, 171.

9. Lloyd, op. cit., n. 8, 33–40; SFBul, Apr. 27, 1876; EMJ, Sept. 9, Oct. 21, 1876, Feb. 10, June 16, 1877; *Frank Leslie's Illustrated Newspaper*, Nov. 24, 1877; Henry G. Langley, *The San Francisco Directory for 1877* (San Francisco: Langley, 1877), 933–934, 996; California Assembly, "Majority Report of the Committee on Corporations," Mar.

7, 1878, *Appendix to the Journals of the Senate and Assembly, 22nd Session* (Sacramento: S.P.O., 1878), v. 4, 3–11; Bancroft Scraps, op. cit., n. 1, v. 52, 3; Lawrence, op. cit., n. 2, 5–21; *SFSR*, Dec. 22, 1879; William R. Balch, *The Mines, Miners and Mining Interests of the United States in 1882* (Philadelphia: Mining Industrial Publishing Bureau, 1882), 475–481; Lord, op. cit., n. 1, 430–435; Hubert H. Bancroft, *History of California, 1860–1890* (San Francisco: History Co., 1891), v. 7, 676–677; King, op. cit., n. 3, 59; Robert Sobel, *The Curbstone Brokers* (New York: Macmillan, 1970), 51–52; Marian V. Sears, *Mining Stock Exchanges, 1860–1930* (Missoula: University of Montana Press, 1973), 40–43, 53–54, 202–203.

10. New York *Sun*, quoted in *Borax Miner* (Columbus, Nev.), Mar. 27, 1875; Peter R. Decker, *Fortunes and Failures* (Cambridge, Mass.: Harvard University Press, 1978), 182–183.

11. *SFCh*, Jan. 9, July 31, Aug. 1, Oct. 10, 1875; *SFAC*, July 29, 31, Aug. 6, 8, 1875, Mar. 1, 30, 1876; *SFBul*, July 30, Aug. 7, Oct. 12, 1875; Langley, op. cit., n. 9, (1878), 495, (1880), 523.

12. "The Bank Convention of 1875," *Bankers' Magazine* 9 (June 1875), 989; *SFCh*, Aug. 1, 1878; *NYT*, Aug. 4, 1878; *BE*, Aug. 4, 1878; United States Census, 1900, Massachusetts, Arlington, 4 June, Enumeration District No. 658, Sheet no. 4, line 80, Conant, Royal B.; *Wall Street Journal*, Oct. 2, 1926.

13. Chicago *Tribune*, Aug. 10, 1878; *NYT*, Oct. 1–2, 1878, Feb. 8, Mar. 6, 1880, Mar. 24, 1882, Aug. 26, 1884; Washington *Post*, Oct. 8, 1878; "Editorial Notes," *The Independent*, Oct. 10, 1878, p. 16; *SFAC*, Feb. 7, 1880; *NYTr*, Feb. 8, 1880, Mar. 22, 1882; United States Census, 1920, Massachusetts, Cambridge, Holy Ghost Hospital for Incurables, 21 January, Enumeration District no. 79, Sheet no. 21, line 35, Conant, Royal B.

14. *SFCh*, May 1–2, Aug. 5–7, 12, 19, 24, 1875; *SFBul*, Aug. 10, Sept. 11, Nov. 8, 26, Dec. 14, 24, 31, 1875, Jan. 3, 1876; *SFAC*, Nov. 9, Dec. 26, 1875, Mar. 16, 1878; *LAT*, July 11, 1892.

15. San Francisco *Post*, July 25, 1877; *SFC*, May 7, 1879; Bancroft Scraps, op. cit., n. 1, v. 53, 584–585, v. 55, 1353–1355.

16. *MSP*, Jan. 30, 1875; James Hezlep Galloway, "Diary, 1853–1883," manuscript, Bancroft Library, University of California, Berkeley, 266–267, 280–281, 288, 298–299, 351; excerpts as "Diary of James Galloway," *University of Nevada Bulletin* no. 45 (1947), 13–21.

17. *SFBul*, Nov. 17, 1877; *NSJ*, Nov. 21, 1877; Reno *Gazette*, July 30, Nov. 15, 1879, June 1, 1881, Mar. 30, 1882; Angel, op. cit., n. 1, 573–574, 618; Lord, op. cit., n. 1, 316; *NYT*, July 28, 1887; Bancroft, op. cit., n. 4, 113–114, 139; Bancroft, op. cit., n. 9, 675–676.

18. Raymond, op. cit., n. 3, 3rd rpt. (1872), 520–521, 4th rpt. (1873), 500–501, 5th rpt. (1874), 529, 532; Mary M. Mathews, *Ten Years in Nevada* (Buffalo, N.Y.: Baker, Jones, 1880; reprinted Lincoln: University of Nebraska Press, 1985), 1–9, 120–122, 175–176, 227–229, 333–337.

19. Alfred Doten, *The Journals of Alfred Doten, 1849–1903* (Reno: University of Nevada Press, 1973), 1157–1161, 1241–1246, 1318–1327, 1336–1347; Angel, op. cit., n. 1, 324–325.

20. Lloyd, op. cit., n. 8, 44–45; San Francisco *Post*, Apr. 14, 18, 1877; *VCTE*, Mar. 13, 1878; Bancroft Scraps, op. cit., n. 1, v. 53, 497–498.

21. John P. Young, *Journalism in California* (San Francisco: Chronicle, 1915), 80–81.

22. Lloyd, op. cit., n. 8, 43–44; *SFCh*, Mar. 4, 1878; Bancroft Scraps, op. cit., n. 1, v. 54, 843; King, op cit., n. 3, 249; famous quote from Henry Clews, *Twenty-eight Years in Wall Street* (New York: Irving, 1888), 416.

23. *SFAC*, Feb. 25, 1877; Bancroft Scraps, op. cit., n. 1, v. 52, 3, v. 53, 444; Paul Prowler, "Glimpses of Gotham," *National Police Gazette* 34 (June 14, 1879), 14–15.

24. *MSP*, Jan. 31, Feb. 7, Mar. 14, 28 1874; *SFAC*, Feb. 14–15, 18, 1874; *The Statutes of California* (Sacramento: 1875), 866–867; Glen A. Vent and Cynthia Birk, "Insider Trading and Accounting Reform: The Comstock Case," *Accounting Historians Journal* 20 (1993), 75–76.

25. William Sharon quoted in Zoeth Eldredge, "Banking in California," *History of California* (New York: Century, 1915), v. 5, 437; Ira B. Cross, *Financing an Empire: History of Banking in California* (Chicago: Clarke, 1927), v. 1, 402–403; Tilton, op. cit., n. 1, 331; George D. Lyman, *Ralston's Ring* (New York: Scribner's Sons, 1937), 221–222; Michael J. Makley, *The Infamous King of the Comstock: William Sharon* (Reno: University of Nevada Press, 2006), 105–106.

26. *SFCh*, Dec. 6, 1874; *MSP*, May 29, 1875; Raymond, op. cit., n. 3, (1875) 503; *NYT*, Dec. 15, 1876, Nov. 9, 1885, Mar. 2, 1909; Lawrence, op. cit., n. 2, 19, 28, 69–72, 228–229; Phelps, op. cit., n. 2, v. 1, (1881), 367–368; *SFC*, Nov. 14, 1885; Hubert H. Bancroft, *Chronicles of the Builders of the Commonwealth* (San Francisco: History Co., 1891), v. 3, 343–344, v. 4, 64–66; Cross, op. cit., n. 25, 404–405; Carl B. Glasscock, *Lucky Baldwin* (Indianapolis: Bobbs-Merrill, 1933), 155–164; Tilton, op. cit., n. 1, 329–332, 385–390; Lyman, op. cit., n. 23, 257–259; Smith, op. cit., n. 1, 163; Makley, op. cit., n. 25, 106–107.

27. *NSJ*, Jan. 5, 1875; *SFCh*, Jan. 11, 13–14, Aug. 27–30, 1875; *SFBul*, Jan. 12, Aug. 27–28, 30–31, 1875; *SFAC*, Aug. 28–30, 1875; *MSP*, Aug. 28, Sept. 4, 1875; *NYT*, Aug. 27–28, Sept. 4, Oct. 2, 1875; *Frank Leslie's Illustrated Newspaper*, Sept. 11, 1875; Lord, op. cit., n. 1, 430–432; Bancroft, op. cit., n. 26, v. 4, 65–66; Theodore H. Hittell, *History of California* (San Francisco: Stone, 1897) v. 4, 553–556; Eldredge, op. cit., n. 25, 437; Samuel P. Davis, *The History of Nevada* (Reno: Elms, 1913), 421; Cross, op. cit., n. 25, 403–405; Tilton, op. cit., n. 1, 338–348; Lyman, op. cit., n. 25, 293–313; *BDAC*, 1587; Smith, op. cit., n. 1, 186–190; Makley, op. cit., n. 25, 107–120.

28. *NYT*, Sept. 13, 1875; Bancroft, op. cit., n. 26, v. 4, 66–69; Cross, op. cit., n. 25, 404–407; Tilton, op. cit., n. 1, 384–390; Smith, op. cit., n. 1, 190–191.

29. *SFAC*, Jan. 13–14, May 15, 1875, Oct. 5, Nov. 2, 1883, June 1, Dec. 26, 1884, Nov. 14, 1885; *NYT*, Jan. 10, Nov. 26, 1880, Jan. 22, 1881, Sept. 9, 1883, May 27, 1884, Jan. 2, Sept. 9, Nov. 14, 1885, May 15, Aug. 15–16, 1889, Jan. 16, 1890, Feb. 15, Oct. 7, 1892; *SFC*, Nov. 14, 1885, Oct. 6, 1892; Bancroft, op. cit., n. 26, v. 4, 71–77; *BDAC*, 873; Milton S. Gould, *A Cast of Hawks* (La Jolla, Calif.: Copley, 1985), 168–305, 333–337; Makley, op. cit., n. 25, 168–209.

30. *SFCh*, Apr. 25, July 16, 1876; Lawrence, op. cit., n. 2, 28; *SFSR*, Dec. 22, 1879;

John P. Mains, *Annual Statistician, 1879* (San Francisco: McCarthy, 1879), 549–552; *NYT*, Jan. 4–5, 1913; *WSJ*, Jan. 4, 1913; Smith, op. cit., n. 1, 191–202; Makley, op. cit., n. 1, 111–113.

31. *MSP*, Aug. 2, 1873, Jan. 3, 1874; *EMJ*, Oct. 17, 24, 1874; Richard E. Lingenfelter, *Death Valley & the Amargosa* (Berkeley: University of California Press, 1986), 113–120.

32. *MSP*, Sept. 5, Dec. 12, 1874; *SFBul*, Sept. 30, Dec. 2, 1874; *SFAC*, Jan. 30, Feb. 1–2, 1875; *The Wonder Consolidated Mining Company and the Wyoming Consolidated Mining Company* (San Francisco: Stock Report, 1875), 1–7; Lingenfelter, op. cit., n. 31, 116–123.

33. *MSP*, Nov. 21, 1874, Aug. 14, Oct. 2, 1875, Jan. 22, May 27, 1876; *SFAC*, Feb. 1, 1875; *NYT*, Mar. 3–May 26, 1876; William M. Stewart, *Reminiscences* (New York: Neale, 1908), 262–263; Lingenfelter, op. cit., n. 31, 131–134.

34. "An Arizona Bonanza," *Scientific American* 36 (Jan. 20, 1877), 38; *MSP*, May 5, July 21, 1877; Richard J. Hinton, *The Hand-book to Arizona* (San Francisco: Payton, Upham, 1878), 138–139, 152; Lawrence, op. cit., n. 2, 40–41, 244; *SFCh*, June 14, 1880; *Arizona Sentinel* (Yuma), June 19, 1880; Phelps, op. cit., n. 2, v. 2 (1882), 128–132; Robert L. Spude, "Mineral Frontier in Transition: Copper Mining in Arizona, 1880–1885" (M.A. thesis, Arizona State University, 1976), 52–54; Jack San Felice, *When Silver Was King* (Mesa, Ariz.: Millsite Canyon, 2006), 9–11, 19–34, 55.

35. *MSP*, Nov. 17, 1877, Oct. 5, 1878, Dec. 13, 1879, Jan. 24, 1880, Jan. 19, 26, 1884; Lawrence, op. cit., n. 2, 244; *SFCh*, June 14, 1880; *Arizona Sentinel* (Yuma), June 19, 1880; Spude, op. cit., n. 34, 53–54.

36. *CMR*, Oct. 1, 1881; Tombstone *Epitaph*, Oct. 4, 1881; *MSP*, Jan. 26, 1884, Jan. 24, 1885, Jan. 16, 1886; O'Leary, op. cit., n. 7, 47; *BDAC*, 929; Spude, op. cit., n. 34, 54.

37. *MSP*, July 8, 1876, Oct. 26, 1878; *SLT*, Jan. 12–13, 30, Feb. 5, 16, 1878; *SFAC*, Jan. 24, 1878; *VCTE*, Feb. 5, 1878; "The Old Telegraph Mine," *Scientific American* 41 (Aug. 30, 1879), 56; "L. E. Holden," *Magazine of Western History* 11 (Feb. 1890), 438–443.

38. *MSP*, Oct. 6, Nov. 3, Dec. 22, 1877, Mar. 2, Oct. 26, 1878; *SLT*, Dec. 9, 1877, Jan. 3–13, 30, Feb. 5, 12–14, 1878; *SFAC*, Jan. 24, 1878; *VCTE*, Feb. 5, 1878.

39. *SLT*, Dec. 9, 13, 1877, Jan. 3–5, 8–13, 30, Feb. 5–8, 12–21, 1878, Feb. 21, Mar. 3, May 10, 1879; Chicago *Inter Ocean*, Dec. 10, 1877; *NYT*, Dec. 10, 1877, May 11, Nov. 8, 23, 24, 1879, July 6, 1880, Mar. 19, 1900 (Philippart obit), Aug. 27, 1913 (Holden obit); *MSP*, Dec. 15, 22, 1877, Mar. 2, Oct. 26, 1878, Feb. 22, Apr. 12, Oct. 25, Nov. 15, 29, 1879, June 20, 1880, Sept. 10, 1881; *SFAC*, Jan. 24, Feb. 21, 1878; *VCTE*, Feb. 5, 21, 1878; *SLGD*, Feb. 24, 1879; *EMJ*, Sept. 3–10, 1881; "L.E. Holden," op. cit., n. 37, 438–443; *NCAB*, v. 22, 230–231.

40. Butte *Miner*, June 3, July 1, 1876, Aug. 17, 28, Sept. 25, 1877; Michael Leeson, *History of Montana* (Chicago: Warner, Beers, 1885), 1326–1328, 1333–1334; Harry C. Freeman, *A Brief History of Butte* (Chicago: Shepard, 1900), 14–15; Christopher P. Connolly, "The Story of Montana, II," *McClure's Magazine* 27 (Sept. 1906), 457; William D. Magnam, *The Clarks* (New York: Silver Bow, 1941), 45–46; Michael P. Malone, *The Battle for Butte* (Seattle: University of Washington Press, 1981), 9–10, 15–16.

41. Butte *Miner*, June 1, 13, 24, July 10, Aug. 17, 24, Sept. 1, Dec. 12, 17, 26, 1876, Mar.

6, 20, Apr. 10, May 1, July 3, 17, Aug. 28, Sept. 25, Oct. 9, 1877; Balch, op. cit., n. 9, 1085; Leeson, op. cit., n. 40, 566, 954, 1078, 1326–1328; *NYT*, Mar. 14, 1890 (Davis obit), June 22, 1895, Aug. 24, 1897 (estate); Joaquin Miller, *An Illustrated History of the State of Montana* (Chicago: Lewis, 1894), 206; Henry Hall, *America's Successful Men of Affairs* (New York: Tribune, 1896), v. 2, 229–230; Freeman, op. cit., n. 40, 14–15, 39–41; Connolly, op. cit., n. 40, 455–457; Magnam, op. cit., n. 40, 43–46; Donald MacMillan, "Andrew Jackson Davis" (M.A. thesis, University of Montana, 1967), 4, 17–20, 24–25; Malone, op. cit., n. 40, 12–16.

42. Butte *Miner*, Sept. 19, Oct. 3, Dec. 12, 1876, Aug. 17, Sept. 25, Oct. 23, 1877, Feb. 16, 1884; *SLT*, Aug. 30, Nov. 17, 1877; Leeson, op. cit., n. 40, 953; Ogden *Standard*, June 16, 1899; Freeman, op. cit., n. 40, 41–45; John Lindsay, *Amazing Experiences of a Judge* (Philadelphia: Dorrance, 1939), 69–70; K. Ross Toole, "Marcus Daly" Ph.D. diss., University of Montana, 1948), 1–6, 12–21, 29; Leonard J. Arrington, "Banking Enterprises in Utah, 1847–1880," *Business History Review* 29 (Dec. 1955), 327–328; H. Miner Shoebotham, *Anaconda: Life of Marcus Daly* (Harrisburg, Pa.: Stackpole, 1956), 38–50; Jonathan Bliss, *Merchants and Miners in Utah: The Walker Brothers* (Salt Lake City: Western Epics, 1983), 183–198.

43. Butte *Miner*, Sept. 19, Oct. 3, Dec. 12, 1876, Aug. 17, Sept. 25, Oct. 23, 1877, Dec. 9, 1883, Feb. 16, 29, 1884; *EMJ*, Mar. 20, 1880, Mar. 5, Dec. 10, 1881, Feb. 4, 1882, Feb. 10, 1883, Mar. 1, May 3, Aug. 16, Dec. 20, 1884; *SLT*, Mar. 21, 1880; *Statement of the Alice G. & S. M. Co.* [New York, 1880], 1–5; *NYT*, Aug. 11, 1880, Apr. 10, 1881; Balch, op. cit., n. 9, 1152–1153; Leeson, op. cit., n. 40, 953; Toole, op. cit., n. 42, 12–15; Patrick Henry McLatchy, "A Collection of Data on Foreign Corporate Activity in the Territory of Montana, 1864–1889" (M.A. thesis, Montana State University, 1961), 93; Bliss, op. cit., n. 42, 194–195.

44. Butte *Miner*, Oct. 3, 1876, Feb. 16, 29, 1884; *EMJ*, Mar. 1, May 3, Aug. 16, Dec. 20, 1884; *Report of the Alice Gold & Silver Mining Co. from May 15, 1880 to January 1, 1884* (Salt Lake City: Tribune, 1884): The company's note for $391,834 (summed from payments and the amount still due) was included in the debits, but no money from it was included in the credits, and although its purpose was not disclosed, it seems quite clear since the amount coincided almost exactly (within 99.9 percent) with the $392,178 value placed on the mills, which were listed as the company's major asset; *MSP*, Jan. 24, 1885; Richard P. Rothwell, *The Mineral Industry* (New York: Scientific, 1893), v. 1, 475, v. 8, 796; Bliss, op. cit., n. 42, 195, 253.

45. *Moulton Mining Company. Capital $10,000,000. 400,000 shares $25 Each*, stock certificate no. 356, New York, May 3, 1881, and . . . *Capital $2,000,000. 400,000 shares $5 Each*, no. 158, New York Apr. 13, 1885; *MSP*, Jan. 28, 1882, Jan. 27, 1883, Jan. 19, Feb. 16, 1884, Jan. 24, 1885; *NYT*, Feb. 29, 1884, Oct. 20, 1899 (last Moulton dividend); Rothwell, op. cit., n. 44, v. 1, (1892), 477, v. 8, (1899), 797; McLatchy, op. cit., n. 43, 95.

46. Butte *Miner*, Aug. 3, 18, 21, 1881, Dec. 8, 1882, Mar. 24, 1885; *EMJ*, Mar. 26, July 9, Sept. 3–10, 1881, Mar. 10, May 5, Dec. 22, 1883, June 21, Dec. 20, 1884; *MSP*, Jan. 28, 1882, Jan. 27, 1883, Jan. 19, Feb. 16, 1884, Jan. 24, 1885; *Société Anonyme des Mines de*

Lexington. Capital Social: Vingt Millions de Francs Divise en 40,000 Actions de 500 Francs, stock certificate no. 21,383, Paris, 5 Janvier 1883; Leeson, op. cit., n. 40, 954; *NYT,* Aug. 13, 15, 1884, Mar. 14, 1890; Miller, op. cit., n. 40, 206–207; Hall, op. cit., n. 41, 229–230; McLatchy, op. cit., n. 43, 108; MacMillan, op. cit., n. 41, 25–33.

47. *MSP,* Mar. 12, 1870, Apr. 22, 1871, Nov. 24, 1877, Feb. 2, Apr. 20, July 20, 1878, May 31, Aug. 30, 1879, Sept. 4, Oct. 2, 1880; Harry L. Wells, *History of Nevada County* (Oakland, Calif.: Thompson and West, 1880), 193.

48. *MSP,* July 14, 1877; Joseph Wasson, *Bodie and Esmeralda* (San Francisco: Spaulding, Barto, 1878), 9–10, 25–31; Lawrence, op. cit., n. 2, 45, 247–248; *Report of the Standard Gold Mining Co.* (San Francisco, 1879), 1–11; "The Bodie Mining District," *Scientific American* 46 (Sept. 6, 1879), 148; *SFEx,* Oct. 6, 1882 (Cook obit); *SFAC,* Oct. 7, 1882, Feb. 27, 1889 (Cook obits); Balch, op. cit., n. 9, 1060–1061, 1145–1146; William B. Clark, *Gold Districts of California* (San Francisco: Division of Mines and Geology, 1970), 13; Warren Loose, *Bodie Bonanza* (New York: Exposition, 1971), 38–39; Michael H. Piatt, *Bodie* (El Sobrante, Calif.: North Bay, 2003), 31–33, 36, 40–42, 85–86.

49. Wasson, op. cit., n. 48, 25–27, 35–36; *MSP,* May 18, Aug. 17, Sept. 14, Nov. 23, 1878; Bodie, Calif., *Standard,* June 5, July 31–Aug. 21, 1878; *SFBul,* Aug. 16, 1878; Lawrence, op. cit., n. 2, 45, 176; *NYT,* Sept. 1, 1878, Jan. 5, 1879, Mar. 17, 1880; *EMJ,* Sept. 6, 1879; Loose, op. cit., n. 48, 40–46, 62–71; Piatt, op. cit., n. 48, 49–57.

50. *NYT,* Aug. 18, Sept. 8, 1878; Bodie, Calif., *Standard,* Aug. 28, 1878; *National Police Gazette* (New York), Sept. 21, 1878; Loose, op. cit., n. 48, 68–69; Piatt, op. cit., n. 48, 52.

51. *SFBul,* Aug. 16, 1878; *MSP,* Aug. 24, 1878, June 14, 1879; *EMJ,* Aug. 24, 1878, June 14, 28, Aug. 23, Sept. 6, 1879, Jan. 24, Feb. 7, 1880, Mar. 6, 1886; *NYMR,* Oct. 23, 1880; Balch, op. cit., n. 9, 1119, 1121; Piatt, op. cit., n. 48, 53–55, 119, 131–134, 144–145, 157–165.

52. *NYT,* Mar. 29, Sept. 8, 1878; *MSP,* Aug. 31, 1878; Bodie *Standard,* Sept. 4, 1878; *SFAC,* Oct. 11, 1878, Oct. 7, 1882, Feb. 27, 1889; *EMJ,* Jan. 24, 1880; Balch, op. cit., n. 9, 1060–1061, 1145–1146; *SFEx,* Jan. 28, 1884; "A Quarter of a Century's Fluctuations in Mining Stocks, 1871 to 1895," in Rothwell, op. cit., n. 44, v. 4, (1896), 69–70; Piatt, op. cit., n. 48, 85–86, 134–144, 165–167.

53. William Bronson and T. H. Watkins, *Homestake: The Centennial History* (San Francisco: Homestake, 1977), 3; Duane A. Smith, *Staking a Claim in History: The Evolution of Homestake Mining Company* (Walnut Creek, Calif.: Homestake, 2001), 2, 187.

54. *NYT,* July 5, Aug. 23, 1874, June 14, 1875, June 12, July 12, 1876; see also Watson Parker, *Gold in the Black Hills* (Norman: University of Oklahoma Press, 1966), 38–54.

55. *Black Hills Pioneer* (Deadwood, S.Dak.), July 8, 1876, Feb. 3, 1877; Moses Manuel, untitled manuscript quoted in John S. McClintock, *Pioneer Days in the Black Hills* (Deadwood, S.Dak.: 1939), 235–237; Lawrence, op. cit., n. 2, 35–36; William B. Vickers, et al., *History of the City of Denver* (Chicago: Baskin, 1880), 323–325; Frank S. Peck, *Map of Ore District of the Northern Black Hills* (Deadwood, S.Dak.: Peck, 1904), map; Rockwell D. Hunt, ed., *California and Californians* (Chicago: Lewis, 1926), v. 3, 522–524; Emma Myron, "The History of the Homestake Mine" (M.A. thesis, University of South Dakota, 1928), 4–7; Mildred Fielder, *The Treasure of Homestake Gold* (Aberdeen,

S.Dak.: North Plains, 1970), 20–40; Joseph H. Cash, *Working the Homestake* (Ames: Iowa State University, 1973), 14–16; Bronson and Watkins, op. cit., n. 53, 24–25; Joel Waterland, *The Spawn & the Mother Lode* (Rapid City, S.Dak.: Grelind, 1987), 123–128; Smith, op. cit., n. 53, 21–22.

56. *Black Hills Times* (Deadwood, S.Dak.), Oct. 13, Nov. 18, 30, Dec. 17, 1877; *Black Hills Champion* (Deadwood, S.Dak.), Oct. 27, 1877; *Black Hills Pioneer* (Deadwood, S.Dak.), Mar. 7, 1878; George Hearst, letter to James Haggin, May 23, 1878, quoted in Fielder, op. cit., n. 55, 57; *Homestake Mining Company. Incorporated Nov. 5th 1877. Capital $10,000,000. 100,000 Shares. 100 Dolls. Each*, stock certificate no. 246, San Francisco, July 18, 1878; George Hearst, *The Way It Was* (New York: Hearst Corp., 1972), 22; Myron, op. cit., n. 55, 8; Cash, op. cit., n. 55, 16; Bronson and Watkins, op. cit., n. 53, 31; Smith, op. cit., n. 53, 23, 26.

57. *Black Hills Times* (Deadwood, S.Dak.), Oct. 18, Nov. 22, Dec. 12, 1877, Jan. 26, Aug. 15, Sept. 5, 18, 27, 1878, Feb. 18, 22, Mar. 26, 1879; Lawrence, op. cit., n. 2, 37, 64–65, 199; *SFBul*, Aug. 21, 1879; Vickers, op. cit., n. 53, 324; Fielder, op. cit., n. 55, 47–63; Mel Gorman, "Financial and Technological Entrepreneurs in the Black Hills," *Huntington Library Quarterly* 45 (Spring 1982), 137–154; Waterland, op. cit., n. 55, 145–146; Smith, op. cit., n. 53, 26.

58. Lawrence, op. cit., n. 2, 204; *Black Hills Pioneer* (Deadwood, S.Dak.), June 26, 1878; *Black Hills Times* (Deadwood, S.Dak.), Jan. 6, 11, 14, 17–18, 24–25, Feb. 5, 1879; *NYT*, Jan. 25, 1879; Jesse Brown and A. M. Williams, *The Black Hills Trails* (Rapid City, S.Dak.: Journal, 1924), 366–368; Parker, op. cit., n. 54, 197; Fielder, op. cit., n. 55, 63–65.

59. *Black Hills Times* (Deadwood, S.Dak.), Jan. 18, 24–25, Feb. 13, 15, Mar. 12–13, 18–22, Apr. 1, 1879; *SLGD*, Mar. 20, Apr. 18, 1879; George Hearst, letters to James Haggin, Mar. 6 and 19, 1879, quoted in Bronson and Watkins, op. cit., n. 53, 32, and Fielder, op. cit., n. 55, 65; Brown and Williams, op. cit. n. 58, 366–368; Parker, op. cit., n. 54, 197; Fielder, op. cit., n. 55, 64–65.

60. *MSP*, Oct. 19, 1878 (new incorp.); George Hearst, letter to James Haggin, Mar. 6, 1879, quoted in Fielder, op. cit., n. 55, 73–75; Lawrence, op. cit., n. 2, 35–36, 199; *SFAC*, July 9, 1881, Sept. 16, 1883; *NYT*, July 9, 1881; *EMJ*, Sept. 22, 1883; Hunt, op. cit., n. 55, 522–524; Smith, op. cit., n. 53, 37–38.

61. *NYT*, Jan. 24, 1879; *Black Hills Times* (Deadwood, S.Dak.), Mar. 6, Aug. 27, 1879; George Hearst, letter to James Haggin, Mar. 9, 1879, quoted in Smith, op. cit., n. 53, 37; McClintock, op. cit., n. 55, 297; *BDAC*, 1347; Fielder, op. cit., n. 55, 66–71; Bronson and Watkins, op. cit., n. 53, 35; Smith, op. cit., n. 53, 37.

62. *Black Hills Pioneer* (Deadwood, S.Dak.), Mar. 4, 1879; *NYT*, Mar. 7, 1880; *EMJ*, Aug. 7, Oct. 2, 1880; *Black Hills Times* (Deadwood, S.Dak.), Mar. 2, 28, May 8, June 2, July 8, Aug. 25, Sept. 8, 12, 21, Dec. 29, 1880, Jan. 5, 1881; Gorman, op. cit., n. 57, 146–149.

63. Hearst, op. cit., n. 56, 21; Lenox H. Rand and Edward B. Sturgis, *The Mines Handbook*. (Suffern, N.Y.: Mines Info. Bureau, 1931), v. 18, 1773–1774; Fremont and Cora

Older, *The Life of George Hearst* (San Francisco: Nash, 1933), 156; Andrew V. Corry and Oscar E. Kiessling, *Grade of Ore* (Philadelphia: WPA., 1938), 28–29; Myron, op. cit., n. 55, 19–20; *Homestake Centennial 1976* (San Francisco: Homestake, 1976), 13; Keith R. Long, John H. DeYoung, and Steve Ludington, "Significant Deposits of Gold, Silver, Copper, Lead, and Zinc in the United States," *Economic Geology* 95 (May 2000), 629, 635–637; Smith, op. cit., n. 53, 187.

64. *SFBul*, Feb. 24, Apr. 25, 27, May 19, 1864; *Charter and By-Laws of The People's Gold and Silver Mining Company, Amador County* (San Francisco, 1864), 1–12; *SacU*, June 17, 1865; *NYT*, June 28, 1873; *VCTE*, July 10, 12, 1873; Butte *Miner*, Oct. 3, 1876; Anthony Comstock, *Frauds Exposed* (New York: Brown, 1880), 132–145; Philip Nordell, "Pattee, the Lottery King," *Annals of Wyoming* 34 (Oct. 1962), 193–204.

65. *NYT*, Aug. 21, 26, Dec. 18, 1876; Comstock, op. cit., n. 64, 125–130; Nordell, op. cit., n. 64, 203–204.

66. *NYT*, Aug. 21, 26, Dec. 18, 1876, June 25, July 2, 1877; *Silver Mountain Mining Company. Shares $100 Each. 100,000 Shares*, stock certificate no. 971, New York, Mar. 5, 1877; *RMN*, July 12, 1877; Comstock, op. cit., n. 64, 114–124.

67. *NYT*, June 25, July 2, 1877, June 4–5, 9, 1878, Feb. 28, 1879, Dec. 29, 1881; Central City, Colo., *Register-Call*, Dec. 20, 1878, Feb. 14, 1879; Comstock, op. cit., n. 64, 122–123, 130–131; Nordell, op. cit., n. 64, 206–210.

68. Sacramento *Bee*, Nov. 11, 1875; *MSP*, Nov. 20, 1875; Columbus, Nev. *Borax Miner*, Dec. 4, 1875; *Inyo Independent* (Independence, Calif.), Dec. 4, 1875.

69. Mains, op. cit, n, 30, (1878), 582–587; *MSP*, Jan. 29, 1881; Angel, op. cit., n. 1, 620; Balch, op. cit., n. 9, 457; California Secretary of State, Division of Archives, "General Index of Corporations Filed and Recorded 1851–1889, Book 1, A–Z," Sacramento; U.S. Bureau of the Census, *Historical Statistics of the United States: Colonial Times to 1970* (Washington, D.C.: G.P.O., 1975), 25, 31, 255; Jung, op. cit., n. 2, 144–145.

70. Lawrence, op. cit., n. 2, 177–178, 209–210, 231, 239–242; Lord, op. cit., n. 1, 419–421.

71. Unsigned letter to "William Sharron, May 10," no year, in the archives of the Bank of California, quoted in David Lavender, *Nothing Seemed Impossible* (Palo Alto, Calif.: American West, 1975), 220.

72. Langley, op. cit., n. 9, (1871), 518, (1875), 585, (1877), 689 (partnership of Pearson and Hearst); *VCTE*, Oct. 4–9, 1874; *NYT*, Oct. 12, 1874; *MSP*, Jan. 23, Nov. 6–13, Dec. 18, 1875; *SFCh*, Oct. 12, 1875; *SFBul*, Nov. 4, 1875, Dec. 12, 1877; *SFAC*, Dec. 12, 1875; Justice Mining Company, *Report of the Committee on Investigation* (San Francisco, 1878), 18–19.

73. *SFCh*, Dec. 3, 1877; *SFBul*, Dec. 4, 26–28, 1877, Jan. 4, 1878; Bancroft Scraps, op. cit., n. 1, v. 53, 718–719; Justice Mining Company, op. cit., n. 70, 1–32; Lawrence, op. cit., n. 2, 61, 177; Dan Plazak, *A Hole in the Ground with a Liar at the Top* (Salt Lake City: University of Utah Press, 2006), 38–39.

74. *SFBul*, Dec. 4–8, 12–13, 26–28, 1877, Jan. 15, Mar. 13, 1878, Apr. 1, 11–12, May 24, June 2, July 3, 18, 29, 31, 1879, Mar. 13, 1880; *SFCh*, Dec. 12–13, 1877, Jan. 11, 1878, Apr. 1, 8, 1879; San Francisco *Post*, Dec. 12, 1877, Mar. 7, 1878, Apr. 5, 8–9, 1879; *SFC*, Dec. 13, 1877; *SFAC*, Dec. 15, 1877, Jan. 15, 18, Dec. 23, 1878, May 4, June 4, 1879; Bancroft

Scraps, op. cit., n. 1, v. 53, 718–719, 722–723, 729, 731, v. 54, 848–849, v. 94, 721–724, 753, 864–874; *VCTE*, Jan. 15–18, Mar. 8–9, 1878; *NYT*, Jan. 28, 1878, June 4, 1879; *RG*, Mar. 25, 1878, Apr. 14, 24, 1879; Justice Mining Company, op. cit., n. 70, 1–32; Lawrence, op. cit., n. 2, 209–210; Richard E. Lingenfelter, *The Hardrock Miners* (Berkeley: University of California Press, 1974), 56–58; Plazak, op. cit., n. 73, 38–39.

75. Squire P. Dewey, *The Fruits of the Bonanza* (San Francisco: Winterburn, 1878) 1–38; *Gleanings From the Inside History of the Bonanza* (San Francisco, 1878), 6–9; *SFBul*, May 25, 1878; *VCTE*, May 26, Dec. 11, 17, 22, 1878; Squire P. Dewey, *The Bonanza Mines of Nevada* (San Francisco, 1879), 13–14; Phelps, op. cit., n. 2, v. 1 (1881), 255–269; *SFAC*, May 1, 1889; *NYT*, May 1, 1889; Smith, op. cit. n. 1, 218–222; Jung, op. cit., n. 2, 170–177, 200–201.

76. *SFBul*, May 25, 1878; *VCTE*, May 26, 1878; *Gleanings*, op. cit., n. 75, 7–11; *The Bonanza Mines*, op. cit., n. 75, 14–17; Plazak, op. cit., n. 73, 36; Makley, op. cit., n. 1, 113–114.

77. *SFAC*, Jan. 12, 1877; *SFCh*, Jan. 12, 1877; *NYT*, Jan. 12, 1877; San Francisco *News Letter*, Jan. 18, 1877; *SFBul*, May 25, 1878; *VCTE*, May 26, 1878; *Gleanings*, op. cit., n. 75, 11–14; *The Bonanza Mines*, op. cit., n. 75, 17–19, 29–34; Makley, op. cit., n. 1, 114–116.

78. *SFAC*, Jan. 12, 1877; *SFCh*, Jan. 12, 1877; *NYT*, Jan. 12, 1877; San Francisco *News Letter*, Jan. 18, 1877; *The Bonanza Mines*, op. cit., n. 75, 19–23, 34–37; Smith, op. cit. n. 1, 203–204, 254–255, 260–261; Makley, op. cit., n. 1, 115. Dewey had confused the standard "dry" weight measures of assays and the "wet" weight measures of rock coming out of the mine, which held roughly 12 percent water. Thus he hadn't corrected the weight of rock sent to the mills and thought the assays implied 12 percent more metal in the rock that wasn't being recovered.

79. *SFAC*, Jan. 12–14, 1877; *SFBul*, Jan. 12, 1877, May 20, 27, 1878; *SFCh*, Jan. 12, 1877; San Francisco *Post*, Jan. 12, 1877; *NYT*, Jan. 12, Feb. 7, 1877; San Francisco *Mail*, Jan. 14, 1877; San Francisco *News Letter*, Jan. 18, 1877; *Gleanings*, op. cit., n. 75, 15–26, 32–37; *The Bonanza Mines*, op. cit., n. 75, 37–42; John H. Burke, "The Bonanza Suits of 1877," in *History of the Bench and Bar of California* (Los Angeles: Commercial, 1901), 95–97; Plazak, op. cit., n. 73, 36–37; Makley, op. cit., n. 1, 116–118, 129–130.

80. *SFCh*, May 19, Sept. 16, 1878, Mar. 31, 1881; *SFBul*, May 20, 25, Sept. 16, Oct. 16, 22, 1878; *VCTE*, May 21, 1878; *NYT*, May 26, 1878; *SFAC*, Aug. 29, 1880; Trehane, op. cit., n. 5, 60–62; Burke, op. cit., n. 79, 95–96.

81. *Gleanings*, op. cit., n. 75, 11–14; *VCTE*, May 28, 1878; *Annual Report*, op. cit. n. 3, (1879), 5–24; *The Bonanza Mines*, op. cit., n. 75, 1–78; *SFAC*, Nov. 29, Dec. 5, 1879, Aug. 29, Nov. 12, 16–19, 1880, Mar. 31, 1881; *SFBul*, Jan. 20, 30, Nov. 12, 16, 18–19, 1880; *NSJ*, Sept. 3, 1880; *NYT*, Nov. 23, 1880; Trehane, op. cit., n. 5, 1–28, 60–62.

82. *SFBul*, Dec. 14–20, 22–25, 27–28, 1880, Mar. 30, 1881; *SFAC*, Dec. 15, 1880, Mar. 31, Aug. 20, 1881; *MSP*, Apr. 2, 1881; Burke, op. cit., n. 79, 96.

83. *NYT*, May 10, 1879, Aug. 29, 1880, May 13, 1881; *SFAC*, Aug. 29, 1880, Mar. 31, 1881; *SFBul*, May 11–12, 1881, July 21, 1882; *SFEx*, May 12–13, 1881; *EMJ*, May 14, 1881; Burke, op. cit., n. 79, 95–97.

84. *MSP*, Oct. 25, 1873; *UMG*, Nov. 1, 1873.

85. Adam Smith, *An Inquiry into the Nature and Causes of the Wealth of Nations* (London: Strahan and Cadell, 1776), v. 2, 154–155; Lord, op. cit., n. 1, 419–421.

86. Raymond, op. cit., n. 3, (1874) 495, (1875) 465; Lawrence, op. cit., n. 2, 37, 182, 188; Lord, op. cit., n. 1, 430.

87. Adolph Sutro, "Mr. Sutro's Argument," *Report of the Commissioners . . . in Regard to the Sutro Tunnel* (Washington, D.C.: G.P.O., 1872), 395–434; *Report of the Superintendent of the Sutro Tunnel Company* (Washington, D.C.: M'Gill and Witherow, 1873), 27–28; *LT*, Jan. 6, 1874; *NYT*, June 20, July 8, 10, 1878; *SFBul*, July 1, 9–10, 17, Sept. 3, 1878; *EMJ*, July 13, 1878; Angel, op. cit., n. 1, 508–509; Lord, op. cit., n. 1, 334–339, 342; *LAT*, Jan. 13, 1887; Bancroft, op. cit., n. 4, 145–147; Robert and Mary Stewart, *Adolph Sutro* (Berkeley, Calif.: Howell-North, 1962), 63, 91, 113, 119–122, 130, 133, 149–151, 169.

88. *SFBul*, Sept. 18, Oct. 2, 1877, June 12, Dec. 9, 1878, Jan. 29–30, Feb. 1, 3, 18, Mar. 20, 27, Apr. 1, July 1, 1879, Oct. 18, 1887; *NYT*, June 9, July 24, 1878, Feb. 1–2, 4, 17, Mar. 7, 19, Apr. 2–3, May 16, 1879; Lord, op. cit., n. 1, 416; Bancroft, op. cit., n. 4, 145–147; Smith, op. cit., n. 1, 297; Stewart, op. cit., n. 87, 153–156, 168.

89. *SFBul*, July 17, Oct. 30, Dec. 9, 1878, July 1, Nov. 7, 1879; *NYT*, July 24, Dec. 13, 1878, July 1, 18, Aug. 11, 1879; weekly stock sales in *EMJ*, Dec. 14, 1878–May 29, 1880; Angel, op. cit., n. 1, 511; Joseph Aron, *History of a Great Work* (Paris: E. Regnault, 1892), 5–72; Joseph Aron, in his Paris *L'Or & l'Argent, Revue Independante*, Aug. 1, 1895; Goodwin, op. cit., n. 2, 243–244; Stewart, op. cit., n. 87, 87, 165–168. The Stewarts (p. 168) conclude that Sutro only got a little over $900,000 from his tunnel stock, based on the secret sales of 300,000 shares by his broker in the spring of 1880, but they ignore the other 300,000 shares of his stock, which must have been sold privately or secretly in 1879, as well as his private sales of the 300,000 appropriated shares, which Aron claims. They say the latter sales can be refuted by the company records, but how else can we account for Aron's intense ire over his "dearly bought experience" (*L'Or & l'Argent*, p. 4)?

90. *SFBul*, Jan. 12, Oct. 18, 1887; *NYT*, Jan. 13, 1887; Aron, *L'Or & l'Argent*, op. cit., n. 89; Stewart, op. cit., n. 87, 161–162, 165, 169, 172–173, 195–196.

91. *SFBul*, Apr. 5, Nov. 21, 1879; *SacU*, Nov. 21, 1879; *SFAC*, Nov. 21, 1879; *NYT*, Nov. 21, 29, 1879; *SFCh*, Nov. 27, 1879; *Inyo Independent* (Independence, Calif.) Dec. 6, 1879; Smith, op. cit., n. 1, 222–227; Makley, op. cit., n. 1, 136–139.

92. *SFCh*, Jul. 20, 1878; *SFBul*, Aug. 8, 10, 20, 1878; *MSP*, Aug. 31, Sept. 7, 1878; *NYT*, Sept. 8, 1878, Aug. 25, 1881, July 18, 1885; Lawrence, op. cit., n. 2, 241–242; Sacramento *Record-Union*, Nov. 23, 1878; Bancroft, op. cit., n. 4, v. 54, 830–831; *National Police Gazette* (New York), Mar. 12, 1881.

93. *SFCh*, Sept. 1, 1878; *SFAC*, Sept. 3, 1878; *SFBul*, Sept. 12, 1878; Lawrence, op. cit., n. 2, 241–242; *MSP*, Oct. 5, 1878; Sacramento *Record-Union*, Nov. 23, 1878; Bancroft, op. cit., n. 4, v. 54, 798, 804, 830–831, 872–873; Doten, op. cit., n. 19, 1324–1325; *NYT*, Dec. 16, 1878; *NSJ*, May 3, 1883.

94. *SFAC*, Nov. 19–21, 24, 1878; *SFBul*, Nov. 19, 21, 1878, July 17, 1885 (obit), Oct. 2, Nov. 8, 1886; *VCTE*, Nov. 19–23, 27, Dec. 1, 1878; *NSJ*, Nov. 21, 1878, May 2, 1883; Reno

Gazette, Nov. 21–22, 26, 1878; *NYT*, Nov. 21, 26, Dec. 16, 1878, Mar. 14, 1880, Aug. 18, 25, 1881, July 18, 1885; Sacramento *Record-Union*, Nov. 23, 1878; *MSP*, Nov. 23–30, 1878; *PO*, Nov. 26, 1878; *SFCh*, Nov. 26, 1878, Aug. 14, 1881, July 17, 1885, July 28, 1903; Bancroft, op. cit., n. 4, v. 54, 830–835; Doten, op. cit., n. 19, 1334–1337; *SFEx*, Aug. 30, 1882; King, op. cit., n. 3, 277.

95. *MSP*, Aug. 24–Nov. 30, 1878 (new incorp); *SFAC*, Nov. 24, 1878; San Francisco *Post*, in *SLGD*, Dec. 26, 1878; *NYT*, July 31, Aug. 1, 1879; *NYMR*, Oct. 9, 1880.

96. *VCTE*, Nov. 20, 22, 26, 1878; Sacramento *Record-Union*, Nov. 23, 1878.

97. *SFCh*, May 20, 1875, Feb. 21, 1878, Apr. 6, July 20, 1879; *SFBul*, Jan. 24, 1877; *SFC*, May 7, 1879; Lloyd, op. cit., n. 8, 43; *MSP*, Aug. 28, 1875, Dec. 14, 1878; Bancroft Scraps, op. cit., n. 1, v. 52, 440, v. 55, 1328–1329, 1354, 1431–1432; King, op. cit., n. 3, 293–294; W. John Carlson "A History of the San Francisco Mining Exchange" (M.A. thesis, University of California, Berkeley, 1941), 94–100.

98. California Assembly, op. cit., n. 9, 3–11; *SFAC*, Feb. 7, 10, 12, 14, 18, 1878; *SFBul*, Feb. 4, 7, 1878; *SFCh*, Feb. 9, 11, 16, 20, 25, 28, Mar. 13, 18, 1878; San Francisco *Post*, Feb. 9, 11–12, 14, 18, 27, Mar. 14, 29–30, 1878; *VCTE*, Feb. 13, 16, 19, Mar. 15, 24, 26, 30, 1878; Reno *Gazette*, Mar. 25, 1878; Bancroft Scraps, op. cit., n. 1, v. 54, 804–823, 828–838, 843, 852–858, 862–863, 873–874, 880–881; *Debates and Proceedings of the Constitutional Convention of the State of California* (Sacramento: State Office, 1880), 808.

99. *Debates*, op. cit., n. 98, 805–811; "Constitution of the State of California," Article IV, Section 26, and Article XII, and "Statutes of California," chaps. CXVIII and CXXI in *The Statutes of California, Passed at the Twenty-Third Session of the Legislature, 1880* (Sacramento: State Office, 1880), xxviii, xxxvi–xxxviii, 131–132, 134–136; *SFBul*, Dec. 23, 1878; *SFAC*, Apr. 10, May 2, 1879; *SFCh*, Feb. 6, 1880; *VCTE*, May 5, 1880; Vent and Birk, op. cit., n. 24, 76–80.

100. *SFAC*, Mar. 19, 29, May 2, 1879; *SFBul*, Mar. 9, 1880; *NYT*, Dec. 5, 1880, Jan. 6, 1903; *Debates*, op. cit., n. 98, 808; Bancroft Scraps, op. cit., n. 1, v. 55, 1577; "Otis v. Parker," 187US606(1903); *LAT*, Jan. 6, 1903; *WSJ*, Jan. 6, 1903; *Securities Exchange Act of 1934* (Washington, D.C.: G.P.O., 1934), 1–42; Richard Budolfson, "Margin Requirements," *Monthly Review, Federal Reserve Bank of Minneapolis* (Oct. 1963), 2–5, 10–11.

101. Angel, op. cit., n. 1, 620; Lord, op. cit., n. 1, 430–435.

102. *MSP*, Dec. 26, 1874; Angel, op. cit., n. 1, 620; Lord, op. cit., n. 1, 430–435; Smith, op. cit., n. 1, 259–262; Jung, op. cit., n. 2, 188–203; Long et al., op. cit., n. 63, 635.

CHAPTER 6. THE SILVER SHARKS

1. *NYT*, Feb. 14, 21, 1880, May 29, 1884; Joseph E. King, *A Mine to Make a Mine: Financing the Colorado Mining Industry* (College Station: Texas A&M University Press, 1977), 89–90.

2. *The New York Mining Directory* (New York: Hollister and Goddard, 1880), 83–95; William R. Balch, *The Mines, Miners and Mining Interests of the United States* (Philadelphia: Mining Industrial Publishing Bureau, 1882), 475–481.

3. *NYT*, Jan. 1, Feb. 14, 21, 27, Mar. 4, 16, Apr. 13, June 2, 19, 1880, May 29, 1884; *EMJ*, Jan. 3, 1880; *Frank Leslie's Illustrated Newspaper*, June 19, 1880; Balch, op. cit., n. 2, 482–487; Frederick H. Smith, *Rocks, Minerals and Stocks* (Chicago: Railway Review, 1882), 211; *NCAB*, v. 5, 478; *WSJ*, May 25, 1906; W. John Carlson, "A History of the San Francisco Mining Exchange" (M.A. thesis, University of California, Berkeley, 1941), 43–46; King, op. cit., n. 1, 101–103.

4. *NYT*, July 31, 1877, Apr. 27, May 25, June 2, 1881, Nov. 17, 1898, Sept. 24, 1905, Jan. 5, 1910 (obit), Apr. 18, 1914 (estate), Feb. 1, 1925, May 1, 1926; Henry Clews, *Twenty-eight Years in Wall Street* (New York: Irving, 1888), 455–459.

5. *LDt*, Apr. 23, May 28, 1880; Alfred Doten, *The Journals of Alfred Doten, 1849–1903* (Reno: University of Nevada Press, 1973), 2276; Charles W. Henderson, *Mining in Colorado* (Washington, D.C.: G.P.O., 1926), 176; Rodman W. Paul, *Mining Frontiers of the Far West, 1848–1880* (New York: Holt, Rinehart and Winston, 1963), 127–129.

6. *EMJ*, Oct. 12, 1878; Frank Fossett, *Colorado, Its Gold and Silver Mines* (New York: Crawford, 1879), 447–448; *NYG*, Mar. 15, 1880; William Vickers, et al., *History of the Arkansas Valley* (Chicago: Baskin, 1881), 217–219, 224–225, 378–380; Hubert H. Bancroft, "Life of Horace Austin Warner Tabor," in *Chronicles of the Builders of the Commonwealth* (San Francisco: History Co., 1891), v. 4, 273–339; Duane A. Smith, *Horace Tabor* (Boulder: Colorado Associated University Press, 1973), 71–74; King, op. cit., n. 1, 84–85; Don and Jean Griswold, *The History of Leadville and Lake County* (Denver: Colorado Historical Society, 1996), 157–158. Duane A. Smith, "The Rise and Fall of Horace Tabor," in *Mining Tycoons in the Age of Empire, 1870–1945*, ed. R. E. Dumett (Burlington, Vt.: Ashgate, 2009), 42–60.

7. *BE*, June 19, 1879; Fossett, op. cit., n. 6 (1879) 448–450, 2nd ed. (1880), 444, 452–455; Charles H. Dow, "Leadville letter" in the Providence *Journal*, June 28, 1879, reprinted in George W. Bishop, *Charles H. Dow and the Dow Theory* (New York: Appleton-Century-Crofts, 1960), 291–295; *NYG*, Mar. 15, 1880; Montezuma, Colo., *Millrun*, Nov. 4, 1884; Smith, op. cit., n. 6, 74–76; King, op. cit., n. 1, 84–85; Steven F. Mehls, *David H. Moffat, Jr.* (New York: Garland, 1989), 56–58; Kathleen P. Chamberlain, "David F. Day and the Solid Muldoon," *Journal of the West* 34 (Oct. 1995), 66; Griswold, op. cit., n. 6, 173–174.

8. *EMJ*, May 24, June 14, 1879; *The Little Pittsburg Consolidated Mining Company* (New York: Evening Post, 1879), 1–23; *United States Annual Mining Review and Stock Ledger* (New York: Mining Review, 1879), 114; Dow, "Leadville letters" in the Providence *Journal*, May 26–July 30, 1879, in Dow, op. cit., n. 7, 248–354; *NYG*, Mar. 15, 1880; *NYTr*, Mar. 23, 1880; Smith, op. cit., n. 6, 109–115; King, op. cit., n. 1, 92–94; Griswold, op. cit., n. 6, 173–174, 231–233; Dan Plazak, *A Hole in the Ground with a Liar at the Top* (Salt Lake City: University of Utah Press, 2006), 134–135.

9. *EMJ*, May 24, June 14, 28, 1879; Dow, op. cit., n. 7, 301–305; *NYTr*, June 21, 1879, Mar. 23, 1880; *NYT*, July 2, 10, 26, 1879; *The Little Pittsburg Consolidated*, op. cit., n. 8, 8, 14–20; *NYB*, Nov. 1, 1879; *NYMR*, Mar. 20, 1880; Philadelphia *Inquirer*, Apr. 29, 1880; Smith, op. cit., n. 6, 115–116; King, op. cit., n. 1, 94–95.

10. *Chrysolite Silver Mining Company* (New York: Gildersleeve, 1880), 3, 16, 23; *LDt*,

Mar. 16, 1880; Fossett, op. cit., n. 6, (1880) 445–449; *NCAB* 365–366; Smith, op. cit., n. 6, 76–77; King, op. cit., n. 1, 91–92; Plazak, op. cit., n. 8, 138–140.

11. *Whereas, George D. Roberts has made a contract for the purchase of the Borden, Tabor & Company Mines* (New York, Sept. 1879), broadside; *EMJ*, Oct. 11, 1879; *Chrysolite*, op. cit., n. 10, 3–6; *Winnebago and O. K. Company* (New York: 1880), 4–7; *NYMR*, Mar. 6, 1880; *LD*, Mar. 16, 1880; Fossett, op. cit., n. 6, (1880) 445–449; Smith, op. cit., n. 6, 116–119; Richard E. Lingenfelter, *The Hardrock Miners* (Berkeley: University of California Press, 1974), 143; King, op. cit., n. 1, 91–92.

12. *Chrysolite*, op. cit., n. 10, 7, 15–16, 21; Smith, op. cit., n. 6, 119.

13. Leadville, Colo., *Chronicle*, Jan. 3, 1880; *NYMR*, Apr. 3, May 22, 1880; *NYT*, Apr. 21, 1880; Fossett, op. cit., n. 6, (1880) 450–452; Ralph M. Hower, "Cyrus Hall McCormick," *Business History Review* 10 (Nov. 1936), 69–76; Griswold, op. cit., n. 6, 174, 427.

14. *EMJ*, July 12, Sept. 20, 1879, Feb. 7, 1880; Leadville, Colo., *Chronicle*, July 14, 1880; Fossett, op. cit., n. 6, (1880) 367–375, 427–429, 450–451, 493–498; *Winnebago*, op. cit., n. 11, 1–8; *NYMR*, Feb. 5, 1881; Griswold, op. cit., n. 6, 648.

15. *EMJ*, Jan. 31, Feb. 14, Mar. 20, Apr. 3, 1880; *NYMR*, Mar. 20, Apr. 3, 1880; *NYTr*, Mar. 23, 1880; *American Exchange* (New York), Mar. 27, 1880; *LD*, Apr. 1, 13, 16, 1880; Fossett, op. cit., n. 6, (1880) 451–454; George T. Ingham, *Digging Gold among the Rockies* (Philadelphia: Edgewood, 1882), 389–390; Smith, op. cit., n. 6, 127–128; Griswold, op. cit., n. 6, 486–490.

16. *EMJ*, Mar. 20, Apr. 3, May 15, June 5, 1880; *NYMR*, Mar. 20, Apr. 3, 1880; *NYTr*, Mar. 23, 1880; New York *American Exchange*, Mar. 27, 1880; New York *Puck*, Mar. 31, 1880; *LD*, Apr. 3, 7, 16, 1880; *RMN*, Apr. 7, 1880; Colorado Springs *Gazette*, Dec. 28, 1880, Jan. 1, Nov. 30, Dec. 3, 1881; Smith, op. cit., n. 6, 128–131; King, op. cit., n. 1, 104–106. A later suit for fraud during the original promotion was tried in 1884–85, but failed, *NYT*, Oct. 14, 16, Dec. 14, 1884, Apr. 23, 28–May 1, 1885.

17. *NYT*, Mar. 30, Apr. 21, 1880, Apr. 2, 1881; *EMJ*, Apr. 17, 24, May 15, 1880; *NYMR*, May 22, 1880; Drake De Kay, *To the Stockholders of the Chrysolite Silver Mining Company* (New York, July 28, 1880), 1–7; *LHd*, July 31, 1880; Smith, op. cit., n. 6, 132–133; Lingenfelter, op. cit., n. 11, 144; King, op. cit., n. 1, 106–107.

18. *EMJ*, Apr. 17–24, June 5, 1880, Sept. 4, 1920, June 1941; *NYT*, Apr. 2, 1881; *NYTr*, Dec. 5, 1880; *MSP*, Oct. 11, 1919; Lingenfelter, op. cit., n. 11, 144–145.

19. Leadville, Colo., *Chronicle*, May 26, 1880; *LD*, May 27–30, June 5, 11, 1880; *LH*, May 29, June 5, 1880; *NYMR*, June 12, 1880; Lundy, Calif., *Homer Mining Index*, Sept. 3, 1881; R. G. Dill, *The Political Campaigns of Colorado* (Denver: Arapahoe, 1895), 50–55; Smith, op. cit., n. 6, 134–139; Lingenfelter, op. cit., n. 11, 145–151; Griswold, op. cit., n. 6, 572–576, 607–608.

20. *EMJ*, June 5–26, 1880; *LD*, June 11–12, 1880; *LH*, June 12, 1880; Vickers, op. cit., n. 6, 287–288; Lingenfelter, op. cit., n. 11, 145–151; Griswold, op. cit., n. 6, 609–630.

21. *EMJ*, July 24–Aug. 7, Sept. 25, Nov. 6, 1880; *RMN*, July 30, 1880; *NYMR*, July 31, Aug. 14, Sept. 4, 1880; *LH*, July 31, Aug. 21, 28, 1880; *NYB*, Aug. 2, 1880; Lingenfelter, op. cit., n. 10, 154–155; King, op. cit, n. 1, 108–109; Griswold, op. cit., n. 6, 681–688.

22. *EMJ*, July 24–Aug. 7, Oct. 9, Nov. 6, 27, 1880, Aug. 13, 27, 1881; *NYB*, July 26, 1880; *LH*, July 31, 1880; Leadville, Colo., *Chronicle*, Aug. 7, 1880; *CMR*, Aug. 7, 1880; *NYTr*, Dec. 5, 1880; *NYT*, Apr. 2, 1881; *MSP*, Apr. 9, 1881; *SFEx*, Apr. 13, 1881; *NYMR*, Apr. 30, 1881; Balch, op. cit., n. 2, 1167; Lingenfelter, op. cit., n. 11, 155; Griswold, op. cit., n. 6, 682–603.

23. *EMJ*, Feb. 21, July 31, Aug. 28, Sept. 18–25, Oct. 2, 1880; *NYTr*, Apr. 27, Sept. 11, Dec. 5, 1880; *NYMR*, May 22, Sept. 18, 1880; Leadville, Colo., *Chronicle*, Sept. 23, 1880; *BDAC*, 476; Griswold, op. cit., n. 6, 684–688.

24. *NYMR*, Apr. 3, 1880, Jan. 15, 1881; *RMN*, Nov. 28, 30, 1880; Central City, Colo., *Register-Call*, Dec. 3, 1880; Fossett, op. cit., n. 5, 493–498; Vickers, op. cit., n. 6, 390–393, 409–410; Stanley Dempsey and James E. Fell, *Mining the Summit* (Norman: University of Oklahoma Press, 1986), 120–127, 137–145; Griswold, op. cit., n. 6, 683–684.

25. Balch, op. cit., n. 2, 475, 488–494, 514; *Mining World* (Las Vegas, N.Mex.), Feb. 1, 1883, advertises the New York *Mining Stock Register & Journal of Finance*, listing over 8,000 mining companies, while Balch, op. cit., n. 2, 1114, cites a total of 10,000 by May of 1882; Henderson, op. cit., n. 5, 176.

26. Thomas B. Corbett, *Colorado Mining Directory* (Denver: Rocky Mountain News, 1879), 440 pp.; Balch, op. cit., n. 2, 1114; and many stock certificates from the period.

27. *EMJ*, May 13, 1876, Apr. 21, 1877, Jan. 5, Feb. 9, Mar. 16, 1878, Oct. 29, 1881; Corbett, op. cit., n. 26; *CMR*, July 30, Aug. 21, Sept. 11, 1880, June 11, July 30, Nov. 12, 1881, Apr. 22, 1882; *NYT*, Nov. 13, 1880, Jan. 11, 12, May 21, 23, Oct. 18, 1882, May 26, 1883; *Mining World* (Las Vegas, N.Mex.), Dec. 1880, Dec. 1881, Nov. 1882; *SFEx*, June 9, 1882; Balch, op. cit., n. 2, 495–506; Jerome C. Smiley, *History of Denver* (Denver: Times, 1901) 572–576; Duane A. Smith, *Rocky Mountain Mining Camps* (Bloomington: Indiana University Press, 1967), 178–179; Marian V. Sears, *Mining Stock Exchanges, 1860–1930* (Missoula: University of Montana Press, 1973), 205–210.

28. *NYB*, Apr. 1, 1880; *LD*, Apr. 23, 1880; *Hadley's Pointers* (Las Vegas, N.Mex.), Sept. 7, 1882.

29. *NYT*, Apr. 28, 1875 (Roudebush), Jan. 23, 1883; *LH*, Nov. 8, 1879; Leadville, Colo., *Chronicle*, Mar. 24, Dec. 20, 1880, July 10, 1882; Fossett, op. cit., n. 6, 456–459; Vickers, op. cit., n. 6, 290–294; Balch, op. cit., n. 2, 1175; Frank Hall, *History of the State of Colorado* (Chicago: Blakely, 1890), v. 2, 440–444; Smith, op. cit., n. 6, 122–123; Griswold, op. cit., n. 6, 356, 427, 457, 737, 985–987.

30. Vickers, op. cit., n. 6, 290–291, 295; *EMJ*, Feb. 7, Mar. 27, Apr. 3, May 22, Dec. 11, 1880, Jan. 22, Mar. 12–26, Apr. 30, July 2, Aug. 13, Dec. 10, 1881; *NYMR*, May 7, 1881; Smith, op. cit., n. 6, 179–181, 270; Griswold, op. cit., n. 6, 570, 689.

31. Fossett, op. cit., n. 6, 459; Vickers, op. cit., n. 6, 294; *NYMR*, Feb. 26–Apr. 9, May 7, 1881; *EMJ*, Sept. 25, Dec. 25, 1880, Jan. 8, 22, Feb. 12, Mar. 5, Apr. 23, May 21, June 4, 1881; Smith, op. cit., n. 6, 184–186.

32. Central City, Colo., *Register-Call*, Mar. 25, 1881; *NYMR*, Mar. 26–Apr. 9, May 7, Sept. 8, 1881, Apr. 8, 1882, May 5, 1883; *LH*, Apr. 30, 1881; *SLGD*, May 1, 1881; *RMN*, July 1, 1881; *EMJ*, Sept. 10, 24, Dec. 24, 1881, Feb. 4, May 20, 1882; Balch, op. cit., n. 2,

1170; *NYT*, Mar. 25, 1883, June 11, 1901, Apr. 21, 1929, Mar. 8, 1935; *ChT*, Apr. 16, 1899, Mar. 8, 1935; *LAT*, June 10, 1901, Aug. 7, 1907, Sept. 28, 1925, Apr. 12, 1929, May 14, 1934, Mar. 8, 1935; Douglas Moore and John Latouche, *The Ballad of Baby Doe: Opera in Two Acts* (New York: Chappell, 1958); Smith, op. cit., n. 6, 184–188, 288–300, 314–315; Judy Nolte Temple, *Baby Doe Tabor: The Madwoman in the Cabin* (Norman: University of Oklahoma Press, 2007).

　　33. *EMJ*, Mar. 12, 1881, Oct. 14, 1882, Oct. 18, 1884; *NYMR*, Apr. 9, 1881; *NYT*, Jan. 23, 25, 1883, Oct. 14, 1884, Apr. 29, 1885; Robert A. Corregan and David F. Lingane, *Colorado Mining Directory* (Denver, 1883), 416; Clark Spence, *British Investments and the American Mining Frontier, 1860–1901* (Ithaca, N.Y.: Cornell University Press, 1958), 75; *BDAC*, 560; Smith, n. 6, 183; Mehls, op. cit., n. 7, 63; Griswold, op. cit., n. 6, 983–984.

　　34. *EMJ*, Dec. 16, 1882, July 14, 1883; *LT*, Jan. 25, 1883; *BE*, Oct. 1, 1884; *NYT*, Oct. 8, 29, 31, 1884, Apr. 29, 1885; Spence, op. cit., n. 33, 75, 265; Smith, op. cit., n. 6, 271.

　　35. *LH*, Nov. 8, 1879; *NYT*, Feb. 16, Mar. 30, 1880, Mar. 11, 1881, Feb. 12, 1882, Oct. 9–10, 1883; *Reports of the Amie Mine* (New York: Kilbourne Tompkins, 1880), 1–6; *EMJ*, Mar. 27, Aug. 28, 1880 (plus weekly sales), Jan. 29, 1881, Feb. 4, 1882; New York *Mining Directory*, op. cit., n. 2, 26; Fossett, op. cit., n. 6, (1880) 455–456; Vickers, op. cit., n. 6, 289; *BDAC*, 1464–1465.

　　36. *EMJ*, Apr. 23, May 28, 1881, Feb. 4, 1882, Feb. 10, 1883, Jan. 1, 1887; Vickers, op. cit., n. 5, 295–296; Corregan and Lingane, op. cit., n. 33, 419; Hubert H. Bancroft, *History of Nevada, Colorado, and Wyoming, 1540-1888* (San Francisco: History Co., 1890), 505; Richard P. Rothwell, *The Mineral Industry* (New York: Scientific, 1893), v. 1, 476, and "A Quarter of a Century's Fluctuations in Mining Stocks, 1871 to 1895," v. 4, (1896), 69–70; Walter H. Weed, *The Mines Handbook* (New York: Stevens, 1920), v. 14, 619–621; Griswold, op. cit., n. 6, 683–685.

　　37. *The New York Mining Directory*, op. cit., n. 2, 17; Fossett, op. cit., n. 6, (1880) 433–434; *NYT*, Nov. 2, 1880, Dec. 8, 1881, May 7, 1884, June 13, Sept. 18, Nov. 10, 1885, Feb. 26, 1890; Vickers, et al., op. cit., n. 6, 279–280, 387–388; *EMJ*, Dec. 17, 1881, July 1, Dec. 30, 1882, June 30, 1883, Jan. 5, 1884; *MSP*, Jan. 28, 1882, Jan. 27, 1883, Feb. 16, 1884; Philadelphia *Inquirer*, Jan. 5, 1885; *BE*, May 6, 1884, Oct. 8, 1885; Baltimore *Sun*, Oct. 9, 1885; Ferdinand Ward, "General Grant an Easy Prey for the Wolves of Finance," *NYH*, Dec. 26, 1909; Griswold, op. cit., n. 6, 220–221, 428, 491, 494.

　　38. *The United States Biographical Dictionary and Portrait Gallery of Eminent and Self-Made Men. Iowa Volume* (Chicago: American Biographical, 1878), 38–40; Fossett, op. cit., n. 6, (1880) 423; Vickers, op. cit., n. 6, 307, 336–337; Samuel F. Emmons, *Geology and Mining Industry of Leadville* (Washington, D.C.: G.P.O., 1886), 639, 668–669; Hall, op. cit., n. 29, v. 2, 451; Bancroft, op. cit., n. 36, 292–297; *Portrait and Biographical Record of Denver and Vicinity* (Chicago: Chapman, 1898), 159–160; *Encyclopedia of Biography of Colorado* (Chicago: Century, 1901), 254–255; Walter Ingalls, *Lead and Zinc in the United States* (New York: Hill, 1908), 78, 155; Henderson, op. cit., n. 5, 476; James E. Fell, *Ores to Metals* (Lincoln: University of Nebraska Press, 1979), 91–94, 103. Based on Emmons's report (pp. 639, 668–669) of 1880 smelting charges of $25 a ton and costs at $15 a ton,

Fell (p. 103) estimates a profit of only $10 a ton, but he neglects the 5 percent deduction on silver and the 75 percent deduction on lead, which for Leadville's average $100-a-ton ores of $86 silver and $14 lead (Fossett, p. 423) added another $15 a ton, raising the profit to $25 a ton. Similarly, for 1885 Ingalls (p. 78) reports an average charge of $15 a ton for such ore, with deductions of 7 percent on silver and 75 percent on lead for a total cost of $32 a ton, by which time the costs were down to $9 a ton (*EMJ*, Oct. 25, 1884) for a profit of $23 a ton.

39. *NYT*, July 26, 1879, Sept. 29, Nov. 9, 1882; Georgetown, Colo. *Courier*, Jan. 8, 1880; Fossett, op. cit., n. 6, (1880) 423–426; Vickers, op. cit., n. 6, 307, 328–329, 348–349; *EMJ*, May 7, 1881, May 27, Sept. 16, 1882; Fell, op. cit., n. 38, 109–113, 143–145.

40. *EMJ*, Nov. 18, 1882, Mar. 24, Aug. 11, Oct. 27, 1883, Apr. 5, Oct. 25, 1884; Ingalls, op. cit., n. 38, 156–157; Fell, op. cit., n. 38, 119–120, 145–148, 155–156.

41. *EMJ*, May 3–June 28, July 26, Aug. 2–16, Sept. 20, 1879; *NYT*, June 17, 1879; Fossett, op. cit., n. 5, (1880) 262–268; Duane A. Smith, *Silver Saga; The Story of Caribou, Colorado* (Boulder: Pruett, 1974), 68–71; Mehls, op. cit., n. 7, 55–56.

42. Boulder *News and Courier*, Apr. 4, Sept. 19, 1879; *RMN*, Sept. 18, 1879, May 15, 1880; *EMJ*, Sept. 27, 1879; Fossett, op. cit., n. 6, (1880) 266–270; Smith, op. cit., n. 41, 71–74; Mehls, op. cit., n. 7, 56.

43. *EMJ*, Jan. 24, Feb. 21, Apr. 24, May 22, June 5, Sept. 18, 1880; Boulder *News and Courier*, May 21, 1880; Corregan and Lingane, op. cit., n. 33, 48; *NYT*, July 15, Nov. 11, Dec. 2, 1900; Smith, op. cit., n. 41, 74–82, 163–172, 206–210.

44. Marcus M. Pomeroy, *When the World Was Young: Cutting the Backbone of the American Continent by the Atlantic-Pacific Tunnel, Opening the Vaults of Gold and Silver* (Denver: Atlantic-Pacific Tunnel Co., 1881), 1; *NCAB*, v. 2, 502–503; *BE*, Apr. 5, 1896.

45. Denver *Republican*, Sept. 14, 1883; Marcus M. Pomeroy, *Early Life of "Brick" Pomeroy* (New York, 1891), 250–251; *NCAB*, v. 2, 502–503; *Dictionary of American Biography*, v. 8, 53–54; Ruth A. Tucker, "M. M. 'Brick' Pomeroy" (Ph.D. diss., Northern Illinois University, 1979), 4–38, 160–174.

46. *RMN*, Mar. 23, 1880; Fairplay *Flume*, Mar. 25, 1880; Colorado Springs *Gazette*, Mar. 28, 1880; *LH*, Apr. 3, 17, May 1, 22, 1880; *NYMR*, Apr. 10, 1880; Boulder *News & Courier*, Apr. 16, 1880; *Colorado Miner* (Georgetown), Oct. 30, 1880; Pomeroy, op. cit., n. 44, 38, 44; Denver *Republican*, Sept. 14, 1883; *The Great West* (Denver), Aug. 23, 1884; Tucker, op. cit., n. 45, 314–321.

47. Georgetown, Colo., *Courier*, Nov. 4, 1880; *Colorado Miner* (Georgetown), Dec. 18, 1880; *The Great West* (Denver), Nov. 14, 1880, Feb. 20, May 14–15, July 16, Sept. 3, Nov. 19, 1881; Pomeroy, op. cit., n. 44, 2–13, 25–26; *RMN*, May 8, 1881; *Freeborn County Standard* (Albert Lea, Minn.), Dec. 1, 1881; *The Atlantic-Pacific Tunnel Co. 700,000 Shares, $10 Each*, stock certificate no. 5417, Denver, Jan. 21, 1882; Marcus M. Pomeroy, *Make Money, A Great and Safe Enterprise* (Denver: Atlantic-Pacific Tunnel Co., 1882), 15; Corregan and Lingane, op. cit., n. 33, 118–119; Erl H. and Carrie S. Ellis, *The Saga of Upper Clear Creek* (Frederick, Colo.: Jende-Hagan, 1983), 179–181. Pomeroy's share of the stock was not explicitly stated, but 400,000 shares were originally reserved for

the purchase of property (*When the World Was Young*, p. 10). By the time of the second annual report (*The Great West*, Oct. 28, 1882, quoted in *NYMR*, Nov. 4, 1882), 350,000 of these shares had been issued in exchange both for unidentified property, presumably the tunnel sites that Pomeroy held, and for shares in Pomeroy's Monte Cristo company, most of which had not been sold.

48. *The Great West* (Denver), June 11, July 16, 1881; Pomeroy, op. cit., n. 44, 37–38; Tucker, op. cit., n. 45, 318–319.

49. Boulder *News & Courier*, Apr. 16, 1880; *The Great West* (Denver), Nov. 14, 1880, Mar. 13, May 14, 15, June 11, July 16, Sept. 3, 1881; Pomeroy, op. cit., n. 44, 12–14, 40–41, 44; Atchison (Kans.) *Globe*, May 18, 21, 1881; "Fortunes in Colorado," *St. Louis Magazine* 21 (Oct. 1881), 375–389; Pomeroy, op. cit., n. 47, 3–11, 20–28.

50. *The Great West* (Denver), Sept. 17, 1881, May 13, 1882; Pomeroy, op. cit., n. 47, 27–28; *NYMR*, Nov. 4, 1882; Denver *Republican*, Sept. 14, 1883; Tucker, op. cit., n. 45, 321–322, 348. The annual reports of the tunnel company published in *Make Money*, pp. 27–28, for 1881, and *NYMR*, Nov. 4, 1882, for 1882, list only the direct sales of the company's shares, amounting to 41,829 shares in the first year and 68,210 shares by the end of the second, which paid all company expenses and much more. But the total sales can only be approximated from the certificate numbers and dates, and the average number of twenty shares per certificate, in the list of shares bought back by the company, published in *The Great West*, May 13, 1882. This list shows that just over 5,000 certificates, representing about 100,000 shares, had been sold during the first year (to Oct. 25, 1881), which indicates that about 60,000 personal shares had also been sold by Pomeroy, who handled all of the stock sales. Similarly another 4,300 certificates, representing about 86,000 shares, had already been issued in the first half of the second year, during which only 26,381 shares of company stock were sold, leaving at least another 60,000 shares in personal stock sales. Nearly all of these were sold for $2.50 a share, yielding a total of more than $300,000 in personal profits in the first two years.

51. *The Great West* (Denver), May 13, 1882; Montezuma, Colo., *Millrun*, Oct. 21, 1882; *NYMR*, Oct. 28, 1882; Colorado Springs *Gazette*, May 23, 1883; Denver *Republican*, Sept. 14, 1883; *NYT*, Oct. 7, 1883; White Pine, Colo., *Cone*, Oct. 3, 1884; Tucker, op. cit., n. 45, 331–333; Ellis, op. cit., n. 47, 182–187.

52. *The Great West* (Denver), Aug. 23, 1884, and previous issue quoted in an unidentified clipping in the Colorado State Historical Society Library, Denver; Tucker, op. cit., n. 45, 334.

53. *The Atlantic-Pacific Railway Tunnel Co. Capital Stock $7,000,000. Par Value Ten Dollars Each. Shares 700,000*, stock certificate no. 5606, Denver, Sept. 14, 1888; Tucker, op. cit., n. 45, 333–335; Ellis, op. cit., n. 47, 182.

54. Aztec Syndicate, *Reports on the Mines of the Aztec Mining District* (San Francisco: Cubery, 1877), 1–42; Prescott *Arizona Miner*, Dec. 14, 1877, Aug. 2, 1878; Denver *Tribune*, July 28, 1878; Philadelphia *North American*, Aug. 13, Nov. 6, 1878; Richard J. Hinton, *The Handbook to Arizona* (New York: American News, 1878), 128–129, 182, 198, 203–206;

Enoch Conklin, *Picturesque Arizona* (New York: Mining Record, 1878), 7–19, 341–342; *SFAC*, Apr. 23, 1880 (Boyle obit); *History of Arizona Territory* (San Francisco: Elliott, 1884), 192; *NYT*, Dec. 17, 1891 (Safford obit); Spence, op. cit., n. 33, 258; *BDAC*, 1041.

55. Edward L. Schieffelin, "Finding Tombstone," Philadelphia *Times*, quoted in *BE*, June 15, 1886; *PO*, Jan. 17, 1889 (Schieffelin interview); Hubert H. Bancroft, "Life of Richard Gird," in *Chronicles of the Builders of the Commonwealth* (San Francisco: History Co., 1891), v. 3, 95–97; Richard Gird, "True Story of the Discovery of Tombstone," *Out West* 27 (July 1907), 39–41; Alice E. Love, "The History of Tombstone to 1887" (M.A. thesis, University of Arizona, 1933), 4–7; Morris Elsing and Robert Heineman, *Arizona Metal Production* (Tucson: University of Arizona, 1936), 91; Odie B. Faulk, *Tombstone: Myth and Reality* (New York: Oxford University Press, 1972), 26–41; Lonnie E. Underhill, ed., "The Tombstone Discovery: The Recollections of Ed Schieffelin and Richard Gird," *Arizona & the West* 21 (Spring 1979), 37–55, 67–69; Richard E. Moore, "The Silver King: Ed Schieffelin, Prospector," *Oregon Historical Quarterly* 87 (Winter 1986), 367–374; William B. Shillingberg, *Tombstone, A.T.* (Spokane, Wash.: Arthur H. Clark, 1999), 21–36; Lynn R. Bailey, *Tombstone, Arizona, "Too Tough to Die"* (Tucson: Westernlore, 2004), 4–9.

56. Schieffelin, op. cit., n. 55; Bancroft, op. cit., n. 55, 95–96; Gird, op. cit., n. 55, 39, 43–46; Love, op. cit., n. 55, 7–9; Faulk, op. cit., n. 55, 162–163; Underhill, op. cit., n. 55, 55–63, 71–73; Moore, op. cit., n. 5, 372–374; Shillingberg, op. cit., n. 55, 38–40; Bailey, op. cit., n. 55, 10–13.

57. *SFBul*, July 1, 1879; *Arizona Miner* (Prescott), Sept. 19, 1879; Schieffelin, op. cit., n. 55; Bancroft, op. cit., n. 55, 96; Love, op. cit., n. 55, 10–11; Faulk, op. cit., n. 55, 53–58; Underhill, op. cit., n. 55, 63–65; Moore, op. cit., n. 55, 374–375; Shillingberg, op. cit., n. 55, 43–51, 54; Bailey, op. cit., n. 55, 13–15.

58. *EMJ*, June 28, 1879, Mar. 20, Apr. 24, May 22–29, June 12–July 3, 1880, June 4, 1881; *MSP*, May 22, June 14, 1880; Thomas R. Sorin, *Hand-book of Tucson and Surroundings* (Tucson: Citizen, 1880), 12–15; *CMR*, June 11, July 30, 1881; Clara S. Brown, "An Arizona Mining District," *Californian* 4 (July 1881), 50–51; *NYT*, Oct. 28, Dec. 30, 1881, Feb. 9, Mar. 20, 24, 26, 1882; Balch, op. cit., n. 2, 1178; Schieffelin, op. cit., n. 55; Bancroft, op. cit., n. 55, 97; Love, op. cit., n. 55, 11–13; Underhill, op. cit., n. 55, 64–66; Moore, op. cit., n. 55, 375–376; Shillingberg, op. cit., n. 55, 106–109; Bailey, op. cit., n. 55, 22–23.

59. *MSP*, Jan. 29, Dec. 17, 1881, Jan. 28, 1882; *PMJ*, May 27, 1882; *EMJ*, Jan. 5, 1884; Bancroft, op. cit., n. 55, 95; Gird, op. cit., n. 55, 49; Love, op. cit., n. 55, 13; Faulk, op. cit., n. 5, 164–165; Underhill, op. cit., n. 55, 76; Moore, op. cit., n. 55, 376; Robert L. Spude, "Mineral Frontier in Transition" (M.A. thesis, Arizona State University, 1976), 57; Shillingberg, op. cit., n. 55, 193; Bailey, op. cit., n. 55, 23.

60. *PMJ*, June 3, 1882; *LAT*, May 9, Oct. 10, 1883, Oct. 14, 1885, Apr. 4, May 16, 1897, June 1, 1910; Schieffelin, op. cit., n. 55; Bancroft, op. cit., n. 55, 97–108; Love, op. cit., n. 55, 13–14; Moore, op. cit., n. 55, 376–384; Shillingberg, op. cit., n. 55, 347–348; Bailey, op. cit., n. 55, 22–25, 214–215.

61. *EMJ*, July 2, 1881, May 6, 27, 1882, May 12, 1883, Jan. 5, 26, Sept. 20, 1884; *MSP*, Jan. 28, 1882, Jan. 27, 1883; *PMJ*, Apr. 29, May 6, June 17, Oct. 7, 1882; June 2, 1883; *SFEx*, May 11, 23, 1882; *NYT*, July 11, 1881, Dec. 17, 1891, Dec. 17, 19, 1892, May 1, 1896, July 12, 1900; *LAT*, Aug. 30, 1890; Frederick W. Dau, *Florida Old and New* (New York: Putnam's Sons, 1934), 251; Spude, op. cit., n. 59, 58; Shillingberg, op. cit., n. 55, 349–350; Bailey, op. cit., n. 55, 48–49.

62. *MSP*, Nov. 9, 1878, June 19, 1880, Jan. 29, Dec. 17, 1881, Jan. 28, 1882, Jan. 27, 1883, Jan. 19, 1884, Jan. 24, 1885; *EMJ*, Jan. 3, Apr. 24, Sept. 25, 1880, June 4, Oct. 22, Nov. 12, 1881, Mar. 10–Aug. 4, 1883, Jan. 5, 1884; Sorin, op. cit., n. 58, 14; Brown, op. cit., n. 58, 51; Myron Angel, *History of Nevada* (Oakland, Calif.: Thompson and West, 1881), between 116–117, 124–225; Balch, op. cit., n. 2, 1167; Spude, op. cit., n. 59, 58; Shillingberg, op. cit., n. 55, 43–44, 339; Bailey, op. cit., n. 55, 25–26.

63. Grand Central claim transfers in Pima County Recorder's Office, Tucson, Mine Deeds, bk. 2, 402, 423, 470, 476, 641, bk. 3, 194, 197, bk. 4, 391, and bk. 5, 215, all courtesy of Robert L. Spude; *Transcript of Record. Peter L. Kimberly vs. Charles D. Arms, et al.* (Washington, D.C., 1889), 113–116, 127–128, 131–133, 145, 163–165, 172–175, 208–214, 697–698, 711; Shillingberg, op. cit., n. 55, 40–41, 47–49; Bailey, op. cit., n. 55, 11–12.

64. *EMJ*, June 4, 1881; *MSP*, Dec. 17, 1881, Jan. 28, 1882, Jan. 27, 1883, Jan. 19, 1884, Jan. 24, 1885; *NYT*, July 27, Oct. 2, 1883, Mar. 28, 1903 (Fairbank obit); Balch, op. cit., n. 2, 1169; *Transcript*, op. cit., n. 63, 5–34, 152, 156–157, 187–188, 192, 215–227, 470, 484–485, 492, 500–502, 516–517, 818–825; Kimberly vs. Arms, 125 U.S. 512; Tombstone *Epitaph*, Dec. 7, 1889; *The Biographical Dictionary and Portrait Gallery of Representative Men of Chicago* (Chicago: American Biographical, 1893), 670–674; Spude, op. cit., n. 59, 58; Shillingberg, op. cit., n. 55, 85–86, 335, 365; Bailey, op. cit., n. 55, 31–32.

65. *NYT*, Dec. 10, 1880, Jan. 1, 1881, May 19, 1882, Dec. 22, 1883, Feb. 17, 1887; *EMJ*, Nov. 12, 1881, Feb. 4, May 20, 1882; *SFBul*, May 30, 1882.

66. George Lee's Quicksilver Claim, Nov. 4, 1875, Miscellaneous Record Book R, 137, and Robert Waterman's and John Porter's Alpha and Omega Claims, Dec. 7, 1879, Miscellaneous Record Book B, 624, 626, San Bernardino County Recorder's Office, San Bernardino, Calif.; San Bernardino *Times*, Aug. 26, 1876, Apr. 6, May 18, 1880, Nov. 4, 1886; Robert W. Waterman, letters to his brother, Oct. 17, Dec. 29, 1880, Apr. 5, 1881, and their agreement of May 14, 1881, in *Supreme Court. October Term 1891. No. 190. Robert W. Waterman, Appellant, vs. James M. Banks, Executor of Abbie L. Waterman, Deceased* (Washington, D.C.: Judd and Detweiler, 1891), 1–2, 169–171, 298–299, 350–357, 376, 410–411; John L. Porter and James S. Waterman's agreement of May 14, 1881, in *Supreme Court. October Term 1891. No. 191. J. L. Porter, Appellant, vs. James M. Banks, Executor of Abbie L. Waterman, Deceased* (Washington, D.C.: Judd and Detweiler, 1891), 1–2, 11–13; *MSP*, Aug. 12, 1882; *Portrait and Biographical Album of DeKalb County, Illinois* (Chicago: Chapman Bros., 1885), 254, 257; San Francisco *Examiner*, Oct. 3, 1886; Theodore S. Van Dyke, *The City and County of San Diego* (San Diego: Leberthon and Taylor, 1888), 199–203; Hubert H. Bancroft, "Life of Robert Whitney Waterman," in *Chronicles of the Builders of the Commonwealth* (San Francisco: History Co., 1891),

v. 7, 566–571; H. Brett Melendy and Benjamin F. Gilbert, *The Governors of California* (Georgetown, Calif.: Talisman, 1965), 227–235.

67. *MSP*, Aug. 11, 1883; *In the Circuit Court of the United States. Ninth Judicial Circuit, in Equity. Abbie L. Waterman, vs. R. W. Waterman, 3418. Abbie L. Waterman, vs. J. L. Porter, 3439. Brief in Support of Respondent's Petition for a Re-Hearing* (San Francisco: Bosqui, 1886), 58–60; San Francisco *Examiner*, Oct. 3, 1886; *Supreme Court, No. 190,* op. cit. n. 66, 1–21, 250–257, 268–273, 356–357.

68. San Bernardino *Times*, July 26–Aug. 1, 8, Sept. 4–7, 1883, Nov. 4, 1886; San Francisco *Examiner*, Sept. 23, Oct. 3, 1886; *Supreme Court. October Term 1889. No. 1286. Robert W. Waterman, Appellant, vs. Philander M. Alden and George S. Robinson, Executors of James S. Waterman, Deceased* (Washington, D.C.: Judd and Detweiler, 1889), 1–10, 48–51; *Supreme Court, No. 190,* op. cit. n. 66, 1–21, 32–36, 98–100, 164–169, 246–248, 268–273, 356–357.

69. San Francisco *Examiner*, Sept. 23, Oct. 3, 1886; *LAT*, May 18, 1888, Apr. 5, 11, 13, Aug. 15–17, 25, 1889, Apr. 13, 1891; *SFBul*, Apr. 13, 1891; *SFCh*, Apr. 13, 1891. The bullion returns and profits, both cash withdrawals and official dividends, are tallied in Waterman's "Account Books," vols. 10–12 for 1881–86, in the Waterman Papers, Bancroft Library, University of California, Berkeley.

70. *EMJ*, Nov. 30, 1878, Mar. 29, 1879, May 6, 1882, Apr. 14, 1883, Mar. 1, 1884; *SLT*, Feb. 9, May 25, 1879; W. A. Hooker, *The Horn Silver Mine* (New York: Tompkins, 1879), 1–4; *United States Annual Mining Review*, op. cit., n. 8, 112–113; *MSP*, Apr. 9, 1881, Jan. 24, 1885; *LAT*, June 21, 1902 (Campbell obit); *NYT*, Feb. 17, 1905 (Cooke obit); Ellis P. Oberholtzer, *Jay Cooke* (Philadelphia: Jacobs, 1907), v. 2, 522–526; Walter Ingalls, *Lead and Zinc in the United States* (New York: Hill, 1908), 198–199; Henrietta M. Larson, *Jay Cooke, Private Banker* (Cambridge, Mass.: Harvard University Press, 1936), 424–425; Leonard J. Arrington and Wayne K. Hinton, "The Horn Silver Bonanza," in *The American West: A Reorientation*, ed. Gene M. Gressley (Laramie: University of Wyoming, 1966), 35–54; Robert Sobel, *Panic on Wall Street* (New York: Macmillan, 1968), 167–170.

71. Hooker, op. cit., n. 70, 21–25; *EMJ*, Mar. 29–Apr. 5, 26, Oct. 25, Nov. 8–Dec. 6, 1879; *United States Annual Mining Review*, op. cit., n. 8, 112–113; *NYT*, Mar. 16, 1880; Balch, op. cit., n. 2, 1171; Arrington and Hinton, op. cit., n. 70, 39–43.

72. *NYT*, Oct. 23, 1881, Jan. 12, 1929 (Francklyn obit); *EMJ*, May 6, 1882, Feb. 10, Apr. 7–14, 1883, Mar. 1, 1884; *MSP*, Apr. 9, 1881, Jan. 24, 1885; *SLT*, Mar. 18, 20, 1883, Jan. 1, Feb. 17, 1885; Salt Lake City *Herald*, Feb. 14, 1885; "Fluctuations in Mining Stocks," op. cit., n. 36, 69–70; Bert S. Butler, *Geology and Ore Deposits of the San Francisco and Adjacent Districts, Utah* (Washington, D.C.: G.P.O., 1913), 164–165; Arrington and Hinton, op. cit., n. 70, 39–47.

73. New York *Sun*, Feb. 22–23, Oct. 21–22, 1887; *NYT*, May 18, Sept. 18, 24, Oct. 21–22, 26, 29, Nov. 13, Dec. 21, 1887, Jan. 7, 18, 22, Oct. 3, 9, 1888, July 15, Nov. 12, 14, 1889, Jan. 14, 1903, Jan. 12, 1929; *LT*, Oct. 22, 29, Dec. 7, 1887; Butler, op. cit., n. 72, 164–165; Salt Lake City *Mining Review*, Oct. 30, 1916; Walter G. Neale, *The Mines*

Handbook (New York: Mines Handbook Co., 1926), v. 17,1494–1495; Arrington and Hinton, op. cit., n. 70, 47–50.

74. *EMJ*, Jan. 24, 1880, Jan. 29, 1881, Jan. 28, 1882, Jan. 27, Oct. 13, 1883, Feb. 16, 1884, Jan. 24, Feb. 21, 1885; *NYT*, Nov. 18, 1884; Rothwell, op. cit., n. 36, v. 1 (1893), 476–478, v. 4 (1896), 69–70; Jack R. Wagner, *Gold Mines of California* (Berkeley, Calif.: Howell-North, 1970), 153–154, 169–171.

75. *SFBul*, Mar. 12, Oct. 15, 1881, Sept. 11, 25, 1883; Balch, op. cit., n. 2, 1160; Robert Kelley, *Gold vs. Grain* (Glendale, Calif.: Arthur H. Clark, 1959), 51–52, 91, 109, 229–239; Powell Greenland, *Hydraulic Mining in California* (Spokane, Wash.: Arthur H. Clark, 2001), 227–255.

76. *SFBul*, Jan. 7–8, 14, 28, 1884; *MSP*, Jan. 12–19, Feb. 2, 1884; Woodruff v. North Bloomfield Gravel Mining Co. and others, 18 Fed. 753; Thomas Egleston, *The Metallurgy of Silver, Gold, and Mercury in the United States* (New York: John Wiley, 1890), v. 2, 89–91, 274–275; Kelley, op. cit., n. 75, 239–242; Greenland, op. cit., n. 75, 255–257.

77. *SacU*, Oct. 12, 1876; Balch, op. cit., n. 2, 1141; *NYT*, Jan. 24, 30, 1885.

78. *EMJ*, Jan. 24, 1880, Jan. 1, 1881, Feb. 4, Dec. 30, 1882, Feb. 10, 1883; *NYT*, Jan. 29, 1880, Jan. 12, 1884, Jan. 24, 30, Feb. 17–18, 1885, Oct. 24, 1911 (obit); Balch, op. cit., n. 2, 1141.

79. *NYT*, Oct. 18, 1878, Sept. 7, 1879, Jan. 13, 27, Dec. 8, 1880, Apr. 18, 1881, Nov. 21, 1886; *Scientific American* 41 (Sept. 20, 1879), 177; *MSP*, Mar. 13, June 26, 1880, Jan. 15, 1881; Baltimore *Sun*, Mar. 30, 1880; Macon, Ga., *Telegraph & Messenger*, Apr. 2, 1880; "Magnetic Ore Separator," United States Patent no. 228,329, Apr. 3, 1880; *SLGD*, Apr. 4, 1880; *EMJ*, Apr. 10, June 5, 1880; *Idaho Statesman* (Boise), Apr. 10, 1880; *Deseret News* (Salt Lake City), Apr. 21, 1880; *SFAC*, Apr. 30, 1880; *SacU*, Apr. 30, 1880; *NSJ*, May 15, 1880; New York *Mining Directory*, op. cit., n. 2, 100; John B. Hodgkins, *Thomas A. Edison and Major Frank McLaughlin* (Chico, Calif.: Assoc. for No. Calif. Records and Research, 1979), 1–4, D1–D5; Paul Israel, *Edison: A Life of Invention* (New York: John Wiley, 1998), 340–341.

80. "List of Subscribers to the capital stock of The Edison Ore Milling Company, Limited," Jan. 5, 1880 (TAEM 97:398), in the Thomas A. Edison Papers, Rutgers University, Piscataway, N.J., online at http://edison.rutgers.edu; *NYT*, Jan. 13–Apr. 18, 1880 (ads); *SFBul*, Mar. 1, 1880; *SLT*, Mar. 21, Apr. 1, 1880; *EMJ*, Apr. 10, June 5, 1880; *Idaho Statesman* (Boise), Apr. 10, 1880; *Deseret News* (Salt Lake City), Apr. 21, 1880; *SFAC*, Apr. 30, 1880; *Proposed Plan of Reorganizing the Edison Ore Milling Company, Limited* (New York: Burgoyne's, 1885), 1–3; Hodgkins, op. cit., n. 79, 4–13; Israel, op. cit., n. 79, 340–341.

81. Sacramento *Record-Union*, April 30, 1880; *SFBul*, May 15, 1880, Mar. 24, 1883; Harry L. Wells, *History of Butte County* (San Francisco: printed by author, 1882), 211–212; *Proposed Plan*, op. cit., n. 80, 3–6; *NYT*, Nov. 21, 1886 (McMahon); *LAT*, July 29, 1890, Mar. 24, 1895, Jan. 1, 1896, Nov. 17–18, 1907 (McLaughlin obit); *The Edison Ore Milling Company, Limited. Capital Stock $2,000,000. Shares $100 Each*, stock certificate no. 1208, New York, 11 May 1892; San Jose *Mercury & Herald*, Nov. 16–17, 1907

(McLaughlin obit); Hodgkins, op. cit., n. 79, iii–iv, 13–16; D12, D15, D19–D20, D22; Michael Peterson, "Thomas Edison, Failure," *American Heritage of Invention & Technology* 6 (Winter 1991), 8–14; Israel, op. cit., n. 79, 340–362, 391.

82. *EMJ*, Apr. 12, Aug. 2, Dec. 6, 1879, Jan. 17–Feb. 21, Aug. 7–21, Dec. 11, 1880, Feb. 4, 1882, Feb. 10, 1883; New York Stock Exchange, Committee on Admissions, Hearing Transcripts, v. 12, June 3, 11, 17, 1880, and Copartnership Directory, v. 1, in NYSE Archives and Corporation Research Center, New York; *SFEx*, May 11, 1881; Galveston *News*, June 2, 1889; *NYT*, Sept. 16, 1897; Duane A. Smith, "The Vulture Mine," *Arizona & the West* 14 (Autumn 1972), 240–245.

83. San Bernardino County Recorder, Deeds Book 22, 446; *EMJ*, Feb. 4, 1882, Feb. 10, 1883; *NYG*, Feb. 11, 1882; *The New York Stock Exchange* (New York: Historical Pub., 1886), 73; *NYT*, Dec. 10, 11, 22, 1886, May 21, 28, June 5, 1889, Apr. 8, 1896, Sept. 16, 1897; *NCAB*, v. 11 (1901), 10–11; Richard E. Lingenfelter, *Death Valley & the Amargosa* (Berkeley: University of California Press, 1986), 158–159.

84. San Bernardino County Recorder, Deeds Book 26, 82–87, bk. 27, 188–189; New York *Indicator*, Nov. 10, 1881, quoted in San Bernardino, Calif., *Times*, June 3, 1882; *EMJ*, Nov. 5, 19, Dec. 24, 31, 1881, Jan. 21, Apr. 22, 1882, Feb. 10, 1883; *NYG*, Jan. 14, 25, Feb. 3, 10, 17, 1882; *NYTr*, Jan. 14, 1882; *NYT*, Jan. 15, Feb. 4, 1882; Lingenfelter, op. cit., n. 83, 158–159.

85. *EMJ*, Nov. 5, 1881; *SFEx*, Nov. 23, 1881, Jan. 28, 31, 1882; *NYMR*, Dec. 10, 1881; New York *Daily Stockholder*, quoted in *NYG*, Feb. 11, 1882; Lingenfelter, op. cit., n. 83, 161.

86. *EMJ*, May 6, 14, 20, 1882, June 9, 1883; San Bernardino, Calif., *Times*, June 3, 1882; San Bernardino County, Superior Court, Case No. 410, Thomas E. Williams et al. v. South Pacific Mining Co., Aug. 4, 1883; *NYT*, Dec. 11, 1886; Copartnership Directory, v. 1, op. cit., n. 82; *BDAC*, 1061; Lingenfelter, op. cit., n. 83, 161.

87. *NYT*, Mar. 24, 1880, Oct. 16, 1883; *SLGD*, July 19, 1887; *Transcript of Record. The United States vs. The Maxwell Land-Grant Company, et al.* (Washington, D.C., 1887), 473–542; William Keleher, *Maxwell Land Grant* (Santa Fe: Rydal, 1942), 118–120; Jim Berry Pearson, *The Maxwell Land Grant* (Norman: University of Oklahoma Press, 1961), 76–77; Maria E. Montoya, *Translating Property: The Maxwell Land Grant* (Berkeley: University of California Press, 2002), 122–127.

88. *NYT*, Nov. 6, 1881, Feb. 5, 9, 1882, Aug. 17, Sept. 30–Oct. 2, 16, 18, 24, Dec. 29, 1883, July 18, Nov. 29, 1884, Nov. 25, Dec. 19, 1885; *SLGD*, July 19, 1887; Pearson, op. cit., n. 87, 94–111.

89. *NYT*, Mar. 11, 1881, Feb. 7, 1882, June 13, 30, July 1, 5, 7, 9, 1885, Oct. 16, 1886; Apr. 19, 30, June 5, 28, 1887, Aug. 26, 1888; New York *Sun*, June 13, 1885, June 12, 1887; *NYH*, in Omaha *Bee*, June 2, 1887; George W. Julian, "Land-Stealing in New Mexico," *North American Review* 145 (July 1887), 17, 25–26; St. Louis *Globe-Democrat*, July 19, 1887; U.S. vs. Maxwell Land Grant Co., 121 U.S. 325; Leon Noel, "The Largest Estate in the World," *Overland Monthly* 12 (Nov. 1888), 480–494; Fort Worth *Gazette*, Feb. 12, 24, 1889; John G. Otis, *Alleged Conspiracy in Connection With the Maxwell Land Grant*

(Washington, D.C.: G.P.O., 1892), 1–5; Gustavus Myers, *History of the Great American Fortunes* (Chicago: Kerr, 1911), v. 3, 328–335; Keleher, op. cit., n. 87, 84–88, 118–119, 127–136; Pearson, op. cit., n. 87, 88–92; Montoya, op. cit., n. 87, 157–190.

90. *EMJ*, Jan. 24, 1880, Jan. 1, July 30, 1881, Feb. 4, Aug. 19, 26, Sept. 2, Oct. 14, 1882, Feb. 10, 1883; *MSP*, Dec. 25, 1880; *NYT*, July 10, Dec. 18, 24, 1881; *SFAC*, Oct. 1, 1881; *LAT*, July 24, 1887.

91. *NYT*, Mar. 24–25, 1870, Oct. 11, Nov. 17, 1877, Mar. 31, Apr. 14, 1878, Sept. 17, 1879, Mar. 21, Apr. 15, 27, July 25, 1880, Apr. 2, 1881, Jan. 2, 1892; *EMJ*, Jan. 24, 1880, Feb. 4, Dec. 30, 1882, Feb. 10, 1883, Jan. 5, 1884; Leonard A. Jones, "Preferred Stock," *American Law Review* 18 (Jan. 1884), 43–50; Rothwell, op. cit., n. 36, v. I, (1893), 477.

92. *NYT*, Nov. 17, 1877, July 25, 1880; Weed, op. cit., n. 36, v. 13, 617–618, v. 14, 477–478.

93. *MSP*, Dec. 16, 1876, Sept. 29, 1877, Jan. 29, 1881, Jan. 28, 1882, Jan. 3, 24, 1885; *NYT*, Oct. 13, 1877, Jan. 5, June 6, 1878; R. S. Lawrence, ed., *Pacific Coast Annual Mining Review and Stock Ledger* (San Francisco: Francis and Valentine, 1878), 170, 173, 200, 205–206, 223; *EMJ*, Jan. 11, 1879, Jan. 24, 1880, Feb. 4, 1882; *SFEx*, Jan. 17, 1882, Jan. 23, 1883; James Gilroy, *The Mining Stock Market; "How They Do It"* (San Francisco: Thomas, 1879), 30–31, 46, 72, 77, 92; Smith, op. cit., n. 3, 208–210; Balch, op. cit., n. 2, 1153, 1169, 1171, 1173; "American Millionaires: The Tribune's List," *The Tribune Monthly* 4 (June 1892), 5, reprinted in *New Light on the History of Great American Fortunes* (New York: Kelley, 1953), 5; Rothwell, op. cit., n. 36, v. I, (1893), 473–474; Bertrand Couch and Jay Carpenter, *Nevada's Metal and Mineral Production (1859–1940)* (Reno: University of Nevada, 1943), 46.

94. *Silver Bend Reporter* (Belmont, Nev.), May 2, 1868; *Inyo Independent* (Independence, Calif.), May 10, 1873, Jan. 15, Nov. 12, 1881; Lawrence, op. cit., n. 93, 74–75; *Esmeralda Herald* (Aurora, Nev.), Sept. 4, 1880; *True Fissure* (Candelaria, Nev.), Oct. 2, Nov. 20, 1880, May 7, 1881; *EMJ*, Oct. 30, 1880, Oct. 29, 1881; *NYMR*, Nov. 6, 20, 1880, Jan. 29, 1881; Lingenfelter, op. cit., n. 83, 104, 108–110, 146.

95. *Inyo Independent* (Independence, Calif.), Sept. 7, Oct. 26, 1872, Jan. 10, 1874, Apr. 9, Nov. 12, 1881; *True Fissure* (Candelaria, Nev.), Oct. 2, 1880, Apr. 2, May 7, 1881; *EMJ*, Oct. 30, 1880; *NYMR*, Nov. 20, 1880, Jan. 29, 1881; New York *Truth*, Mar. 13, 1882, quoted in the *SFEx*, Mar. 24, 1882; Lingenfelter, op. cit., n. 83, 109, 147–150.

96. *NYB*, Nov. 1, 1880; *Inyo Independent* (Independence, Calif.), Jan. 15, Feb. 5, May 28, 1881; *Esmeralda Herald* (Aurora, Nev.), Jan. 29, Apr. 16, 1881; *True Fissure* (Candelaria, Nev.), Feb. 26, May 7, 1881; *EMJ*, May 28, 1881; *NYG*, May 28, 1881; Lingenfelter, op. cit., n. 83, 149–150.

97. *EMJ*, Oct. 23, 1880, Mar. 26, May 7–28, June 18, July 2, 1881; *NYMR*, Nov. 6, 1880, Apr. 30, June 4, 1881; *True Fissure* (Candelaria, Nev.), Nov. 20, 1880; *Inyo Independent* (Independence, Calif.), May 14, June 11, July 23, 1881; *SFEx*, Mar. 24, 1882; Lingenfelter, op. cit., n. 83, 149–152.

98. *NYT*, Apr. 27, 1881; *NYG*, May 28, 1881; *EMJ*, May 28, June 18, July 2, Aug. 6, 1881; *NYTr*, May 9, 13, June 4, 13, 1881; *Inyo Independent* (Independence, Calif.), June 11, July 9, 1881; *True Fissure* (Candelaria, Nev.), July 16, 1881; Charles Kaufman, letter to

George Roberts, June 23, 1881, Asbury Harpending Collection, California Historical Society, San Francisco, courtesy of Jeremy Mouat; *SFEx*, Nov. 16, 1881, Mar. 24, 1882; Lingenfelter, op. cit., n. 83, 149–151.

99. *EMJ*, June 18, July 2–Aug. 6, 1881; *Inyo Independent* (Independence, Calif.),July 16, 1881; Lingenfelter, op. cit., n. 83, 151–152.

100. *EMJ*, July 30–Aug. 27, 1881; *Inyo Independent* (Independence, Calif.), Sept. 17, Nov. 12–19, 1881; *NYMR*, July 23, 1881; Henry Hall, *America's Successful Men of Affairs* (New York: Tribune, 1896), v. 2, 22; *BDAC*, 474; Lingenfelter, op. cit., n. 83, 152; Maury Klein, *Union Pacific* (Garden City, N.Y.: Doubleday, 1987), passim.

101. *NYMR*, July 23, 1881; *EMJ*, Aug. 6–27, Nov. 12, 1881; *Inyo Independent* (Independence, Calif.), Sept. 17, Nov. 12–19, 1881; *SFEx*, Nov. 14, 18, 1881, Mar. 24, 1882.

102. *NYT*, Dec. 14, 1876, Mar. 5, 1886, Oct. 19, 1889 (Selover); *EMJ*, Nov. 12, Dec. 3–17, 1881, Feb. 18, Apr. 29, 1882, Feb. 10, 1883; *Inyo Independent* (Independence, Calif.), Nov. 12–19, Dec. 3, 17, 1881, Jan. 7, July 22, 1882; *True Fissure* (Candelaria, Nev.), Jan. 14, July 22, 1882; *Esmeralda Herald* (Aurora, Nev.), Feb. 18, 1882; *SFEx*, Mar. 24, May 11, 1882; Lingenfelter, op. cit., n. 83, 152; Homer E. Milford, *History of the Lake Valley Mining District* (Santa Fe: New Mexico Mining and Minerals Division, 2000), 68.

103. *EMJ*, July 2, 16, 1881, Feb. 4, 1882, Feb. 10, 1883; *NYMR*, Jan. 7, 1882; Lingenfelter, op. cit., n. 83, 151–152.

104. *NYT*, Nov. 30, 1880, Oct. 30, 1881, Jan. 4, 1882; Vickers, op. cit., n. 4, 393, 400–403; *EMJ*, June 18–25, 1881; *NYMR*, Dec. 24, 1881; Dempsey and Fell, op. cit., n. 24, 152–158; David A. Remley, *Bell Ranch* (Albuqerque: University of New Mexico Press, 1993), 112–113; Milford, op. cit., n. 102, 51, 61; Plazak, op. cit., n. 8, 141–143.

105. *NYT*, Oct. 30, 1881; *EMJ*, Aug. 13–Dec. 10, 1881; *NYMR*, Dec. 3–10, 24–31, 1881; Dempsey and Fell, op. cit., n. 24, 158–164.

106. *EMJ*, Dec. 3–10, 1881, Jan. 7, Feb. 4, Apr. 8, May 13, 1882, Feb. 10, 1883; *NYMR*, Dec. 3–10, 24–31, 1881, Feb. 4, Mar. 11–18, May 13, June 17, July 8, 1882; *CMR*, Jan. 28, 1882; *MSP*, Mar. 25, 1882; *WSJ*, July 21, 1923; Dempsey and Fell, op. cit., n. 24, 158–168; Remley, op. cit., n. 104, 113–114.

107. *NYT*, Oct. 4, 1881, May 28, 1882, July 13, Aug. 7, 1883, July 28, 1887, July 17–18, 1888; *NSJ*, Mar. 18, 1882; *SFEx*, Mar. 24, 1882; *NYMR*, Apr. 15, 1882; *BE*, July 27, Dec. 29–30, 1887, July 17, 1888; *EMJ*, July 30, Dec. 31, 1887.

108. *NYT*, Oct. 4, 1881, May 28, 1882, July 13, Aug. 7, 1883, July 17–18, 1888; *BE*, July 27, Dec. 29–30, 1887, Jan. 3, July 17, 1888; *EMJ*, July 30, Dec. 31, 1887.

109. *SFAC*, July 21, 1867 ("Podgers" identified by Mark Twain); Reno *NSJ*, Mar. 18, 1882; *NYT*, May 7, 14, 21, 28, June 3, 1882; Roberts's young superintendent, John Hays Hammond's "bulling" statements about the Santa Maria are mentioned in a letter of Donald B. Gillette to Henry Janin, dated Jan. 16, 1882, from the James D. Hague Papers in the Huntington Library, quoted in Milford, op. cit., n. 102, 61, but Hammond in his entertaining but highly unreliable *Autobiography* (Murray Hill, N.Y.: Farrar and Rinehart, 1935), 110–143, twists around his role as superintendent, pretending he didn't even see it until April!

110. *NYT*, July 13, Aug. 7, 1883, July 28, 1887, Feb. 11, July 17–18, 1888, June 26, 1896; *BE*, July 27, Dec. 29–30, 1887, Jan. 3, 8, July 17, 1888, Feb. 13, 24, 1889; *EMJ*, July 30, Dec. 31, 1887.

111. *SFEx*, May 10, 1881; *MSP*, Aug. 27, 1881; *CMR*, Jan. 14, 1882; *NYMR*, Apr. 15, 1882; Milford, op. cit., n. 102, 41–47.

112. *MSP*, Aug. 27, 1881; *CMR*, Jan. 14, 1882; *PMJ*, Jan. 14, Apr. 22–29, May 20, June 17, 1882; *NYMR*, June 10–17, 1882; *SFEx*, July 9, 1883; Bernard MacDonald, "Genesis of the Lake Valley, New Mexico, Silver-Deposits," *Transactions of the American Institute of Mining Engineers* 39 (1909), 851–853; *Dictionary of National Biography*, v. 3, 711–713; Dale Collins, "Frontier Mining Days in Southwestern New Mexico" (M.A. thesis, Texas Western College, 1955), 108, 209; David J. Jeremy, ed., *Dictionary of Business Biography* (London: Butterworths, 1984), v. 5, 901–904; Milford, op. cit., n. 102, 47–51, 57, app. ii, 5–6, 24–26; Jeremy Mouat, "Looking for Mr. Wright," *Mining History Journal* 10 (2003), 6–8; Jeremy Mouat, "Whitaker Wright, Speculative Finance, and the London Mining Boom of the 1890s," in *Mining Tycoons in the Age of Empire, 1870–1945*, ed. R. E, Dumett (Burlington, Vt.: Ashgate, 2009), 126–149.

113. Colorado Springs *Gazette*, Aug. 21, 1881; *MSP*, Aug. 27, 1881, Aug. 19, 1882; *LD*, Sept. 3, 1881; *Homer Mining Index* (Lundy, Calif.), Sept. 3, 1881; *Mining World* (Las Vegas, N.Mex.), Sept. 15, 1881; *NYMR*, Dec. 17, 1881; *SFEx*, Dec. 23, 1881, July 9, 1883; Tombstone, Ariz., *Epitaph*, Dec. 24, 1881; *CMR*, Jan. 14, 1882; *PMJ*, Mar. 11, 25, Apr. 1, 1882; MacDonald, op. cit., n. 112, 855–856; Griswold, op. cit., n. 6, 688–689; Milford, op. cit., n. 102, 50–60, app. ii, 16–19.

114. *SFEx*, Jan. 25, Feb. 4, Apr. 5, 1882; *NYMR*, Feb. 25, 1882, Aug. 4, 1883; *PMJ*, Apr. 1, June 17, 1882; Milford, op. cit., n. 102, 57–76.

115. *PMJ*, Mar. 4, May 20, June 10, July 1, Sept. 9, 1882; *SFEx*, Apr. 5, July 5, 20, 1882; *NYMR*, Apr. 8–15, 1882; Collins, op. cit., n. 112, 109–111; Milford, op. cit., n. 102, 65–71.

116. *NYMR*, Feb. 25, June 10–17, 1882; *SFEx*, Mar. 24, Apr. 5, July 5, 20, 1882; *PMJ*, June 3–July 15, Aug. 26, 1882; *SFBul*, June 28, 1882; *Mining World* (Las Vegas, N.Mex.), July 15, 1882; Milford, op. cit., n. 102, 71–75.

117. *NYMR*, Apr. 15, June 3, July 1, 1882; *SFEx*, July 5, 1882; *EMJ*, Apr. 8–May 27 (end page ads), July 15, 1882; *PMJ*, July 22, 1882.

118. *PMJ*, June 10, July 1, 1882; *SFEx*, July 7, 20, Aug. 9, 30, Sept. 11–12, 1882; *NYMR*, July 1, 1882; *EMJ*, June 17, Aug. 1, 9, 1882; Balch, op. cit., n. 2, 501–502; Milford, op. cit., n. 102, 78.

119. *SFEx*, Sept. 5, 12, 25, Oct. 14, 1882; *PMJ*, Aug. 26, Sept. 9, Oct. 21, Nov. 18, Dec. 23, 1882; Las Vegas, N.M. *Mining World*, Sept. 15, Nov. 1882; *Black Range* (Chloride, N.Mex.), Jan. 19, 1883; *EMJ*, Apr. 14, 1883; *MSP*, Nov. 11, 1882; *NYMR*, Dec. 16, 30, 1882, Jan. 6, 1883; Milford, op. cit., n. 102, 84–95.

120. *PMJ*, Jan. 13, Mar. 23–30, Apr. 6, June 2, 23–30, July 21, Aug. 4–18, 1883; *Black Range* (Chloride, N.Mex.), Jan. 19, 1883; *SFEx*, July 9, 1883, Feb. 19, 1884; *EMJ*, Sept. 22, 1883, Feb. 16, Sept. 27, 1884; *NYMR*, Dec. 16, 1882, June 16, Aug. 4, 1883; Milford, op. cit., n. 102, 96–121.

121. *NYT*, Nov. 2, 1878, June 12, 1882, Mar. 4, June 28, Aug. 15–16, Dec. 2, 1883, Mar. 20, July 18, 27, Sept. 13, 1884, July 1, 1887, Jan. 14, 1892, July 21, 1902 (Mackay obit); *SFEx*, Aug. 25, 1883; *Statement of George D. Roberts, Representing the Postal Telegraph and Cable Company, Before the Committee on Post-Offices and Post-Roads of the U.S. Senate* (Washington, D.C.: Committee, 1884) 1–7; *BE*, July 21, 1902; *Nevada Mining News* (Reno), June 18, 1908; Michael J. Makley, *John Mackay* (Reno: University of Nevada Press, 2009), 157–173.

122. *NYT*, Dec. 29–31, 1900, Mar. 16, 1903, Jan. 27, 1904 (Wright obit); *EMJ*, Jan. 24, 1903, Jan. 28, 1904; *LT*, Jan. 12–16, 19–22, 26–28, 1904; Atlanta *Constitution*, Mar. 17, 1903, Jan. 29, 1904; *WSJ*, Jan. 27–28, 30, 1904; "Whitaker Wright Finance," *Blackwood's Magazine* 175 (Mar. 1904), 397–409; Lee, op. cit., n. 112, 711–713; Aylmer Vallance, *Very Private Enterprise* (London: Thames and Hudson, 1955), 52–65; Jeremy, op. cit., n. 112, 901–904; Michael F. Gilbert, *'Fraudsters'* (London: Constable, 1986), 17–45; Jeremy Mouat, *Roaring Days* (Vancouver: UBC Press, 1995), 47–66, 167; Jeremy Mouat, "The Great Tycoon," *The Beaver* 75 (Jan. 1996), 19–25; Mouat, op. cit., n. 112, 6–17; Mouat, op. cit., n. 112, 126–149.

123. *NYT*, Jan. 13, 1890, Apr. 13, 1897 (Cope); Milford, op. cit., n. 102, 111–118, app. ii, 5–7.

124. *NYT*, Nov. 7, 1884, Feb. 4, 1885, Dec. 25, 1888.

125. *NYT*, Dec. 23, 1880; "My Mining Investments," *Lippincott's Magazine* 1 (Jan. 1881), 86–96.

126. "My Mining Investments," op. cit., n. 125, 86–87.

127. *Mining World* (Las Vegas, N.Mex.), Oct. 15, 1882; *CMR*, Oct. 21, 1882, May 24, 1883; *MSP*, July 14, 1883; *NYMR*, May 26, 1883; Henry B. Clifford, *Years of Dishonor* (New York: Brown, 1883), 5–21; Henry B. Clifford, *Rocks in the Road to Fortune* (New York: Gotham, 1909), i.

128. *SLT*, Dec. 12, 1879, Mar. 26, May 5, Aug. 12, 1880, Feb. 28, Mar. 2, 1883; *EMJ*, Feb. 14, 1880, Jan. 29, 1881; *NYT*, Sept. 1, 1881, Feb. 27–28, Mar. 2, 1883, Nov. 15, 1884; Ogden *Herald*, Feb. 27–28, 1883; *Park Mining Record* (Park City, Utah), Mar. 3–17, 1883; Edward W. Tullidge, *Tullidge's Histories* (Salt Lake City: Juvenile Instructor, 1889), v. 2, 513–514.

129. *NYT*, Dec. 8, 1881, May 7, 1884, Apr. 8, Nov. 10, 1885, Feb. 26, 1890; *BE*, Oct. 8, 31, 1885; New York *Sun*, Oct. 9, 1885; Baltimore *Sun*, Oct. 9, 1885; Ward, op. cit., n. 37; William S. McFeely, *Grant* (New York: Norton, 1981), 493.

130. *NYT*, May 19, 22, 1884, Jan. 22, Mar. 28, 31, June 27, Nov. 10, 1885; "Monaco on Wall Street," *Harper's Weekly* 28 (May 24, 1884), 326; Philadelphia *Inquirer*, Dec. 29, 1884; New York *Sun*, Feb. 1, Oct. 9, 1885; Clews, op. cit., n. 4, 215–221; Robert Sobel, *Panic on Wall Street* (New York: Macmillan, 1968), 210–211; McFeely, op. cit., n. 129, 491.

131. *NYT*, May 7, 14–16, 19, 28, June 18, July 8, 19, Dec. 28, 1884, Jan. 22, Mar. 28, Apr. 6, June 17, 27, July 24, Sept. 9, 29–30, Oct. 29, 1885, Mar. 18, 1886, Feb. 26, 1890; Philadelphia *Inquirer*, May 7, 10, Dec. 29, 1884; New York *Sun*, May 7–10, 15–17, 24, 30, 1884, Mar. 28, Oct. 9, 29, 1885; *NYTr*, May 8–10, 30, 1884, July 23–24, Oct. 29, 1885, Mar. 18,

1886; "The Wall Street Scandal," *The Nation* 38 (May 15, 1884), 420; "General Grant and James D. Fish," *Harper's Weekly* 28 (June 7, 1884), 359; *BE*, July 31, Oct. 8, 31, 1885, Apr. 30, 1892; *National Police Gazette* (New York), Aug. 8, 1885; Baltimore *Sun*, Oct. 9, 1885; Adam Badeau, *Grant in Peace* (Hartford, Conn.: Scranton, 1887), 418–422; Ward, op. cit., n. 37; Hamlin Garland, "A Romance of Wall Street. The Grant and Ward Failure," *McClure's Magazine* 10 (Apr. 1898), 498–505; Sobel, op. cit., n. 130, 214–229.

132. *CMR*, Aug. 12, 1882; *PMJ*, Sept. 16, 1882; *SFEx*, Feb. 7, Apr. 26, 1883, Jan. 2, 29, Mar. 11, Aug. 11, 1884.

133. *MSP*, Jan. 24, 1880, Jan. 29, 1881, Jan. 28, 1882, Jan. 27, 1883, Feb. 16, 1884, Jan. 24, 1885; Balch, op. cit., n. 2, 619, 989; S. F. Emmons and G. F. Becker, *Statistics and Technology of the Precious Metals* (Washington, D.C.: G.P.O., 1885), 105, 111–114, 156–157, 185; Neale, op. cit., n. 73, 43, 54.

134. *EMJ*, Feb. 4, 1882; Balch, op. cit., n. 2, 502; *SFEx*, Aug. 11, 1884; Emmons and Becker, op. cit., n. 133, 111; *NYT*, Jan. 25, 1886; U.S. Bureau of the Census, *Historical Statistics of the United States: Colonial Times to 1970* (Washington, D.C.: G.P.O., 1975), 12, 22, 255.

INDEX

References to illustrations appear in italics.

Ada Elmore Gold and Silver Mine, 130

Ahtna Indians, 36–37

Ajo mine, *44*, 79–80, 137

Albany *Journal*, 111

Albuquerque, N.Mex., 136, 290

Alice Gold & Silver Mining Co., 236–237, 357

Alley, John B., 339–340

Allison Ranch mine, 124, 356

Allsop, Thomas, 60

Almagro, Diego de, 26–27

Alta, Utah, 156, 163–168

Alta California, 95, 115, 198, 202, 223

Alta Silver Mining Co., 254–255

Alvarado, Juan Bautista, 58, 67

Amador City, Calif., *84*, 150, 189

Amador Consolidated Mining Co., 190

Amador Mining Co., 65–66, 189–190

American Bankers Association, 216

American Mining Stock Exchange, 274–275, *275*

American River, 43, 55

Ames, Adelbert, 286

Amie Consolidated Mining Co., 294

Anglo-Californian Gold Mining Co. Ltd., 50

Anker, Moses, 178

Antonio, Frank, 92

Apache Indians, 33–36, 78–79, 82, 136–137, 230, 232, 306, 341

Apex law, 91, 159–160, 295

Appleton, Daniel S., 280

Argenta, Mont., 84, 132

Argonaut companies, 45–48

Arizona, Cerro de la San Antonio de la, 33

Arizona and Sonora (Mowry), 136

Arizona Copper Mining Co., 137

Arizona Land & Mining Co., 81

Arizona Mining & Trading Co., 79–80

Arizona Mining Co., 136

Arkansas River, 32

Arms, Charles D., 311

Arnold, Philip, 199–204, 407n75

Aron, Joseph, 263–265

Arrowhead Springs, Calif., 312

Arthur, Chester A., 218, 326

Ashburner, William, 337

Assessments (stock), 54–55, 62, 90, 98, 101, 123, 131, 133, 141–142, 144–146, 149, 152–153, 173, 192, 195, 197, 206–209, 211–212, 215, 233, 240–241, 250, 252–254, 260–262, *262*, 265–266, 288, 311, 318, 327–330, 347–348

Atahuallpa (Inca emperor), 26–27

Atchison, Topeka & Santa Fe Railroad Co., 326, 340

Atlantic-Pacific Railway Tunnel Co., 301, 304

Atlantic-Pacific Tunnel Co., 300–304

Atwell, Cordelia, 348
Aurora, Nev., 84, 100–101, 103
Austin, Frederick, 310
Austin, Nev., 84, 103, 108–109, 330
Avila, Pedro Arias de, 26
Aztec empire, 23–25
Aztec Syndicate of Arizona, 305, 307

Badger-Eureka mine (Hayward's), 65,
    355
Baffin Island, 30–31
Bailey, John W., 243, 245
Bailey, Robert, 98–99, 172–173
Bailey Silver Mining Co., 98–99
Balbach, Charles and Leopold, 186, 297
Balbach, Edward S., 186
Balboa, Vasco Núñez de, 25–26
Baldwin, Elias J. "Lucky," 225, 255
Baldwin, Joseph, 68
Ballad of Baby Doe (opera), 203
Balzac, Honoré de, 7
Bancroft, Hubert Howe, 204
Bankers, 14, 59, 71–72, 74, 116, 124,
    128–129, 132, 163, 175, 180, 191, 209,
    235–236, 241, 256, 263, 267, 274,
    295–296, 312, 315; monopolies, 71,
    74, 76–77, 141, 143–145, 147, 194, 197,
    209–211, 263; practices, 109, 116,
    128–129, 174, 235–237, 258–260, 348;
    speculations, 59, 67, 88, 109–110, 132,
    146, 158, 161, 163–164, 174, 191, 200,
    202, 216–218, 225–227, 227, 235, 267,
    280, 295–296, 305, 327, 349–351
Bank of California, 116–117, 141,
    144–147, 202–203, 226–227, 227, 256
Bank of England, 174, 305
Bank of the United States, 39
Bank Ring, 144–147, 194, 197, 210, 254,
    263
Banque Générale, 32
Baranov, Aleksandr, 37
Barlow, Samuel L. M., 200–201, 241

Barney, James M., 230–232, 305
Barron, Forbes & Company, 70
Barron, William E., 71–74, 97, 112–119,
    151, 193, 208, 328, 355
Bartlett, Washington M., 314
Bates, Edward, 113–115, 119
Baum, Frederick, 100–101
Baxter, Henry, 163, 166
Beachy, Hill, 172
Bear River & Auburn Water &
    Mining Co., 74–76, 356
Bechtel, Frederick K., 126, 241
Belcher Silver Mining Co., 195, 196,
    196–197, 201, 205, 224, 355
Bell, Thomas, 116, 151, 193–195, 208, 243,
    246, 328
Belle Isle Mining Co., 329
Belmont, Calif., 226
Benton, Thomas Hart, 57–58, 67
Big Bonanza, 207–212, 215–221, 228,
    251–252, 261, 264–265, 270–271, 328,
    353
Big Horn Mountains, 248–249
Big Pittsburgh Consolidated Silver
    Mining Co., 291–292
Billing, Gustav, 186, 296
Bingham Canyon, Utah, 84, 129,
    133–134, 232–234, 237
Bishop, Jack, 177
Bissell, Charles, 277–278, 283
Black, Jeremiah S., 73, 112–115, 118, 120
Black Butte Station, Wyo., 156, 201
Black Hawk, Colo., 84, 127, 187–188
Black Hills, 156, 228, 241–248
Black Hills Times, 245–246
Blacks, 24, 28, 39, 215, 226, 238, 302
Blaine, James G., 293
Blair, Francis P., 119
Blake, William P., 99, 164, 348
Blake, Williston, 305
Blue Gravel mine, 151–152, 320
Blue Gravel Mining Co., 151

Bobtail Lode, 127–128, 187

Bodie, Calif., 84, 126–127, 238–241, 284

Bodie Bluff Consolidation Mining Co., 126–127, 127

Bodie Mining Co., 239–241, 356

Bodie *Standard*, 239–240

Boggs, Biddle, 68–69

Boise, Idaho, 84, 129

Boissevain, Gideon Maria, 180–181, 181

Bolton, Barron & Co., 71–74, 355

Bolton, James R., 71–72, 115

Bonanza Firm, 208–212, 228, 254–260

Bonanza Suit (*Burke et al. v. Flood et al.*), 258–260, 269–270

Bonaparte, Louis Lucien, 51

Bonaparte, Napoleon, 33

Bonds, 68, 110–111, 116, 133, 135, 152, 180–181, 192, 309, 315, 318, 325–326, 350, 385, 392

Borden, William, 279–280, 285

Borland, Archibald, 212, 244–248, 267

Boston, Mass., 55; stock exchanges, 41, 138, 288, 290; stockholders, 45, 122, 138, 187, 216–218, 267–268, 325, 334, 338, 340, 353

Boston & California Joint Stock Mining & Trading Co., 45

Boston & Colorado Smelting Co., 187, 355

Boston *Commercial Bulletin*, 140–141

Bourn, William B., 153, 157

Bowie, Augustus J., 248

Boyer, Oliver, 310

Boyle, William G., 305

Bradford, Putnam F., 130

Bradshaw Mining Co., 311–312

Breed, Abel D., 177–178, 281, 331, 346

Brennan, Michael, 62, 124

Bridal Chamber mine, 340–344

British & Colorado Mining Bureau, 174

British stockholders, 30–32, 40, 49–52, 56, 59–61, 64, 139, 157–158, 160, 162–169, 174, 180, 190–192, 199, 205, 229–230, 234, 263, 293, 305, 344–346

Broadway (New York City), 274–275, 275, 279, 323

Brown, George, 93

Browne, John Ross, 2, 105, 116, 126, 136–137, 305

Brown mine, 174

Brown University, 187

Brumagim, Mark, 109, 191–192, 327

Brunchow, Frederick, 80, 306

Brunchow mine, 306, 311

Brydges-Willyams, Edward, 166

Buchanan, James, 73

Bullion Gold & Silver Mining Co., 249

Bullion Mining Co., 252, 254

Bunker Hill mine, 126, 238

Burke, John H., 258–260

Burnett, Peter, 7, 14

Burning Moscow Mining Co., 91

Burr, Thomas, 175

Burro Mountains, 199

Butler, Samuel, 60

Butte, Mont., 15, 156, 228, 234–238, 263, 290

Butterworth, Samuel F., 73, 115–120, 124, 136–137, 151, 193, 305

Byers, William N., 302

Cabeza de Vaca, Alvar Núñez, 28, 33

Cahuilla Indians, 119

Cajalco mine, 119–120

Cajamarca, Peru, 26

Calafia (fictional queen), 28

Calhoun, John C., 40

Calico, Calif., 272, 312–313

California Civil Code, 258

California constitution of 1879, 13, 260, 268–270, 273; margin sales banned, 13, 270

California Geological Survey, 110
California gold rush, 43–53, 83
California Mining Co., 210–212, 216–217, 221, 224, 254, 259–260, 263, 266, 271, 282, 355
California Quicksilver Co., 194
California Stock Board (New York), 107
California Stock Exchange, 205, 213
California Street (San Francisco), 88
California Supreme Court, 68, 92, 227
Calumet & Hecla mine, 42
Calvo y Muro, Blas, 34
Campbell, Allen G., 315–316
Camp Huachuca, Ariz., 310
Canadian mines, 250, 345
Canadians, 129–130, 163, 179–182, 265–267, 336–337, 346
Capitancillos grant, 72–73, 113–114
Capps mine, 39–40
Cap-Rouge, Quebec, 30
Caribou Consolidated Mining Co., 298
Caribou mine, 156, 173, 177–178, 298–300
Carleton, James H., 82, 136–137
Carlos I of Spain, 25, 27, 30
Carpentier, Horace W., 120
Carrasco, José Manuel, 33
Carroll, Daniel, 39
Carson Hill mine, 44, 63–65, 121
Carsons Creek Consolidated Mining Co., 64, 356
Cartier, Jacques, 30
Castillero, Andrés, 70, 112–113
Cataract & Wide West Gravel Mining Co., 319
Catron, Thomas Benton, 180–182
Central Arizona Mining Co., 323
Central City, Colo., 84, 127–129, 282
Central Pacific Railroad Co., 125–126, 155–156, 156, 315

Central Park (New York City), 110, 348
Cerrillos, N.Mex., 29, 33, 77
Cerro Colorado, Ariz., 80, 84
Cerro de la San Antonio de la Arizona, 33
Cerro Rico de Potosí, Bolivia, 27, 29
Chaffee, Jerome B., 182–183, 189, 273, 291–299, 350; British promotions, 174–176, 178, 205; early career, 128–129; Little Pittsburg mine, 277–279, 282–283, 297; Maxwell Grant scam, 178–180; U.S. Senate seat, 129, 189
Chamberlain, Oscar, 65
Charlotte, N.C., 40
Chavanne, Andre, 124, 238
Cherokee Nation, 39
Chicago, 186, 322; stock exchanges, 288, 290; stockholders, 176, 279, 281, 311, 314, 325
Chihuahua, Mex., 33, 78, 134; stockholders, 34–36, 78
Chino Rancho, Calif., 308
Chinese, 74, 184, 192
Chippewa Indians, 40
Chisholm, Robert, 163
Chisholm, William, 163
Chitina River, 36
Chollar-Potosí Mining Co., 221, 355
Chrysolite Silver Mining Co., 279, 280, 280–288, 295, 332–333, 347, 356
Cibola quest, 28–29
Cincinnati, Ohio, stockholders, 80–81, 136, 177
Civil War, 82, 85, 107, 109, 112–113, 120, 127, 133, 139, 155, 176, 201, 207, 220, 232, 297, 300, 315
Clark, Frederick, 174
Clark, William Andrews, 235–238
Clay, Henry, 40
Clear Creek, Colo., 84, 173
Clemens, Samuel L. (Mark Twain), 95, 99–100, 103, 141

Clergymen. *See under* Stockholders

Cleveland, Grover, 133, 293, 326–327

Cleveland, Ohio, stockholders, 232–234

Cleveland *Plain Dealer*, 232, 234

Clever, Charles P., 137

Clews, Henry, 338

Cliff mine, 41–42

Clifford, Henry B., 347–348

Clifton, Ariz., 156, 183–185

Clifton, Henry, 183

Coffin, James, 210, 259

Cohen, Abraham, 245

Coleman, John and Edward, 123, 149, 318

Coleman, William Tell, 81

Colfax, Schuyler, 97

Collinson, John, 179–181, 325

Colombo, Cristoforo (Christopher Columbus), 22–23; first European gold mine in Americas, 23

Colorado Real-Estate & Mining Bureau, 303

Colorado River, 33

Colorado Terrible Lode Mining Co. Ltd., 173–174

Colorado United Mining Co., 174

Colt, Samuel, 80–82, 132, 136

Colton, Walter, 43, 45

Columbia, Calif., 44, 76

Columbia School of Mines, 235, 296, 315

Commercial Cable Co., 345

Como Miners' Stock Board, 103

Compagnie des Indes, 32

Compagnie d'Occident, 32–33

Compagnie Française & Américaine de San-Francisco, 50, 51

Companye of Kathai, 30

Company of Adventurers for Exploring the Gold Districts of New California, 51

Comstock, Anthony, 250

Comstock, Henry T. P., 94–97, 129–130

Comstock Lode, 15, 85–99, 102–106, 146–148, 195–198, 204–227, 258–261, 263–271, 276

Comstock Mill & Mining Co., 356

Comstock Mill Ring, 144–146, 196, 201, 205, 209, 226, 274

Conant, Royal B., 216–218

Conklin, Enoch, 305

Connelly, Henry, 134–135

Connor, Patrick Edward, 133

Conquistadores, Spanish, 19, 29

Consolidated Gregory Co., 128, 138, 163

Consolidated Virgina Mining Co., 207–210, 210, 211–212, 217–221, 224, 226, 254–260, 263, 266, 271, 282, 355

Contention Consolidated Mining Co., 306, 309–310, 356

Continental Stereoscope Co., 305

Converse, James W., 187

Conway, Edward, 120

Cook, Dan and Seth, 239, 241, 318

Cooke, Jay, 315–316

Cooper, James B., 199

Cope, Edward D., 340–346

Copper: market prices, 21, 34, 82, 120–122, 182–184, 234, 405n50; mining, 19, 32–42, 78–80, 120–123, 133–134, 137–138, 182–184, 228, 234–238; smelting, 34, 79, 121–122, 182–185

Copperopolis, Calif., 84, 121–123

Copper Queen Mining Co., 357

Copper River, Alaska, 36

Corbet, Michael, 124

Corbin, Elbert and Frank, 307–309

Corbin Mill and Mining Co., 307

Coronado, Francisco Vásquez de, 28–29, 33

Corruption, 101, 144; Congress, 147, 173, 194, 263; courts, 59, 68, 91, 102, 109, 111, 140, 161, 176, 179, 195, 246;

Corruption (*continued*), executive, 73,
   83, 112, 114–115, 176, 179; experts,
   89–90, 164, 201, 331; press, 166, 204,
   335–336, 342; stock market, 14–15, 67,
   83, 104–105, 140, 144–146, 155, 168, 171,
   197–198, 201, 209, 215, 219, 222, 224,
   240, 251, 254, 264, 268–269, 274, 276,
   285, 288, 292, 294, 323–324, 328, 337,
   344, 347, 352
Cortés, Hernan, 24–28, 363n11
Corydon Mining Co., 128, 138
Coulter, George, 190, 263–265
Courcier, Esteban. *See* Curcier,
   Andrew and Stephen
Cox, Jacob D., 194
Crédit Mobilier, 334
Crocker, Charles, 125
Crowell, Gilbert L., 348
Crown Point Gold & Silver Mining
   Co., 195–198, 201, 204–205, 207, 221, 355
Cuba colony, 24
Cunard, Bache, 317
Cunard Steamship Co., Ltd., 316–317
Curcier, Andrew and Stephen, 35–36
Curtis, Allen A., 108–109, 330
Curtis California Mining Co., 49
Custer, George Armstrong, 242
Cuzco, Peru, 27

Daly, George, 284–287, 291, 336, 340–341
Daly, Marcus, 171, 236–238
Daniel, John and William, 124
Davis, Andrew Jackson, 235, 237–238
Davis, David, 113
Davis, Erwin, 163, 167–168, 234, 237
Davis, Henry G., 182
Dawson, Thomas M., 218–219
Deadwood, Dak. Terr., 156, 243–248
Deadwood Mining Co., 246
Dean, Walter E., 309
Deane, Coll, 244, 265
Death Valley, Calif., 84, 98, 172,
   228–230, 323–325, 330–334

Defoe, Daniel, 32
DeGroot, Henry, 159
Deidesheimer, Philipp, 212, 220, 338
De Kay, Drake, 286
Delano, Alonzo "Old Block," 62
Delavan, James, 55–56, 62
Delmonico, Charles, 291
Delmonico's Restaurant, 274
Dene Indians, 33, 36
Denham Jail, 218
Denver, 84, 129, 188–189, 243, 290,
   300–304, 344; smelting, 189, 297–298
Detroit, 169, 183
Detroit Copper Co., 183
Dewey, Squire P., 255–260
Diamond Drill Co., 199
Diamond drilling, 146, 199, 211
Diamond swindle. *See* Great Diamond
   Swindle
Dickens, Charles, 65, 195
Dickerman, Watson B., 295–296
Dillon, Sidney, 315
Disston, Hamilton, 307–308
Dividends, 95, 103, 109, 138, 145–146,
   154, 169, 211, 221, 257, 263, 271, 288,
   347, 349; earned, 36, 41–42, 76, 87, 92,
   107, 117, 122–124, 130–131, 133, 138–139,
   142, 149–151, 153, 157–158, 160–161, 165,
   170–171, 173–174, 189–190, 195, 197, 205,
   208, 228, 231–232, 237–239, 241, 246–
   248, 252, 259, 261, 271, 278, 280–281,
   286, 294–295, 298, 308, 310–311, 316,
   318–319, 327–330, 336; fraudulent, 56,
   107–108, 116, 140, 165–168, 231, 236,
   240–241, 292, 299, 309, 320, 337, 344;
   promised, 51, 56–57, 88, 141, 158, 164,
   166, 168–169, 192, 212, 249, 254, 282,
   286, 293, 301–303, 319–320; total, 54,
   261, 352, 355–357
Doane, George, 232–233
Doctors. *See under* Stockholders
Dodge, George S., 200–201, 204
Dodge, William Earl, Jr., 131

Doe, Elizabeth McCourt "Baby Doe" Tabor, 292–293

Dollars: 19th-century vs. 15th- and 16th-century values, 363n9; 19th-century vs. 21st-century values, 15, 363n9

Donnell, Robert W., 237

Doten, Alfred, 221

Dow, Charles H., 278

Downs, Robert C., 97, 125

Drake, Francis, 31

Drake, Frank, 157–158

Drew, Daniel, 193, 327

Duero, Andrés de, 24–25

Dun, Robert G., 299–300

Dunbar, Edward Ely, 79–80, 137

Duncan, William Butler, 241

Dutch stockholders, 178–181, 325–326

Eberhardt & Aurora Mining Co., 156–158, 161

Eberhardt Mining & Milling Co., 157

Eddy, Edward, 297

Edison, Thomas Alva, 135, 320–322

Edison Electric Light Co., 321

Edison Ore Milling Co. Ltd., 321, 322

Edison Reduction Works, 321

Eilers, Frederic Anton, 186, 296

Eldredge, James, 72–73

Elguea, Francisco Manuel, 34–35, 78, 137, 182

Eliot National Bank, 216

Elizabeth I of England, 30–31

Elizabethtown, N.Mex., 156, 179

Elkins, John, 179, 294, 325

Elkins, Stephen Benton, 179–182, 273, 278, 293–294, 325–327; Maxwell Grant scam, 179–181; Santa Fe Ring, 179–180, 182, 327; U.S. Senate seat, 182

Elkins, W.Va., 182

El Paso, Texas, 44, 78, 82, 290

Ely, John H., 160

Emma mine swindle, 163–168, 186, 191–192, 209, 229–230, 234

Emma Silver Mining Co. Ltd., 164–165, 165, 166–167, 357

Emma Silver Mining Co. of New York, 163

Empire Gold & Silver Mining Co., 126, 238, 241

Empire mine (Grass Valley), 63, 124, 152–153

Empire mine (Hayward's), 189

Empire mine (Park City), 348

Empire Mining Co., 152–153

Engineering & Mining Journal, 139, 241, 278, 286, 316, 343, 348

Envy, 14, 72, 189, 215, 218, 251, 267

Escalante, Felipe de, 29

Eshington, Peter, 238–239

Espinosa, Gaspar de, 26–27

Eureka, Nev., 156, 159–160, 278, 280

Eureka, Utah, 156, 168–170

Eureka Consolidated Mining Co., 160, 318, 355

Eureka Hill Mining Co., 170

Eureka mine (Hayward's), 65–66, 355

Eureka Mining Co., 168–170

Eureka Quartz Mining Co., 149–150, 189, 356

Evening Star Mining Co., 295–296, 349–350, 356

Ewer, Warren B., 139; on corrupt management, 140, 145–146, 260–261; on stock jobbing, 206, 269

Ewing, Thomas, 336–337

Excelsior Water & Mining Co., 319–320

Extralateral rights, 91, 159–160, 295

Fair, James G., 208–209; "Big Bonanza," 208, 228, 258, 265; U.S. Senate seat, 226

Fairbank, Nathaniel K., 311, 325

Fairplay (Colo.) Flume, 301

Fall, Albert B., 182
Farish, William A., 334–335
Farlin, William L., 234
Farwell, John V., 281
Father DeSmet Consolidated Gold
    Mining Co., 247–248
Father DeSmet mine, 243–244, 247
Faull, John A., 191
Fear, 14, 96, 256, 282, 341, 350
Field, David D., 68, 111
Field, Marshall, 279, 285, 325
Field, Stephen J., 68–69, 111, 120, 160,
    170, 227, 327
"Fighting companies," 90–91
Finnegan, James, 64–65
First Choice Diamond Mining Co.,
    202
First National Bank of Denver, 129
Fish, James Dean, 295–296, 349–350
Fisk, James, 111, 350
Flagstaff Silver Mining Co., 167–168
Flood, James C., 208, 226, 255–259,
    265–267
Fly By Night Gold & Silver Mining
    Co., 89
Fogus, David H., 172
Foord, John, 242, 260, 270
Forbes, Alexander, 70–71, 112–114
Forbes, James, 70–71, 73, 112
Forbes, William H., 107–108
Forney, John W., 119
Fort Yuma, Ariz., 44, 79–82
Francklyn, Charles G., 316–317
François I of France, 30
Franco-Prussian War, 263
Frank, Moses, 100–101
Frank Silver Mining Co., 93
Frankenberg, Alex, 245
Freeborn, Isaac, 97
Freeman, George A., 107–108
Freiberg School of Mines, 278, 297
Fremont, Jessie Benton, 57–58, 111

Fremont, John Charles, 57–60, 66–69,
    73, 109–111, 134
French stockholders, 32, 49–52, 59–61,
    234, 237–238, 263
Fricot, Jules, 149
Frisco, Utah, 272, 314–315
Frobisher, Martin, 30–31
Fryer, George, 277
Fryer Hill, Leadville, 277

Gadsden Purchase, 44, 79
Gage, Eliphalet B., 310–311
Galaktionov, Konstantin, 37
Galloway, James, 219–220
Gambling (stock), 105–106, 141, 198, 216,
    219, 224, 262, 268–269
Garfield, James A., 232, 283
Garrison, William, 251
Gashwiler, John W., 243, 245–246
Gates, Thomas, 31
Georgetown, Colo., 156, 173–177, 301
Germania Separating & Refining
    Works, 186, 356
Gila Copper Mining Co., 183
Gila River, 33, 79
Gilbert, Humphrey, 31
Gilded Age, The (Twain and Warner), 315
Gilpin, William, 326
Gilpin County (Colo.), 127–129, 138,
    185–188, 282
Gird, Richard, 306–308
Godkin, Edwin L., 111, 167, 203
Gold, 13, 22–26, 31–32, 38, 43–45; lodes,
    20–21, 39, 53–69, 77–83, 86–90,
    109–112, 123–131, 134–136, 138–139, 142,
    149–152, 179, 187–192, 207–221, 238–
    249, 251–268, 318, 322–325, 330–335;
    placers, 20, 23, 33–34, 38–40, 43–53,
    74–77, 83, 129, 135, 151–152, 318–322;
    price, 15, 20–21
Golden Chariot Mining Co., 171–173
Golden Terra Extension claim, 245

Golden Terra Mining Co., 243–246
Gold Hill, Nev., 84, 93, 103, 195–198, 204–205, 221
Gold Hill *News*, 221
Gold Mines, Lands & Rivers of California Co., 51
Gómez, Vicente Perfecto, 118, 194
Goodfellow, Dr., 158
Goodwin, Charles C., 221–222, 268
Gopher Hills, Calif., 120–121
Gordon (Cummings), George, 48, 69, 74
Gordon's California Association, 47–48
Gould, Jason "Jay," 111, 211, 276, 345
Gould & Curry Silver Mining Co., 86–87, 87, 93–96, 136, 141–142, 197, 220, 355
Grand Central Mining Co., 306, 310–311, 357
Grand Prize Mining Co., 329
Grant, Albert (Abraham Gottheimer), 162–167, 199, 204, 209
Grant, Frederick D., 176, 338
Grant, James B., 296–298
Grant, Ulysses S., 15, 176, 178, 194, 296
Grant, Ulysses S. "Buck," Jr., 176, 295–296, 349–350
Grant & Ward, brokers, 296, 349–351, 351
Grant Smelting Co., 297, 356
Grass Valley, Calif., 44, 54, 57 62–63, 76, 86, 91, 94, 103, 123–124, 139, 149–152, 189, 238
Grass Valley Silver Mining Co., 91–92
Gray, Elisha, 345
Grayson, George Washington, 329–330
Great Diamond Swindle, 155, 199–204, 239, 332, 340, 406n69, 407n75
Great West, 301–304
Greed, 14–15, 20, 25, 27, 37, 64, 68, 71, 83, 99, 142, 157, 193, 196, 209, 215–216, 223, 225, 260, 267, 292, 339, 347
Greeley, Horace, 57, 112, 276, 333

Green, Edward H., 190
Green, Henrietta "Hetty," 190
Green, Norvin, 177
Greenbacks, 106–107, 111
Gregory, John H., 127
Grenier, John, 77
Grosh, E. Allen and Hosea B., 92–96
Grosh Consolidated Gold & Silver Mining Co., 92, 94, 94–95
Grubstakes, 55, 93, 277, 310
Guerra, Francisco Pedro de, 34–35
Gunnell Gold Co., 127–128, 138

Hadley's Pointers, 290
Haggin, Ben Ali (James's son), 320
Haggin, James Ben Ali, 171, 241, 243, 245–246, 306, 319–320
Hale & Norcross Silver Mining Co., 208, 356
Halleck, Henry W., 112, 115
Hamill, William A., 174–177
Hance, William, 64
Hanover mine, 78
Hardy, Thomas (miner and legislator), 121–122
Harmon, Albion K. P., 309–310
Harpending, Asbury, 162–163, 199–204, 330, 334, 338–339, 346
Harper's New Monthly Magazine, 2, 105, 116, 126, 136
Harper's Weekly, 351
Hart, Henry, 310
Hart, Simeon, 78
Harte, Francis Bret, 195
Hartley, William M. B., 136
Hastings, Serranus Clinton, 92, 95–97
Hauser, Samuel T., 131–133, 235
Hay, John Milton, 201
Hayes, Martin, 182
Hayes, Rutherford B., 219
Hays, John (druggist), 41
Hays, John Coffee "Jack," 67

Hayward, Alvinza, 65, 83, 123, 220;
Comstock mines, 196–198, 205, 212;
Mother Lode millionaire, 65–66, 125,
150–151, 189–191, 318, 355
Hearst, George, 14, 244, 253, 278, 306;
early career, 63, 131, 159, 161–162,
171–173, 191, 193, 241; Homestake
bonanza, 242–248, 271, 274, 318;
Ontario bonanza, 171, 274, 315, 317;
Ophir bonanza, 86, 94
Heintzelman, Samuel, 80–81, 136
Heintzelman mine, 80–82, 136
Helena, Mont. Terr., 84, 131–132, 235
Hendricks, Thomas A., 305
Henkel, Sofio, 78
Henriett Mining & Smelting Co.,
293–294
Hermes Silver Mining Co., 161, 246
Hibernia Consolidated Mining Co.,
291–293
Hickman, Ky., 251
Hill, Nathaniel P., 187–189
Hill, Sarah Althea, 226
Hispaniola colony, 23–24
Hite, Calif., 272
Hite's Cove Gold Quartz Mining Co.,
356
Hittell, Theodore H., 154
Hobart, Garret A. 323, 325
Hobart, Walter Scott, 189
Hoffman, Lewis, 313–314
Holden, Liberty E., 232–234
Holmes, Oliver Wendell, 270
Holy Ghost Hospital for Incurables, 218
Homestake Mining Co., 241–247, 247,
248, 271, 274, 318–319, 353, 356
Honest Miner Gold & Silver Mining
Co., 99
Honore, Henry H., 176, 338
Hook, George, 277
Hope Mining Co., 132–133
Hopkins, Mark, 125
Horn Silver Mines Co., 317

Horn Silver Mining Co., 314–317, 355
How, John, 132
Hoy, James, 110–111
Hughes, Hiram, 121
Hughes, Napoleon, 121
Humbert, Pierre Jr., 338–339
Huntington, Collis, 125
Hussey, Curtis G., 41
Hussey, Warren, 163
Hydraulic mining, 76–77, 135, 151–152,
318–320

Ida Elmore Mining Co., 172–173
Idaho Quartz Mining Co., 149, 189,
318, 355
Imperial Mining Co., 356
Inca empire, 23, 26–28
Independence Mining Co., 329
Indian mines, 23, 29, 33, 36–37, 40, 70,
77, 119
Indians, 33–37, 39–40, 43, 45, 70, 77–79,
82, 119, 129, 136–137, 155, 228, 230, 232,
242–243, 246, 306, 313, 334, 341. See
also names of individual tribes
Initial public offerings: gutted mines,
110, 162, 164–165, 173, 177–178, 190,
234, 239, 271, 281–285, 343, 347;
inflated prices, 158, 166, 190, 232, 309;
misrepresentations, 56–57, 59–60,
77, 89–90, 107–108, 132, 135–136, 140,
154, 164, 169, 174, 191, 228–229, 234,
240, 249, 282–283, 302, 305, 311–312,
320–321, 323–324, 331, 339; pools,
89, 104, 323, 332, 338, 340–341; total
offerings, 52, 61, 88, 138, 141–142,
153–154, 157, 202, 205, 251–152, 267, 271,
288, 352–353; underwriters, 279–280,
285, 287; watered stocks, 126, 154, 241,
246, 291; without mines, 14, 42, 62,
89–90, 98–99, 142, 249, 324, 352. See
also Mining promotion
Inquiry into the Nature and Causes of the
Wealth of Nations (Smith), 7, 38

Interest rates, 27, 87, 170, 258–259, 349;
  banks, 116, 129, 132, 144; brokers, 104,
  274
Investments. *See* Bonds; Grubstakes;
  Mining stocks
Investors. *See* Stockholders; *names of*
  specific countries and cities
"Irish dividends." *See* Assessments
Iron Silver Mining Co., 281, 285, 287,
  295–296, 356
Irwin, William, 269
Isabella and Ferdinand of Spain, 22
Ives, Brayton C., 278, 298, 325, 336–337

Jackson, Andrew, 39
Jackson, David H., 341–344
James, Charles, 68
James, William H., 297
James I of England, 31
Jamestown, Va., 32, 45
Janin, Henry, 190–191, 200–203
Jerome, Leonard, 128
Jesuit missionaries, 40, 305
Johnson, John "Quartz," 55, 66
Johnson, Reverdy, 95
Jones, John Percival, 195–198, 205, 208,
  212, 221, 228–230, 265–267, 280, 291
Jones, William Carey, 58
Jordan Silver Mining Co., 133
Journal des Débats, 52
Julian, George W., 326–327
Justice Mining Co., 252–255, 255

Kalamazoo, Mich., stockholders,
  232–233
Kalamazoo Theological Seminary, 232
Kansas City smelter, 186
Kaufman, Charles, 333
Keene, James R., 225, 228, 291, 311, 345
Kellogg, L. D., 243
Kelly, David B., 238
Kelly, Eugene, 191–193, 327
Kennett, Luther M., 132

Kentuck Mining Co., 356
Kerens, Richard C., 294
Ketchum, Morris, 110–111
Keweenaw Peninsula, 40–42
Keyes, William S., 160, 278, 280–286
Keystone Consolidated Mining Co.,
  150–151, 189, 191, 355
Kimberly, Peter L., 311
Kindleberger, Charles P., 14
King, Clarence, 166, 202–203
King, Thomas Butler, 55
King, Thomas Starr, 86
Kirker, James, 34–35
Kuchel, Charles Jr., 215–216

LaCoste, Jean Batiste, 78, 82
La Crosse *Democrat*, 300
Ladies' Mining & Stock Exchange, 290
Lake Valley, N.Mex., 272, 340–346
Lares, Amador de, 24–25
Larkin, Thomas, 58, 72
Las Cruces, N.Mex., 156, 183
Las Nueve Minas de Santa Maria
  Gold & Silver Mining Co., 338
La Société des Mines d'Argent et
  Fonderies de Bingham, 234
Last Chance Gulch (Helena), 129, 131
Last Chance Silver Mining Co., 168
Latham, Milton S., 201
La Toison d'Or Cie. des Placers, 51
L'Aurifère Cie. Universelle des Mines
  d'Or de la Californie, 51
Law, John, 32
Lawrence, Joab, 168–170
Lawyers, 15, 73, 95, 137, 145, 154, 182, 216,
  218, 257–259, 304, 313; investments, 61,
  171, 243; practices, 68, 73, 91, 100, 112–
  113, 120, 161, 246, 310; speculations,
  73, 80, 88, 92, 97, 109, 118, 128, 133, 143,
  163–168, 170, 179, 200, 241
Lead, Dak. Terr., 245
Lead mining, 32–33, 80, 133–134, 155, 159,
  163, 168, 171, 230, 232–233, 277,

Lead mining (*continued*), 281, 314–315, 340; smelting, 185–187, 296–298

Leadville, Colo., 272–273, 276–296, 349, 353; smelters, 296–298

Leadville Mining & Stock Exchange, 276

Lecompton mine, 63, 86

Lee, George, 312–314

Lee, Stephen W., 151–152

Leighton, John, 138, 178

Leiter, Levi Z., 281, 287, 295, 325

Le Nouveau Monde Cie., 51, 60, 66–67

Lent, William M., 162–163, 200–204, 239–240, 291

Lesinsky, Henry, 183–185

Leverage. *See* Margin trading

Lexington mine, 235, 237–238

Lightner, Charles, 161

Limantour, José Yves, 73

Lincoln, Abraham, 69, 109, 112–115, 119, 194, 315

Little Chief Mining Co., 279, 281, 285–287, 291

Little Dorritt (Dickens), 65

Little Pittsburg Consolidated Mining Co., 278–279, 281–283, 287–288, 290, 293–295, 297–299, 347, 356

Little Pittsburg Mining Co., 277

Lockberg, Louis, 238

Lode & Gulch (N.Y. mining journal), 290

Logan, John A., 176

Lok, Michael, 30–31

Lolor, Charles P., 121–122

London & California Mining Co., 191

London & Globe Finance Corp., 345

London *Mining Journal*, 164, 166

London Stock Exchange, 345

London stockholders, 30–32, 40, 49, 52, 56, 59–61, 64, 114, 139, 158, 160, 162, 164–168, 174, 178, 180, 190–191, 199, 204–205, 229, 234, 257, 263, 293, 305, 344, 346

London *Times*, 52, 162, 191, 204

Longet, William, 93, 97

Longfellow Mining Co., 183–185

Los Angeles, Calif., 84, 118–119, 229, 309

Los Angeles *Times*, 314

Louis XIV of France, 32

Louisiana colony, 32–33

Lounsbery, Richard P., 320

Lovell, William "Chicken Bill," 237

Lucky Cuss mine, 306–308

Lyon, James E., 138, 163, 167, 187

MacDonald, Bernard, 341

Mackay, John W., 14, 209, 221, 330, 345; "Big Bonanza," 208–211, 228, 257–258

MacWillie, Marcus H., 137, 182

Malet, Alexander, 167

Management. *See* Mining management

Manhattan Quartz Mining Co., 57

Manhattan Silver Mining Co., 108–109

Manuel, Moses and Fred, 242–243

Margin trading, 104–105, 213, 215, 252, 265, 274; cause of losses, 217, 220–221, 223, 228, 267; denounced, 222–223, 268; easily wiped out, 215; outlawed in California, 13, 224, 269–270; profitable, 240

Marine National Bank, 295, 349–350

Mariposa, Calif., 44, 54–55, 172

Mariposa Co., 110, 110–111, 112, 127, 274, 385n50

Mariposa grant, 58–60, 66–69, 73, 109–112, 114, 125–126, 135, 163, 178, 181, 191–192, 229, 327

Mariposa Land & Mining Co., 192, 327

Mariposa Mining Co., 59

Market regulation, 13, 15, 146, 216, 221–222, 224, 254, 256, 260, 269–270

Marshall, James W., 43

Marysville, Calif., 74, 103

Mason, Charles, 230
Mason, Richard, 43
Matchless mine, 290–293
Mathews, Mary M., 220–221
Mattison, Edward E., 76
Maxwell, Lucien Bonaparte, 179
Maxwell Grant, 156, 178–182, 278, 311, 325–327
Maxwell Land Grant & Railway Co., 179–181, 181
Maxwell Land Grant Co., 325–327
McCalmont, Hugh, 263
McCalmont, Robert, 190–191, 263–265
McClellan, George B., 201
McCook, Edward M., 176
McCormick, Cyrus Hall, 281, 286, 325–326
McCunniff, Thomas, 175
McDonald, James Monroe, 150–151
McGarrahan, William, 118–119, 194–195
McKnight, Robert, 35
McLaughlin, Frank, 321–322
McLaughlin, Patrick, 93
McMahon, William, 321–322
McMaster, Samuel, 244–246
McMurdy, John H., 175–177
Meader, Charles T., 121–122
Meadow Valley Mining Co., 161, 356
Mecklenburg Gold Mining Co., 40
Medhurst, Frank H., 238
Mendoza, Antonio de, 28–29
Merced Mining Co., 55, 66–69
Merced River, 55
Mercury, 75; gold recovery, 21, 69, 83; market prices, 70–71, 73–74, 116–117, 192–194; mines, 69–74, 112–119, 192–195, 327–329; monopoly, 69, 75, 114, 144, 151, 192–194
Metcalf, Robert, 183
Mexican land grant frauds. See Capitancillos; Mariposa; Maxwell; Panoche Grande; San Jacinto

Mexican mine grants. See Ortiz; Santa Clara; Santa Rita del Cobre
Mexican mines, 29, 34–36, 39, 70–71, 77–79, 115, 134, 305, 338–339
Mexicans, 78–79, 81–82, 115, 184, 339
Mexico City, 34–35, 70
Miller Gold Mining Co., 331–332, 346
Milling: charges, 144–145, 237; processes, 53–54, 61, 138, 322; rings, 144–146, 197, 205, 211, 254, 257–259
Mills, Darius Ogden, 14, 71, 72, 72–76, 83, 135, 152, 345; Bank of California, 116–117, 226; Comstock mill ring, 144–147, 196–197; early banking, 71, 76; mercury monopoly, 71–74, 116, 193–194, 328; New York real estate, 274–276; water companies, 74–76
Mills Building, "first skyscraper," 15, 276
Mimbres Indians, 33
Minas de Santa Clara, 70
Mineral Hill Silver Mines Co., Ltd., 156, 162–163
Miners: cooperative ventures, 45–48, 53–54, 74–75; earnings, 76, 78, 81, 83, 115, 148, 153, 184, 197, 206–207, 352; grubstakes, 55, 93, 277, 310; stock speculation, 148, 220–221; strikes, 76, 284–285; unions, 148, 255, 284
Mining. See Copper; Gold; Lead; Mercury; Silver; Tin
Mining & Scientific Press, 139–140, 145–146, 260–261, 269
Mining companies: average capitalization, 45, 49, 51, 53–55, 88–90, 141–142, 261, 288, 353; British, 49–52, 59–60, 157–169, 179–181, 190–191, 263; chartered, 22, 37, 40–41; close corporations, 239, 247, 294; cooperatives, 45–48, 53–54, 74–75; Dutch, 178–181, 325–326; estimated total numbers, 40, 46, 49, 88–89,

Mining companies (*continued*), 138, 141–
142, 154, 205, 251–252, 288, 353; failure,
46, 48, 61–62; French, 49–52, 59–60,
234, 237–238, 263; incorporation,
22, 41, 46, 86, 88, 99, 106, 228, 288;
joint stock, 22, 39, 49–50, 53–54, 74;
litigation, 55, 64, 67–69, 72–74, 90–97,
111, 138, 167, 174–175, 182, 255–260,
281–282, 327–328; Mexican, 33–36,
70; overcapitalization, 135, 138, 142,
162, 206; partnerships, 24, 26, 171,
182, 191, 311; placer vs. lode, 53–55,
317–318; productive lifetimes, 355–357;
profits and losses, 52–53, 88–89, 154,
157, 205–206, 261, 352–353; Spanish,
23; wildcats, 14, 62, 87, 89–91,
98–100, 105–107, 123, 128, 138, 141–143,
153–155, 157, 198–199, 205, 213, 215, 228,
249, 251–252, 261, 267, 348. *See also*
Assessments; Dividends; Copper;
Gold; Lead; Mercury; Silver; Tin
Mining "experts," 60, 66, 89, 91,
110, 112, 136, 158, 164, 169, 180, 189,
199–202, 236, 278, 316, 323–324, 331,
333–334, 337, 342
Mining Investigator, 290
Mining Law of 1872, 91, 137, 200–201,
295
Mining management: admirable,
35–36, 117, 124–125, 131, 149–150, 160,
171–172, 172, 238, 241–242, 247–248,
295; bribery, 59, 68, 73, 89, 91, 101–102,
109, 112, 114, 134, 144, 161, 164, 166,
179, 195, 201, 204, 246, 255, 331; costly
cost-cutting, 66, 316–317; directors,
14, 52, 54, 62, 66, 112, 115–116, 158,
161, 165–166, 168, 192–193, 196, 203,
253–254, 269, 287, 295, 307–309,
347; gutted mines, 110, 162, 164–165,
178, 239, 271, 281–285, 299, 343, 347;
incompetent, 40, 61–62, 138–139, 347;
insider manipulation, 14–15, 83, 140,

144–146, 168, 171, 197–198, 201, 209,
215, 219, 222, 240, 251, 254, 269, 274,
288, 294, 299, 328–330, 332–335, 344,
347, 352; other fraudulent practices,
14, 20, 71–72, 100–101, 108–109,
144–148, 154, 161, 163–168, 174, 195–198,
232–234, 236–237, 253–255, 257–260,
269–270, 292, 294, 308, 329
Mining promotion: advertising, 52,
57, 61, 140, 249, 290, 300–303, 343;
bonanzas, 117, 149–150, 159–161, 171,
190, 195–196, 208–210, 238–239, 246,
278–285, 290–291, 309–310, 314–316;
borrascas, 20, 38–39, 41–42, 49–52,
54, 56, 59–62, 80, 89–92, 110–111,
127–129, 131, 155–157, 162–166, 171–173,
193, 201–203, 229–230, 248–250,
323–324, 330–333, 340–343, 352–53;
"dummy" directors, 51, 89, 111, 144,
164–166, 174, 199, 201, 254, 280–281,
286, 293, 325, 331, 333–334, 336, 338,
340, 346; excursions, 278, 342;
"expert" reports, 66, 89–91, 110,
136–137, 158, 164, 169, 180, 191–192,
199–202, 278, 315–316, 323–324, 331,
333–334, 337, 341–343; gimmicks, 51,
249–250; misrepresentations, 56–57,
59–60, 77, 89–90, 107–108, 132, 135–136,
140, 154, 164, 166, 169, 174, 191, 195, 198,
228–229, 234, 240, 249, 282–283, 302,
305, 311–312, 320–321, 323–324, 331, 337,
339; "picture rock," 55–56, 59–60, 90,
98, 102, 130, 178, 199–200, 202, 342–44;
pools, 89, 104, 323, 332–334, 338,
340–341; prospectuses, 39, 50, 56–57,
61, 89, 99, 126, 164, 174, 179, 199, 303,
316; publicity, 116, 120, 126, 136–137,
202, 305, 324, 332, 339, 342–344. *See
also* Initial public offerings
Mining stock: bearing, 95, 101, 104,
114, 144–145, 154, 166, 193, 197, 228,
266, 292, 320, 334, 344; bulling, 145,

154, 158, 166, 239, 333–334, 341, 344;
common, 116, 193, 328; elaborately
decorated, 74–75, 75, 99, 147–147, 148,
249–250, 301; gambling, 105–106, 141,
198, 216, 224, 226, 268–269; "lottery
tickets," 14, 40, 51, 95, 249, 261,
347; manipulations, 14–15, 83, 140,
144–146, 168, 171, 197–198, 201, 209,
215, 219, 222, 240, 251, 254, 269, 274,
288, 328, 344, 347, 352; markets, 13–15,
19, 41–42, 88–89, 102, 104, 110–111, 131,
136, 139, 141, 143, 145, 154, 161, 197–198,
205–207, 213, 223–224, 228, 251–252,
260, 266, 268, 270–271, 273–274, 276,
300, 304, 320, 322, 328, 337, 349, 353;
options, 89, 166, 201, 279; par value,
45, 50, 52, 88, 98, 117, 133, 141–142, 150,
160, 169, 215, 237, 239, 241, 288, 294,
317, 323, 332; preferred, 116, 132, 180,
193, 327–328; prices, 41–42, 88–89,
102–104, 122, 126, 138–139, 142, 145,
154, 158, 161, 173, 204–206, 210, 212,
216–217, 222–226, 241, 263, 271, 323,
332, 352; proxies, 210–211, 258–259,
270; sales volume, 205, 213, 270, 288,
324, 335; short sales, 66, 104–105, 175,
196, 216, 224–225, 228, 236, 269, 287,
292, 294, 334–335, 337, 344, 352–353;
"street names," 210, 259, 336; used for
bribery, 73, 112–113, 194–195; used like
banknotes, 106; wash sales, 104, 224,
276, 288, 292, 294, 323–324, 332–334;
watering, 126, 154, 241, 246. See also
Assessments; Dividends; Margin
trading
Mining stock bubbles, 19, 21–22, 32,
39–43, 49, 51, 57, 61–62, 88, 102–105,
107, 112, 123, 128–129, 139–143,
153–155, 159, 166, 181, 195, 197, 199,
201, 203–206, 212–213, 226, 228, 230,
234, 261, 267–269, 271, 273, 285, 288,
324–325, 353

Mining stock exchanges:
Albuquerque, N.Mex., 290; Aurora,
Nev., 103; Austin, Nev., 103; Boston,
288; Chicago, 288; Como, Nev., 103;
Denver, 290; Gold Hill, Nev., 103;
Grass Valley, Calif., 103; Kansas
City, Mo., 290; Leadville, Colo., 276;
Marysville, Calif., 103; New York
City, 61, 107, 138, 213, 274–275, 275,
290, 292, 331, 336, 353; Philadelphia,
288, 308; Portland, Ore., 103;
Sacramento, 103; Salt Lake City,
290; San Francisco, 102–103, 142, 161,
205, 212–214, 214, 215, 225–226, 252,
274; Silver Mountain, Calif., 103;
Socorro, N.Mex., 290; St. Louis, 290;
Stockton, Calif., 103; Tombstone,
Ariz., 290, 397; Topeka, 290;
Victoria, B.C., 103; Virginia City,
Nev., 103
Mining Trust Co., 274
Miocene Mining Co., 321
Misinformation. See under Initial
public offerings; Mining promotion
Mississippi Bubble, 32, 51, 203
Missouri & Montana Mining Co., 132
Mitchell, Robert B., 137
Moctezuma (Aztec emperor), 25, 29
Moffat, David H., 174–175, 178, 273,
277–279, 282–283, 293, 298–300
Moffat, John L., 55
Mogollon Mountains, 33
Mohawk & Montreal Mining Co., 90
Mojave Desert, 312
Mono Gold Mining Co., 240–241
Montalvo, Garci de, 28
Montana Silver & Copper Co., 132
Monte Cristo Mining & Milling Co.,
300
Monterey, Calif., 43–44
Montgomery Street (San Francisco),
102, 140, 155

Moody, Gideon C., 246–247
More, J. Marion (John N. Moore), 172
Morenci, Ariz., 184
Morgan, Alfred Grey, 64
Morgan, Henry, 199
Morgan, Johnny, 92
Morse, Anthony, 128, 136
Mortgages, 24, 62, 82, 109, 111, 145, 152, 159, 162, 170, 180–181, 192, 232, 237, 265, 283, 325, 392
Mother Lode, 54, 63, 65, 121, 125, 127, 149–150, 192
Moulton Mining Co., 235–237, 413
Mount Hope Mining Co., 62
Mowry, Sylvester, 81–83, 136–137
Mowry mine, 44, 136–137
Mowry Silver Mining Co., 136
Mutual Trust Co., 274
Myncompagnie Nederland, 178

Napias Jim, 155
Napoleon Copper Mining & Smelting Co., 123
Napoleonic Wars, 19, 33
Nation, The, 111–112, 167, 203
National Academy of Sciences, 89
National Mining & Exploring Co., 131
National Police Gazette, 240
Navajo Mining Co., 329
Naylor, Edward, 175–176
Naylor, Peter, 185
Neagle, David, 227
Nevada Bank, 209, 226, 259, 267
Nevada City, Calif., 44, 63, 86, 91, 94, 139, 249
Nevada Gold, Silver & Copper Mining Co., 153
Nevada Legislature, 146
Nevada Mill & Mining Co., 197, 356
Nevada Stock Exchange Board, 213
Nevada Supreme Court, 91
New, John C., 305

New Almaden grant, 70–73, 113–115
New Almaden mine, 44, 69–74, 112–117, 193–194, 327–328, 356
New Britain, Conn., 307
Newbury, John S., 315
New Idria Mining Co., 84, 117–118, 194–195, 355
Newlands, Francis G., 227
New Mexico Mining Co., 77, 134, 134–135
Newport, Christopher, 32
New Racket (steamer), 308
New York American Exchange, 290
New-York & Nevada Gold & Silver Mill & Mining Co., 107–108
New York & Owyhee Gold & Silver Mining Co., 130
New-York & Reese River Gold & Silver Mill & Mining Co., 107
New-York & Santa Fe Gold & Silver Mill & Mining Co., 107
New York & Utah Prospecting & Mining Co., 134
New-York & Washoe Gold & Silver Mill & Mining Co., 107
New York City, 15, 40, 61, 135, 139, 147, 167, 200, 228; stockholders, 55, 62, 107, 110–111, 123, 126–128, 131, 134, 136, 138, 147, 163, 182, 190, 237, 241, 273, 282, 285, 291, 298–299, 316, 319–320, 323, 327–328, 353
New York Daily Stockholder, 324
New York Herald, 138, 142, 176
New York Mining & Mineral Exchange, 61
New York Mining News, 290
New York Mining Record, 292
New York Mining Share Board, 61
New York Mining Stock Exchange, 274, 292, 331, 336
New York's "first skyscraper," 15, 276
New York Society for the Prevention of Vice, 250

New York Stock Exchange, 15, 61, 107, 336, 349; mining stocks, 61, 110–111, 113, 171, 192, 236, 241, 246–247, 264, 273–274, 278–279, 298, 320, 323, 336–337

New York *Stock Report*, 334

New York *Sun*, 176

New York Supreme Court, 317

New York *Times*, 108, 128, 148, 202, 217, 240, 242, 250, 260, 270, 339

New York *Tribune*, 56–57, 113, 201, 333, 335, 336, 342

New York *World*, 104

New York Yacht Club, 241

Nickerson, Benjamin R., 92, 94–97

Nicolai (Ahtna *tyone*, or chief), 37

No Name mine, 299–300

North Bloomfield Gravel Mining Co., 84, 151–152, 318, 356

North Carolina Gold-Mine Co., 38–39

Northern Belle Mill & Mining Co., 356

Northern Pacific Railroad Co., 315

North Ophir salting, 99

North Star Gold Mining Co., 123–124, 149

Oberlin College, 345

O'Brien, William S., 208

Ogden, Richard L. "Podgers," 339

Ogilvie, George B., 133

O'Hara, William "Uncle Billy," 238

Old, Robert Orchard, 174

Old Abe Mining Co., 243, 245–246

Old Telegraph Mining Co., 232–234, 237

Olmsted, Frederick Law, 110–111

Olney, James N., 102

Omaha, Neb., 185–187

Omaha & Grant Smelting & Refining Co., 297–298, 357

Omaha Smelting & Refining Co., 186, 355

Oñate, Juan de, 29–30

Onis, Juan de, 34

Ontario Silver Mining Co., 168, 171–172, 172, 274, 315, 317–319, 347–348, 353, 355

Opdyke, George, 110–111

Ophir Silver Mining Co., 86–87, 91–92, 94–96, 98, 103, 127–128, 141–142, 207, 220, 224–226, 256, 356

Ord, Pacificus, 118, 194

Ore, 19–21

Oriental Gold Mining Co., 331–332, 346

Original Amador Mining Co., 191

Original Hidden Treasure Mining Co., 157

O'Riley, Peter, 93–94

Orphans, 61

Ortiz, Jose Francisco, 77

Ortiz Mine Grant, 44, 77, 134–135, 181–182

Osborn, John J., 107–108

Otero, Miguel A., 77, 134

Otis, Harrison Gray, 314

Overman Silver Mining Co., 220, 252

Owyhee mining district, 129–131, 171–172

Pacific Mill & Mining Co., 209, 355

Pacific Mining Co., 189

Pacific Stock Exchange, 213, 225

Packard, John Q., 169–170

Palmer, Cook & Co., 59, 67, 73, 109

Panamint, Calif., 156, 228–230

Panic of 1873, 207, 273, 315

Panic of 1884, 15, 350–351, 351

Panoche Grande Quicksilver Mining Co., 118–119, 194–195

Paris, France, 32, 168; stockholders, 49–52, 59–61, 234, 263

Paris International Exposition 1867, 130

Park, Trenor W., 109–111, 125–126, 135, 163–168, 206, 229–230, 234, 238, 335

Park City, Utah, 156, 171, 348
Patagonia mine, 81–82
Pattee, James Monroe, 248–250
Pattie, Sylvester, 34–35
Paul, Almarin B., 269
Pearce, Edwin M., 183
Pearce, Richard, 188
Pearson, John W., 253–254
Pecos & Placer Mining & Ditch Co.,
    135
Pelican & Dives Mining Co., 175–177
Pels, Martinus, 326
People's Gold & Silver Mining Co.,
    249
People's Mining Co., 64
Perdue Gold & Silver Mining & Ore
    Reducing Co., 175
Perley, Duncan W., 161
Phelps, Bethuel, 135, 323
Phelps, Dodge & Co., 131
Philadelphia, 45, 56, 288, 290, 308, 315;
    stockholders, 40, 48, 73, 305, 307, 338,
    340–346, 353
Philadelphia & California Mining
    Co., 59
Philadelphia Mining & Stock
    Exchange, 308, 343
Philadelphia *Mining Journal*, 341
"Philadelphia Syndicate," 340–346
Philippart, Simon, 234
Phillips, John Arthur, 66
Phillipsburg, Mont., 84, 132–133
Phillpotts, Thomas, 158
Picturesque Arizona, 305
Piedmont gold rush, 39–40, 45
Pierce, Franklin, 67
Pike's Peak rush, 85, 127–129, 154–155
Pine Street (San Francisco), 215, 259,
    268
Pine Tree and Josephine mines, 55, 66,
    110, 192
Pioche, François L. A., 161

Pioche, Nev., 156, 160–161
Pitts, Henry, 117–118
Pittsburgh & Boston Mining Co.,
    41–42
Pizarro, Francisco, 24–27
Pizarro, Hernando, 24–27
Placerville, Calif., 92
Planchas de plata, 33, 79
Pleasant, Mary Ellen "Mammy," 226
Plumas-Eureka Mine Ltd., 190–191, 356
Plumb, Preston B., 294
Plymouth Consolidated Mining Co.,
    184, 357
Polk, James, 43, 45, 48
Pomeroy, Marcus M. "Brick," 300–304,
    424–425n47, 425n50
Poor, Daniel W., 218
Poorman Gold & Silver Mining Co.,
    130–131
Poorman mine, 130, 173
Porter, Fitz-John, 138
Porter, John, 312
Portland, Ore., 84, 103, 130
Post, George B., 276
Postal Telegraph Co. 345,
Poston, Charles D., 80–82, 136
Potosí, Cerro Rico de, Bolivia, 27, 29
Potosí Mining Co., 92
Price-to-earnings ratio, 41, 168, 173, 221,
    231, 240–241, 246, 263, 278, 280, 283,
    298, 308, 318, 320, 327, 336, 344
Pride of the West claim, 245
Pride of the West Mining Co., 141
Princeton mine, 55, 110
Probert, Edward, 160
Prospectors, 20, 37, 53, 122–123, 155, 159,
    242. *See also names of individuals*
Pryor, Nathaniel, 34–35
Puck, 282, 289
Pueblo Indians, 29–30, 77
Puleston, John H., 164
Pullman, George M., 325–326

Pyramid Range Silver Mountain Co., 199–200, 204

Quandary claim, 300
Quicksilver. *See* Mercury
Quicksilver Mining Co., 73, 112–117, 192–194, 274, 327–329, 329, 356
Quivira quest, 29

Railroads, 132, 179, 278, 297, 304, 314, 342; short line, 108, 120, 145, 149, 179, 182, 184, 189, 229, 276, 298, 315–316; stock, 128, 200, 273, 291, 326, 352; transcontinental, 21, 80, 125–126, 134, 155, 157, 183–186, 201, 211, 230, 271, 305, 307, 315, 326, 334, 340
Raleigh, Walter, 31
Ralston, William Chapman: Bank of California, 143–146, 226–227, 227; Diamond swindle, 200–204; profligate speculator, 151–152, 162–163, 191, 196–197, 224–225; suicide, 226
Rancho Santa Anita, 225
Randol, John Butterworth, 117, 193
Raton, N.Mex., 156, 179, 326
Rattlesnake lode, 132
Raymond, Rossiter W.: exposed mining frauds, 139–140, 24, 294, 332–333, 335–336, 343, 348; Leadville mines, 278, 282–284, 286
Raymond, William H., 160
Raymond & Ely Mining Co., 160–161, 218, 355
Reagan, Benjamin, 230–231
Redington, John H., 117, 192
Redington Quicksilver Mining Co., 84, 117, 356
Reed, Conrad, 38–39
Reed, William K., 121
Reese, Michael, 189
Reese River, 88, 107–108, 189
Reid, Whitelaw, 276, 333, 335

Reis, Christian, Ferdinand, and Gustave, 124–125, 190, 356
Rencher, Abraham, 77, 134
Richmond Consolidated Mining Co. Ltd., 160, 318, 355
Riggs, Elisha, 128
Rio Grande, 29, 33, 82
Rische, August, 277
Rising Star Silver Mining Co., 131
Rivafinoli, Vincent de, 40
Robbins, George Collier, 130
Robbins, James J. (pseud. "Mary Jane Simpson") 148, 304n109
Robert E. Lee Mining Co., 290–291, 356
Roberts, George D., 152–153, 243, 294, 297, 305, 345; British promotions, 159–160, 162–163; Comstock mines, 91–92, 228, 253–254, 330; Death Valley scam, 330–335, 346; Diamond swindle, 199–205, 406nn69–70; early career, 63, 131, 139, 149, 157, 193; Leadville bonanzas, 278–288, 295, 336–337, 346; Mexican scams, 337–339, 32n109; New Mexico bonanza, 340–346; New York mining exchanges, 271, 273–275; Postal Telegraph Co., 345
Roberval, Jean-François de la Rocque de, 30
Robinson, George B. 281, 287, 336
Robinson, Lester L., 120
Robinson, Thomas C., 65
Robinson Consolidated Mining Co., 281–282, 287, 336–337, 346
Rocky Bar Mining Co., 55–57, 62, 124, 238, 357
Rocky Mountains, 85, 141
Rocky Mountain News, 302
Roosevelt, Nicholas I., 37–38
Rose, Albert, 150
Rothschild, Lionel, 201
Roudebush, Lorenzo, 291

Roughing It (Twain), 141
Royal stockholders, 22–24, 30–33
Rubery, Alfred, 201–204
Rusk, Thomas Jefferson, 78
Russian-American Company, 37

Sacramento, 44, 55, 71; stockholders, 74, 103, 106–107, 125–126
Sacramento *Record-Union*, 268
Sadleir, John, 64–65
Safford, Anson P. K., 307
Sage, Russell, 345
Salt Lake City, 84, 129, 133; smelters, 185, 315–316; stockholders, 163, 186, 235–237
Salt Spring, Calif., 272, 323–324
Samoa, 219
Sampson, Marmaduke, 52, 162, 191, 199, 204
San Antonio Silver Mining Co., 136–137
San Diego, Calif., 35, 44, 48, 314
San Francisco, 44, 48, 86; smelters, 185–186, 230; stockholders, 2, 55, 71, 74, 79, 81, 86–88, 95, 100, 102, 105, 109, 117, 123, 131, 150, 157, 161, 163, 171, 173, 189, 196, 203, 205, 207–208, 211–212, 215–216, 218–219, 221, 228–229, 231, 251–252, 256–257, 268–271, 309, 328
San Francisco *Alta California*, 95, 115, 198, 202, 223
San Francisco & New York Mining & Commercial Co., 200–204
San Francisco *Bulletin*, 95, 98, 101
San Francisco *Call*, 140, 219, 268
San Francisco *Chronicle*, 145, 198, 202–203, 208, 211, 216, 218, 258, 268
San Francisco Stock & Exchange Board, 142, 161, 173, 205, 214, 226, 244, 265, 274; described, 102–103, 212–215; formed, 102; restricted by new laws, 269–270; sales volume, 213, 252, 261,

270; stock gambling assailed, 105–106, 224, 268–269
San Francisco *Stock Report & California Street Journal*, 145, 239
San Jacinto grant, 120
San Jacinto Tin Co., 120
San Quentin, Calif., 102, 216
Santa Anna, Antonio Lopéz de, 79
Santa Clara Mining Co., 72
Santa Fe, N.Mex., 34, 44, 77, 134, 290
Santa Fe Ring, 179–182, 327
Santa Rita del Cobre mine, 44, 78–79, 123, 137, 183, 340, 356; first western bonanza, 33–36, 42
Santa Rita Mining Assoc., 137
Santa Rita Silver Mining Co., 80
Santa Rosalia mine (Ortiz's), 77
Santiago Mining Co., 97–98
Santillan grant fraud, 73–74
Sargent, Aaron A., 200
Savage Mining Co., 96, 197–199, 212, 221, 252, 263–264, 330, 355
Sawyer, Lorenzo, 318
Schenck, Robert C., 164–165, 167
Schieffelin, Edward, 306–308
Schultz, George, 254–255
Sea Serpent Co., 45
Selby, Prentiss, 186
Selby, Thomas H., 185–186
Selby Lead & Silver Smelting Works, 185
Selby Smelting & Lead Co., 186
Selover, Abia, 385n50
Selover, John, 335
Serrano, Leonardo, 119–120
Seward, William Henry, 113–114, 120
Sexton, Thomas B., 135
Seymour, James Madison, 322–325, 337
Sharon, William, 143, 152, 221–222, 226; Comstock mill ring, 144–146, 196–198, 205, 224–226; death, 227; U.S. Senate seat, 198, 225–226

Shaw, Thomas Jefferson, 330

Sherwin, Frank Remington, 325–326

Sherwood, Robert, 266

Shoshone Indians, 155

Shryock, Lee R., 331

Sieger, Henry, 186

Sierra Apache Mining Co., 340, 343

Sierra Bella Mining Co., 340, 343

Sierra Buttes Gold Mining Co. Ltd., 190–191, 356

Sierra Buttes mine, 124, 156, 190–191, 263, 357

Sierra City, Calif., 44, 54

Sierra Grande Mining Co., 340–345

Sierra Madre Mining Co., 340

Sierra Nevada, 43, 54, 58

Sierra Nevada Deal, 265–268

Sierra Nevada Silver Mining Co., 252, 265–267

Sierra Plata Mining Co., 340, 343–344

Silliman, Benjamin Jr., 89–90; endorsed frauds, 164, 166–167, 192, 331, 342–343

Sills, Beverly, 293

Silver: demonitization, 207; market prices, 21, 207, 403n51; mining, 27, 29–30, 33, 80–83, 85–90, 92–101, 108–109, 131–134, 142–148, 155–178, 185–187, 195–198, 207–221, 228–238, 250–268, 276–317, 329, 336–346; smelting, 80, 82, 132–133, 138, 155, 160, 164, 174, 185–189, 278, 296–298, 316, 337, 355–357

Silver City, Idaho, 84, 129–131

Silver City, Nev., 84, 92

Silver King Mining Co., 156, 230–232, 305, 356

Silver Mountain Mining Co., 250

Sing Sing prison, 350

Sioux Indians, 228, 242–243, 334

Siqueiros, Leonardo, 78

Skae, John, 239, 265–267

Slack, John, 200–204, 406nn69–70, 408n75

Smelting: charges, 121, 185, 188, 298, 423–424n38; companies, 185–189, 296–298; copper ores, 34, 79, 121–122, 182–185; recovery fraction, 188, 315; refractory silver-lead ores, 80, 82, 132–133, 138, 155, 160, 164, 174, 185–189, 278, 296–298, 316, 337, 355–357; smoke, 185, 188, 288

Smiley, George, 274

Smith, Adam, 7, 38, 45, 53, 99, 261

Smith, Eben L., 128, 298

Smith, John, 32

Snider, Jacob, 175–177

Société Anonyme des Mines de Lexington, 235, 237–238

Sonora Exploring & Mining Co., 80

Sonorans, 79, 81–82

Sopori Land & Mining Co., 81

South Aurora Consolidated Mining Co., 158

South Aurora Silver Mining Co., 157–158

Southern Pacific Railroad Co., 184, 231, 272, 305, 307, 312

South Pacific Mining Co., 323–324

Spanish mines, 23–25, 27, 29–30, 33–35, 39

Speculators, 2, 13–15, 22, 24, 27–30, 32, 34, 42, 50, 72–73, 79, 91–92, 97, 103, 120–121, 129, 134–135, 143, 146, 150, 161, 163, 171, 174, 189, 192, 202, 208, 210, 225, 229, 231–232, 242–246, 253, 256, 270–273, 278–279, 287, 306, 309–311, 325, 328–329, 332–335, 338–339, 341–347; estimated numbers, 40–41, 49, 88–89, 138, 141–142, 154, 205, 252, 267–268, 288, 353; margins, 104, 213, 215–224, 228, 240, 252, 265–268, 274. See also Stockholders; names of individual cities

Standard Consolidated Mining Co.,
  241, 318–319, 353
Standard Gold Mining Co., 238–241,
  347, 355
Stanford, Asa Philip, 127
Stanford, Leland, 125–127, 157–158, 211
Stanton, Edwin M., 73, 112–115, 117
Star Route scandal, 182, 293–294, 326
State Line Gold Mining Cos., 330–336,
  340–341, 343–344
Stearns, Abel, 119–120
Stevens, William H., 281, 287, 295
Stevenson, Robert Louis, 212–213
Stewart, William M., 91, 146–148;
  Comstock litigation, 91–92; Emma
  mine scandal, 163–167, 198; Panamint
  scam, 228–230; U.S. mining law, 91,
  137, 159, 295; U.S. senator, 91
Stewart's Wonder Mining Co., 229
St. Louis, Mo., 35, 109, 188, 250
St. Louis & Montana Mining Co., 132
Stockbrokers, 102–106, 128, 136, 143, 205,
  213, 259, 268–270, 273–274, 295–296,
  320, 325, 330–333, 335, 347, 349–350;
  bucketed orders, 223; commissions,
  103–104, 122, 162, 166, 215, 222–223;
  described, 102–106, 213–215; estimated
  numbers, 102, 213; fraudulent
  practices, 104, 222–224
Stockholders, 13–14, 22, 24–26, 31, 46,
  48–49, 55, 71, 74–75, 90, 98, 100–101,
  166–167; average investment, 88–89,
  205–206, 252, 353; clergymen, doctors,
  etc., 38–39, 41, 61, 88, 218–219, 222,
  277; eastern vs. western, 141–142, 271,
  352–353; estimated numbers, 21, 89,
  102, 154, 205, 213, 252, 353; investigating
  committees, 66, 162, 203, 236–237,
  254–255; lawsuits, 31, 52, 64, 71, 92, 108,
  161, 167, 193, 230, 233, 255–260, 269–270,
  283, 293, 303, 311–312, 317, 320, 337,
  339, 346; liability, 49–50, 54–55; men

vs. women, 31, 60–61, 105, 223, 302;
  profits vs. losses, 52, 83, 88–89, 154,
  159, 205–206, 261, 271, 317–318, 352–353;
  proxies, 210–211, 258–259, 270; sample
  holdings, 90, 212, 215–221, 225, 240,
  347–348; suicides, 94, 131, 153, 161, 219,
  226, 268, 348. See also Margin trading
Stock jobbing, 67, 105, 140, 155, 191, 264,
  268–269, 285, 294, 337; insiders, 14,
  20, 72, 145, 154, 171, 205–206, 208–209,
  212, 219, 266, 270, 288, 328, 336, 339,
  344, 346–347, 352; pools, 89, 104, 323,
  332–334, 338, 340–341, 353
Stock market regulation, 13, 15, 146, 216,
  221–222, 224, 254, 256, 260, 269–270
Stock speculation, 14, 19, 22, 39–40, 49,
  83, 85, 103, 105–107, 109, 113–114, 122,
  126, 128–129, 141, 148, 155, 193, 198,
  201, 205, 213, 215–216, 218, 220–221,
  224–226, 228, 268–269, 272, 276, 305,
  346–347, 354; denounced, 105–106, 159,
  197–198, 206, 218–219, 224, 268–270
Stone & Donner (bankers), 267
Streeter, Eli S., 175
"Street names." See under Mining stock
Struthers, Thomas, 310
Surprise Valley Mill and Water Co., 229
Sutro, Adolph, 146–148, 148, 192,
  263–265, 418n89
Sutro, Nev., 147–148, 156
Sutro Tunnel Co., 146–148, 148, 263–264
Sutter Creek, Calif., 44, 65, 125
Swamp Lands, 200, 309
Swansea, Wales, 174, 187–188, 229
Sweet, James R., 78, 82, 137
Swett, Leonard, 113–115

Tabor, Augusta, 277, 292
Tabor, Elizabeth McCourt "Baby
  Doe," 292–293
Tabor, Horace A. W., 276; early career,
  277; Little Pittsburg mine, 277–283;

subsequent scams, 284–285, 290–293;
U.S. senator, 292
Tahiti, 99, 172
Tallman, Caroline and Catherine, 348
Taney, Roger B., 67
Taylor, Bayard, 139
Taylor, George, 219
Taylor, Isaac M., 330–331, 333
Taylor (John) & Sons, 66, 162, 293
Taylor, John, Jr., 162
Tecoma Silver Mining Co., 168
Teller, Henry M., 128, 176–177
Temescal grant, 84, 119–120
Tenochtitlán, 25
Tepic, Mexico, 70
Terrible Lode Mining Co., 173–175
Territorial Enterprise (Virginia City,
Nev.), 95, 145, 221–222, 258, 268
Terry, David S., 68, 227, 314
Tevis, Lloyd, 171, 243
Thornton, William, 38–39
Tidal Wave (sloop), 241
Tiffany, Charles L., 200
Tin mining, 119–120
Tintic mining district, 168, 183
Toltec Syndicate of Mines, 305
Tombstone, Ariz., 272, 290, 305–312, 314
Tombstone Gold & Silver Mill &
Mining Co., 307
Tombstone Mill & Mining Co.,
308–309, 340, 353, 356
Tough Nut mine, 306–308
Treasure Hill, Nev., 155, 156, 156–158
Trehane, John, 258–260
Trollope, Anthony, 167
Tubac, Ariz., 33, 44, 80–81, 305–306
Tubac *Arizonian*, 80
Tubac Mining & Milling Co., 305
Tucker, Thomas T., 39
Tucson, Ariz., 272, 290, 305–306, 310
Tunnel companies, 146–148, 148,
263–265, 301, 301–304, 320

Tuolumne County Water Co., 74–75,
75, 76, 356
Tuscarora, Nev., 272, 328–329
Tuthill, Franklin, 98, 101
Twain, Mark. *See* Clemens, Samuel L.
Tweed, William M. "Boss," 260, 300

U.S. Congress, 37–38, 40, 43, 58, 73, 81,
119, 134–135, 146–148, 176, 180, 194,
200, 207, 230, 263, 345
U.S. congressmen, 55, 77, 119, 164, 195,
200, 232, 333
U.S. presidents, 15, 39, 43, 48, 67, 69, 73,
119, 133, 176, 178, 182, 194, 201, 218–219,
232, 296, 326, 338, 349
U.S. Senate, 38, 43, 58, 119, 126, 129, 182,
194, 225, 227, 255, 265, 323
U.S. senators, 40, 57, 58, 72, 78, 91, 95,
128, 129, 137, 146–147, 163, 176, 179, 182,
189, 198, 201, 226, 228, 247, 265, 273,
280, 286, 291, 293, 294, 305, 337
U.S. Supreme Court, 67, 69, 73–74, 91,
111–115, 118–120, 179, 227, 270, 311, 314,
327
Union Consolidated Silver Mining
Co., 265–267
Union Copper Mining Co., 121, 356
Union Mill & Mining Co., 144, 355
Union Pacific Railroad Co., 156, 186,
201, 315, 334
Union Quartz Mining Co., 125
United Mining and Investment
Company of North and South
America, 348
United States Democrat, 304
University of California, 310
Upson, Lauren, 120
Usher, John P., 113–114
Utah Enterprize Co., 93
Utah Mining Co., 100

Van Bosse, Pieter Philip, 178

Van Ness, John P., 39
Vanderbilt, William Henry, 291, 345
Velasquez, Diego, 24
Victoria, B.C. (Canada), 103
Virginia & Truckee Rail Road Co., 145
Virginia City, Nev., 84, 86–87, 103, 146–148, 218, 228, 256, 267, 270, 273, 276; stockholders, 148, 219–221
Virginia City Miners' Union, 148
Virginia Company, 31–32
Vosburg, John, 306–307
Vulture Mining Co., 135, 323, 356

Waddingham, Wilson, 129–130, 163, 179–182, 336–337, 346
Waddingham Gold & Silver Mining Co., 130
Waite, Catherine V., 133
Waite, Davis H., 327
Walker brothers (Frederick, Matthew, Robinson, and Sharp), 163, 235–237, 357
Walker, Robert J., 72–74, 112–119, 194
Wall-Street Journal, 164
Wall Street Journal, 228, 274, 278
Wall Street (New York), 15, 48, 56, 61, 110, 114, 127, 190, 193, 215, 259, 274, 292–293, 338, 349, 351
Ward, Eber B., 169–170, 183
Ward, Ferdinand, 15, 295–296, 349–352
Ward, William S., 295–296
Warner, William S., 350
Warren, Marion E., 290
Washington, D.C., 51, 77, 81, 118, 180, 200, 226, 242, 265. See also U.S. *entries*
Washington *Chronicle*, 114, 119
Washington Mining Co., 311
Washington *Post*, 217
Washoe Stock & Exchange Board, 103
Washoe Stock Exchange, 102
Washoe United Consolidated Gold & Silver Mining Co., 139

Wash sales. *See under* Mining stock
Water ditch companies, 53, 69, 74–75, 76–77, 83, 135, 144, 152
Waterman, Abbie L., 313–314
Waterman, James S., 312–313
Waterman, Robert W., 312–314
Waterman mine, 272, 312–314
Watt, John S., 182
Watt, Robert and William, 149, 238
Way We Live Now, The (Trollope), 167
Webster, Daniel, 43
Webster Gold & Silver Mining Co., 89
Weller, John B., 68
Wells, Fargo & Co., 170–171
Western Electric Manufacturing Co., 345
Western Mining Co., 309
Western Union Telegraph Co., 177, 345
Whang Doodle Mining Co., 157
White, Stephen V. "Deacon," 274, 331–332, 336
White Pine mining district, 155, 157–163
White Pine News, 165
Whitlatch, James W., 131
Whitlatch Union mine, 131, 356
Whitney, John N., 168–170
Whitney, Josiah D., 62, 110
Whittlesey, Elisha, 77
Wickenburg, Ariz., 84, 135
Wickenburg, Henry, 135
Wildcats. *See under* Mining companies
Williams, Henry, 306, 310
Willing, George M., 135
Wilson, Alex, 251
Wilson, J. Downes, 97–98
Wiss, Dr. G. (fictitious), 324
Witherell, W. Frank, 310–311
Wolcott, Henry R., 297
Women, 29, 34, 78, 122, 177, 183, 215–216, 226, 241, 249, 263, 265, 276–277, 292, 304, 322, 334, 348; investors, 22, 31, 60–62, 88, 120, 133, 190, 205, 220–221,

232, 240, 292–293, 302, 313–314, 348, 353; speculators, 22, 88, 105, 159, 205, 213, 221, 223, 290, 353; stock exchanges, 213, 290; widows, 34, 61, 77, 119–120, 159, 177, 215, 220, 223, 232, 292, 303, 308, 313–314, 348

Wonder Consolidated Mining Co., 228–230

Woodman, James, 163

Woodville Consolidated Mining Co., 254

Woodville Mining Co., 253–254

Workingmen's Party, 269

Wright, James Whitaker, 340–346

Wrightson, William, 80

Wyoming Consolidated Mining Co., 229

XIT Ranch, 281

Yale University, 89, 164, 192

Yankie, Joseph M., 183

Years of Dishonor (Clifford), 348

Yellow Jacket Silver Mining Co., 356

York Mining Co., 53

Yuba River, 151, 318, 320

Yukon River, 308

Yuma, Ariz., 44, 79–82, 230, 305–306

Zacatecas mines, Mexico, 29–30

Zaldívar, Vincente de, 29

Zuni Indians, 29